# horizons

Exploring the Universe

# Horizons

## Exploring the Universe

**Michael A. Seeds**
Joseph R. Grundy Observatory
Franklin and Marshall College
Lancaster, Pennsylvania

Wadsworth Publishing Company
Belmont, California
A Division of Wadsworth, Inc.

Astronomy Editor: Marshall Aronson

Production Editor: Jeanne Heise

Managing Designer: Detta Penna

Designer: Carol Kummer

Art Editor: Wendy Palmer

Copy Editor: Bob McNally

Page Layouts: Cheryl Carrington

Technical Illustrators: J & R Associates, Christine Dorsaneo



Printed in the United States of America

1 2 3 4 5 6 7 8 9 10—85 84 83 82 81

Library of Congress Cataloging in Publication Data

Seeds, Michael A

Horizons, exploring the universe.

Bibliography: p.

Includes index.

1. Astronomy. I. Title.

QB45.S44 520 80-21250

ISBN 0-534-00888-7

A study guide has been specially designed to help students master the concepts presented in this textbook. Order from your bookstore.

for Janet and Katie

# CONTENTS

# PREFACE

The fundamental question of our existence is, "What am I?" We go through our lives trying to answer that question, but few ever succeed in getting even a hint. That is why an introductory astronomy course is an exciting experience for both instructor and student. On the physical level, at least, astronomy gives us insight into what we are and where we came from.

If our goal is to describe our place in the universe, our role in nature, then we cannot think of astronomy as an accumulation of facts. Facts are merely data and can only be meaningful when they are synthesized into a consistent description of nature. Consequently, this book views astronomy as a small number of basic physical processes that are responsible for a wide variety of phenomena and that explain a diverse assortment of objects. For example, galaxies, star clusters, individual stars, and planets are all expressions of the same process—gravitational contraction. Only the scale is different. Emphasizing processes presents astronomy not as a collection of unrelated facts, but as a unified body of knowledge.

The most important astronomical process is evolution—the gradual, natural, irrestible transformation of a celestial body as it passes through stages in its formation, development, and ultimate dissolution. Stellar evolution is the principal example of astronomical change, and the student must understand stellar evolution in order to understand the evolution of other celestial bodies. Of course, the ultimate evolutionary question is the origin of things, including the origin of the universe, galaxies, stars, planets, and life. The study of astronomical origins is an important part of this book.

The use of physical processes as a unifying theme and the importance of stellar evolution dictate an organization that deals with stars, galaxies, cosmology, and the solar system, in that order. Stars must come first because the star is the fundamental unit of structure on the astronomical scale. However, since most students cannot leap into a discussion of star formation in Chapter 1, the introduction and first three chapters provide a transition from an earth-bound frame of reference to an astronomical view of nature.

An Historical Supplement, appearing between Chapters 2 and 3, discusses the history of astronomy up to the time of Newton. This is an important part of most astronomy courses, but, because instructors vary in their scheduling of this material, it could be introduced at any point after Chapter 1.

Because this book is intended for nonscience majors, it does not rely on mathematical reasoning. However, because science is fundamentally a quantitative discipline and because students should see that aspect of science, some mathematical discussions are included in boxes. None of these mathematical boxes is necessary to the development of the astronomical principles, and they can be deleted. For those who wish to include mathematics, examples have been worked in the boxes, and problems requiring mathematical skills have been placed at the end of each chapter.

## TEACHING AIDS

Each chapter ends with a group of aids to the student and instructor. The final section of several chapters is called a Perspective to set it off from the rest of the text. It serves to introduce a new and interesting idea that will allow the student to review and apply the principles covered in the chapter. Some Perspectives discuss the development of a theory, the synthesis of hypotheses from data, the testing of theories by observation, and the meaning of statistical evidence. Others describe some of the controversies of modern astronomy, showing that clashes between contending theories are an important and valuable part of science.

Additional end-of-chapter materials include a chapter summary, a list of new terms, questions, problems, and recommended readings. The first time the new terms appear in the text they are in **boldface.** These same terms appear in the glossary. The recommended reading is intended for the student and ranges from *National Geographic* to *Science*. Instructors may wish to guide students in selecting appropriate reading material. The questions are nonquantitative and would lead to essay answers or could be used to stimulate class discussion. The problems are quantitative or involve mathematical reasoning. Answers to even-numbered problems appear at the end of the book, and answers to odd-numbered problems are given in the *Instructor's Resource Manual.*

The end-of-book material consists of a glossary, four appendices, answers to even-numbered problems, an index, and star charts. The glossary contains all of the new terms that appear in boldface in the text, plus some other words that may arise in classroom discussions.

Appendix A describes right ascension, declination, and sidereal time. Instructors who wish to use this material could assign it at the end of Chapter 1.

Appendix B and C contain supplementary tables. Appendix B discusses exponential notation, the metric system, units of measurement, temperature scales, and fundamental constants. Appendix C contains astronomical data, including lists of the Greek alphabet, the constellations, the nearest stars, the brightest stars, the properties of main-sequence stars, the properties of planets and satellites, meteor showers, and the Messier catalogue.

Finally, the star charts included are adaptations of the monthly charts from the *Griffith Observer.* I would like to thank E. C. Krupp, the editor of the *Griffith Observer,* for making the charts available, and encourage both instructors and students to use them and enjoy the sky.

## ACKNOWLEDGMENTS

Among the reviewers who helped in the development of this book were David R. Alexander, Wichita State University; DeWayne Backhus, Emporia State University; Stephen Hill, Michigan State University; H. W. Ibser, California State University at Sacramento; Stephen Lattanzio, Orange Coast College; L. A. Mink, Arkansas State University; James Mullaney, Community College of Allegheny County; C. W. Price, Millersville State College; Michael M. Shurman, University of Wisconsin; Harding E. Smith, University of California at San Diego; Michael Stewart, San Antonio College; Walter G. Wesley, Moorhead State University; Raymond E. White, University of Arizona; and Benjamin Zellner, University of Arizona. In addition, I want to thank Roger Thomas and Ira Feit for reading large parts of the material on planetary astronomy and the origin of life, John McDermott for his helpful comments on biological questions, Joseph Holzinger for his assistance with astronomical and mathematical points, Peter Usher for his comments on QSOs, and the Franklin and Marshall Evolution Roundtable for their interesting discussions. Thanks are due Satoshi Matsushima and the members of the department of astronomy at Pennslyvania State University for their helpful comments and for making me welcome during the sabbatical when I developed the overall design for the book.

I wish to thank those people listed in the illustration credits for kindly providing photographs and diagrams. I would especially like to thank Philip Bedient and William Langs for locating some of the stamps used for illustrations, and Dean Richard Traina and Prof. Angela Jeannet for bringing Italian banknotes all the way from Sicily and Rome. Also, special thanks to Martin Burkhead for providing two photographs that were otherwise unavailable.

My appreciation also goes to the staffs of the following institutions for their assistance in providing photographs and diagrams: AstroMedia Corporation, Ames Research Center, Bell Laboratories, Brookhaven National Laboratories, Celestron International, Hale Observatories, High Altitude Observatory, Jet Propulsion Laboratories, Johnson Space Center, Kitt Peak National Observatory, Lowell Observatory, Lunar and Planetary Laboratory, Martin Marietta Aerospace Corporation, National Aeronautics and Space Administration, National Radio Astronomy Observatory.

Special thanks go to the editorial staff at Wadsworth for their assistance: to Mike Snell for seeing this project underway, to Autumn Stanley for extensive help in writing development, and to Marshall Aronson for bringing it all to a successful conclusion.

Finally, I must thank my wife Janet for putting up with a basement author and for being my most valuable reviewer.

Michael A. Seeds
Lancaster, Pennsylvania
31 January 1980

# INTRODUCTION

Facts are like uncut gems—of value because they can be cut and polished and placed in a pleasing and meaningful relation to each other. For example, here is an astronomical fact. Orbiting telescopes have detected x-rays coming from a faint blue star in the summer sky. Not very interesting. But later in this book we will polish this fact and set it in a meaningful relationship with hundreds of other facts involving the nature of stars and their lives and deaths, and we will conclude that the x-rays may be coming from a black hole that is ripping matter from its companion star and swallowing it in a whirlpool of intensely hot gas. Facts are only interesting when they are properly interpreted.

We need facts, of course, as the basis for our study, and we will therefore spend some time discussing how astronomers discover the properties of celestial bodies. In this respect, astronomy is a series of complicated puzzles in measurement whose solutions are intriguing in themselves. Thus one of the things you should gain from an astronomy course is some insight into the way astronomical measurements are made.

However, the properties of stars, galaxies, and planets are only the cold raw materials on which we will base astronomical theories. Our most important goal is to piece together the available data and learn how astronomical objects change with time—how they evolve. We will discover that every object in the universe is temporary. The earth is changing and so is the sun: In only a few billion years the sun will die and life on earth will end. All of the stars we see in the sky are evolving and will, sooner or later, die. In fact, some theories predict that the entire universe may someday come to an end. Thus we want to know what rules govern the evolution of celestial bodies and how those bodies will change in the future.

In addition, we want to know the origins of things. How did the earth form, and how did the sun and other stars take shape? We will even discuss the origin of life on earth and the possibility that it exists on other planets. An astronomy course should give you some insight into how matter and energy combined to create the universe in which you live.

*Astronomy as a Science.* Astronomy is a science, which means that we should base our conclusions on evidence and test those conclusions whenever possible. When we meet something we don't understand, we must consider the available evidence, form a hypothesis, and then test that hypothesis by making further measurements or observations. We must then keep, change, or abandon the hypothesis as the evidence indicates. This is sometimes called the scientific method, but as you can see, it is simply a logical way to solve a problem.

The existence of a method does not give astronomers a foolproof recipe for moving from raw data to completed theory. Human ingenuity and creativity are absolute necessities. Forming a hypothesis from raw data requires insight and imagination, and only a truly creative person can devise an experiment to test a generally accepted hypothesis. Astronomy, as mentioned above, is a series of puzzles, and solving puzzles is one of the most human of activities.

Clearly, measurements are important in a science, nowhere more so than in astronomy; and if we measure things we must use some system of units. For most quantities we will use the metric system, giving the English equivalent in parentheses when useful. We will make temperature measurements on the Kelvin temperature scale. The metric system and the Kelvin temperature scale are summarized in Appendix B.

Astronomical numbers are often enormous. (In fact, *astronomical* has come into the language as a synonym for "excessively large," as in "an astronomical rise in the cost of living.") For example, the nearest star is about 41 trillion km (25 trillion mi) away. We will cope with these numbers by expressing them in scientific notation. That is, instead of writing 41 followed by 12 zeros, we will write $41 \times 10^{12}$. This handy system of notation is summarized in Appendix B.

Another way of dealing with large numbers is to invent new units. One such unit is the light-year (ly), the distance light travels in a year. One light-year is about $5.8 \times 10^{12}$ mi, and the nearest star is about 4.3 ly away. We will define a few other special units later as we need them.

*Four Questions.* Although there are many ways of organizing a study of astronomy, we will concentrate on the answers to four fundamental questions. First, *What does the sky look like and how do its motions arise?* When we look at the night sky, we can see the stars, the moon, and five of the planets in our solar system. Though the stars move very slowly, the moon and planets move relatively quickly across the sky as they follow their orbits. In addition, we live on a spherical planet that rotates on its axis once a day and orbits the sun once a year. These motions produce corresponding motions in the sky as we see it. Our view of the universe corresponds to the view of an amusement park seen from the spinning car of a carnival ride. As we try to answer our first question, we will gain an insight into the relationship of the earth to the rest of the universe.

Our second question is, *What are stars?* The universe is filled with stars and we must understand how they work if we are to understand how nature acts on the astronomical scale. Modern astronomers have deduced that stars are enormous, hot balls of gas. Our sun is a perfectly normal star. It is 109 times larger in diameter than the earth, and its surface temperature is about 6000°K. The temperature at its center is about 14,000,000°K.

Stars are held together by their own gravity, producing very high pressure and temperature near their centers. Under these extreme conditions, atoms fuse together in nuclear reactions, releasing tremendous amounts of energy. Since most stars are about 80 percent hydrogen, we can think of a star as a giant hydrogen bomb continuously exploding. The star is caught in a tug of war between its gravity, which tries to make it collapse into itself, and the nuclear reactions at its center, which try to blow it apart. When the fuel runs out, the nuclear reactions stop, gravity wins, and the star collapses. How stars die is a subject we will save for a later chapter.

We must study stars not only as individuals, but in larger groupings. All of the stars we see in the sky are part of our Milky Way galaxy, a great wheel of stars 100,000 ly in diameter. The sun is just one of the 100 billion stars in the galaxy, each moving in its own orbit around the center. In addition, our galaxy is merely one of millions of galaxies scattered through space. Some, like our own, are disk-shaped and show beautiful spiral patterns of bright stars. Others are plain, featureless clouds of stars.

We understand only part of the story of galaxies. Their evolution is connected to the way stars live and die, but many of the details are still unknown.

The third question we will consider is, *How did the universe itself begin?* Modern astronomers have found some exciting clues to the origin of the universe, but the clues don't quite fit together. At the moment we have a theory called the big bang. According to this theory, the universe began 10 to 20 billion years ago with a violent explosion. As the fragments of the explosion flew away from each other, they became gas clouds and then formed galaxies of stars.

We will find a number of important clues to the origin of the universe. For example, when we look at galaxies we will find they appear to be moving away from each other, suggesting that the universe is still expanding from the original explosion.

Finally, our fourth question: *What is the origin of the solar system?* The nine planets that circle the sun should be easier to study than distant galaxies, but planets are very complicated objects. We can't be sure of the details, but we can infer that the planets formed with the sun about 5 billion years ago. Since that time, the planets have evolved in complicated ways to reach their present state. If we can unravel the processes that affected the ancient earth, we may uncover clues to the origin of life and the future of our planet. That may even help us consider the possibility that life exists on other planets orbiting other stars.

*Why Bother?* Why should you study astronomy? One reason is that the space age has begun and you will live well into the 21st century. You may see the colonization of the moon and Mars; your children may even be colonists. If you are to understand what is happening, you must know something about astronomy.

Another reason is that you probably have some natural curiosity about astronomy. You have probably heard about black holes, neutron stars, and the rings of Saturn, and you may want to know more about them. Satisfying your own curiosity is the most noble of reasons for studying astronomy. Curiosity leads some individuals to become astronomers, though you will probably not do so—the profession is small and jobs are scarce. But if you find astronomy interesting, consider adopting it as a hobby. The magazines listed here will keep you up to date with the rapid advances in the field and give you some ideas for further projects such as telescope building and photography.

**Nontechnical Magazines**

*Astronomy* 411 East Mason St., P.O. Box 92788, Milwaukee, WI 53202

*The Griffith Observer* 2800 East Observatory Road, Los Angeles, CA 90027

*Mercury* Astronomical Society of the Pacific, 1290 24th Ave., San Francisco, CA 94122

*Sky & Telescope* Sky Publishing Corporation, 49 Bay State Rd., Cambridge, MA 02238

*Star & Sky* 44 Church Ln., Westport, CN 06880

One reason for taking a college course is to prepare for a particular profession. If you become a lawyer, your government courses will be valuable, and if you go into business, your mathematics and accounting courses will pay dividends. In a sense, you are taking these courses now and will later trade the knowledge you gain to an employer in return for a good job. Astronomy is different because it is just for you. It will show you our tiny planet spinning in space amid a vast cosmos of stars and galaxies. It will take you from the first moment of creation to the end of the universe. You will see our planet form, life develop, and our sun die. This knowledge has no monetary value, but it is priceless if you are to appreciate your existence as a human being.

Astronomy will change you. It will not just expand your horizons, it will do away with them. You will see humanity as part of a complex and beautiful universe. If by the end of this course you do not think of yourself and society differently, if you don't feel excited, challenged, and a bit frightened, then you haven't been paying attention.

*The four-day-old crescent moon and Venus setting. Exposures were made every eight minutes, illustrating how the eastward rotation of the earth causes celestial objects to move westward across the sky. Note also the motion of the moon relative to Venus. (Photo by William P. Sterne, Jr.)*

# Chapter 1 THE EARTH AND SKY

From a city, the night sky is unimpressive; a murky haze of air pollution and city lights hides the stars. Only a few miles away, however, in the open countryside, the sky is a mysterious star-studded ceiling. It is not surprising that the ancients held the sky in awe and peopled it with their greatest gods and heroes. We begin our study of astronomy as ancient astronomers began, by identifying stars, estimating their brightness, and grouping them into constellations.

However, our goal in this chapter is not to admire, but to expand our perception of the earth and sky. We must learn to think of earth as a planet turning on its axis and moving in its orbit around the sun. More important, we must abandon a mystical view of the sky and see it as the natural appearance of the universe seen from our planet.

The sky as a whole looks to us like a great sphere surrounding the earth and rotating slowly, carrying the stars westward across the sky. Actually, of course, it is the earth that turns, making the sky seem to rotate around us, and the thousands of stars are really other suns scattered through space—some nearby and some far away.

This chapter will not answer all of our questions about the universe. We are merely setting the scene for the challenging explorations to follow. Clearly before we can begin, we must take this first look at the sky.

## THE STARS

*Constellations.* Gazing at the night sky on a clear, moonless evening, we can see thousands of stars scattered in random groups. Some of these groups have been named and are called **constellations** (Figure 1-1). Though we identify these patterns by name, we must keep in mind that the stars in a constellation are usually not physically associated with each other. Some may be many times farther away than others and moving in different directions. The only thing they have in common is that they lie in approximately the same direction from earth.

About half of today's 88 constellations come from the ancient Greeks. Many of the constellation names—Scorpio, Libra, Leo, Hercules, Pegasus—are familiar from Greek mythology. Nevertheless, the names are Latin, the language of science from the Renaissance into the 19th century. However, as we will see later, the names of many of the brighter stars come from Arabic.

To the ancients, a constellation was a loose grouping of stars that represented a certain figure. A star could even belong to more than one constellation, as in the case of Alpheratz in the constellations of Andromeda and Pegasus (Figure 1-2a). Modern astronomers have given each constellation definite boundaries. Thus a constellation represents not just a group of stars, but an area of the sky. Any star within the

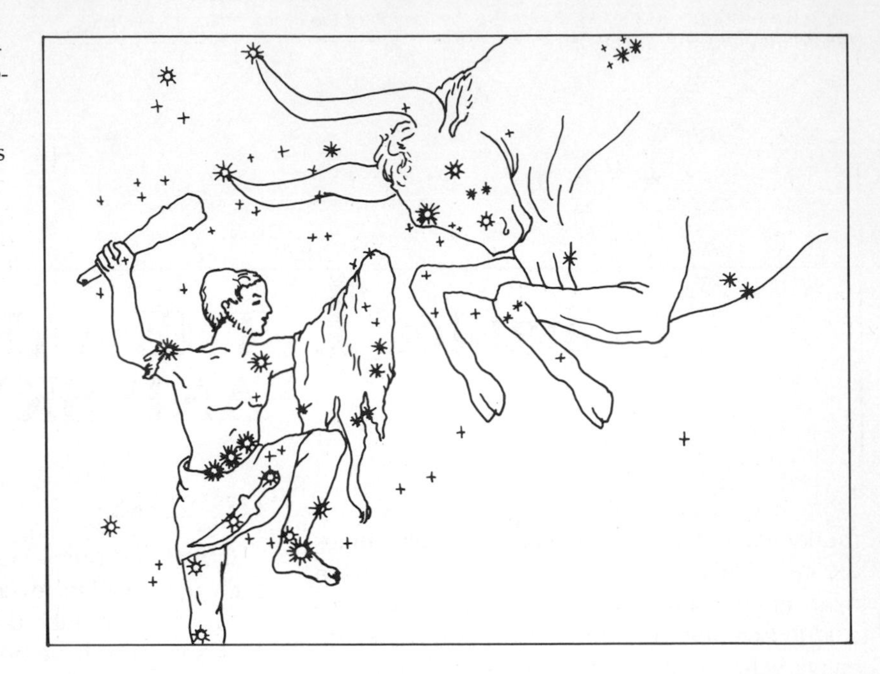

*Figure 1-1. The constellations Orion and Taurus represent figures from Greek mythology. (Adapted from Duncan Bradford,* Wonders of the Heavens, *Boston: John B. Russell, 1837.)*

specified area is a part of the particular constellation. This solves the problem of Alpheratz, making it an official member of Andromeda (Figure 1-2b).

Unfortunately, the ancient astronomers ignored parts of the sky where there were few bright stars, and those who devised our constellations could not see parts of the southern sky from their northern latitudes. When modern astronomers began to redraw the star charts, they found empty spaces where there were no constellations. To fill these spaces, they invented over 40 modern constellations. Although given Latin names, the objects represented are sometimes strangely modern, such as Telescopium (the telescope), Microscopium (the microscope), and Antlia (the air pump). Box 1-1 introduces the brighter constellations.

Modern astronomers still use the constellations as a convenient way of referring to areas of the sky. But merely naming the constellation in which a certain star is found does not identify it uniquely. Thus we must discuss ancient and modern ways of naming stars.

*The Names of the Stars.* Greek and Arab astronomers named the brightest stars thousands of years ago, and modern astronomers still use many of these names. Although the names of the constellations are in Latin, most star names come from ancient Arabic. Such names as Sirius (the Scorched One) and Vega (the Falling Eagle) are beautiful additions to the mythology of the sky.

However, there aren't enough names for all of the thousands of stars we can see, and these names do not help us locate the star in the sky. Another way to identify stars is to assign Greek letters to the bright stars in a constellation in approximate order of brightness. Thus the brightest star is usually designated $\alpha$ (alpha), the second brightest $\beta$ (beta), and so on. For almost all constellations the letters follow the order of brightness (Figure 1-3). To identify a star in this

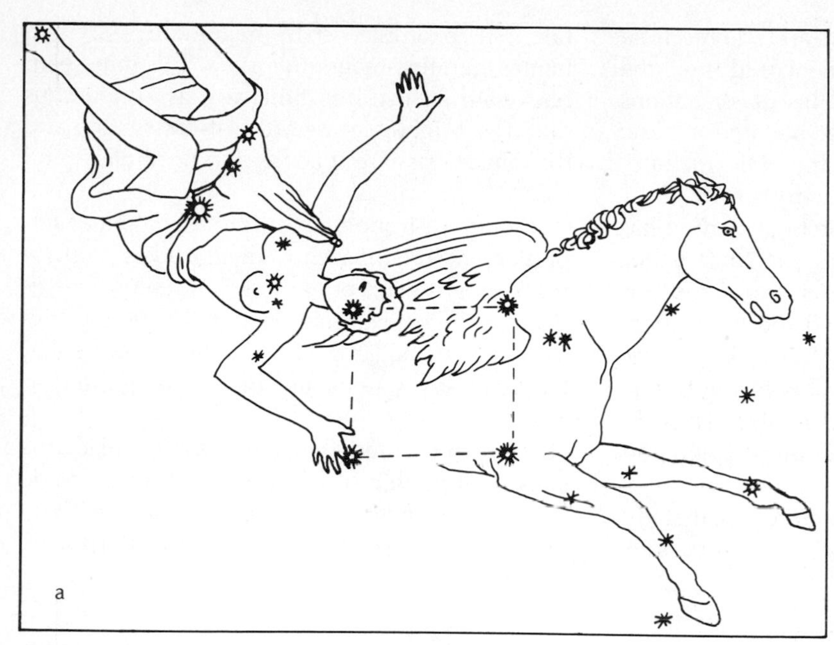


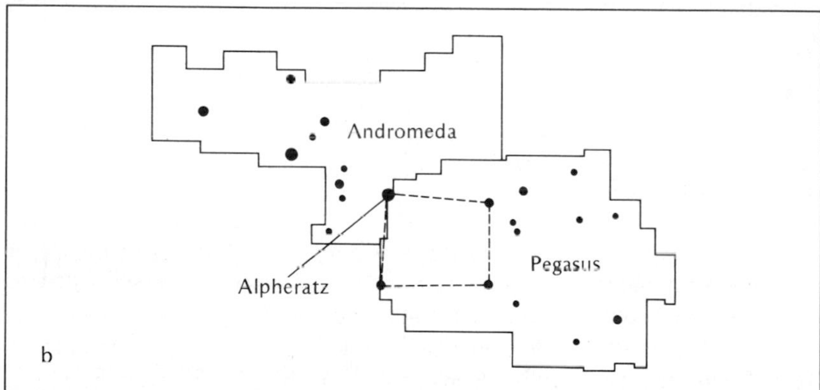


*Figure 1-2.* *(a) The ancient constellations Andromeda and Pegasus share the star Alpheratz at the upper left corner of the great square of Pegasus. (Adapted from Duncan Bradford,* Wonders of the Heavens, *Boston: John B. Russell, 1837.) (b) The modern constellation boundaries assign Alpheratz to Andromeda.*

way, we give the Greek letter followed by the genetive form of the constellation name, such as $\alpha$ Canis Majoris. This both identifies the star and constellation and gives us a clue to the brightness of the star. Compare this with the ancient name for this star, Sirius, which tells us nothing about location or brightness.

This method of identifying a star's brightness is only approximate. In order to discuss the sky with precision, we must have an accurate way of referring to the brightness of stars, and for that we must consult one of the first great astronomers.

*The Brightness of Stars.* Hipparchus, a Greek astronomer who lived about 2100 years ago, divided the stars into six classes. The brightest were first-class stars, and those slightly fainter were second-class stars. Continuing down to the faintest stars he could see, the sixth-class stars, he recorded his classifications in a great star catalogue that became a basic reference in ancient astronomy. His method, slightly modified, is still in use today.

In spite of its value, Hipparchus' method has confused astronomy students for 2000 years. First, when early astronomers translated the catalogue into Latin they used the word *magnitudo,* meaning size. In English, this became magnitude, even though it refers to the brightness of the stars and not to their size. Thus the **magnitude scale** is the astronomers' brightness scale.

The second source of confusion is that the fainter the star, the larger the magnitude number. For example, sixth-magnitude stars are fainter than first-magnitude stars. This may seem backward at first, but think of it as Hipparchus did. The brightest stars are first-class stars, and the fainter stars are second- and third-class, and so on.

Modern astronomers have made a major improvement in Hipparchus' magnitude system by measuring stellar brightness with sensitive instruments. For example, instead of merely saying that $\theta$ (theta) Leonis is a third-magnitude star, they can say specifically that its magnitude is 3.34.

If we measured the brightness of all of the stars in Hipparchus' first brightness class, some would be brighter than 1.0. For instance, Vega ($\alpha$ Lyrae) is so bright its magnitude is almost zero

## BOX 1-1 KEY CONSTELLATIONS

Constellations are difficult to learn because the constellations above the horizon change with the seasons and with the time of night. To simplify the process, use the descriptions in this box in conjunction with the monthly star charts at the end of this book to find one of the key constellations. We assume you are observing in the evening a few hours after sunset.

In the summer, soon after sunset, look for a very bright star nearly overhead. This star is Vega in the key constellation Lyra (the Lyre). Consult the appropriate star chart from the back of the book to find Hercules, Corona Borealis (the Northern Crown), and Bootes (the Bear Driver) to the west of Lyra. East is Cygnus (the Swan—also known as the Northern Cross), and southeast is Aquila (the Eagle).

Early in the autumn Lyra is nearly overhead as darkness falls, but by about October it is in the western sky and Pegasus (the Winged Horse) becomes the key constellation high in the east. Once you've found Pegasus, look for Andromeda.

As winter comes, Pegasus moves into the western sky in the evenings. Starting in December look for Orion (the Hunter) in the southeast sky. When you have found Orion, you can find the surrounding constellations, including Canis Major (the Big Dog), which contains the brightest star in the sky, Sirius.

As winter passes and spring approaches, Orion moves into the southwestern quadrant of the evening sky. Beginning in late March look for the sickle shape of the key constellation Leo (the Lion). West of Leo is Cancer (the Crab), a faint constellation that is hard to find. East of Leo, look for the kite shape of Bootes (the Bear Driver). By late spring and early summer, Leo is in the western sky in the evening. Beginning in June look again for the summer key constellation, Lyra, in the east.

A few constellations of the northern sky appear in Figure 1-4. To use this chart face north soon after sunset and hold the chart directly in front of you. Turn it until the current date is on top and the chart will represent the sky in front of you. Locate the Big Dipper and note that it is not a constellation. It is actually part of Ursa Major (the Great Bear), just as the Little Dipper is part of Ursa Minor (the Small Bear). At the tip of the handle of the Little Dipper is the North Star, Polaris.

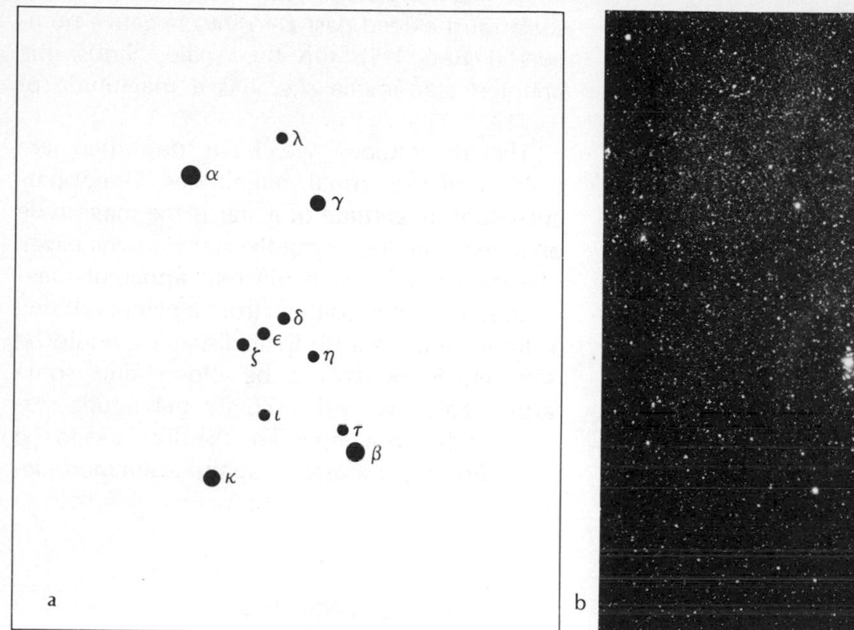


*Figure 1-3.* *(a) The brighter stars in each constellation are assigned Greek letters in approximate order of brightness. In Orion, κ is brighter than its Greek letter suggests. (b) A long exposure photograph reveals the many faint stars that lie within Orion's constellation boundaries. These are members of the constellation but they do not have Greek letter designations. (Hale Observatories.)*

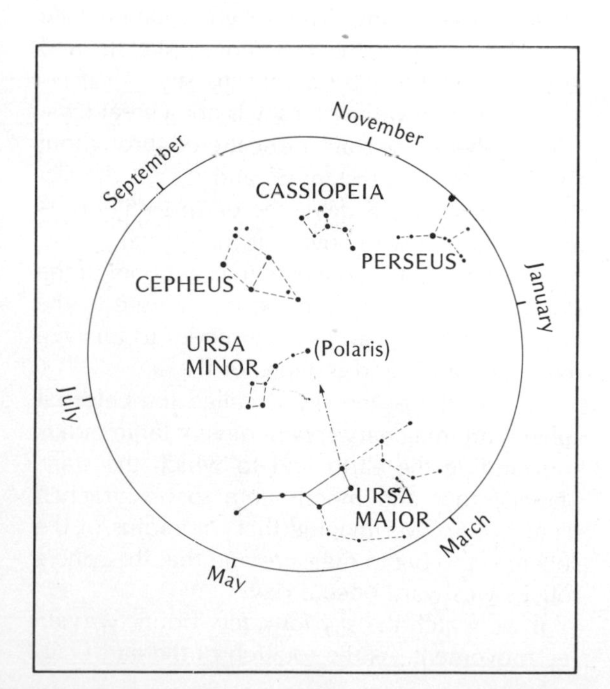


*Figure 1-4.* *The northern constellations. To use the chart, face north soon after sunset and hold the chart in front of you with the current date at the top.*

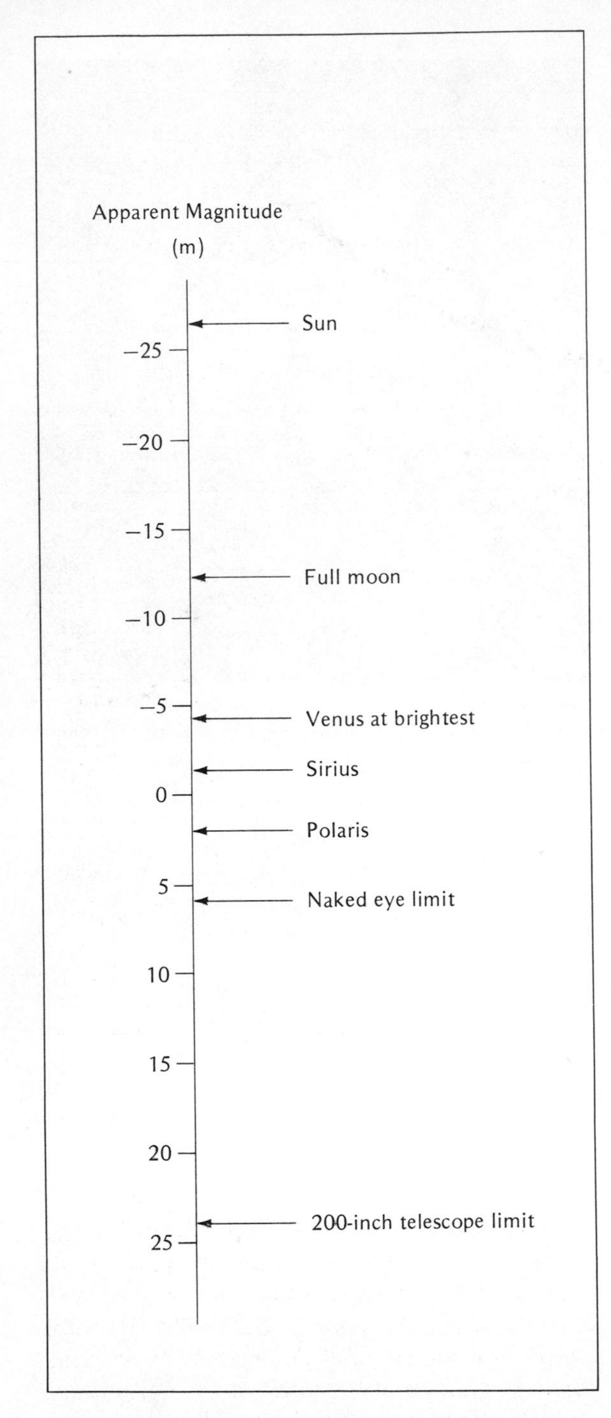


*Figure 1-5. The scale of apparent visual magnitudes extends into negative numbers to represent the brighter objects.*

at 0.04. A few stars are so bright the magnitude scale must extend past zero into negative numbers (Figure 1-5). On this scale, Sirius, the brightest star in the sky, has a magnitude of −1.42.

The magnitudes we have described are called apparent visual magnitudes. The **apparent visual magnitude** of a star is the magnitude obtained from looking at the star from the earth. The stars would have different apparent magnitudes if we viewed them from a planet orbiting a distant star because their distances would be different. Some would be closer and some farther. Later we will discuss a magnitude system that does not depend on the distances to the stars. (Box 1-2 discusses magnitudes in more detail.)

## THE CELESTIAL SPHERE

*A Model of the Sky.* Ancient astronomers thought of the sky as a great crystalline sphere surrounding the earth. The stars, they imagined, were attached to the sphere like thumbtacks stuck in the ceiling. The sphere rotated once each day, carrying the sun, moon, planets, and stars from east to west across the sky.

We know now that the sky is not a great crystalline sphere. The stars are scattered throughout space at different distances, and it isn't the sky that rotates once a day—the earth turns on its axis. Although we know that the crystal sphere does not exist, it is convenient as a model of the sky. As long as we keep the true nature of the sky in mind, we can use the model to analyze the appearance and motions of the sky.

Our model of the sky is called the **celestial sphere,** an imaginary sphere of very large radius surrounding the earth and to which the stars, planets, sun, and moon seem to be attached (Figure 1-6). We imagine that the radius of the sphere is too big to measure and that the sphere rotates westward once a day.

If we watch the sky for a few hours, we can see movement. As the rotation of the earth car-

## BOX 1-2 MAGNITUDES

The amount of visible light we receive from a star can be called its brightness. If we represent the brightness of two stars A and B as $I_A$ and $I_B$, then the ratio of their brightnesses is $I_A/I_B$. This brightness ratio tells us how many times brighter star A is than star B.

Modern astronomers have defined the magnitude scale so that two stars whose magnitudes differ by 5 magnitudes have a brightness ratio of exactly 100. That is, if star A is 5 magnitudes brighter than star B, then star A must be 100 times brighter than star B. Then two stars that differ by 1 magnitude have a brightness ratio of 2.512—that is, one star is about 2.5 times brighter than the other. Two stars that differ by 2 magnitudes will have a brightness ratio of 2.512 × 2.512 or about 6.3, and so on (Table 1-1).

A table of brightness ratios makes magnitude calculations simple. For example, consider two stars C and D. If star C is third magnitude and star D is ninth magnitude, how many times brighter is star C? It is 6 magnitudes brighter, and from the table we find that the brightness ratio is 250. Thus star C looks 250 times brighter to us than star D.

This is a simple example because the magnitude difference is a whole number, 6. If it were not a whole number, we would have to estimate the proper brightness ratio from the table, or calculate the brightness ratio directly. The brightness ratio $I_A/I_B$ is equal to 2.512 raised to the power of the magnitude difference $m_B - m_A$.

$$\frac{I_A}{I_B} = (2.512)^{(m_B - m_A)}$$

For example, if the magnitude difference was 6.32 magnitudes, the brightness ratio would be

$$\frac{I_A}{I_B} = (2.512)^{6.32}$$

A pocket calculator shows that the brightness ratio is 337.

**Table 1-1 Magnitude and Brightness**

| Magnitude Difference | Brightness Ratio |
|---|---|
| 0 | 1 |
| 1 | 2.5 |
| 2 | 6.3 |
| 3 | 16 |
| 4 | 40 |
| 5 | 100 |
| 6 | 250 |
| 7 | 630 |
| 8 | 1600 |
| 9 | 4000 |
| 10 | 10,000 |
| . | . |
| . | . |
| . | . |
| 15 | 1,000,000 |
| 20 | 100,000,000 |
| 25 | 10,000,000,000 |
| . | . |
| . | . |
| . | . |

ries us eastward, the sun moves across the sky and sets in the west. As it gets dark, we can see the stars, and in a few hours it becomes obvious that the rotation of the earth is making the sky rotate westward. As some constellations set in the west, others rise in the east.

*Reference Marks on the Sky.* The pivots about which the sky seems to rotate are called the celestial poles. The **north celestial pole** is the point on the sky directly above the earth's north pole, and the **south celestial pole** is the point directly above the earth's south pole. Stars located near the celestial poles seem to describe small circles about the poles as the earth turns. A time exposure photograph of the sky shows curved streaks made by the stars as the sky rotates (Figure 1-7).

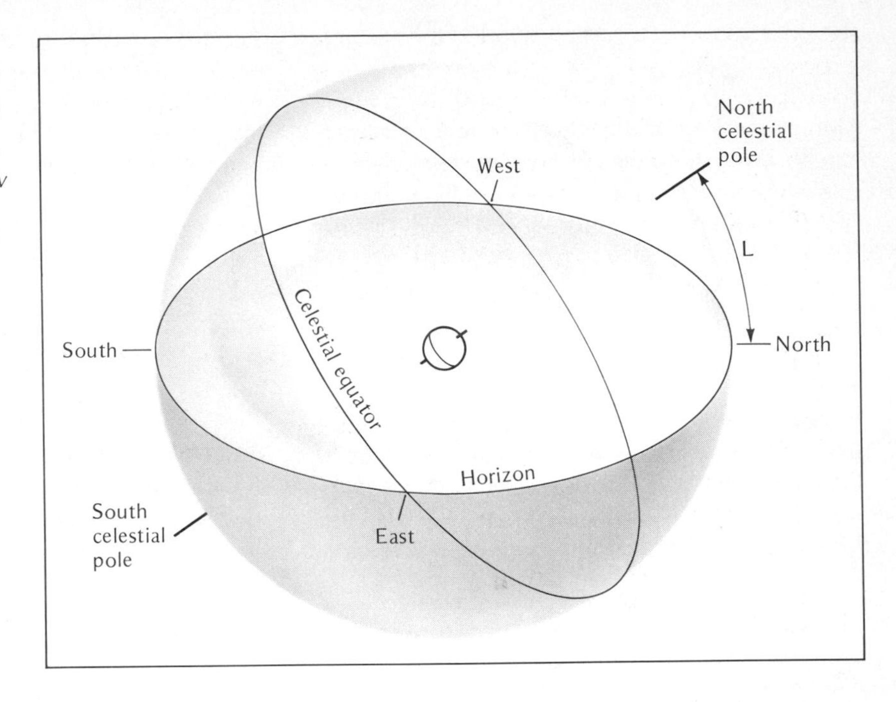


*Figure 1-6.* *The modern celestial sphere models the appearance of the sky. The poles mark the pivots, and the equator divides the sky in half. Those objects below our horizon are invisible. The angle L is equal to the observer's latitude.*

*Figure 1-7.* *A time exposure taken with a camera pointed at the north celestial pole shows star trails that demonstrate the rotation of the sky. (Lick Observatory photograph.)*

The location of the north celestial pole depends on the latitude of the observer. For example, if we stood in the ice and snow at the earth's north pole, the north celestial pole would be directly overhead. If we stood on the earth's equator, the north celestial pole would lie on our northern horizon. At intermediate latitudes, such as those of the United States, the north celestial pole lies about halfway between overhead and the northern horizon, as in Figure 1-6. To be precise, the angular distance from the horizon up to the north celestial pole equals the latitude of the observer. This relationship, by the way, makes it simple for navigators to find their latitude by simply measuring the angle between the northern horizon and the north celestial pole.

The south celestial pole is not visible from the United States, since it lies below our horizon, but we can locate the north celestial pole easily by looking for the Big Dipper and following the pointer stars, as shown in Figure 1-4.

The star Polaris happens to lie very near the north celestial pole, and thus hardly moves as the sky rotates. This is shown in the 8-hour time exposure in Figure 1-7 where the bright arc near the center was made by Polaris. At any time of night, in any season of the year, Polaris always stands above the northern horizon and is consequently known as the North Star. Later we will see that other stars have occupied this location in the past.

Another important reference mark on the celestial sphere is the **celestial equator,** an imaginary line around the sky directly above the earth's equator (Figure 1-6). The celestial equator divides the sky into two equal halves, the northern and southern celestial hemispheres.

These reference marks on the sky will be useful trail markers as we explore. They will help us understand the motion of the sun and moon, the seasons, eclipses, and so on. Before we can explore further, however, we must see how to measure angles in the sky.

*Angles in the Sky.* Astronomers often use angles to describe distance across the sky. They might say, for instance, that the moon was 8° north of a certain star, meaning that if we point one arm at the moon and the other arm at the star, the angle between our arms is 8°. With this system astronomers can describe the extent and location of any celestial object.

We measure angles in degrees, minutes of arc, and seconds of arc. There are 360° in a circle, and 90° in a right angle. Each degree is divided into 60 **minutes of arc** (sometimes abbreviated 60′). If you view a 25¢ piece from the

---

BOX 1-3 THE SMALL ANGLE FORMULA

One important use of angles in astronomy is the determination of the linear (or true) diameters of celestial objects from their angular diameters. Linear diameter is just the distance between an object's opposite sides, commonly measured in meters or kilometers. The angular diameter of an object is the angle formed by lines extending from opposite sides of the object and meeting at our eye. Clearly, the farther away an object is, the smaller its angular diameter.

The relationship between the angular diameter **a,** the linear diameter **D,** and the distance **d** to the object is given by the small-angle formula:

$$D = d\frac{a}{206265}$$

The number 206265 is the constant needed because **a** is measured in seconds of arc. Both **D** and **d** must be expressed in the same unit of length—meters, kilometers, or whatever is convenient.

Consider an earthly example. Suppose we saw an approaching automobile a mile away and noted that its angular diameter was about 5 minutes of arc. We can find its linear diameter from the small-angle formula, but we must express **a** in seconds of arc. Since each minute of arc contains 60 seconds of arc, 5 minutes of arc equals 300 seconds of arc. We can express the distance in any unit we wish, but the linear diameter will be in the same units. For convenience, we can use 5280 feet instead of 1 mile. Putting these into the formula we find the diameter.

$$D = 5280\frac{300}{206265} = 7.7 \text{ feet}$$

In astronomy, we might use the small-angle formula to find the diameter of a planet. For example, we might observe that when Jupiter is $7.48 \times 10^8$ km from earth, it has an angular diameter of 39 seconds of arc. Then its linear diameter can be found from the small angle formula.

$$D = (7.48 \times 10^8)\frac{39}{206265} = 141{,}000 \text{ km}$$

The answer is in kilometers because the distance we used was in kilometers. Thus Jupiter is about 141,000 km in diameter, over eleven times earth's diameter.

length of a football field, it has an angular diameter of about 1 minute of arc. Each minute of arc is divided into 60 **seconds of arc** (sometimes abbreviated 60″). If you view a dime edgeways from the length of a football field, it is about 7 seconds of arc thick.

We can establish some helpful angles on the sky. The sun and the moon are each about 0.5° in diameter. The pointer stars in the Big Dipper are about 5° apart, and the Big Dipper is about 30° from the north celestial pole.

Astronomers locate objects on the sky by measuring their angular distance from reference marks such as those we have established. This forms a celestial coordinate system that works much like the system of latitude and longitude on earth (see Appendix A). And that makes it possible to be precise in locating objects in the sky.

We now have a model of the sky—the celestial sphere—complete with reference marks to guide us. In addition, we can measure angles on the sky to give the locations and diameters of celestial objects. Box 1-3 explains how we could use these angular measurements to determine the actual diameters of objects in kilometers. With these tools, we are ready to investigate the motions of objects in the sky.

## SUMMARY

Astronomers divide the sky into 88 areas called constellations. Although the constellations originated in Greek mythology, the names are Latin. Even the modern constellations, added to fill in between the ancient figures, have Latin names. The names of stars usually come from ancient Arabic, though modern astronomers often refer to a star by constellation and Greek letters assigned according to brightness within each constellation.

The magnitude system is the astronomer's brightness scale. First magnitude stars are brighter than second magnitude stars, which are brighter than third magnitude stars, and so on. The magnitude we see when we look at a star in the sky is its apparent visual magnitude.

The celestial sphere is a model of the sky, carrying the celestial objects around the earth. Because the earth rotates eastward, the celestial sphere appears to rotate westward on its axis. The northern and southern celestial poles are the pivots on which the sky appears to rotate. The celestial equator, an imaginary line around the sky above the earth's equator, divides the sky in half.

## NEW TERMS

constellation
magnitude scale
apparent visual magnitude
celestial sphere
north and south celestial poles
celestial equator
minute of arc
second of arc

## QUESTIONS

1. Why have astronomers added modern constellations to the sky?
2. What information does a star's Greek letter designation often contain?
3. Give two reasons why the magnitude scale might be confusing.
4. Describe the difference between the ancient and the modern conception of the celestial sphere.
5. How do we define the locations of the celestial poles and the celestial equator?

## PROBLEMS

1. If one star is 40 times brighter than another star, how many magnitudes brighter is it?
2. If two stars differ by 8.6 magnitudes, what is their brightness ratio?
3. If star A is fourth magnitude and star B is sixth magnitude, which is brighter? By what factor?
4. By what factor is the sun brighter than the full moon? (Hint: See Figure 1-5.)
5. Sketch the celestial sphere and label the poles, equator, and horizon.
6. When Mars is $80 \times 10^6$ km from earth, it has an angular diameter of 17.5 seconds of arc. What is its linear diameter in kilometers? Is it larger or smaller than earth? (Hint: earth's diameter = 12,756 km.)

## RECOMMENDED READING

Holzinger, J. R., and Seeds, M. A. *Laboratory Exercises in Astronomy*. Ex. 5, 17, 23, and Appendix A. New York: Macmillan, 1976.

Menzel, D. H. *A Field Guide to the Stars and Planets*. Boston: Houghton Mifflin, 1964.

Norton, A. P. *A Star Atlas*. Cambridge, Mass.: Sky Publishing, 1964.

*A total eclipse of the sun, the most spectacular of earth-moon-sun phenomena, occurs when the moon crosses in front of the sun. The sun's upper atmosphere is visible as bright streamers protruding beyond the moon's edge. (G. Newkirk, Jr., High Altitude Observatory.)*

# Chapter 2 THE EARTH, MOON, AND SUN

The sky is the rest of the universe as seen from our planet. In the previous chapter, we studied the appearance of the sky as a whole. Now we are ready to look more carefully and study the motions of individual celestial bodies.

The brightest object in the sky is the sun, the star around which the earth revolves. In this chapter we will discover that the earth's orbital motion makes the sun appear to move eastward against the background of stars. Because the earth's axis of rotation is inclined to its orbit, the sun's path is inclined to the celestial equator. Thus the sun spends half of the year in the Northern Hemisphere and half in the Southern, giving rise to the seasons.

The sun is not the only celestial object that appears to move against the background of stars. The moon orbits the earth in only 27.3 days and thus appears to travel eastward across the sky following a path nearly the same as the sun's. In addition, five planets—Mercury, Venus, Mars, Jupiter, and Saturn—are bright enough to be easily visible, and Uranus can sometimes be detected by the naked eye. Their orbital motions carry the planets eastward across the sky along paths that differ only slightly from that of the sun.

One of the results of the motion of the moon and sun around the sky is the cycle of lunar phases. We will see that the sequence of phases from new moon to full moon and back to new moon is simply explained by the relative positions of the earth, moon, and sun. Other aspects of the earth-moon-sun system are the ocean tides that sweep the beaches of the world and the spectacular eclipses of the sun and moon.

Finally, we will study the motion of the earth and discover that the pleasant climate that makes life possible on this planet is delicately balanced. Small, periodic variations in the earth's motion may cause the advance and retreat of the glaciers and trigger the onset of ice ages. If this is true, then much of the northern United States may be covered by glacier ice within the next 20,000 years.

## THE MOTION OF THE SUN

Everything in the sky is moving. The sun, moon, planets, and even the stars move along their various orbits. Because the stars are so distant, their motion is not obvious to us even over decades, but the sun, moon, and planets are closer and move noticeably against the background of stars. In this chapter we will discuss the motion of the earth-moon-sun system, concentrating first on the motion of the sun.

*The Ecliptic.* We are familiar with the daily motion of the sun as it rises, moves across the sky, and sets. That motion, due to the eastward rotation of the earth, is not what we will examine here.

The motion we wish to study is the apparent motion of the sun due to the orbital motion of the earth. To see how this works, suppose we were riding in an automobile driving in a circle around a tree. As we begin we would see the tree against a background of other more distant trees, but as we drove we would see the tree from different directions against different backgrounds. If we did not know that our automobile was moving, we would imagine that the tree was moving around us. This is precisely what happens when we observe the sun from the moving earth. The earth moves so smoothly along its orbit we feel motionless, and it appears that the sun moves around the sky.

In January, we see the sun in the direction of the constellation Sagittarius (Figure 2-1). We can't see the stars of the constellation, of course, because the sun is too bright, but we can observe that the sun is located in that part of the sky merely by noting the time of sunset and the constellations in the evening sky. As the earth moves through space, we observe the sun from a different part of earth's orbit, and the sun appears to be in Capricornus. Thus as the earth moves along its orbit, the sun seems to move eastward through the constellations, taking a year to circle the sky one time.

The apparent path of the sun around the sky is called the **ecliptic.** Because the apparent mo-

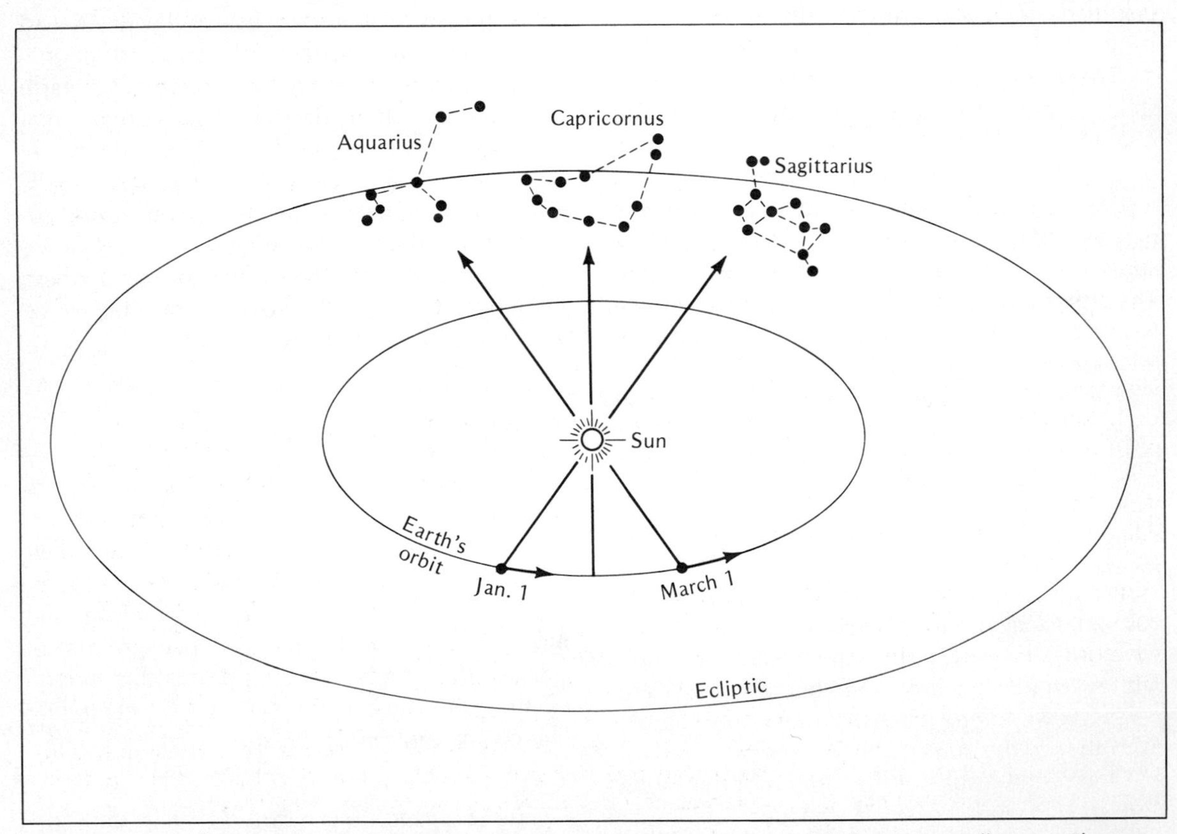


*Figure 2-1. As the earth moves around its orbit, we see the sun in front of different constellations. The sun appears to move around the ecliptic, the projection of the earth's orbit on the sky.*

tion of the sun is due to the orbital motion of the earth, it is easy to see that the ecliptic is the projection of the earth's orbit on the sky. If the celestial sphere were a great screen illuminated by the sun at the center, then the shadow cast by earth's orbit would be the ecliptic (Figure 2-1).

Because of the rotation of the earth, this slow eastward motion of the sun is not easy to visualize. Since the earth spins on its axis once each day, we see the sun, stars, moon, and planets rise in the east and set in the west. While this daily motion is taking place, the sun is moving slowly eastward along the ecliptic about 1° per day, which is about twice its own angular diameter.

An additional complication is that the earth does not rotate perpendicular to the plane of its orbit. Its axis of rotation is tipped 23½° from perpendicular. The spinning earth, like a spinning top, holds its axis fixed in space as it moves around the sun (Figure 2-3). Although the direction of the axis does drift slowly (see Box 2-1), we will not notice any change during our lifetimes.

Because the earth is tipped 23½° in its orbit, the ecliptic is tipped 23½° from the celestial equator. Recall that the celestial equator is the projection of the earth's equator, and that the ecliptic is the projection of the earth's orbit. Since the earth is tipped 23½°, its equator is tipped 23½° from the plane of its orbit. When we project this on the sky, we find that the ecliptic and celestial equator meet at an angle of 23½° (Figure 2-4).

The ecliptic and celestial equator cross at two places on the sky called equinoxes (Figure 2-5). The **vernal equinox** is the place where the sun crosses the celestial equator moving northward

## BOX 2-1 PRECESSION

If we could watch the sky for a few hundred years, we would discover that the north celestial pole is moving slowly with respect to Polaris. The celestial poles and the celestial equator, our supposedly fixed reference marks, are moving very slowly because of the slow change in the direction of the earth's axis of rotation. This slow toplike motion is called **precession.** The earth's axis sweeps around in a cone, taking almost 26,000 years for each sweep (Figure 2-2).

Because the earth's precession has such a long period, it has little effect over a few hundred years. During our lifetimes the north celestial pole will draw slightly closer to Polaris, pass its nearest point about 2100 AD and then begin to move away. Only by making careful observations could we detect this motion.

Precession is caused by the gravitational pull of the sun and moon. Because the earth is not a perfect sphere it has a slight bulge around its equator—the sun and moon pull on it, trying to make it spin upright in its orbit. This forces the axis of the earth to precess. The same thing happens to a child's toy top. Gravity tries to make it fall over, and the spinning top precesses.

One result of precession is the change of the pole star. As the north celestial pole moves around the sky, it sometimes comes close to one star or another. It just happens to be near Polaris now. Ancient Egyptian records show that 5000 years ago the north celestial pole was near the bright star Thuban (α Draconis). In about 13,000 years the pole will have moved away from Polaris and will be near Vega (α Lyrae).

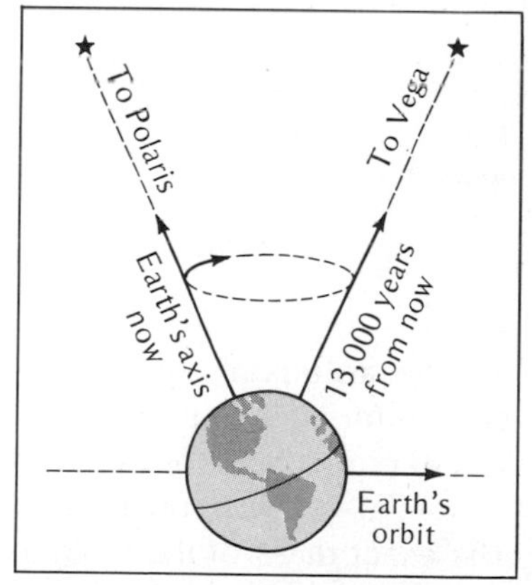


*Figure 2-2. Precession causes the earth's axis to wobble like a top. Although the axis now points toward Polaris, in 13,000 years it will point toward Vega.*

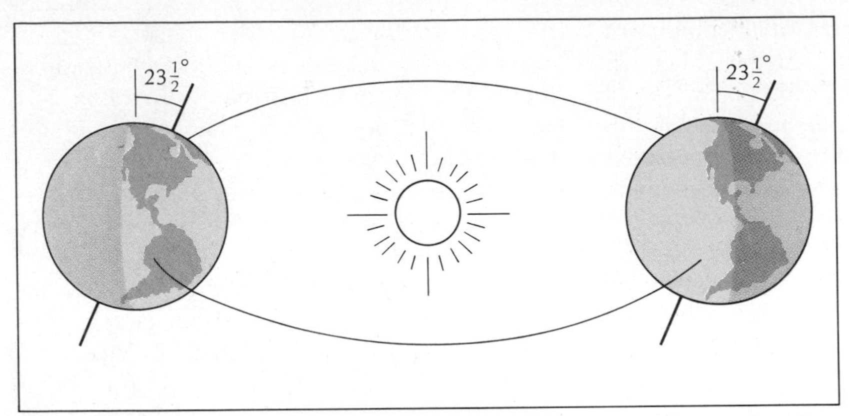


*Figure 2-3. The earth, spinning like a top, holds its axis fixed as it orbits the sun. The earth's northern hemisphere is tipped toward the sun in June and away in December.*

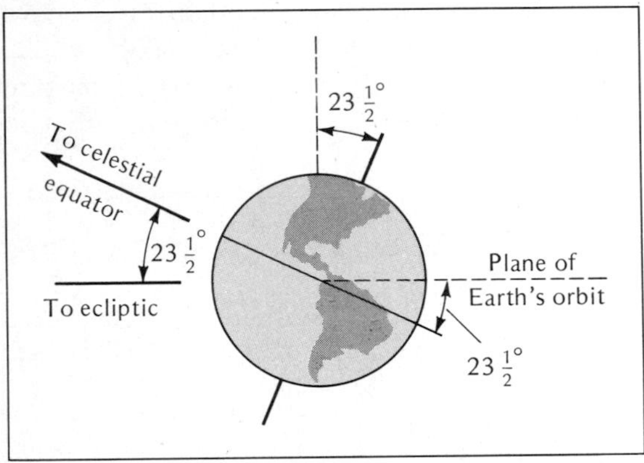


*Figure 2-4. Because the earth's equator is tipped 23½° to its orbit, the ecliptic is tipped 23½° to the celestial equator.*

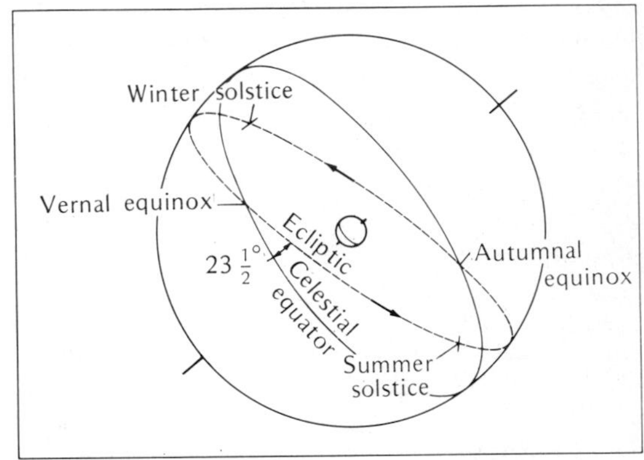


*Figure 2-5. The ecliptic (dashed line), the sun's apparent path around the sky, crosses the celestial equator at the equinoxes. The solstices mark the most northerly and most southerly points.*

and the **autumnal equinox** is the place where it crosses moving southward. The sun crosses the vernal equinox on or about March 21, and the autumnal equinox on or about September 22. The exact dates of the equinoxes can vary by a day or two because of leap year and other factors.

We can identify two other reference marks on the ecliptic by noting where the sun is farthest from the celestial equator. About June 22 the sun is farthest north at the point called the **summer solstice.** The **winter solstice** is the point where the sun is farthest south. The sun passes the winter solstice about December 22. These four reference points on the ecliptic are important because they mark the beginning of each of the seasons.

*The Seasons.* The seasonal temperature depends on the amount of heat we receive from the sun. To hold the temperature constant, there must be a balance between the amount of heat we gain and the amount we radiate to space. If we receive more heat than we lose, we get warm; if we lose more than we gain, we get cooler.

The motion of the sun around the ecliptic tips the heat balance one way in summer and the opposite way in winter. Because the ecliptic is inclined with respect to the celestial equator, the sun spends half the year in the northern celestial hemisphere and half the year in the southern celestial hemisphere (Figure 2-5). When the sun is in the northern celestial hemisphere, the northern half of the earth receives more direct sunlight—and therefore more heat—than the southern half. This makes North America, Europe, and Asia warmer.

The seasons are reversed in the southern half of the earth. While the sun is in the northern celestial hemisphere warming North America, South America becomes cooler. Southern Chile has warm weather on New Year's Day and cold in July.

To see how the sun can give us more heat in summer, think about the path the sun takes across the sky between sunrise and sunset. Figure 2-6 shows these paths when the sun is at the summer solstice and at the winter solstice as seen by a person living at latitude 40°, a good average latitude for most of the United States. Notice in Figure 2-6a that at the summer solstice the sun rises in the northeast, moves high across the sky, and sets in the northwest. But at the winter solstice, Figure 2-6b, the sun rises in the southeast, moves low across the sky, and sets in the southwest. Two features of these paths tip the heat balance.

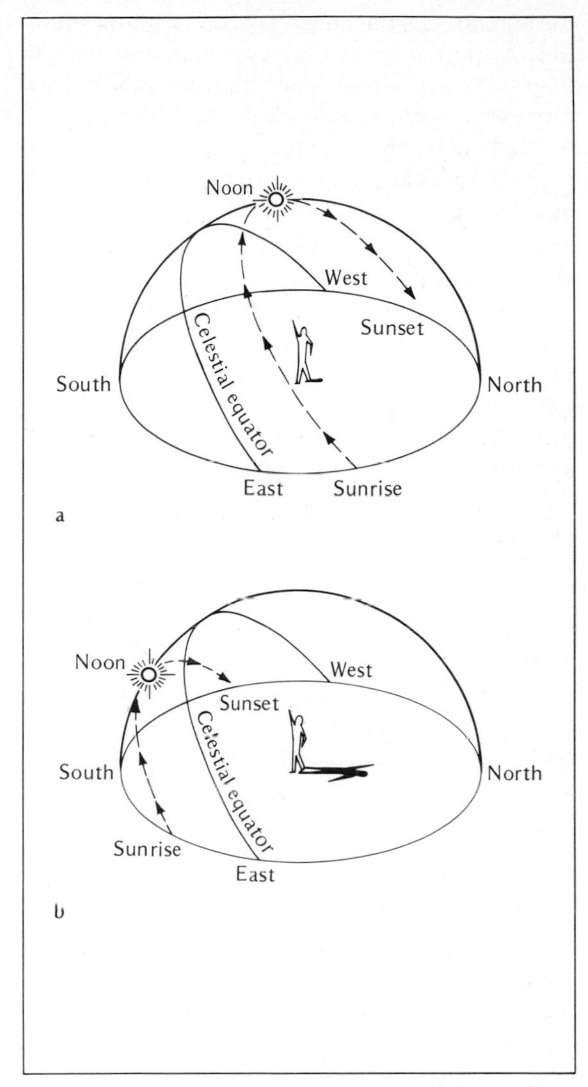


*Figure 2-6. (a) The path of the sun across the sky at the summer solstice. (b) The sun's path at the winter solstice.*

First, the summer sun is above the horizon for more hours of each day than the winter sun. Summer days are long and winter days are short. Since the sun is above our horizon longer in summer, we receive more energy each day.

Second, the sun stands high in the sky at noon on a summer day. It shines almost straight down, as shown by our small shadows. On a winter day, however, the noon sun is low in the southern sky. The ground gains little heat from the winter sun because the sunlight strikes the ground at an angle and spreads out, as shown by our longer shadows. These two effects work together to tip the heat balance and produce the seasons.

We mark the beginning of the seasons by the position of the sun. Spring begins at the moment the sun crosses the celestial equator going north (the vernal equinox). Summer begins at the moment the sun reaches its most northerly point (the summer solstice), and autumn begins when the sun crosses the celestial equator going south (the autumnal equinox). We mark the official beginning of winter when the sun reaches its most southerly position (the winter solstice).

Of course, the weather does not turn warm the instant spring begins. The ground, air, and oceans are still cool from winter, and they take a while to warm up. Likewise, in the fall the earth slowly releases the heat it has stored through the summer. Because of this thermal lag, the average daily temperatures lag behind the solstices by about one month. Although the sun crosses the summer solstice on about June 22, the hottest months are July and August. The coldest months are January and February, even though the sun passes the winter solstice earlier, about December 22.

The ecliptic is important to our daily lives because of its connection with the seasons, but it may also be familiar in a different guise. The ecliptic marks the center line of the zodiac, a band 18° wide that encircles the sky. The twelve constellations that lie on or near the zodiac correspond to the astrological signs. Astrology was once an important part of astronomy, but the two are now almost exact opposites—astronomy is a science that depends on evidence, and astrology is a superstition that depends on faith. Thus the signs of the zodiac are no longer important in astronomy. However, the zodiac itself is still of interest because it is the path followed by the planets as they move around the sky.

*The Motion of the Planets.* The planets of our solar system produce no visible light of their own; we see them by reflected sunlight. Mercury, Venus, Mars, Jupiter, and Saturn are all easily visible to the naked eye, but Uranus is usually too faint to be seen and Neptune is never bright enough. Pluto is even fainter, and we need a large telescope to find it.

All of the planets of our solar system move in nearly circular orbits around the sun. If we were looking down on the solar system from the north celestial pole, we would see the planets moving in the same counterclockwise direction around their orbits. The farther from the sun, the more slowly the planets move.

When we look for planets in the sky, we always find them near the ecliptic because their orbits lie in nearly the same plane as the orbit of the earth. As they orbit the sun, they appear to move eastward along the ecliptic. Mars moves completely around the ecliptic in slightly less than 2 years, but Saturn, being farther from the sun, takes nearly 30 years.

As seen from earth, Venus and Mercury can never move far from the sun because their orbits are inside earth's orbit. They appear near the western horizon just after sunset or near the eastern horizon just before sunrise. Venus is easier to locate because its larger orbit carries it higher above the horizon than Mercury (Figure 2-7). Mercury's orbit is so small that it can never get far from the sun. Consequently it is hard to see against the sun's glare, and is often hidden in the clouds and haze near the horizon.

By tradition, any planet visible in the evening sky is an **evening star,** even though planets are not stars. Similarly, any planet visible in the sky shortly before sunrise is a **morning star.** Perhaps the most beautiful is Venus, which can become

*Figure 2-7. The orbits of Venus and Mercury sometimes carry them far enough from the sun to be visible in the evening sky soon after sunset (a) or in the morning just after sunrise (b).*

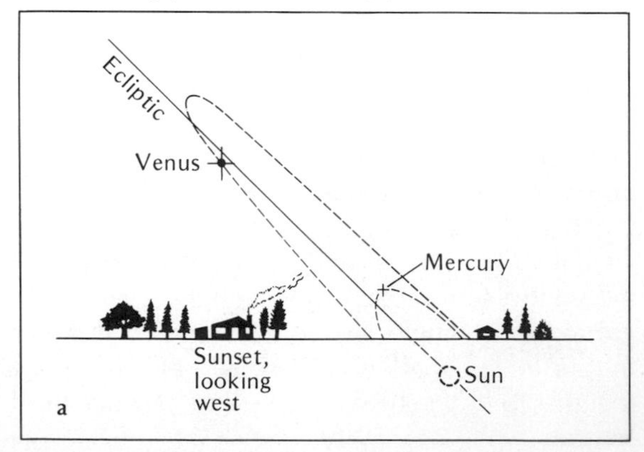

as bright as minus fourth magnitude. As Venus moves around its orbit, it can dominate the western sky each evening for many weeks, but eventually its orbit carries it back toward the sun and it is lost in the haze near the horizon. In a few weeks it reappears in the dawn sky, a brilliant morning star.

## THE EARTH-MOON-SUN SYSTEM

Among the changing aspects of the sky, none are so striking as those that involve the earth-moon-sun system. We are most familiar with the phases of the moon, but other phenomena include tides and eclipses of the moon and sun.

*The Phases of the Moon.* The moon orbits eastward around the earth in 27.32 days. Thus it moves rapidly across the sky. In 24 hours it moves 13°, about 26 times its apparent diameter. If you compare the position of the moon to stars in the background, you will see it move slightly more than its diameter in an hour. Because the moon's orbit is inclined only about 5° to the plane of earth's orbit, the moon's motion never takes it farther than 5° from the ecliptic.

Because the moon does not produce visible light of its own, it is visible only by the sunlight it reflects, and we can see only that portion illuminated by the sun. As the moon moves around the sky, the sun illuminates different amounts of the side of the moon facing earth,

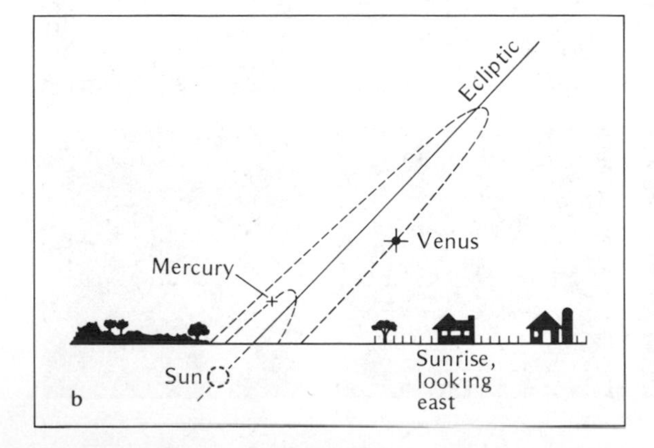


and the moon passes through a sequence of phases.

Figure 2-8 shows how the phases of the moon are related to its orbital position, and Figure 2-9 gives close-up photographs of the phases. When the moon is approximately between the earth and sun, the side toward us is in darkness. The moon is invisible, and we refer to it as new moon. A few days after new moon, it has moved far enough along its orbit to allow the sun to illuminate a small sliver of the side toward us and we see a thin crescent. Night by night this crescent moon waxes (grows), until we see half of the side toward us illuminated by sunlight and refer to it as first quarter. The moon continues to wax, becoming gibbous, and then, when it is nearly opposite the sun, the side toward earth is fully illuminated and we see a full moon.

The second half of the lunar cycle reverses the first half. After reaching full, the moon wanes (shrinks) through gibbous phase to third quarter, then through crescent to new moon. To distinguish between the gibbous and crescent phases of the first and second half of the cycle, we refer to gibbous waning and crescent waning when the moon is shrinking, and gibbous waxing and crescent waxing when it is growing.

Although the moon orbits the earth in 27.32 days, it takes slightly longer to go through one cycle of phases. While the moon moves around the sky in 27.32 days, the sun moves eastward about 1° per day. In 27.32 days the moon returns to the same place in the sky where it was last new, but the sun has moved about 27° east, so the moon needs slightly more than two days to catch up with the sun and reach new moon again. Thus one cycle of lunar phases takes 29.53 days, about four weeks, and new moon, first quarter, full moon, third quarter, and new moon occur at nearly one week intervals.

To summarize, let us follow the moon through one cycle of phases. At new moon, the moon is nearly in line with the sun and sets in the west with the sun. Thus we see no moon at new moon. A few days after new moon, we see the waxing crescent above the western horizon

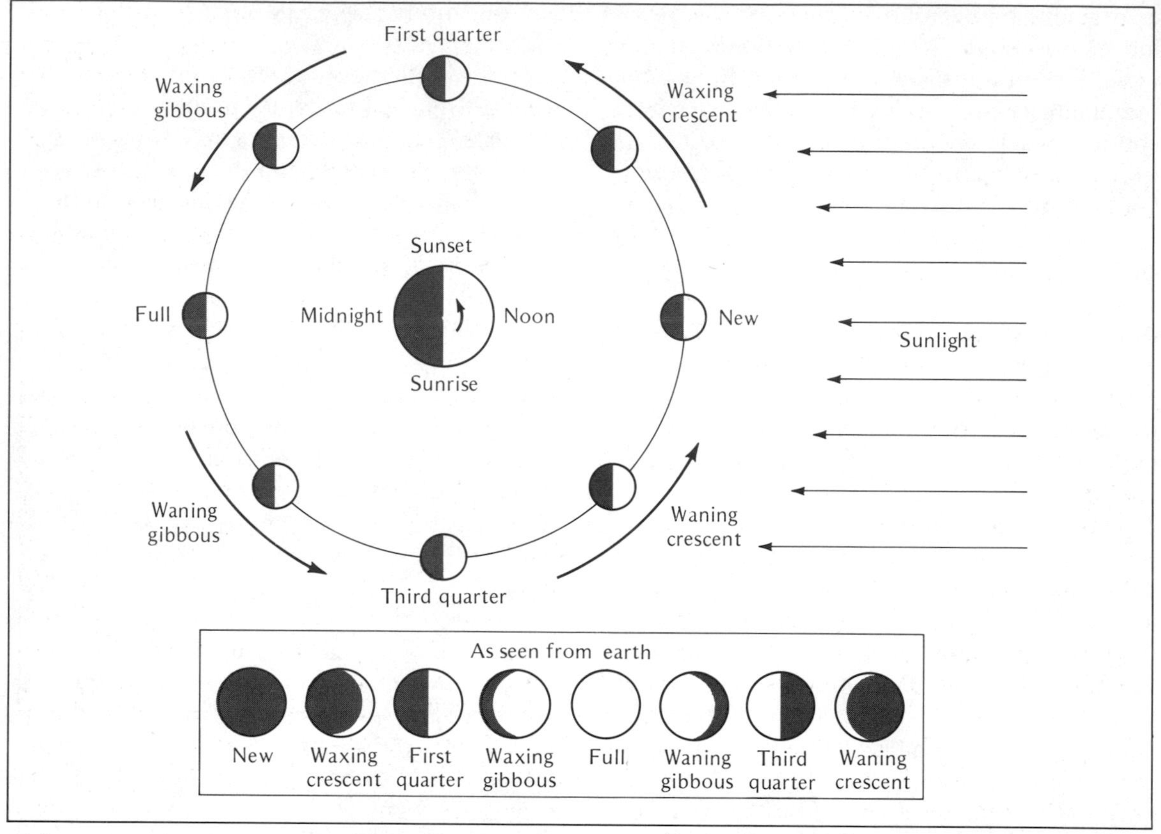


*Figure 2-8. The phases of the moon are produced by the varying amounts of the illuminated surface we can see. The box shows the moon as it appears from earth.*

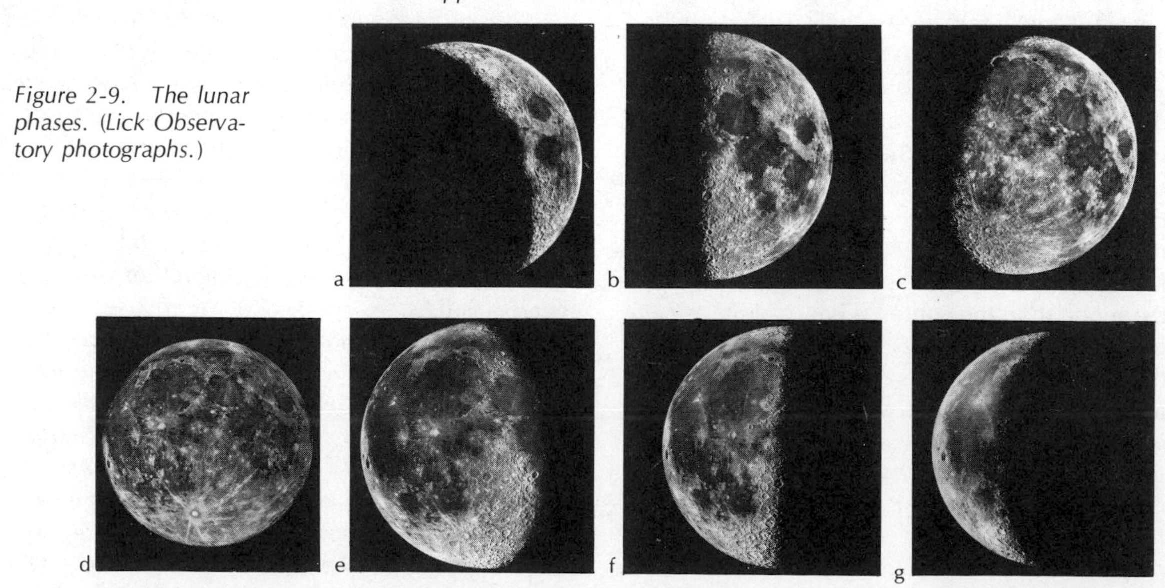


*Figure 2-9. The lunar phases. (Lick Observatory photographs.)*

soon after sunset, and each evening it is fatter and higher above the horizon, until, about one week after new moon, it reaches first quarter and stands high in the southern sky at sunset. The first quarter moon does not set until about midnight. In the days following first quarter, the moon waxes fatter, becoming gibbous waxing. Each evening we find it farther east among the stars and it sets later and later. About two weeks after new moon, the moon reaches full, rising in the east as the sun sets in the west. The full moon is visible all night, setting in the west at sunrise.

The waning phases of the moon may be less familiar because the moon is not visible in the early evening sky. As it wanes through gibbous, it rises later and later and by the time it reaches third quarter, it does not rise until midnight. The waning crescent does not rise until even later, and if we wish to see the thin waning crescent just before new moon, we must get up before sunrise and look for the moon above the eastern horizon.

Almost everyone is familiar with the changing phases of the moon, but those who live near the seashore are probably familiar with another phenomenon related to the earth-moon-sun system—the periodic advance and retreat of the ocean tides.

*Tides.* We feel the earth's gravity drawing us downward with a force we refer to as our weight, but that is not the only gravity acting on us. The moon is less massive and is farther away than the center of the earth is, but its gravity measurably affects earth. The side of the earth facing the moon is about 4000 miles closer to the moon than the center of the earth is, and the moon's gravity pulls on it more strongly than on the earth's center. Though we think of the earth as solid, it is not perfectly rigid, so the moon's gravity draws the rocky surface of the near side up into a bulge a few inches high.

We don't notice the mountains and plains rising and falling by a few inches, but we do notice the moon's influence over the oceans. Sea water is fluid and can respond to the small force of the moon's gravity, flowing into a bulge of water on the side of the earth facing the moon. There is also a bulge on the side away from the moon, which develops because the moon pulls more strongly on the earth's center than on the far side. Thus the moon pulls the earth away from the oceans, which flow into a bulge on the far side (Figure 2-10).

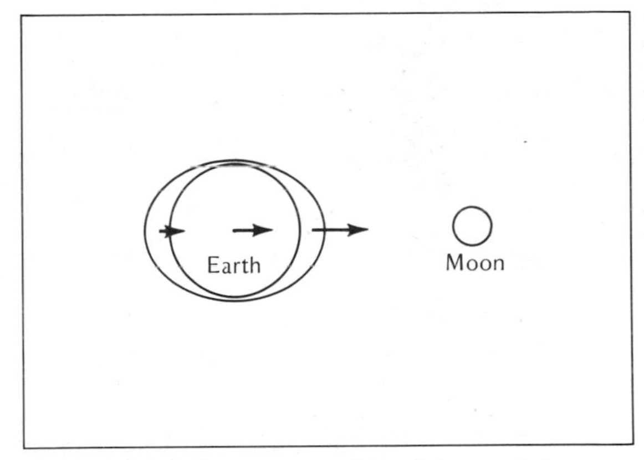


*Figure 2-10. Because one side of the earth is closer to the moon than the other side, the moon's gravitational force produces tidal bulges in the oceans.*

We can see dramatic evidence of this effect if we watch the ocean shore for a few hours. Though the earth rotates on its axis, the tidal bulges remain fixed along the earth-moon line. As the turning earth carries us into a tidal bulge, the ocean water deepens and the tide crawls up the beach. Later when the earth carries us out of the bulge, the water becomes shallower and the tide falls. Because there are two bulges on opposite sides of the earth, the tides rise and fall twice a day.

The sun, too, produces tidal bulges on the earth. At new moon and at full moon, the moon and sun produce tidal bulges that add together (Figure 2-11a) and produce extreme tidal changes; high tide is very high, and low tide is very low. Such tides are called **spring tides** even though they occur at every new and full moon and not just in the spring. **Neap tides** occur at

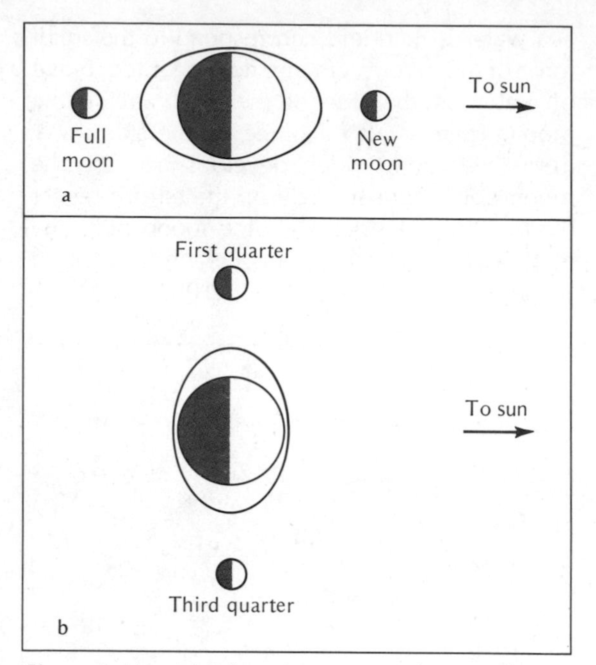


*Figure 2-11. (a) When the moon and sun pull in the same direction, their tidal forces add and the tidal bulges are larger. Thus spring tides occur at new moon and full moon. (b) When the moon and sun pull at right angles, their tidal forces do not add and the tidal bulges are smaller. Such neap tides occur at first and third quarter moon.*

first and third quarter moon, when the moon and sun pull at right angles to each other (Figure 2-11b). Then the tides do not add together and are less extreme than usual.

Tidal forces can have surprising effects. The friction of the ocean waters with the seabeds slows the rotation of the earth by 0.001 seconds per day per century. Fossils of marine animals confirm that only 400 million years ago the earth's day was 22 hours long. In addition, the earth's gravitational field exerts tidal forces on the moon, and, although there are no bodies of water on the moon, friction within the flexing rock has slowed the moon's rotation to the point that it now keeps the same face toward the earth.

Tidal forces can also affect orbital motion. Friction with the rotating earth drags the tidal bulges eastward out of a direct earth-moon line (Figure 2-12). These tidal bulges contain a large amount of mass, and their gravitational field pulls the moon forward in its orbit. As a result, the moon's orbit is growing larger and it is receding from earth at about 3 cm per year, an effect that astronomers can measure by bouncing laser beams off reflectors left on the lunar surface by the Apollo astronauts.

These and other tidal effects are important in many areas of astronomy. In later chapters we will see how tidal forces can pull gas away from stars, rip galaxies apart, and melt the interiors of satellites orbiting near massive planets. For now, however, we must consider further aspects of the earth-moon-sun system. The stately progression of the lunar phases and the ebb and flow of the ocean tides are commonplace, but occasionally something peculiar happens. The moon darkens and turns angry red in a lunar eclipse.

*Lunar Eclipses.* A **lunar eclipse** occurs at full moon when the moon moves through the shadow of the earth. Since the moon shines only by reflected sunlight, we see the moon gradually darken as it enters the shadow.

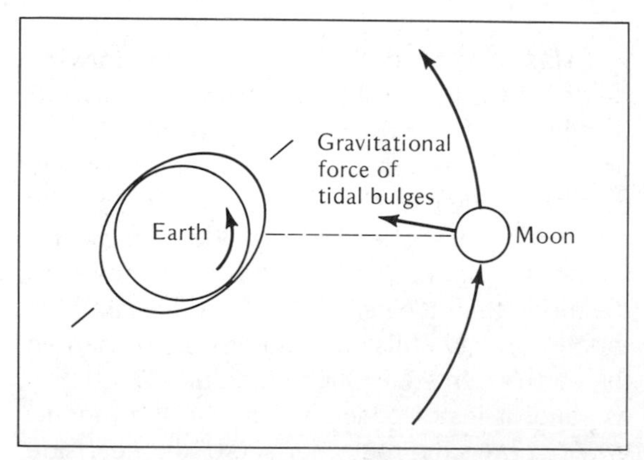


*Figure 2-12. The rotation of the earth drags the tidal bulges ahead of the earth-moon line (exaggerated here). The gravitational attraction of these masses of water pulls the moon forward in its orbit, forcing its orbit to grow in size.*

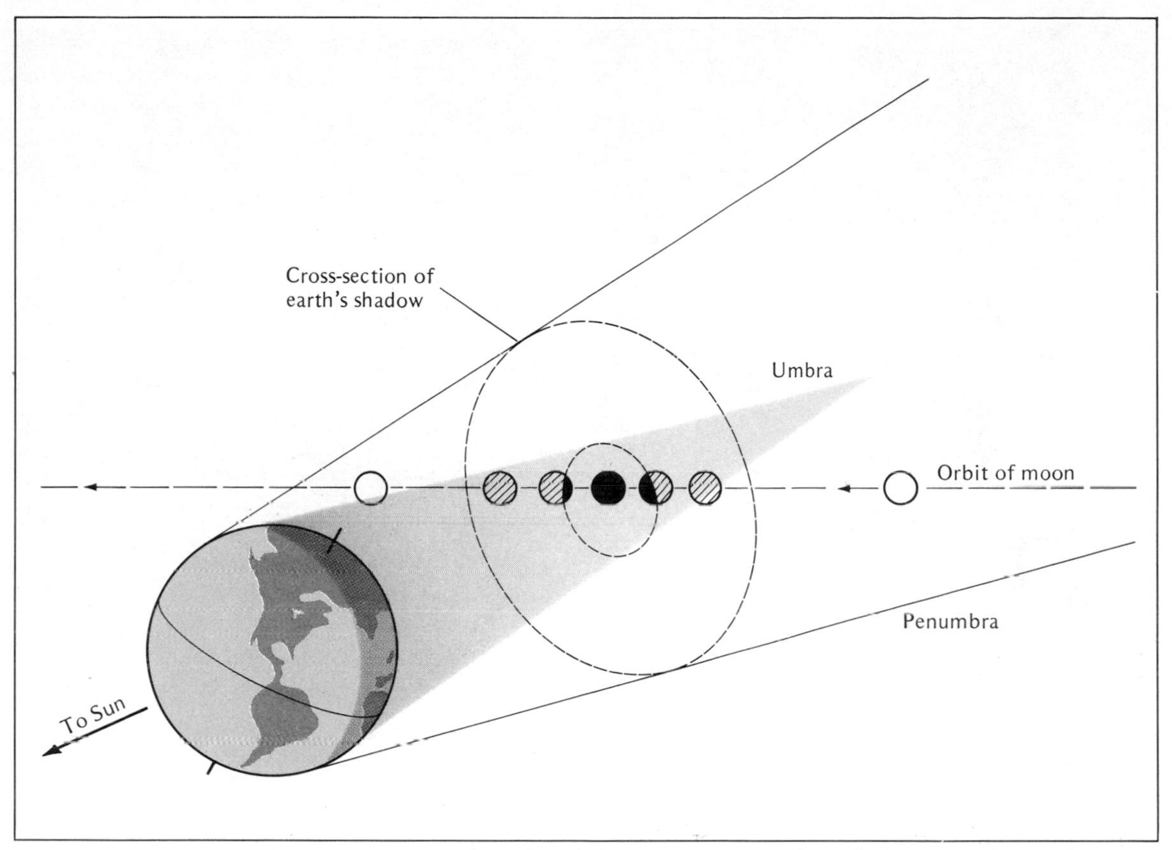


*Figure 2-13. During a total lunar eclipse the orbit of the moon carries it through the penumbra and completely into the umbra.*

The earth's shadow consists of two parts. The **umbra** is the region of total shadow. If we were in the umbra of the earth's shadow, we would see no portion of the sun. However, if we moved into the **penumbra,** we would be in partial shadow and would see part of the sun peeking around the edge of the earth. Thus in the penumbra the sunlight is dimmed but not extinguished.

If the orbit of the moon carries it through the umbra, we see a total lunar eclipse (Figures 2-13 and 2-14). As we watch the moon in the sky, it first moves into the penumbra and dims slightly, and the deeper it moves into the penumbra the more it dims. In about an hour the moon reaches the umbra, where we see the dark umbral shadow darken part of the moon. It takes about an hour for the moon to enter the umbra completely and become totally eclipsed. Totality, the period of total eclipse, may last as long as an hour and 40 minutes, though the timing of the eclipse depends on where the moon crosses the shadow.

When the moon is totally eclipsed, it does not disappear completely. While it receives no direct sunlight, it does receive some sunlight from the earth's atmosphere. If we were on the moon during totality, we would not see any part of the sun because it would be entirely hidden behind the earth. However, we would be able to see the earth's atmosphere illuminated from behind by the sun. The red glow from this "sunset"

*Figure 2-14. A sequence of exposures taken at five-minute intervals shows the moon entering the umbra, becoming totally eclipsed, and then moving out of the shadow. (Edward E. Robinson.)*

illuminates the moon during totality and makes it glow coppery red. (See Color Plate 2.)

If the moon does not move completely into the umbra, we see a partial lunar eclipse (Figure 2-15). The part of the moon that remains outside the umbra receives some direct sunlight, and the glare prevents our seeing the faint coppery glow of the part of the moon in the umbra.

A penumbral lunar eclipse occurs when the moon passes through the penumbra but misses the umbra entirely. Since the penumbra is a region of partial shadow, the moon is only partially dimmed. A penumbral eclipse is not very impressive.

While there are usually no more than one or two lunar eclipses each year, it is not difficult to see one. We need only be on the dark side of the earth when the moon enters the earth's shadow. (Table 2-1 lists some future lunar eclipses.) A total solar eclipse, however, is an event that few people ever witness. (See Table 2-2, page 32.)

*Solar Eclipses.* A **solar eclipse** occurs when the moon passes directly between the earth and sun, blocking our view of the sun. If the moon completely covers the sun, the eclipse is total, but if the moon covers only part of the sun, the eclipse is partial.

Whether we see a total or partial eclipse depends on whether we are in the umbra or the penumbra of the moon's shadow (Figure 2-16). The umbra of the moon's shadow barely reaches the earth and casts a small circular shadow never larger than 269 km (168 mi) in diameter. If we are standing in that umbral spot, we are in total shadow, unable to see any part of the sun's surface and the eclipse is total. But if we are located outside the umbra, in the penumbra, we see part of the sun peeking around the edge of the moon and the eclipse is partial. Of course, if we are outside the penumbra, we see no eclipse at all.

Because of the orbital motion of the moon

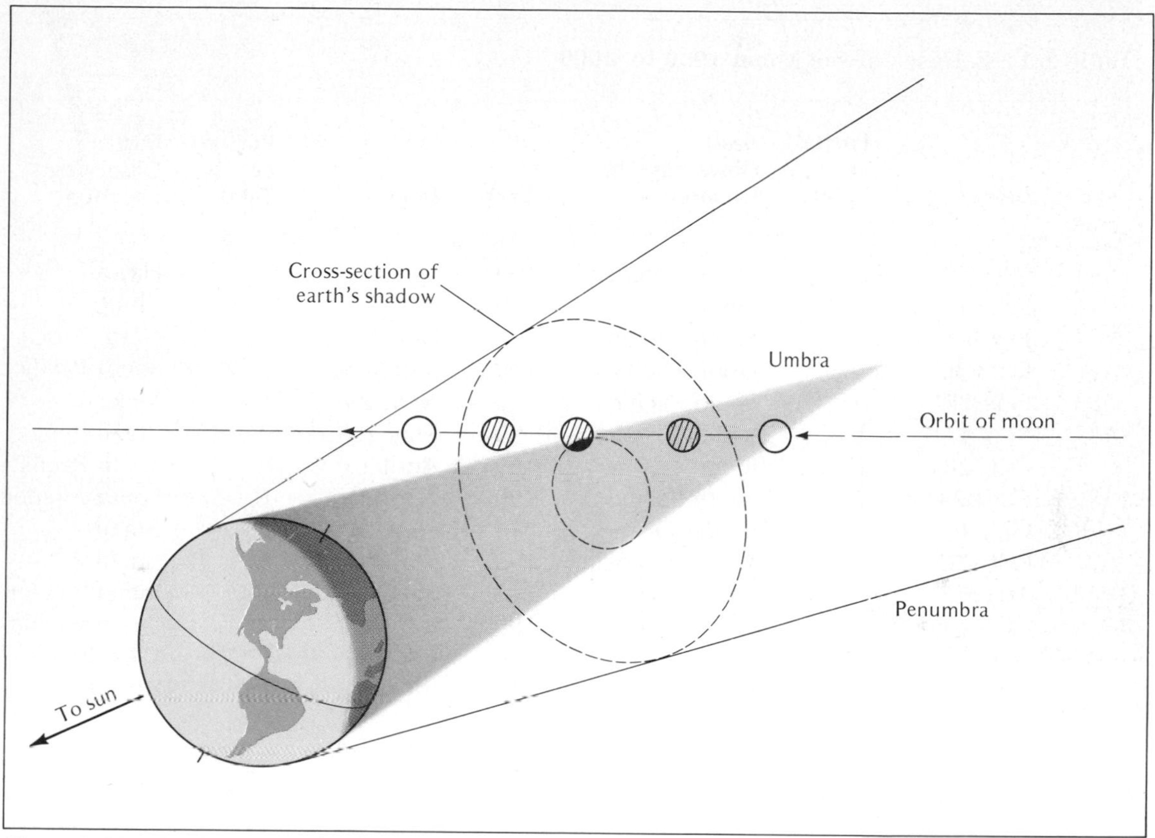


*Figure 2-15. During a partial eclipse the orbit of the moon carries it through the penumbra and only partially into the umbra.*

and the rotation of the earth, the moon's shadow sweeps rapidly across the earth in a long, narrow path of totality. If we want to see a total solar eclipse, we must be in the path of totality. When the umbra of the moon's shadow sweeps over us, we see the most dramatic sight in astronomy, the totally eclipsed sun.

The eclipse begins as the moon slowly crosses in front of the sun. It takes about an hour for the moon to cover the solar disk, but as the last sliver of sun disappears, dark falls in a few seconds. Automatic street lights come on, drivers of cars turn on their headlights, and birds go to roost. The sky becomes so dark we can even see the brighter stars.

The darkness lasts only a few minutes because the umbra is never more than 168 mi in diameter and sweeps across the earth's surface at about 1000 mph. The sun cannot remain totally eclipsed for more than 7.5 minutes, and the average period of totality lasts only 2 or 3 minutes.

When the moon covers the bright surface of the sun, called the **photosphere,** we can see the sun's faint outer atmosphere, the **corona,** and the bright gases called the **chromosphere** (Figure 2-17 and Color Plate 1) just above the photosphere. The corona is low-density, hot gas that glows with a pale white color. Streamers caused by the solar magnetic field streak the corona. The chromosphere is often marked by eruptions on the solar surface called **promi-**

**Table 2-1 Eclipses of the Moon 1980 to 2000**

| Year | Date | Partial or Total | Best Observing Location | Year | Date | Partial or Total | Best Observing Location |
|---|---|---|---|---|---|---|---|
| 1981 | July 17* | P | South America | 1991 | Dec. 21* | P | Hawaii |
| 1982 | Jan. 9 | T | India | 1992 | June 15* | P | Chile |
| | July 6* | T | South Pacific | | Dec. 9* | T | North Africa |
| | Dec. 30* | T | North Pacific | 1993 | June 4 | T | South Pacific |
| 1983 | June 25* | P | South Pacific | | Nov. 29* | T | Mexico |
| 1985 | May 4 | T | Indian Ocean | 1994 | May 25* | P | Brazil |
| | Oct. 28 | T | India | 1995 | April 15 | P | South Pacific |
| 1986 | April 24* | T | South Pacific | 1996 | April 4* | T | South Atlantic |
| | Oct. 17 | T | Indian Ocean | | Sept. 27* | T | Brazil |
| 1987 | Oct. 7* | P | South America | 1997 | March 24* | P | Brazil |
| 1988 | Aug. 27* | P | South Pacific | | Sept. 16 | T | Indian Ocean |
| 1989 | Feb. 20 | T | Philippines | 1999 | July 28* | P | South Pacific |
| | Aug. 17* | T | Brazil | 2000 | Jan. 21* | T | West Indies |
| 1990 | Feb. 9 | T | India | | July 16 | T | Australia |
| | Aug. 6 | P | Australia | | | | |

* Visible in at least part from the United States.

**nences.** The corona, chromosphere, and prominences are visible only while the moon covers the brilliant photosphere. As soon as part of the photosphere reappears, the fainter corona, chromosphere, and prominences vanish in the glare, and totality is over. The moon moves on

*Figure 2-16. Observers in the path of totality see a total solar eclipse when the umbral shadow sweeps over them. Those in the penumbra see a partial eclipse.*

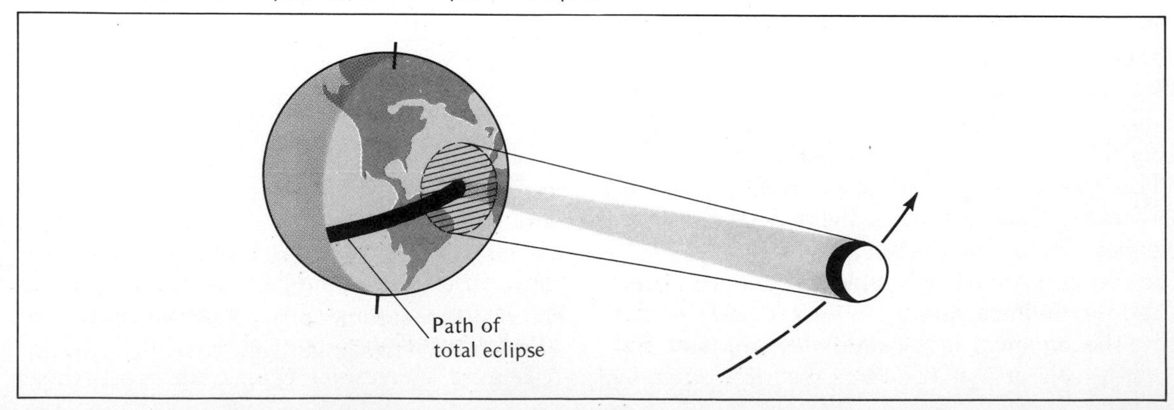

*Figure 2-17. The sun's extended atmosphere is visible during a total solar eclipse. This photograph was taken November 12, 1966, with a special filter to enhance the outer portions of the corona. The planet Venus is visible near the left edge. (G. Newkirk, Jr., High Altitude Observatory.)*

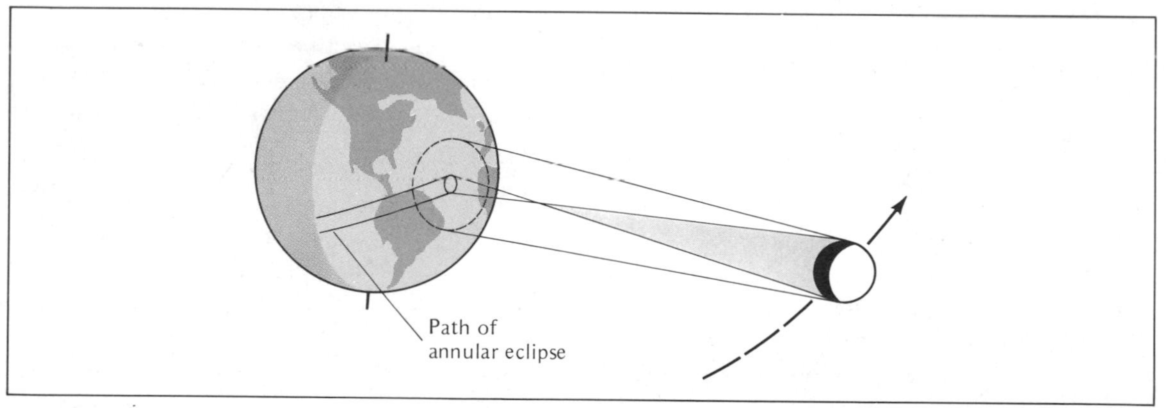


*Figure 2-18. If the moon is near the farther part of its orbit, the umbral shadow does not reach the earth, resulting in an annular eclipse.*

in its orbit and in about an hour the sun is completely visible once again.

Because the moon's orbit is slightly elliptical, it is sometimes nearer and sometimes farther from earth. If it crosses between the sun and earth while it is in the farther part of its orbit, its umbra does not reach all the way to the earth (Figure 2-18). If the umbra does not touch the earth, there can be no path of totality and no total eclipse. Under these circumstances we see an **annular eclipse.** (See Table 2-2.)

During an annular eclipse the moon looks

**Table 2-2 Total and Annular Eclipses 1980 to 2000**

| Year | Date | Total or Annular | Best Observing Location |
|---|---|---|---|
| 1980 | Feb. 16 | T | Africa, India, China |
| | Aug. 10 | A | Pacific, South America |
| 1981 | Feb. 4 | A | Pacific, Australia |
| | July 31 | T | Russia, North Pacific |
| 1983 | July 11 | T | Indian Ocean, Indonesia |
| | Dec. 4 | A | Atlantic, central Africa |
| 1984 | May 30 | A | Mexico, southeast U.S. |
| | Nov. 22 | T | Indonesia, South Pacific |
| 1985 | Nov. 12 | T | South Pacific |
| 1986 | Oct. 3 | T | North Atlantic |
| 1987 | March 29 | A-T* | South Atlantic, Africa |
| | Sept. 23 | A | Asia, China, Pacific |
| 1988 | March 18 | T | Indian Ocean, Indonesia |
| | Sept. 11 | A | Indian Ocean |
| 1990 | Jan. 26 | A | Antarctica, South Atlantic |
| | July 22 | T | Finland, northern Siberia |
| 1991 | Jan. 15 | A | Australia, New Zealand |
| | July 11 | T | Hawaii, Mexico, South America |
| 1992 | Jan. 4 | A | Pacific |
| | June 30 | T | South America, Africa |
| 1994 | May 10 | A | Pacific, central U.S. |
| | Nov. 3 | T | South America, Atlantic |
| 1995 | April 29 | A | South Pacific, South America |
| | Oct. 24 | T | Iran, India, Southeast Asia |
| 1997 | March 9 | T | Asia, Siberia, Arctic |
| 1998 | Feb. 26 | T | Pacific, South America |
| | Aug. 22 | A | Indian Ocean, Indonesia |
| 1999 | Feb. 16 | A | Indian Ocean, Australia |
| | Aug. 11 | T | Europe, Asia, India |

* Begins annular but becomes total.

slightly smaller because it is slightly farther away. It is too small to completely cover the sun, and we see the sun's bright photosphere around the edge of the moon in a brilliant ring, or annulus (Figure 2-19). Annular eclipses are less impressive than total solar eclipses because the bright annulus of the photosphere blinds us to the fainter corona, chromosphere, and prominences.

Predicting when an eclipse will occur and

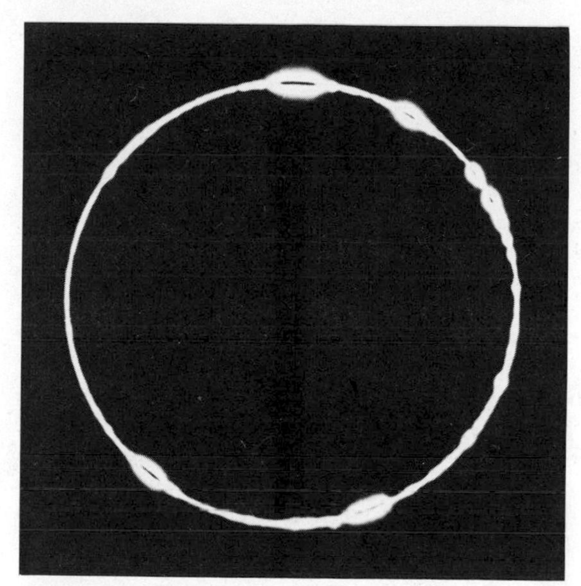

*Figure 2-19. The annular eclipse of April 28, 1930. A bright ring of photosphere remains visible at the edge of the moon. (Lick Observatory photograph.)*

the location from which it will be visible involves difficult calculations. However, eclipses do not occur at random, but with a regular pattern. Box 2-2 explains how that pattern arises and how ancient astronomers used it to predict eclipses.

Eclipses are a product of the relationships between the earth, moon, and sun. Similarly, all of the phenomena discussed in this chapter—the ecliptic, seasons, lunar phases, tides—are connected with the motions of the earth, moon, and sun. The Perspective that follows discusses the theory that earth's climate is also affected by small changes in its motion.

## PERSPECTIVE: CLIMATE AND ICE AGE

*Short-Term Variation in Climate.* We know from bitter experience that the earth's climate can change. In the early 1970s a terrible drought in North Africa caused seven years of suffering and death. Shorter, less costly droughts periodically affect the West Coast of the United States. But these are only local variations, and although they may cause death and hardship, they do not affect the earth's hospitality to human life.

We cannot apply astronomical data to such short-term, local fluctuations, but we can search for astronomical causes of longer-term, planet-wide variations in climate.

*Long-Term Variation in Climate.* The earth has gone through periods called ice ages, when the worldwide climate was cooler and dryer and thick layers of ice covered the higher latitudes. The earliest known occurred roughly 570 million years ago, and the next occurred about 280 million years ago. The latest began only 3 million years ago and may not have ended yet. That is, we may even now be in an ice age. Dating these periods is difficult, so the timing of the ice ages is uncertain. Nevertheless, some earth scientists believe they occur about every 250 million years.

From plant and animal fossils, paleontologists conclude that the normal climate between ice ages is about 10°C warmer than it is today. Tropical forests extended about 1100 km (700 mi) farther north than they do today, and the temperate forests of North America reached up to 2200 km (1400 mi) farther north. In addition, the chemical composition of fossil shells suggests that in the past the oceans were warmer. The most recent variation began roughly 45 million years ago, when the earth's climate began to cool, and culminated in the beginning of an ice age a few million years ago.

*Glacial Periods.* During an ice age, water freezes to form large polar caps, and the atmosphere becomes cooler and dryer. Ice sheets may cover as much as 30 percent of the land, but the remainder of the earth remains ice free.

In the course of an ice age, the ice alternately advances and melts back. A **glacial period** is an interval when the ice sheets engulf huge areas of the land; an **interglacial period** is the time when the ice sheets melt back. Glacial and interglacial

## BOX 2-2 ECLIPSE SEASONS

We see a solar eclipse when the moon passes between the earth and sun, that is, when the lunar phase is new moon. We see a lunar eclipse at full moon. However, we don't see eclipses at every new moon and every full moon. Why not?

Figure 2-20 is a scale drawing of the umbral shadows of the earth and moon. Notice that they are extremely long and narrow. The earth, moon, and sun must line up almost exactly or the shadows miss their mark and there is no eclipse.

To be eclipsed the moon must enter earth's shadow. However, because its orbit is tipped (Figure 2-21a), the moon often misses the shadow, passing north or south of it, and no lunar eclipse occurs. Also, in order to produce a solar eclipse, the moon's shadow must sweep over the earth. However, the inclination of the moon's orbit means that it often reaches new moon with its shadow passing north or south of the earth, and there is no solar eclipse.

For an eclipse, the moon must reach full or new moon at the same time it passes through the plane of the earth's orbit, otherwise the shadows miss (Figure 2-21b). The points where it passes through the plane of the earth's orbit are called the **nodes** of the moon's orbit, and the line connecting these is called the line of nodes. Twice a year this line of nodes points toward the sun, and for a few weeks eclipses are possible at new moon and full moon (Figure 2-21c). These intervals when eclipses are possible are called eclipse seasons, and they occur about six months apart.

If the moon's orbit were fixed in space, the eclipse seasons would always occur at the same time each year. However, the moon's orbit precesses slowly because of the gravitational pull of the sun on the moon, and the precession slowly changes the direction of the line of nodes. The line turns westward, making one complete rotation in 18.61 years. As a result, the eclipse seasons occur about three weeks earlier each year. The motion of the line of nodes, combined with the periodicity of the lunar phases, means that every 18 years 11.3 days the eclipse seasons start over and the same pattern of eclipses repeats. Because this cycle contains about one-third of a day more than an integer number of days, an eclipse visible in North America will reoccur, after 18 years 11.3 days, about one-third of the way around the world, in this case in the eastern Pacific. Many ancient peoples recognized this pattern from their records of previous eclipses and were able to predict when eclipses would occur, even though they did not understand what the sun and moon were or what alignments produced eclipses.

*Figure 2-20. This scale drawing of the umbral shadows of the earth and moon shows how easy it is for the shadows to miss their mark at full moon and new moon.*

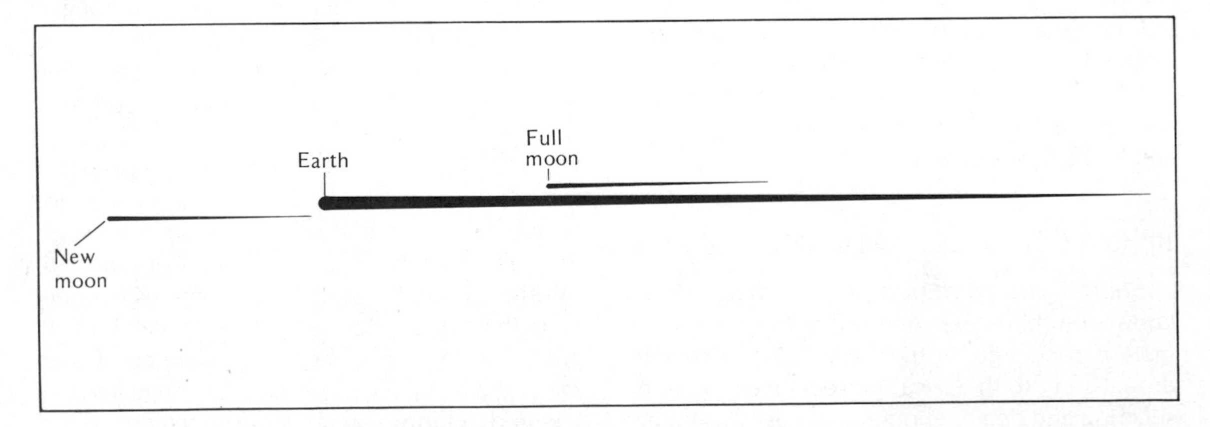

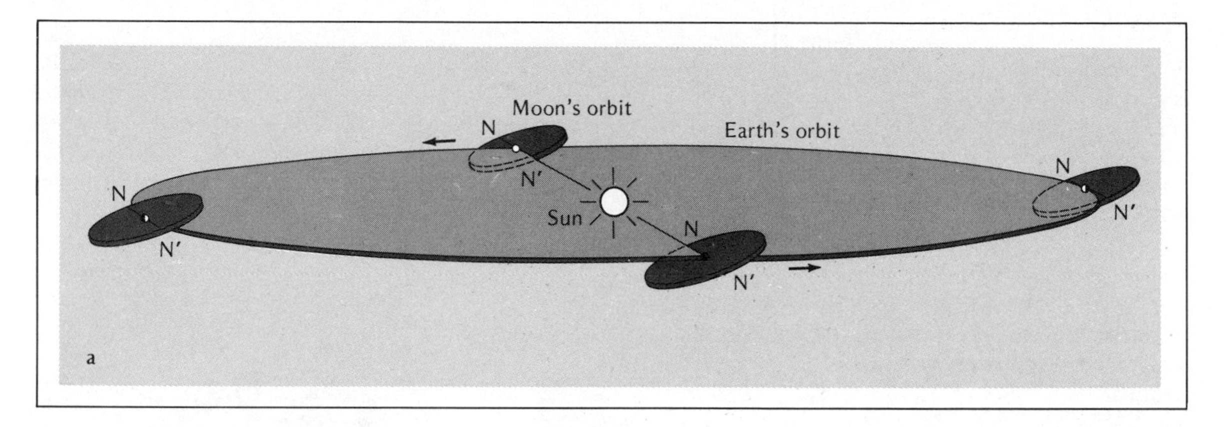


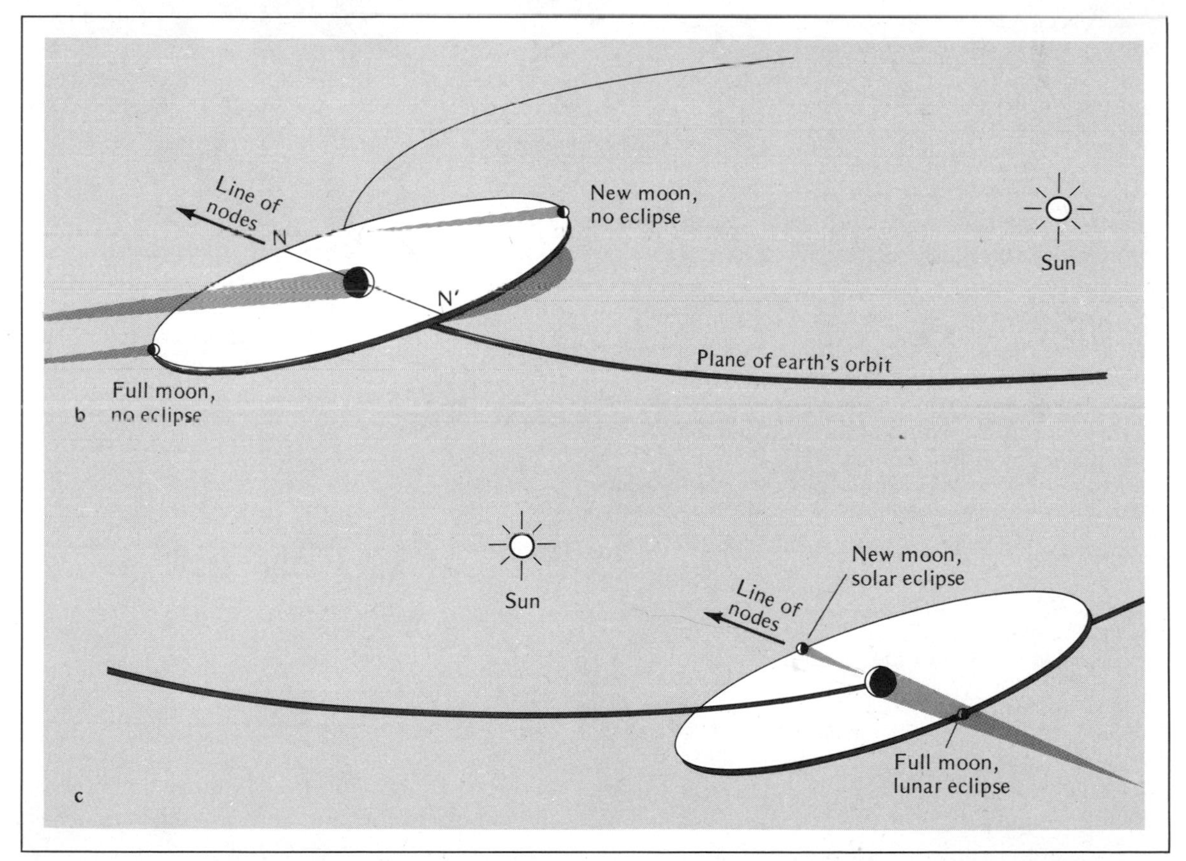


*Figure 2-21.* *(a) The moon's orbit is tipped about 5° to the earth's orbit. The nodes N and N' are the points where the moon passes through the plane of earth's orbit. (b) If the line of nodes does not point at the sun, the shadows miss and there are no eclipses at new moon and full moon. (c) At those parts of earth's orbit where the line of nodes points toward the sun, eclipses are possible at new moon and full moon.*

*Figure 2-22. Geologists can locate evidence that much of North America was once covered by glacier ice. In this scene from Wyoming's Beartooth Mountains, the small boulders and debris in the foreground were washed from a glacier that once filled the valley in the background. (M. Kauffmann, Franklin and Marshall College.)*

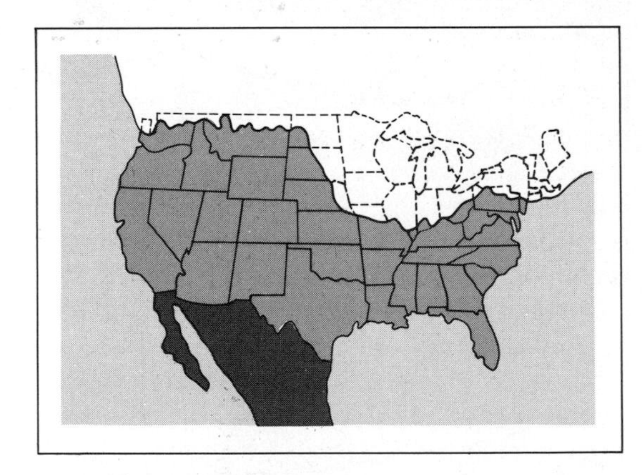

*Figure 2-23. Only 20,000 years ago much of the northern United States was covered by glacier ice. The solid line indicates the farthest advance of the ice.*

periods alternate during an ice age, in a cycle lasting roughly 40,000 years.

We are living in an interglacial period. Geological evidence, such as large boulders transported hundreds of miles from their place of origin, deposits of rock debris washed off of glaciers (Figure 2-22), and striations in rock surfaces cut by moving ice all show that Canada and much of the northern United States were repeatedly covered by sheets of ice up to 2 km thick (Figure 2-23). The last of these ice sheets melted about 20,000 years ago. If the advance of the glaciers is really periodic, and if the current ice age is not coming to an end, we have another 20,000 years before the ice begins its next crushing advance.

The advance and retreat of the glaciers seems to depend on how warm the summers are in the northern half of the earth. If the summers are not warm enough, the previous winter's snow and ice fail to melt completely and accumulate year after year, building into advancing glaciers. However, if the summers are warm enough to melt the previous winter's snow and ice, the ice sheets do not grow.

The northern half of the earth is more important to the worldwide climate than the southern half, because most of the land mass is north of the earth's equator. Cooling or warming this land mass directly affects the ice sheets and the global climate.

No one knows what causes ice ages and glacial advance, but many theories exist. The one we will examine suggests that small changes in the orbit and the axial rotation of the earth affect the heat balance and modify the seasons.

*The Changing Shape of the Earth's Orbit.* The earth orbits the sun at an average distance of 93,000,000 miles ($1.5 \times 10^8$ km). Astronomers use this as a unit of distance called the **astronomical unit** (**AU**). Like all planetary orbits, the earth's is not a circle, but an ellipse. This means that the earth is slightly less than 1 AU from the sun at one part of its orbit and slightly farther away six months later. The total change, about 2 percent, is so small that we hardly notice the difference.

The orbital point of closest approach is called **perihelion,** meaning "close to the sun." The earth passes perihelion about January 2 each year, but the slight increase in heat that we get by being 2 percent closer to the sun is wiped out by the seasonal temperature variation from the inclination of the earth's axis. January is a cold month for the northern half of the earth, even though the earth is slightly closer to the sun.

By studying the motion of the earth and the other planets astronomers have discovered that the shape of the earth's orbit changes with a period of about 93,000 years. When the orbit gets more elliptical the variation in the distance from earth to sun is more than 2 percent and may be enough to make our winters milder and the summers cooler. If the summers in the northern half of the earth are cooler, ice and snow accumulate and glaciers advance. When the orbit becomes more circular, summers are warmer and ice sheets melt back.

This sounds like a good theory, but glacial periods occur every 40,000 years, not every 93,000 years. There must be more to the ice ages than just variation in the shape of the earth's orbit.

*Precession.* When we discussed the rotation of the earth, we saw that the earth's axis precesses like a spinning top. This precession is slowly changing the seasons. The earth is now tipped toward the sun in June, producing summer in the northern half of the earth. But in 13,000 years precession will have tilted the earth the other way, away from the sun in June, producing winter in the northern half of the planet. We of course will adjust our calendar to move the months with the seasons, keeping June a summer month. The important point is not the month but the place in earth's orbit where winter occurs.

It happens that winter in the northern half of the earth now occurs when the earth is near perihelion. Because it is 2 percent closer to the sun, our winters are very slightly less severe than they would be if earth's orbit were circular. However, in 13,000 years precession will have moved winter to the other side of earth's orbit, and we will be 2 percent farther from the sun in winter, and it will be slightly colder. Summer will occur near perihelion, we will be 2 percent closer to the sun, and summers will be warmer. If summer in the earth's northern half is warmer, the ice should melt faster, and the glaciers should recede. Thus we might expect earth's climate to change with the same period as precession, about 26,000 years.

It is not that simple, however. The effects of

precession combine with the effects of the changing shape of earth's orbit. If the orbit is almost circular, then it doesn't matter when perihelion occurs because the earth won't be significantly closer to the sun. But if the orbit is more elliptical, then the time of perihelion is important. The fact that the two variations work together with different periods complicates the problem. A third variation in the motion of the earth adds another complication.

*The Inclination of the Earth's Axis.* Not only can astronomers study the changing shape of the earth's orbit and its axial precession, they can also study its changing inclination. The earth has not always been tipped at an angle of 23½°. The inclination varies with a period of about 41,000 years from 22° to 28°. Since the ecliptic is the projection of earth's orbit onto the sky, a change in the earth's axial tilt changes the inclination of the ecliptic to the celestial equator (see Figure 2-4). If this inclination becomes smaller, the sun does not travel as far north in the sky during summer, producing cooler summers and favoring the accumulation of ice sheets.

It is tempting to point to the 40,000-year glacial cycle and identify it with the 41,000 year variation in the earth's inclination, but we would probably be wrong to do so. We must remember that there are at least three factors working to change the climate, each with a different period. When the three processes work together to produce cool summers, ice sheets may accumulate, producing a glacial period. But the pattern of the advance and retreat of the glaciers is very complex because of the three effects at work.

Scientists can find the temperature of the ancient earth by studying microscopic fossils in ocean-floor sediments. A group of these scientists say they are convinced that these three astronomical effects cause the earth's climatic changes. Other scientists, analyzing the same data, are equally certain that variations in the motion of the earth have little effect on climate. Most agree that the orbital changes could affect the climate, but they are reserving their decision as to whether the ice ages and glacials really have an astronomical cause.

Our study of the earth and sky has given us a secure base from which to explore the universe. That exploration is based on a search for information about the things we see in the sky, so we must consider the way astronomers gather data with telescopes, analyze the content of the data, and reach conclusions about stars, galaxies, and so on. These three steps will carry us through the next three chapters.

## SUMMARY

Because the earth orbits the sun, the sun appears to move eastward around the sky following the ecliptic. Since the ecliptic is tipped 23½° to the celestial equator, the sun spends half the year in the northern celestial hemisphere and half the year in the southern celestial hemisphere, producing the seasons. The seasons are reversed south of the earth's equator.

Because we see the moon by reflected sunlight, its shape appears to change as it orbits the earth. The lunar phases wax from new moon to first quarter to full moon, and wane from full moon to third quarter to new moon. A complete cycle of lunar phases takes 29.53 days.

The moon's gravitational field exerts tidal forces on the earth that pull the ocean waters up into two bulges, one on the side of the earth facing the moon and the other on the side away from the moon. As the rotating earth carries the continents through these bulges of deeper water, the tides ebb and flow. Friction with the sea beds slows the earth's rotation, and the gravitational force the bulges exert on the moon force its orbit to grow larger.

When the moon passes through the earth's shadow, sunlight is cut off and the moon darkens in a lunar eclipse. If the moon only grazes the shadow, the eclipse is partial, or penumbral, and not total.

If the moon passes directly between the sun and earth, it produces a total solar eclipse. During such an eclipse, the bright photosphere of the sun is covered and the fainter corona, chromosphere, and prominences become visible. An observer outside the path of totality sees a partial eclipse. If the moon is in the farther part of its orbit, it does not cover the photosphere completely, resulting in an annular eclipse.

The motion of the earth may change in ways that can affect the climate. Changes in orbital shape, in precession, and in axial tilt can alter the planet's heat balance and may be responsible for the ice ages and glacial periods.

## NEW TERMS

ecliptic
precession
vernal equinox
autumnal equinox
summer solstice
winter solstice
morning and evening stars
spring and neap tides
lunar eclipse
umbra
penumbra
solar eclipse
photosphere
corona
chromosphere
prominence
annular eclipse
node
glacial period
interglacial period
astronomical unit (AU)
perihelion

## QUESTIONS

1. Why is Venus sometimes called a "morning star" and sometimes an "evening star"?
2. Explain why the summer solstice is not the hottest day of the year.
3. Explain how the moon is slowing the rotation of the earth.
4. Why isn't there an eclipse at every new moon?
5. Why are we normally unable to see solar prominences?
6. Describe three astronomical factors that may affect the earth's climate.

## PROBLEMS

1. Draw a diagram like Figure 2-6 and show the path of the sun across the sky at the vernal equinox.
2. If the earth is about 5 billion ($5 \times 10^9$) years old, how many precessional cycles have occured?
3. Identify the phases of the moon if on March 21 the moon is located at (a) the vernal equinox, (b) the autumnal equinox, (c) the summer solstice, or (d) the winter solstice.
4. Identify the phases of the moon if at sunset the moon is (a) near the eastern horizon, (b) high in the south, (c) in the southeast, or (d) in the southwest.
5. About how many days must elapse between first quarter moon and third quarter moon?
6. Draw a diagram showing the earth, moon, and shadows during (a) a total solar eclipse, (b) a total lunar eclipse, (c) a partial lunar eclipse, and (d) an annular eclipse.
7. The average distance from Jupiter to the sun is 5.2 AU. How far is this in miles?

## RECOMMENDED READING

Evans, D. L., and Freeland, H. J. "Variations in the Earth's Orbit: Pacemaker of the Ice Ages?" *Science* 198 (Nov. 4, 1977), p. 528.

Gribbin, J. "Why Does Earth's Climate Change?" *Astronomy* 6 (February 1978), p. 18.

Hawkins, G. S. *Stonehenge Decoded.* Garden City, N.Y.: Doubleday, 1965.

Hays, D. D., Imbrie, J., and Shackleton, N. J. "Variations in the Earth's Orbit: Pacemaker of the Ice Ages." *Science* 194 (Dec. 10, 1976), p. 1121.

Imbrie, J., and Imbrie, K. P. *Solving the Mystery.* Short Hills, N.J.: Enslow, 1979.

Menzel, D. H. *A Field Guide to the Stars and Planets.* Boston: Houghton Mifflin, 1964.

Shipman, H. L. "Megaliths and the Moon: Eclipse Prediction in the Stonehenge Era." *The Griffith Observer* 38 (Feb. 1974), p. 7.

# HISTORICAL SUPPLEMENT

# THE 99 YEARS THAT CHANGED ASTRONOMY

The story of modern astronomy begins with the death of the great Polish astronomer Nicolaus Copernicus (Figure H-1) in May 1543 and the almost simultaneous publication of his theory of the universe. That theory revolutionized not only astronomy, but all science, and inspired a new consideration of our place in nature. We will trace this story over the 99 years from 1543 and Copernicus' death to 1642 and the death of another great astronomer, Galileo, and the birth of one of the greatest scientists in history, Isaac Newton.

## PRE-COPERNICAN ASTRONOMY

To understand why the Copernican theory was important, we must backtrack to ancient Greece and meet the two great authorities of ancient astronomy, Aristotle and Ptolemy. Aristotle, a Greek philosopher who lived from 384 to 322 BC, taught and wrote on philosophy, history, politics, ethics, poetry, drama, and so on. Because of his sensitivity and insight, he became the great authority of antiquity. Unfortunately, his scientific work was almost entirely wrong. He wrote on biology, physics, and astronomy and in every case made serious blunders, but because of his authority in other areas of knowledge, his teachings in science were accepted unquestioningly.

He believed the universe was divided into two parts—the earth, corrupt and changeable, and the heavens, perfect and immutable. He knew the earth was spherical, but concluded that it sat immobile at the center of the universe. It could not move around the sun, as some of his contemporaries suggested, because he saw no parallax in the position of the stars. Actually, he saw no parallax because the stars are much farther away than he supposed, and the parallax is much too small to be visible to the naked eye.

In his astronomy, Aristotle adopted a belief originated by Pythagoras, a Greek mathematician and philosopher who had lived about 150 years before. Pythagoras taught that the seven heavenly bodies—five planets, the sun, and moon—were carried by separate crystalline spheres whose various rotations gave rise to the motions of the sun, moon, and planets. Above these spheres was the sphere of fixed stars. Aristotle thought the earth was immobile, and consequently he believed all of the spheres whirled westward around the earth each day. Thus most Greek philosophers viewed the sky as a perfect, heavenly machine not many times larger than the earth itself.

One reason for the survival of Aristotle's views was the work of the mathematician Claudius Ptolemaeus, usually referred to as Ptolemy (Figure H-2). His nationality and birth date are unknown, but he lived and worked in

*Figure H-1. Nicolaus Copernicus (1473–1543) proposed that the sun and not the earth was the center of the universe. These stamps were issued in 1973 to commemorate the 400th anniversary of his birth. Note the sun-centered model on the United States stamp and the pages from his book* De Revolutionibus *on the East German stamp (DDR).*

*Figure H-2. The muse of astronomy guides Ptolemy (about 140 AD) in his study of the heavens. (Courtesy Owen Gingerich and The Houghton Library.)*

the Greek settlement at Alexandria about 140 AD. There he studied mathematics and astronomy and developed a model of the universe based on the teachings of Aristotle.

The Ptolemaic model was **geocentric** (earth centered) in agreement with Aristotle. In addition, it incorporated the Greek belief that the heavenly bodies moved perfectly. Since the only perfect motion is uniform motion and the only perfect curve is a circle, Ptolemy assumed the planets moved with **uniform circular motion**. But simple, circular paths centered on the earth do not account for the motions of the planets in the sky. The planets sometimes move faster and sometimes slower, and occasionally they appear to slow to a stop and move backwards for a time, tracing a **retrograde loop** (Figure H-3).

To describe the complicated planetary motions and yet preserve a geocentric model with uniform circular motion, Ptolemy adopted a system of wheels within wheels. The planet moved in a small circle called an **epicycle,** and the center of the epicycle moved along a larger circle around the earth called a **deferent** (Figure H-4). By adjusting the size of the circles and the rate of their motion, Ptolemy could account for most planetary movement. But as a final adjustment, he placed the earth off center in the deferent circle and specified that the center of the epicycle would only appear to move at constant speed if viewed from a point called the **equant** located on the other side of the deferent's center. Thus by using a few dozen circles of various sizes rotating at various rates, Ptolemy's system predicted the positions of the planets (Figure H-5).

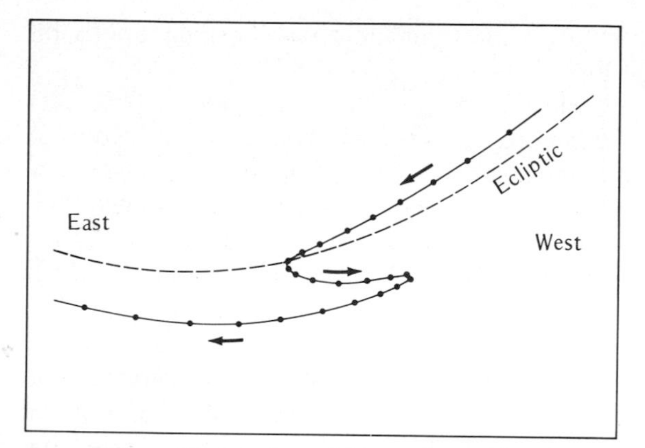


Figure H-3. *The motion of Mars along the ecliptic is shown at 10-day intervals. Though it usually moves eastward, it sometimes slows to a stop and moves westward in retrograde motion.*

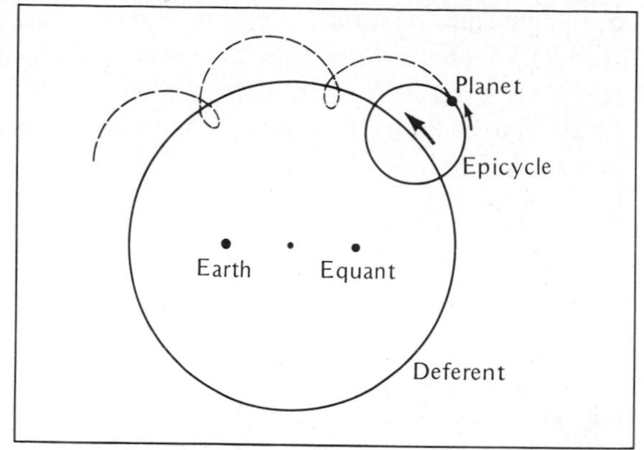


Figure H-4. *Ptolemy accounted for a planet's motion by placing it on a small circle (epicycle) that moved along a larger circle (deferent). Viewed from the equant, the center of the epicycle would have moved at constant speed.*

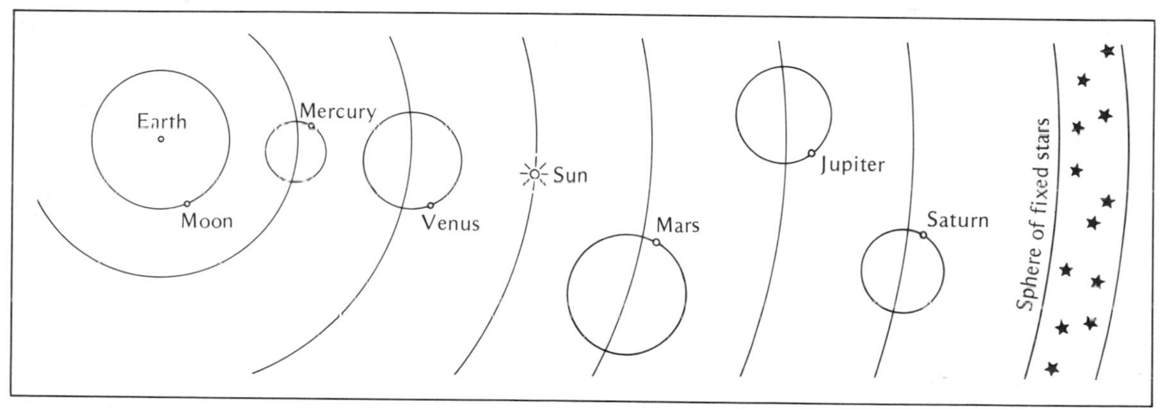


Figure H-5. *The Ptolemaic system was geocentric (earth centered) and based on the uniform circular motion of epicycles.*

About 140 AD, Ptolemy included this work in a book now known as *Almagest*, a title that Ptolemy never knew. He called his work *Great Composition*, but that title changed. With Ptolemy's death the classical astronomy of Greece ended forever. The breath of civilization passed to invading Arabs who dominated most of the known world for 1000 years. Arabian astronomers translated, studied, and preserved many classical manuscripts, and in Arabic Ptolemy's book became *Al Magisti* (Greatest). Beginning in the 1200s, European Christians drove the Arabs out of Spain and recovered their classical heritage through Arabic translations. In Latin, Ptolemy's book became *Almagestum* and thus our modern *Almagest*. For 1000 years Arab astronomers studied and preserved Ptolemy's work, but they made no significant improvement in his theory.

At first the Ptolemaic system predicted the positions of the planets with fair accuracy, but as centuries passed, errors accumulated and Arabian and later European astronomers had to update the system, computing new constants and sometimes adding more epicycles. In the middle of the 13th century a team of astronomers supported by King Alfonso X of Castile worked for 10 years revising the Ptolemaic system and publishing the result as the *Alfonsine Tables*. It was the last great adjustment of the Ptolemaic system.

## COPERNICUS

By 1543, the year Copernicus died, the Ptolemaic system, in spite of many revisions, was still a poor predictor of planetary positions. Yet, because of the authority of Aristotle, it was the officially accepted theory of the universe. The Catholic Church had adopted the teachings of Aristotle as part of church dogma, and anyone who questioned the Ptolemaic system risked a charge of heresy.

Throughout his life Copernicus had been associated with the church. His uncle, by whom he was raised and educated, was an important bishop in Poland, and after studying canon law and medicine in some of the major universities in Europe, Copernicus became a canon of the church at the age of 24. He served as secretary and personal physician to his powerful uncle for 15 years. When his uncle died, Copernicus went to live in quarters adjoining the cathedral in Frauenburg. Because of this long association with the church and his fear of persecution, he hesitated to publish his revolutionary ideas in astronomy.

In fact, his theory was already being discussed long before the publication of his book. His interest in astronomy had begun during his college days, and he apparently doubted the Ptolemaic system even then. About 1507, at the age of 34, he wrote a short pamphlet that discussed the motion of the sky and outlined his theory that the sun, not the earth, was the center of the universe and that the earth rotated on its axis and revolved around the sun. To avoid criticism and possible charges of heresy, he distributed his pamphlet in handwritten form to scientific friends. By 1515 it was well known, and by 1530 church officials were asking about his theory.

He evidently wrote his book *De Revolutionibus Orbium Coelestium* (On the Revolution of the Celestial Orbs) in the early 1530s, finishing it 10 years before it was finally printed. Perhaps Copernicus was wise to delay publication until he was safely beyond the reach of critics. This was a time of rebellion in the church—Martin Luther was speaking harshly about many church teachings, and others, both scholars and scoundrels, were questioning the authority of the church. It may be understandable that church officials were not open to more criticism, even on something as abstract as astronomy, and the penalty for heresy was often burning at the stake. Thus Copernicus delayed publication and probably never saw his book in print, though tradition has it that the book was put into his hands just before he died.

*De Revolutionibus* did not prove that the Ptolemaic theory was wrong. It placed the sun at the center of the universe and thus reduced the earth to a mere planet orbiting the sun with the others. Copernicus quoted a number of criticisms of the Ptolemaic theory while defending his own, but his arguments were not conclusive. His theory did not predict the positions of the planets more accurately than Ptolemy's. Since Copernicus retained epicycles and deferents much like Ptolemy's, the *Prutenic Tables* (1551), tables of planetary position based on the Copernican system, were no more accurate than the *Alphonsine Tables*. Both could be in error by as much as 2°, four times the angular diameter of the moon.

The Copernican *system* was wrong, but the Copernican *theory*, that the universe was **heliocentric** (sun centered), was right. Why that theory gradually won acceptance in spite of the inaccuracy of the epicycles and deferents is a question historians still debate. There are probably a number of reasons, including the revolutionary temper of the times, but the most important factor may be the elegance of the theory. Placing the sun at the center of the universe produced a symmetry among the motions of the planets that was pleasing to the eye as well as to the intellect (Figure H-6). No longer did Venus and Mercury revolve around empty points located between the earth and sun. Now they, like the rest of the planets, moved in orbits around the sun.

NICOLAI COPERNICI

net, in quo terram cum orbe lunari tanquam epicyclo contineri diximus. Quinto loco Venus nono menſe reducitur. Sextum denique locum Mercurius tenet, octuaginta dierum ſpacio circũ currens. In medio uero omnium reſidet Sol. Quis enim in hoc

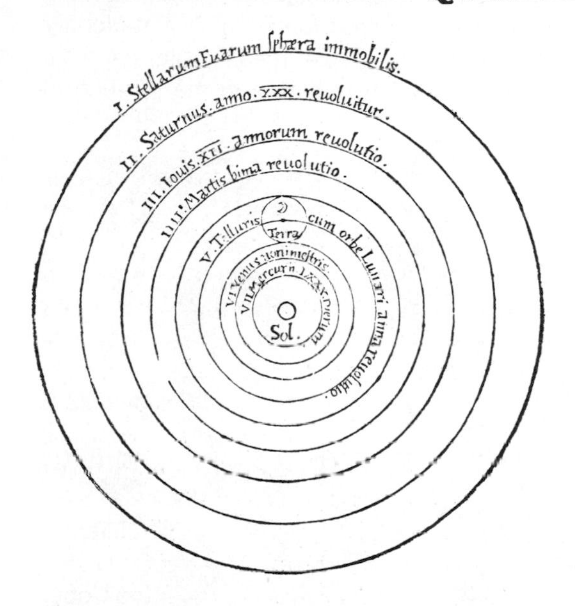


pulcherrimo templo lampadem hanc in alio uel meliori loco poneret, quàm unde totum ſimul poſsit illuminare? Siquidem non inepte quidam lucernam mundi, alij mentem, alij rectorem uocant. Trimegiſtus uiſibilem Deum, Sophoclis Electra intuentẽ omnia. Ita profecto tanquam in ſolio regali Sol reſidens circum agentem gubernat Aſtrorum familiam. Tellus quoque minime fraudatur lunari miniſterio, ſed ut Ariſtoteles de animalibus ait, maximam Luna cum terra cognationẽ habet. Cõcipit interea à Sole terra, & impregnatur annno partu. Inuenimus igitur ſub hac

*Figure H-6. The Copernican universe as reproduced in* De Revolutionibus. *The earth and all of the known planets moved in separate orbits centered on the sun, surrounded by an outer, immobile sphere of fixed stars. (Yerkes Observatory photograph.)*

In addition, the Copernican theory explained the retrograde (westward) motion of the planets in a straightforward way. The earth moves faster along its orbit than the planets that lie farther from the sun. Consequently, the earth periodically overtakes and passes these planets. Imagine that you are riding in a race car driving rapidly along the inside lane of a circular race track. As you pass slower cars driving in the outer lanes, they fall behind, and, if you did not know you were moving, it would seem that the cars in the outer lanes occasionally slowed to a stop and then backed up for a short interval. The same thing happens as the earth passes a planet such as Mars. Although Mars moves steadily along its orbit, as seen from earth, it appears to slow to a stop and move westward (retrograde) as the earth passes it (Figure H-7). Because the planetary orbits do not lie in precisely the same plane, a planet does not resume its eastward motion in precisely the same path it followed earlier. Consequently, it describes a loop whose shape depends on the angle between the orbital planes. This simple explanation of retrograde motion did not prove that the Copernican theory

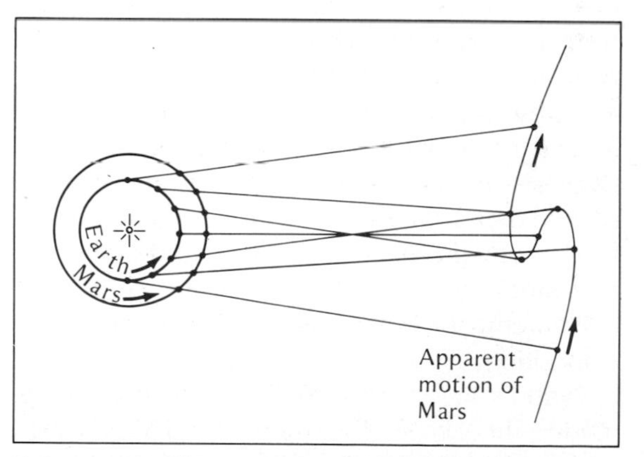


*Figure H-7. The positions of the earth and Mars are shown at equal intervals of one month. As the earth overtakes and passes the slower-moving Mars, the latter appears to slow to a stop and move westward (retrograde) against the background of stars. In this way, the Copernican theory accounted for retrograde motion.*

was correct, but it was a much simpler explanation than that provided by the Ptolemaic theory.

However interesting astronomers found *De Revolutionibus*, it did not become an instant best seller. It was written in Latin, the language of science at that time, and it included mathematical and philosophical arguments that were beyond the grasp of most people. Thus the church took little notice at first, probably assuming that with the author dead the theory would wither in a few years. The book was not officially condemned until 1616.

## TYCHO BRAHE

The great observational astronomer of our 99-year story was a Danish nobleman, Tycho Brahe (Figure A-8), born December 14, 1546, only three years after the publication of *De Revolutionibus*. In histories of astronomy the great astronomers are often referred to by their last name, but Tycho Brahe is usually called Tycho. Were he alive today he would no doubt object to such familiarity from his obvious inferiors. He was well known for his vanity and lordly manners.

Tycho's college days were eventful. He was officially studying law with the expectation that he would enter Danish politics, but he made it clear to his family that his real interest was astronomy and mathematics. It was also during his college days that he became involved in a duel over some supposed insult and received a wound that disfigured his nose. For the rest of his life he wore false noses made of gold and silver and stuck on with wax (see Figure H-8). The disfigurement probably did little to improve his disposition.

Tycho's first astronomical observations were made while he was a student. In 1563 Jupiter and Saturn passed very near each other in the sky, nearly merging into a single point on the night of August 24. Tycho found that the *Alphonsine Tables* were a full month in error and that the *Prutenic Tables* were in error by a number of days. These discrepancies dismayed Tycho and sparked his interest in the motions of the planets.

Then in 1572 a "new star" (now called Tycho's supernova) appeared in the sky, shining brighter then Venus. Tycho carefully measured its position and concluded that it displayed no parallax. To understand the significance of this observation, we must note that the rotation of the earth carries us eastward and that the positions of celestial objects near the earth, such as the moon, change slightly through the night because the observer is in different positions. Tycho believed the earth was stationary and that the heavens were rotating westward, but that makes no difference. Objects near a stationary earth should still have measurable parallax. That he detected no change in the position of

*Figure H-8. Tycho Brahe (1546–1601), a Danish nobleman, established an observatory at Hveen and measured planetary positions with high accuracy. His artificial nose is evident in this engraving.*

the new star through the night proved it was farther away than the moon. Therefore, he concluded, it was a new star in the heavens. Since Aristotle and Ptolemy held that the heavens were perfect and unchanging, the new star led Tycho to question the Ptolemaic system. He summarized his results in a small book, *De Stella Nova* (The New Star) published in 1573.

The book attracted the attention of astronomers throughout Europe, and soon Tycho was summoned to the court of the Danish king Frederik II and offered funds to build an observatory on the island of Hveen just off the Danish coast. Tycho also received a steady source of income as landlord of a coastal district from which he collected rents. (He was not a popular landlord.) On Hveen, Tycho constructed a luxurious home with four towers especially equipped for astronomy, and populated it with servants, assistants, and a dwarf to act as jester. Soon Hveen was an international center of astronomical study (Figure H-9).

Tycho's great contribution to the birth of modern astronomy was not theoretical. Because he could measure no parallax for the stars, he concluded the earth had to be stationary, thus rejecting the Copernican theory. However, he also rejected the Ptolemaic theory because of its inaccurate predictions. Instead, he devised a complex theory in which the earth was the immobile center of the universe around which the sun and moon moved. The remaining planets orbited the sun. This Tychonic theory was popular for a short time.

The true value of Tycho's work was observational. Because he was able to devise new and better observing instruments, he was able to make highly accurate observations of the positions of the stars, sun, moon, and planets. Tycho had no telescopes—they were not invented till the next century—so his observations were made by the naked eye peering along sights on his large instruments (Figure H-10). In spite of these limitations he measured the positions of 777 stars to better than 4 minutes of arc and measured the positions of the sun, moon, and planets almost daily for the twenty years he stayed on Hveen.

*Figure H-9. Tycho's palatial observatory and grounds at Hveen. (Yerkes Observatory photograph.)*

Unhappily for Tycho, King Frederik II died in 1588, and his young son took the throne. Suddenly Tycho's temper, vanity, and noble presumptions threw him out of favor. In 1596, taking most of his instruments and books of observations, he went to Prague, the capital of Bohemia, and became imperial mathematician to the Holy Roman Emperor Rudolph II. His assignment there was to revise the *Alphonsine Tables* and publish the revision as a monument to his new patron. It would be called the *Rudolphine Tables*.

Tycho did not intend to base the *Rudolphine Tables* on the Ptolemaic system, but rather on his own Tychonic system, proving once and for all the validity of his theory. To assist him, he hired a few mathematicians and astronomers, including one Johannes Kepler. Then in November 1601, Tycho overate at a nobleman's home, an internal organ ruptured, and he collapsed. Before he died, nine days later, he asked Rudolph II to make Kepler imperial mathematician. Thus the newcomer, Kepler, became Tycho's replacement (though at half Tycho's salary).

*Figure H-10. Much of Tycho's success was due to his skill in designing large, accurate instruments. In this engraving of his mural quadrant, the figure of Tycho at the center, his dog, and the scene in the background are a mural painted on the wall within the arc of the quadrant. The observer at the extreme right peers through a sight out the loophole in the wall at the upper left and thus measures the object's altitude above the horizon. (Yerkes Observatory photograph.)*

## JOHANNES KEPLER

No one could have been more different from Tycho Brahe than Kepler (Figure H-11). He was born December 27, 1571 to a poor family in a region now included in southwest Germany. His father was unreliable and shiftless, principally employed as a mercenary soldier, fighting for whoever paid enough. He finally failed to return from a military expedition, either because he was killed or because he found circumstances more to his liking. Kepler's mother was apparently an unpleasant and unpopular woman. She was accused of witchcraft in later years, and Kepler had to defend her in a trial that dragged

*Figure H-11. Johannes Kepler (1571–1630) derived three laws of planetary motion from Tycho Brahe's observations of the positions of the planets. This Romanian stamp commemorates the 400th anniversary of Kepler's birth. Ironically it contains an error—the orientation of the moon.*

on for three years. She was finally acquitted but died the following year.

Kepler was the oldest of six children and his childhood was no doubt unhappy. The family was not only poor and often lacked a father, but it was also Protestant in a predominantly Catholic region. In addition, Kepler was never healthy, even as a child, so it is surprising that he did well in the pauper's school he attended, eventually winning a scholarship to the university at Tübingen, where he studied to become a Lutheran pastor.

During his last year of study, Kepler accepted a job in Graz teaching mathematics and astronomy, a job he resented because he knew little of the subjects. Evidently he was not a good teacher. He had few students his first year, and none at all his second. His superiors put him to work teaching a few introductory courses and preparing an annual almanac that contained astronomical, astrological, and weather predictions. Through good luck in 1595 some of his weather predictions were fulfilled and he gained a reputation as an astrologer and seer, and even in later life he earned money from his almanacs.

While still a college student, Kepler had become a believer in the Copernican theory, and at Graz he used his extensive spare time to study astronomy. By 1596, the same year Tycho left Hveen, Kepler was ready to solve the mystery of the universe. That year he published a book called *The Forerunner of Dissertations on the Universe, Containing the Mystery of the Universe*. The book was, like nearly all scientific works, in Latin, and is now known as *Mysterium Cosmographicum*.

The book contained almost nothing of value. It began with a long appreciation of Copernicanism and then went on to speculate on the reasons for the spacing of the planetary orbits. Kepler felt he had found the underlying architecture of the universe in the five regular solids*—the cube, tetrahedron, dodecahedron, icosahedron, and octahedron. Because these five solids were the only regular solids, he supposed they were the spacers between the planetary orbits (Figure H-12). Kepler advanced astrological, numerological, and even musical arguments for his theory.

The second half of the book was no better than the first, but it had one virtue—as Kepler tried to fit the five solids to the planetary orbits, he demonstrated that he was a talented mathematician and that he was well versed in astronomy. He sent copies to Tycho and to Galileo in Rome, and both recognized his talent in spite of the mystical content of the book.

Life was unsettled for Kepler because of the persecution of Protestants in the region, so when

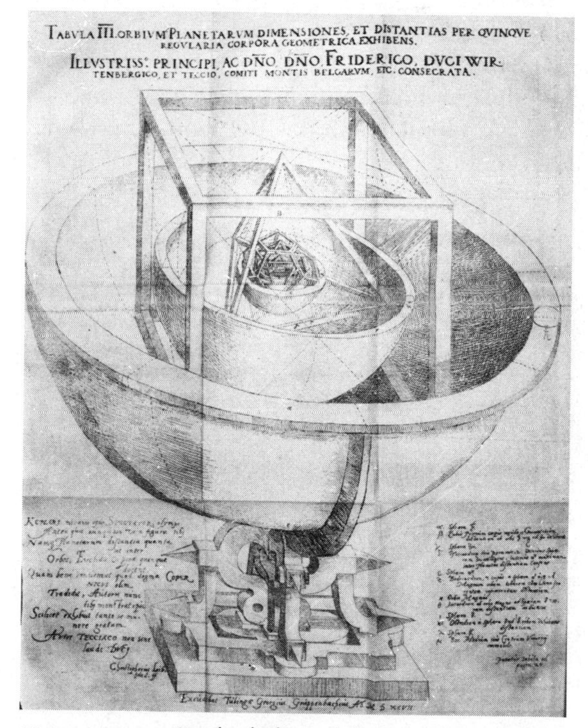


*Figure H-12.* *Kepler believed the five regular solids were the spacers between the spheres containing the planetary orbits. His book* Mysterium Cosmographicum *contained this fold-out illustration of the spheres and spacers. (Courtesy Owen Gingerich and The Houghton Library.)*

* A regular solid is a three-dimensional body each of whose faces is the same. A cube is a regular solid, each of whose faces is a square.

Tycho invited him to Prague in 1600, he came readily, anxious to work with the famous astronomer. Tycho's sudden death in 1601 left Kepler in a position to use the observations from Hveen to analyze the motions of the planets and complete the *Rudolphine Tables*. Tycho's family, recognizing that Kepler was a Copernican and guessing that he would not follow the Tychonic system in completing the *Rudolphine Tables*, sued to recover the instruments and books of observations. The legal wrangle went on for years. They did recover the instruments Tycho had brought to Prague, but Kepler had the books and he kept them.

Whether Kepler had any legal right to Tycho's records is debatable, but he put them to good use. He began by studying the motion of Mars, trying to deduce from the observations how the planet moved. By 1606 he had solved the mystery, this time correctly. The orbit of Mars (and all planets) was an ellipse with the sun at one focus. Thus he abandoned the 2000-year-old belief in circular motion. But the mystery was even more complex. The planets did not move at constant speed along their orbits—they moved faster when close to the sun and slower when farther away. Thus Kepler abandoned both uniform motion and circular motion.

Kepler published his results in 1609 in a book called *Astronomia Nova* (New Astronomy). Like Copernicus's book, *Astronomica Nova* did not become an instant success. It was written in Latin for other scientists and was highly mathematical. In some ways the book was surprisingly advanced. For instance, Kepler discussed the force that holds the planets in their orbits and came within a paragraph of discovering the principle of mutual gravitation.

In spite of the abdication of Rudolph II in 1611, Kepler continued his astronomical studies. He wrote about a supernova that had appeared in 1604 (now known as Kepler's supernova) and about comets, and he authored a textbook about Copernican astronomy. In 1619 he published *Harmonice Mundi* (The Harmony of the World) in which he returned to the cosmic mysteries of *Cosmographicum Mysterium*. The only thing of note in *Harmonice Mundi* is his discovery that the radii of the planetary orbits are related to the planet's orbital period. That and his two previous discoveries are now recognized as Kepler's three laws of planetary motion (Table H-1).

Kepler's first law states that the orbits of the planets around the sun are ellipses with the sun at one focus. An ellipse is defined as a figure drawn around two points called the foci such that the distance from one focus to any point on the ellipse back to the other focus equals a constant. This makes it very easy to draw ellipses with two thumbtacks and a loop of string. Press the thumbtacks into a board, loop the string about them, and place a pencil in the loop as in Figure H-13a. If you keep the string taut as you move the pencil, it traces out an ellipse. The closer together the thumbtacks, the more circular the ellipse. Though Kepler was able to determine the elliptical shape of the planetary orbits, they are nearly circular. Of the planets known to Kepler, Mercury has the most elliptical orbit, but even it differs only slightly from a circle (Figure H-14).

Kepler's second law states that a line from the planet to the sun sweeps over equal areas in equal intervals of time. This means that when the planet is closer to the sun and the line connecting it to the sun is shorter, the planet must move more rapidly if the line is to sweep over

**Table H-1 Kepler's Laws of Planetary Motion**

I. The orbits of the planets are ellipses with the sun at one focus.
II. A line from the planet to the sun sweeps over equal areas in equal intervals of time.
III. A planet's orbital period squared is proportional to its average distance from the sun cubed:

$$P_{yr}^2 = a_{AU}^3$$

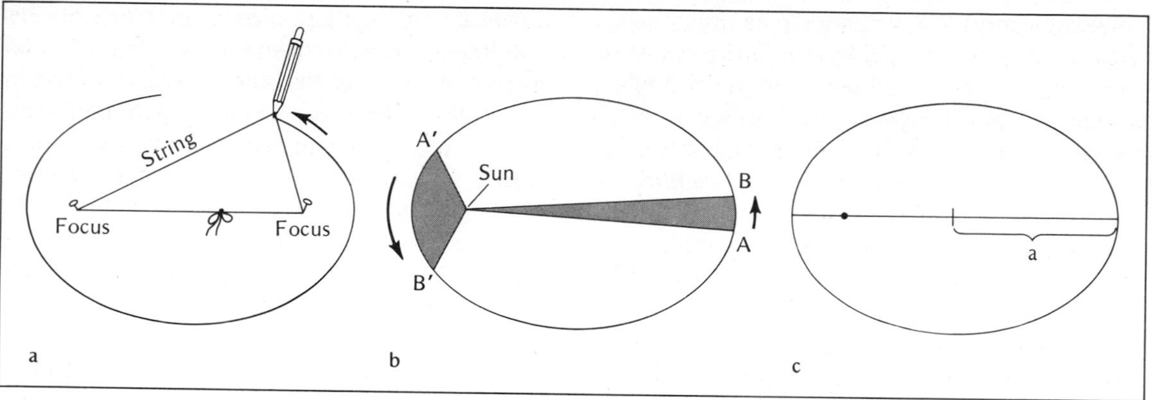


*Figure H-13.* *(a) Drawing an ellipse with two tacks and a loop of string. (b) "A line from a planet to the sun sweeps out equal areas in equal intervals of time." (c) The average distance from a planet to the sun equals* **a,** *the semimajor axis of its orbit.*

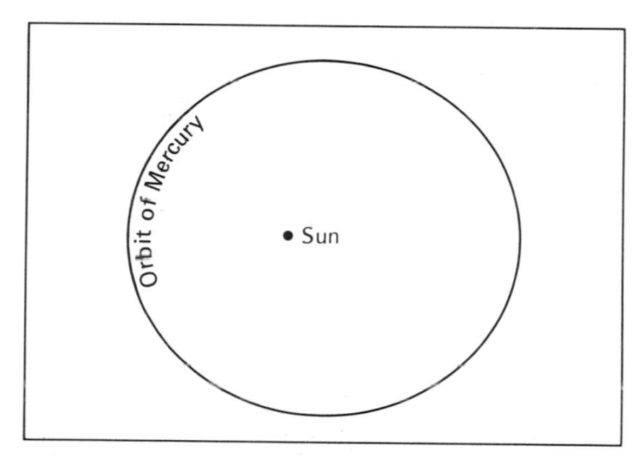


*Figure H-14.* *The orbits of the planets are nearly circular. Of the planets known to Kepler, Mercury has the most elliptical orbit. Pluto's orbit is only slightly more elliptical.*

the same area. Thus the planet in Figure A-13b would move from point A′ to point B′ in one month sweeping out the area shown. But when it is farther from the sun, one month's motion would carry it from A to B.

Kepler's third law states that a planet's orbital period squared is proportional to its average distance from the sun cubed. Because of the way the planet moves along its orbit, its average distance from the sun is equal to half of the long diameter of the elliptical orbit, the so-called semimajor axis **a** (Figure H-13c). If we measure this quantity in astronomical units and the orbital period in years, we can summarize the third law as:

$$P_{yr}^2 = a_{AU}^3$$

For example, Jupiter's average distance from the sun is roughly 5 AU. The semimajor axis cubed would be 125, so the period must be the square root of 125, about 11 years.

It is important to notice that Kepler's three laws are empirical. That is, they describe the phenomenon without explaining why it occurs. Kepler derived them from Tycho's extensive observations, not from any fundamental assumption or theory. In fact, Kepler never knew what held the planets in their orbits or why they continued to move around the sun. His books are a fascinating blend of careful observation, mathematical analysis, and mystical theory.

In spite of Kepler's recurrent affairs with astrology and numerology, he continued to work on the *Rudolphine Tables*. At last in 1628 they were ready, and he financed their printing himself, dedicating them to the memory of Tycho Brahe. In fact, Tycho's name appears in larger type on the title page than Kepler's own. This is especially surprising when we recall that the tables were based on the heliocentric theory of Copernicus and the elliptical orbits of Kepler, and not on the Tychonic system. The reason for Kepler's care was Tycho's family, still powerful and still intent on protecting the memory of Tycho. They even demanded a share of the profits and the right to censor the book before publication, though they changed nothing but a few words on the title page.

The *Rudolphine Tables* was Kepler's masterpiece, the final proof that the Copernican theory was an accurate description of the heavens. It was the proof that Copernicus had sought but failed to find.

Kepler died November 15, 1630. During his life he had been a Copernican and had written a number of books proclaiming his belief, but he had never been seriously persecuted. Others were less lucky. Giordano Bruno had been tried and condemned by the Inquisition in Rome and burnt at the stake in 1600. Bruno was an outspoken critic of the church in many respects, but one of his offences was Copernicanism. Kepler escaped such a fate by living in northern Europe beyond the reach of the Inquisition. The last astronomer in our story was a Copernican too, but he had the misfortune to be born in southern Europe where the Inquisition held sway.

## GALILEO GALILEI

Copernicus, Tycho, and Kepler were all mathematicians and astronomers and are remembered for their work in astronomy. Galileo (Figure H-15) is most famous as the defender of Copernicanism. This is, to a certain extent, unfair because Galileo was a mathematician and scientist of deep insight. He made significant discoveries in the physics of motion and invented a calculating device that was a forerunner of the slide rule. However, the astronomical thrust of his life was directed at the defense of Copernicanism and not at the improvement of astronomical theory.

Galileo was born in Pisa in 1564. Thus he was a contemporary of Tycho and Kepler. At the university he studied medicine, although his true interest was mathematics, mechanics, and astronomy. We can be sure he was talented in mathematics because, even though family finances forced him to leave school without a degree, he obtained a position as a professor of mathematics at the University of Pisa only four years later.

In Pisa his teaching assignment included astronomy, and he must have taught the Ptolemaic system. He may even then have been a Copernican, but he would have been foolish to try to teach it. Italy was no place to introduce unorthodox ideas that might challenge church teachings. Perhaps while he taught astronomy he became convinced that the universe was Copernican and not Ptolemaic.

In 1609 Galileo obtained some lenses and built a telescope (Figure H-16) following descriptions of similar instruments built by Dutch lens makers. He did not invent the telescope, nor was he the first to turn one toward the sky, but he was the first to study, night after night, the telescopic appearance of the heavenly bodies, record his observations, and report them publicly. In 1610 he published a book called *Sidereus Nuncius* (The Sidereal Messenger), in which he described what he saw through his telescope and showed how his observations supported the Copernican theory.

One of his important discoveries was that Venus goes through phases like the moon. In the Ptolemaic theory Venus revolves around an epicycle located between the earth and sun. Thus an observer on earth would never see the planet fully illuminated by the sun. That is, it would always be seen as a crescent. But Galileo saw Venus go through a complete set of phases, which proved that it went around the sun (Figure H-17).

In addition, he discovered four satellites orbiting the planet Jupiter. Some critics of the Copernican theory had said the earth could not move because the moon would be left behind. But Jupiter moved yet kept its satellites, so Galileo's discovery proved that the earth could move and

Figure H-15. *Galileo Galilei (1564–1642), remembered as the great defender of Copernicanism, also made important discoveries in the physics of motion. He is honored here on an Italian 2000-lira note. The reverse side shows one of his telescopes at lower right and a modern observatory above.*

Figure H-16. *Galileo's telescopic discoveries generated intense interest and controversy. Some critics refused to look through a telescope lest it deceive them. (Yerkes Observatory photograph.)*

keep its moon. Also, the accepted teaching was that everything revolved around the earth. Galileo's telescope revealed that some objects orbited Jupiter. Thus there could be other centers of motion.

Finally, the telescope challenged Aristotle's contention that the heavens were perfect and unchanging. Galileo observed spots on the sun, raising the suspicion that the sun was less than perfect. By noting the movement of the spots, he

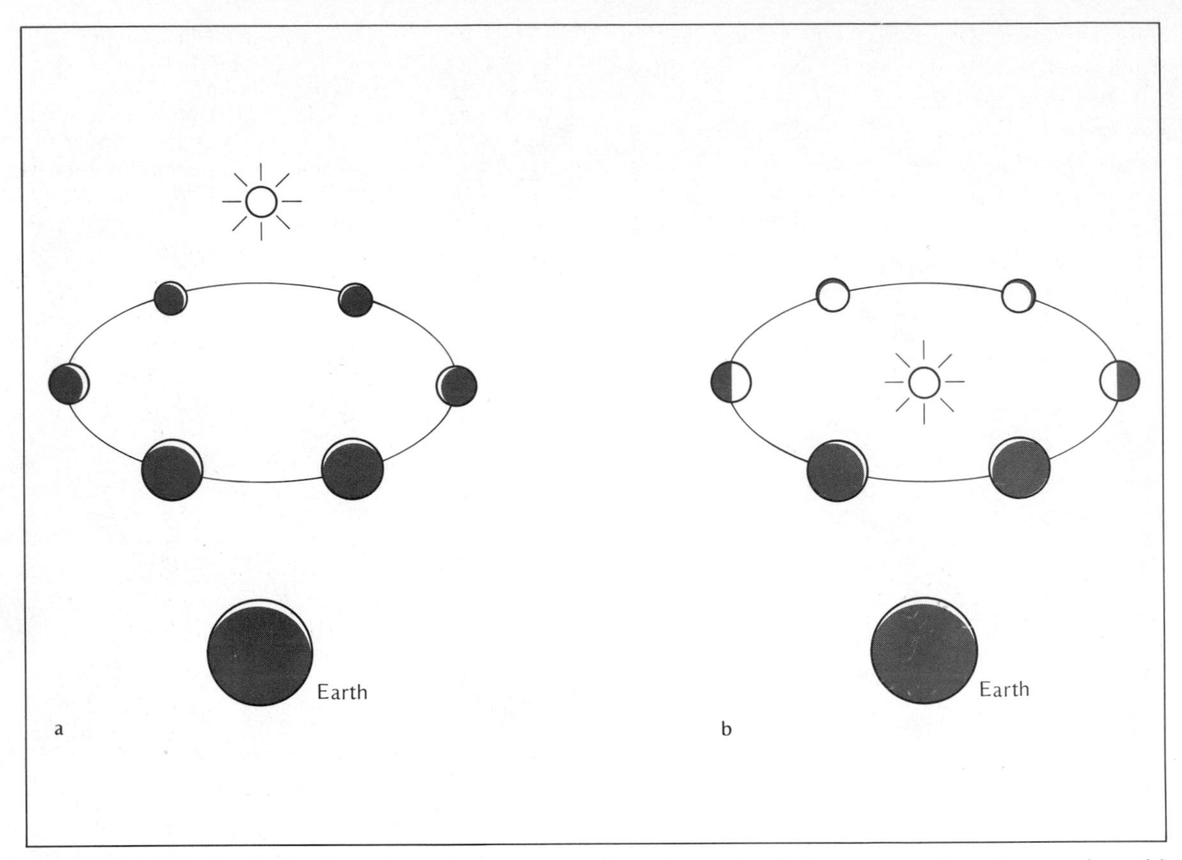


*Figure H-17.* *In the Ptolemaic universe Venus moves around an epicycle between earth and sun and would always appear as a crescent (a). Galileo's telescope showed that Venus went through a full set of phases, proving that it must orbit the sun as in the Copernican universe (b).*

concluded that the sun rotated on its axis, just as Copernicus said the earth did. When he observed the moon he found not a polished perfect sphere but a craggy, mountainous terrain surrounding flat areas he mistook for seas.

*Sidereus Nuncius* was a popular book and it made Galileo famous. He soon left his teaching position and went to Florence where he became personal philosopher and mathematician to the Grand Duke of Tuscany. When Galileo visited Rome in 1611 he gave well-attended lectures and had long and friendly discussions with the powerful Cardinal Barberini. But he also made enemies. Personally, Galileo was outspoken, forceful, sometimes tactless, and he enjoyed debate. Most of all he enjoyed being right. Thus in the years following publication of *Sidereus Nuncius*, in lectures, debates, and letters, he offended important people who questioned his telescopic discoveries.

By 1616 Galileo was the center of a storm of controversy. Some said he was mistaken (or worse), and others refused even to look through a telescope. Pope Paul V decided to end the disruption and so, when Galileo visited Rome in 1616, Cardinal Bellarmine interviewed him and ordered him to end his astronomical work. Books relevant to Copernicanism were banned, including *De Revolutionibus,* and Galileo rode back to Florence.

In some respects the years that followed gave Galileo hope that he could resume his work. His interviews with the pope and Cardinal Bellarmine in 1616 had been stern but not disrespectful, and in a last interview before he left Rome the pope had shown open friendship. Then in 1621 Pope Paul V died and his successor, Pope Gregory XV, died in 1623. The next pope was Galileo's friend Cardinal Barberini, who took the name Urban VIII. Galileo visited the new pope and, though the order of 1616 was not revoked, Urban VIII was friendly.

Thus in 1624 Galileo began to write a book defending the Copernican theory. It was a long project that was finally completed in the last days of 1629. After some delay the book was approved by the local censor in Florence and by the head censor of the Vatican in Rome. It was printed in 1632.

The book was called *Dialogo Dei Due Massimi Sistemi* (Dialogue Concerning the Two Chief World Systems) (Figure H-18). It was written as a conversation between three friends who debate the Copernican and Ptolemaic theories. Salviati, a swift-tongued defender of Copernicus, dominates the book. Sagredo is a reasonably intelligent but uninformed believer in the Ptolemaic theory who is gradually convinced to adopt Copernicanism. Simplicio, the third character in the book, is the dismal defender of Ptolemy. In fact, Simplicio does not seem very bright.

Galileo made three mistakes. First, the book was not an unbiased account of the two systems, though he later claimed it was. Second, he wrote the book in Italian, appealing to the public, not to scientists. Third, either intentionally or unintentionally, Galileo made Simplicio resemble Pope Urban VIII. The pope's friendship vanished overnight and Galileo was ordered to Rome to meet the Inquisition.

Upon Galileo's arrival he was interrogated by the Inquisition four times. He was not tortured or, so far as is known, threatened with torture, but the interviews were probably unpleasant enough. Galileo must have thought often of Bruno, burnt at the stake only 32 years before. However, the Inquisition had a problem in that

*Figure H-18.* *Aristotle, Ptolemy, and Copernicus discuss astronomy in this frontispiece from Galileo's book* Dialogue Concerning the Two Chief World Systems. *(Courtesy Owen Gingerich and The Houghton Library.)*

the book had been approved by the censors, so in the end they had to return to the orders given Galileo in 1616. The official record of the interview between Galileo and Cardinal Bellarmine included the statement that Galileo was "not to hold, teach, or defend in any way" the principles of Copernicus.* Galileo's assertion that his dialogue was an unbiased discussion and not a defense of any one theory was worthless even if it had been true.

On June 22, 1633, at the age of 70, Galileo knelt before the Inquisition and read a recanta-

*Because the document is unsigned, some scholars suspect it was forged and that Galileo never received those instructions in 1616.

tion admitting his errors. Tradition has it that as he rose he whispered *"Eppur si muove"* (still it moves) referring to the earth, but it is unlikely that anyone with any sense would risk such defiance.

Galileo was sentenced to life imprisonment. Perhaps through the intervention of the pope, he was held in confinement at his villa where he could meet his family, though other visitors were forbidden. During these years he studied mechanics and physics and even wrote a book on the subject that was published in Holland in 1638. During his last few years he was allowed a few visitors, and two young scientists came to stay with him. As he was blind by then, he no doubt enjoyed their discussions. At last, on January 8, 1642, after 10 years' imprisonment, 99 years after the death of Copernicus, Galileo died.

## MODERN ASTRONOMY

We date the origin of modern astronomy from the 99 years between the deaths of Copernicus and Galileo because it was an age of transition. That period marked the transition between the Ptolemaic theory and the Copernican, but it also marked a transition in the nature of astronomy in particular and science in general. Before the events of our story, scientific principles were drawn not from observation but from philosophical judgments of what the universe should be like. Thus Aristotle believed the heavenly bodies were perfect because he felt they should be. In such an atmosphere scientific discoveries and observations had to be bent to fit expectations.

The discoveries of Kepler and Galileo were accepted in the 1600s because the world was in transition. Astronomy was not the only thing changing during this period. The Renaissance is commonly taken to be the period between 1300 and 1600, and thus the 99 years of this history lie at the culmination of the reawakening of learning in all fields (Figure H-19). The world was open to new ideas and new observations. Martin Luther remade religion, and other philosophers and scholars reformed their areas of human knowledge. Had Copernicus not published his theory, someone else would have suggested that the universe was heliocentric. History was ready to shed the Ptolemaic theory.

In addition, this period marks the beginning of the modern scientific method. Beginning with Copernicus, Tycho, Kepler, and Galileo, scientists depended more and more on evidence, observation, and measurement. This, too, is coupled to the Renaissance and its advances in metalworking and lens making. Before our story began, no astronomer had looked through a telescope because one could not be made. By 1642, not only telescopes, but also other sensitive measuring instruments made science into something new and precise. Also, the growing number of scientific societies increased the exchange of observations and theories among scientists and stimulated more and better work. However, the most important advance was the application of mathematics to scientific questions. Kepler's work demonstrated the power of mathematical analysis, and as the quality of these numerical tools improved, the progress of science accelerated. Thus our story is not just the story of the birth of modern astronomy, but that of modern science as well.

But our history is not quite over. Galileo died in January 1642. Eleven months later, on Christmas day 1642,* a baby was born in the English village of Woolsthorpe. His name was Isaac Newton (Figure H-20), and his life was the first flower of the seeds planted by the four astronomers of our story.

Newton was a quiet child from a farming family, but his work at school was so impressive that his uncle financed his education at Trinity College, where he studied mathematics and physics. In 1665 the black plague swept through England and the colleges were closed. During

---

*Because England had not yet reformed its calendar, December 25, 1642, in England was January 4, 1643, in Europe. It is only a small deception to use the English date and thus include Newton's birth in our 99-year history.

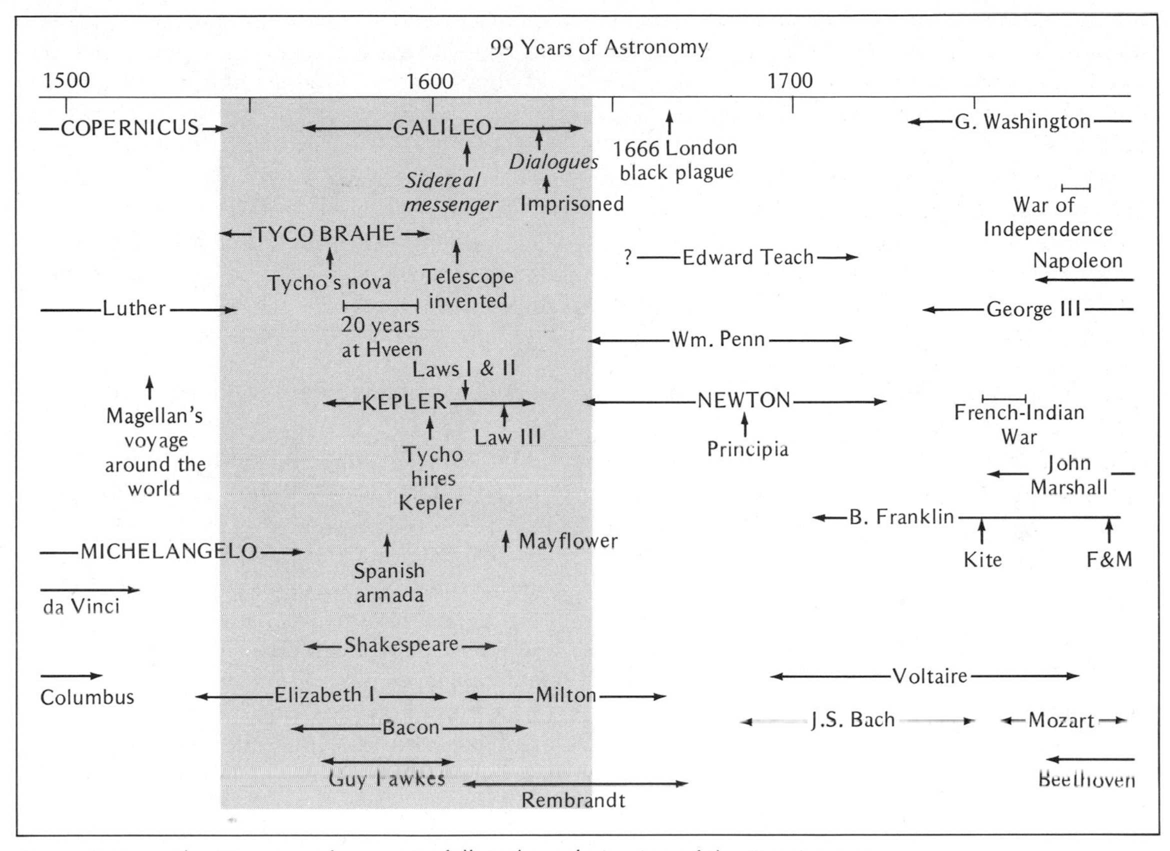


*Figure H-19.* *The 99 years of our story fell at the culmination of the Renaissance.*

*Figure H-20. Isaac Newton (1641–1727), working from the discoveries of Galileo and Kepler, derived three laws of motion and the principle of mutual gravitation. He and some of his discoveries are honored on this English one-pound note.*

1665 and 1666 Newton spent his time in Woolsthorpe, thinking and studying. It was during these years that he made most of his discoveries in optics, mechanics, and mathematics. Among other things, he studied optics, developed three laws of motion, divined the nature of gravity, and invented differential calculus. The publication of his work in his book *Principia* in 1687 placed science on a firm analytical base.

It is beyond the scope of this book to analyze all of Newton's work, but his laws of motion and gravity had an important impact on the future of astronomy. From his study of the work of Galileo, Kepler, and others, Newton extracted three laws that related the motion of a body to the forces acting on it. These laws (Table H-2) made it possible to predict exactly how a body would move if the forces were known. In addition, Newton realized that bodies attract each other with a force called gravity, and he devised a law that described how the force depends on the masses of the bodies and the distance between their centers. These discoveries remade astronomy into an analytical science in which astronomers could measure the positions and motions of celestial bodies, calculate the gravitational forces acting on them, and predict their future motion.

Were we to trace the history of astronomy after Newton, we would find scientists predicting the motion of comets, the gravitational interaction of the planets, the motions of double stars, and so on. Astronomers built on the discoveries of Newton, just as he had built on the discoveries of Copernicus, Tycho, Kepler, and Galileo. It is the nature of science to build on the discoveries of the past, and Newton was thinking of that when he wrote, "If I have seen farther than other men, it is because I stood upon the shoulders of giants."

**Table H-2 Newton's Three Laws of Motion**

| | |
|---|---|
| I. | A body continues at rest or in uniform motion in a straight line unless acted upon by some force. |
| II. | A body's change of motion is proportional to the force acting on it and is in the direction of the force. |
| III. | To every action there is an equal and opposite reaction. |

## NEW TERMS

geocentric universe
uniform circular motion
retrograde motion
epicycle
deferent
equant
heliocentric universe

## REVIEW QUESTIONS

1. Why did Aristotle conclude that the earth did not move?
2. Draw and label a diagram showing an epicycle, deferent, and equant.
3. Why did Copernicus delay publication of his book?
4. In what ways were the systems of Ptolemy and Copernicus similar?
5. Why did the Copernican theory win gradual acceptance?
6. When Tycho observed the new star of 1572, he could detect no parallax. What did he conclude from this?
7. How were Tycho and Kepler similar? How were they different?
8. How did the *Alfonsine Tables*, the *Prutenic Tables*, and the *Rudolphine Tables* differ?
9. What are Kepler's three laws of planetary motion?
10. Review Galileo's telescopic discoveries and explain why they supported the Copernican theory.
11. What three mistakes did Galileo make when he wrote his *Dialogues*?
12. How did Newton's discoveries change astronomy?

## RECOMMENDED READING


Armitage, A. *Copernicus: The Founder Of Modern Astronomy*. New York: A. S. Barnes, 1962.

Berry, A. *A Short History of Astronomy*. New York: Dover, 1961.

Bronowski, J. *The Ascent of Man*. Chapters 6 and 7. Boston: Little, Brown, 1973.

Christianson, J. "The Celestial Palace of Tycho Brahe." *Scientific American* 204 (Feb. 1961), p. 118.

Fermi, L., and Bernardini, G. *Galileo and the Scientific Revolution*. Greenwich, Conn.: Fawcett Publications, 1965.

Gingerich, O. "Johannes Kepler and the Rudolphine Tables." *Sky and Telescope* 42 (Dec. 1971), p. 328.

———. "Copernicus and Tycho." *Scientific American* 229 (Dec. 1973), p. 86.

———. "Tycho Brahe and the Great Comet of 1577." *Sky and Telescope* 54 (Dec. 1977), p. 452.

Koestler, A. *The Watershed: A Biography of Johannes Kepler*. Garden City, N.Y.: Doubleday, 1960.

Koyre, A. *From the Closed World to the Infinite Universe*. New York: Harper & Row, 1958.

Kuhn, T. S. *The Structure of Scientific Revolutions*. 2nd ed. Chicago: University of Chicago Press, 1970.

Moore, P. *Watchers of the Sky*. New York: G. P. Putnam's Sons, 1973.

Ravetz, J. "The Origins of the Copernican Revolution." *Scientific American* 215 (Oct. 1966), p. 88.

Shea, W. R. *Galileo's Intellectual Revolution*. New York: Macmillan, 1972.

Wilson, C. "How Did Kepler Discover His First Two Laws?" *Scientific American* 226 (March 1972), p. 93.

*The 5-m (200-in) Hale telescope on Mount Palomar is the largest astronomical telescope in the Western Hemisphere. The telescope and support weigh 500 tons. Locate the human figure just to the right of the lower end of the telescope tube. (Hale Observatories.)*

# Chapter 3 ASTRONOMICAL TOOLS

Almost all of the data we have about celestial bodies comes to us as light or one of its related forms—infrared, radio, ultraviolet, x-ray, or gamma-ray radiation. The only exceptions are missions to the moon that have returned samples to earth, bits of matter called meteorites that have fallen into earth's atmosphere and reached its surface, and planetary probes that have made direct observations and radioed back the results. In all other cases we can only observe. That limitation requires us to study the various forms of radiation bringing us information.

Telescopes are important because they gather radiation and concentrate it for study. If we wish to examine visible light, a normal telescope will do, but to extract information from other forms of radiation requires specialized telescopes. Radio telescopes, for example, give us an entirely different view of the sky.

Because visible light and radio waves are the only radiation that can freely penetrate our atmosphere, infrared, ultraviolet, x-ray, and gamma ray observations cannot be made from the earth's surface. Some infrared telescopes observe from high mountain tops or high-flying aircraft, but x-ray telescopes can function only above the atmosphere in space.

A telescope alone tells us little about astronomical bodies—it merely gathers radiation. To extract information from the radiation, we need analytic instruments attached to the telescope. A camera that can take photographs through a telescope is an obvious example, but more specialized instruments can analyze the radiation in detail.

Although analytic instruments and specialized telescopes are important in modern astronomy, we begin by studying light and traditional telescopes. Visible light is the key to the sky.

## RADIATION: INFORMATION FROM SPACE

We are familiar with light since we see it every day, but if we are to use it as a source of information, we must know what it is and how it works. Unfortunately, we can't look at a sunbeam and tell immediately what it is. To understand light, we must borrow a few ideas from physics.

*Electromagnetic Radiation.* Light is merely one form of radiation called **electromagnetic** because it is associated with changing electric and magnetic fields that travel through space and transfer energy from one place to another. When light enters our eye, the fluctuating electric and magnetic fields carry energy that stimulates nerve endings, and we see what we call light.

The oscillating electric and magnetic fields

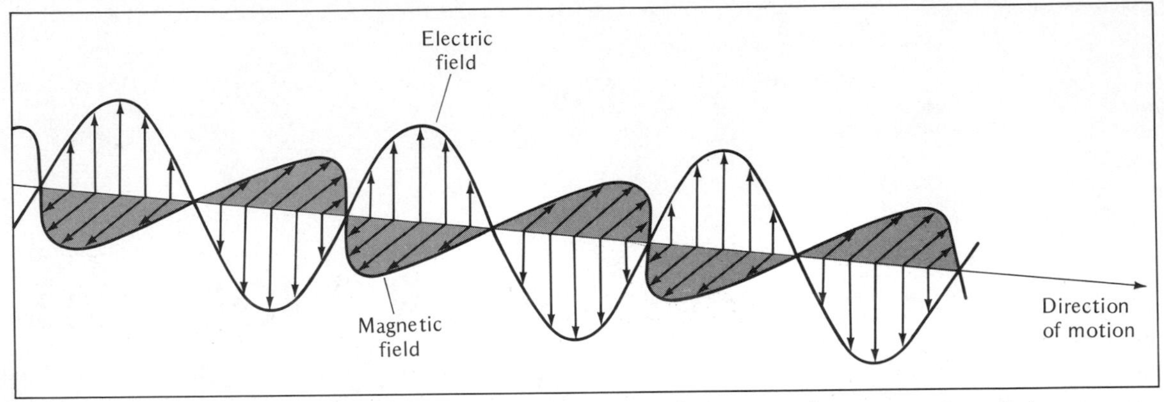


*Figure 3-1. Electric and magnetic fields travel together through space as electromagnetic radiation.*

that constitute electromagnetic radiation move through space at about 300,000 km/sec (186,000 mi/hr). This is commonly referred to as the speed of light **c,** but it is in fact the speed of all such radiation. If we represent the fluctuating electric and magnetic fields as arrows, electromagnetic radiation might resemble Figure 3-1.

Electromagnetic radiation is a wave phenomenon—that is, it is associated with a periodically repeating disturbance, or wave. We are familiar with waves in water. If we disturb a quiet pool of water, waves spread across the surface. Imagine that we use a meter stick to measure the distance between the successive peaks of a wave. This distance is the **wavelength,** usually represented by $\lambda$. If we were measuring ripples in a pond, we might find the wavelength was a few centimeters, while the wavelength of ocean waves might be a hundred meters or more. There is no restriction on the wavelength of electromagnetic radiation. Wavelengths can range from smaller than an atom to larger than the earth.

It is incorrect, or at least incomplete, to say that electromagnetic radiation is a wave, because it has sometimes the properties of a wave and sometimes the properties of a particle. For instance, the beautiful colors in a soap bubble arise from the wave nature of light. On the other hand, when light strikes the photoelectric cell in a camera's light meter, it behaves like a stream of particles carrying specific amounts of energy. We will refer to "a particle of light" as a **photon,** and we can recognize its dual nature by thinking of it as a bundle of waves.

The amount of energy a photon carries depends on its wavelength. The shorter the wavelength, the more energy the photon contains; the longer the wavelength, the less energy it contains.

*The Electromagnetic Spectrum.* A spectrum is an array of electromagnetic radiation in order of wavelength. We are most familiar with the spectrum of visible light, which we see in rainbows for instance, but the visible spectrum is merely a small segment of the much larger electromagnetic spectrum (Figure 3-2).

The average wavelength of visible light is about 0.0005 mm. We could put fifty light waves end to end across the thickness of a sheet of household plastic wrap. It is too awkward to measure such short distances in millimeters, so we will measure the wavelength of light in **Angstrom** units. One Angstrom (Å) is $10^{-10}$ m. The wavelength of visible light ranges from 4000 Å to 7000 Å.

Just as we sense the wavelength of sound as pitch, we sense the wavelength of light as color.

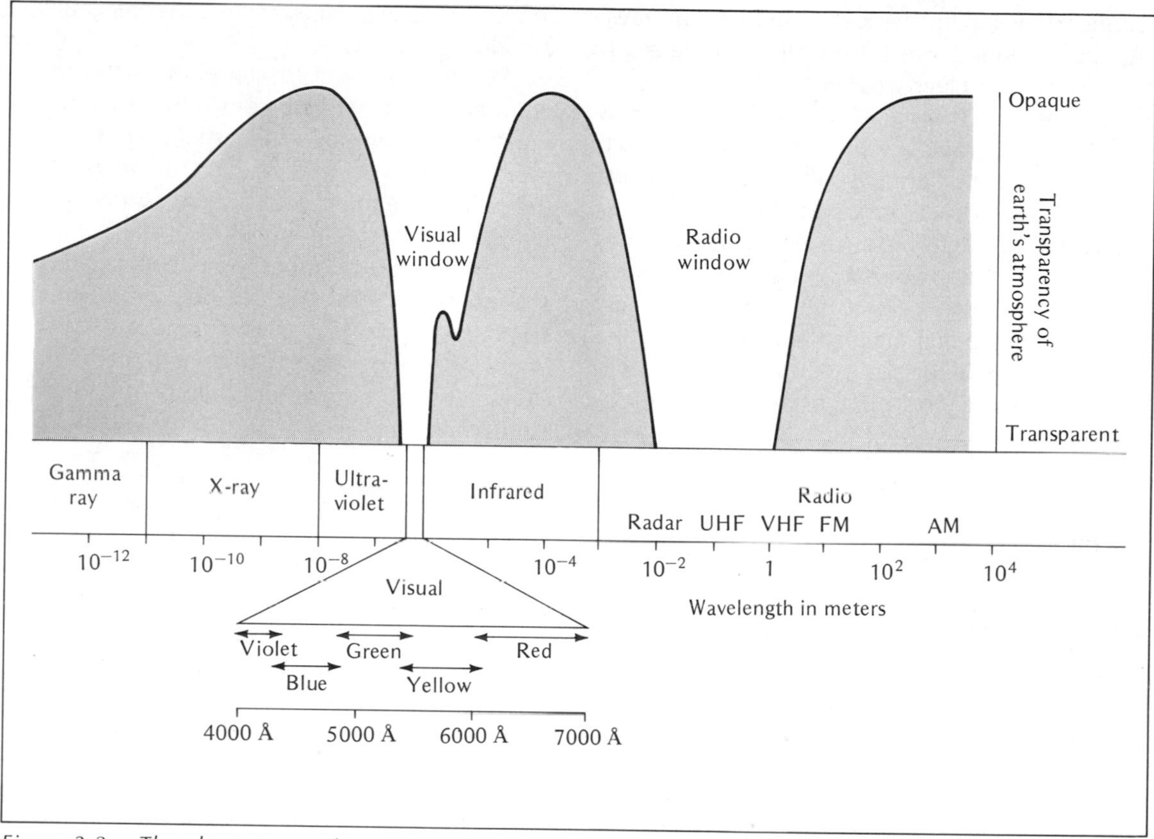


*Figure 3-2. The electromagnetic spectrum includes all wavelengths of electromagnetic radiation. Earth's atmosphere is relatively opaque at most wavelengths. Visual and radio windows allow light and some radio waves to reach earth's surface.*

Light near the short wavelength end of the visible spectrum (4000 Å) looks violet to our eyes, and light near the long wavelength end (7000 Å) looks red (Figure 3-2).

Beyond the red end of the visible spectrum lies infrared radiation, where wavelengths range from 7000 Å to about 1 mm. Our eyes are not sensitive to this radiation, but our skin senses it as heat. A "heat lamp" is nothing more than a bulb that gives off infrared radiation.

Beyond the infrared part of the electromagnetic spectrum lie radio waves. The radio radiation used for AM radio transmissions has wavelengths of a few kilometers, while FM, television, military, governmental, and ham radio transmissions have wavelengths that range down to a few meters. Microwave transmissions, used for radar and long-distance telephone communication for instance, have wavelengths from a few centimeters down to about 1 mm.

The distinction between the wavelength ranges is not sharp. Long-wavelength infrared radiation and the shortest microwave radio waves are the same. Similarly, there is no clear division between the short-wavelength infrared and the long-wavelength part of the visible spectrum. It is all electromagnetic radiation.

At wavelengths shorter than violet, we find ultraviolet radiation, with wavelengths ranging from 4000 Å down to about 100 Å. At even shorter wavelengths lie x-rays and gamma rays. Again, the boundaries between these wavelength ranges are not clearly defined.

X-rays and gamma rays can be dangerous and even ultraviolet photons have enough energy to do us harm. Small doses produce a suntan; larger doses can cause sunburn, and extreme doses might produce skin cancers. Contrast this to the lower-energy infrared photons. Individually they have too little energy to affect skin pigment, a fact that explains why you can't get a tan from a heat lamp. Only by concentrating many low-energy photons in a small area, as in a microwave oven, can we transfer significant amounts of energy.

We are interested in electromagnetic radiation because it brings us clues to the nature of stars, planets, and other celestial objects. However, only a small part of this radiation can get through the earth's atmosphere. Only visible light and some radio waves can reach the surface of the earth; other wavelengths are absorbed. The highest parts of the atmosphere absorb x-rays, gamma rays, and some radio waves, and a layer of ozone ($O_3$) at an altitude of about 30 km absorbs ultraviolet radiation. In addition,

---

BOX 3-1 RADIATION FROM A HEATED OBJECT

Whenever we produce a changing electric field, we produce electromagnetic radiation. If we run a comb through our hair, we disturb electrons (negatively charged subatomic particles) in both hair and comb, producing static electricity. Since each electron is surrounded by an electric field, any sudden change in the electron's motion gives rise to electromagnetic radiation. Running a comb through your hair while standing near an AM radio produces radio static. This illustrates an important principle: Whenever we change the motion of an electron we generate electromagnetic waves.

To see what this has to do with a heated object, think of what we mean by heat. When we say an object is hot, we mean that its atoms are vibrating rapidly. The hotter the object is, the more motion among the atoms. The vibrating atoms collide with electrons in the material, and each time the motion of one of these electrons gets disturbed, it emits a photon. Consequently, we should expect a heated object to emit electromagnetic radiation. Such radiation is called **black body radiation** and is quite common. In fact, it is responsible for the light emitted by the hot filament in an incandescent light bulb.

In the heated filament, gentle collisions produce low-energy photons with long wavelengths, and violent collisions produce high-energy photons with short wavelengths. If we graph the energy emitted at different wavelengths, we get a curve like those shown in Figure 3-3. The curve shows that gentle collisions and violent collisions are rare. Most collisions are intermediate in violence, producing photons of intermediate wavelength.

The wavelength at which an object emits the maximum amount of energy, called the **wavelength of maximum** ($\lambda_{max}$), depends on the object's temperature. If we heat the object, the average collision should be more violent, producing higher-energy, shorter-wavelength photons. The hotter the object is, the shorter $\lambda_{max}$ (Figure 3-3). Basic physics shows that $\lambda_{max}$ in Angstroms equals 30 million divided by the temperature in degrees Kelvin.

$$\lambda_{max} = \frac{30{,}000{,}000}{T}$$

This is a powerful tool in astronomy because it means we can determine the temperature of a star from its light.

In fact, we can estimate a star's temperature from its color. For a hot star, $\lambda_{max}$ lies in the ultraviolet and we cannot see most of the radiation, but in the visible range the star emits more blue than red. Thus a hot star looks blue. In contrast, a cool star radiates its maximum energy in the infrared. However, in the visible part of the spectrum, it radiates more red than blue and thus looks red.

The total amount of radiation emitted at all wavelengths depends on the number of collisions per second. If an object's temperature is high, there

water vapor in the lower atmosphere absorbs infrared radiation. The two wavelength regions in which our atmosphere is transparent are called **atmospheric windows.** Obviously, if we wish to study the sky from the earth's surface, we must look out through one of these windows.

In later chapters we will discuss various ways of interpreting the information this radiation brings to earth, but one deserves mention here because it is closely related to the nature of light. Because of the way objects emit electromagnetic radiation, we can tell a star's temperature from its color. Hot stars look blue and cool stars red (see Box 3-1). This is some of the most important information in starlight.

Having described the nature of electromagnetic radiation and the electromagnetic spectrum, we can now study the tools astronomers use to analyze radiation. The first we will consider is the optical telescope—the basic tool of astronomy.

## OPTICAL TELESCOPES

In this section we will examine the two types of telescopes that work with visible light. They are

---

are many collisions and it emits more light than a cooler object of the same size. We measure energy in units called **ergs;** one erg is about the energy a house fly needs to climb over a toothpick. The total radiation given off by 1 $cm^2$ of the object in ergs per second equals a constant number, represented by $\sigma$, times the temperature raised to the fourth power:*

$$E = \sigma T^4 \text{ (ergs/sec/cm}^2\text{)}$$

If we doubled an object's temperature, for instance, it would radiate $2^4$, or 16, times more energy. Thus we can expect hot stars to radiate large amounts of energy from each square centimeter, and most of their radiation is at short wavelengths.

* For the sake of completeness we should note that the constant $\sigma$ equals $5.67 \times 10^{-5}$ ergs/(cm$^2$ sec degree$^4$). We will not need it in our calculations here.

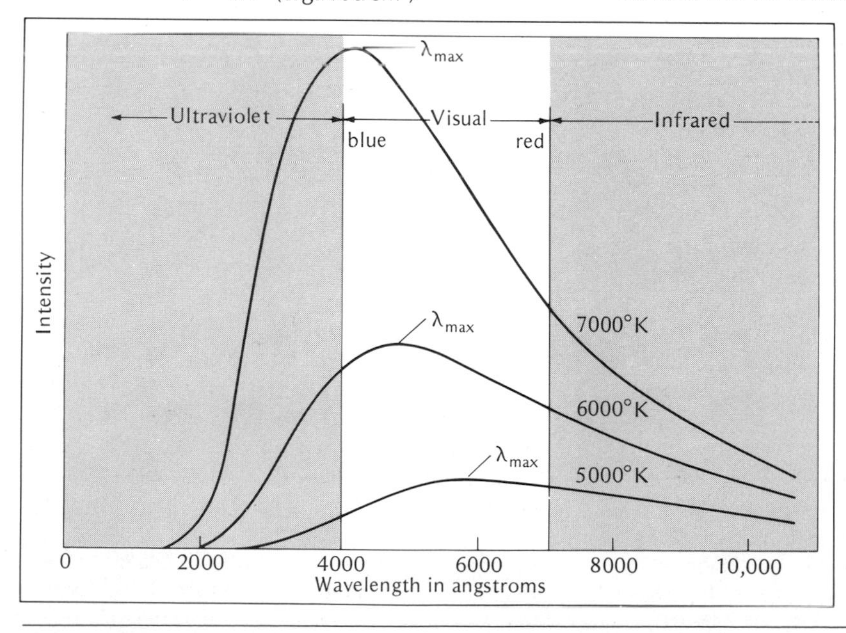


*Figure 3-3. The intensity of radiation emitted by a heated body depends on wavelength. $\lambda_{max}$ designates the wavelength of maximum intensity. Hotter objects radiate more energy and have shorter $\lambda_{max}$ than cooler objects. A hot object radiates more blue light than red and therefore looks blue. A cool object radiates more red than blue and therefore looks red.*

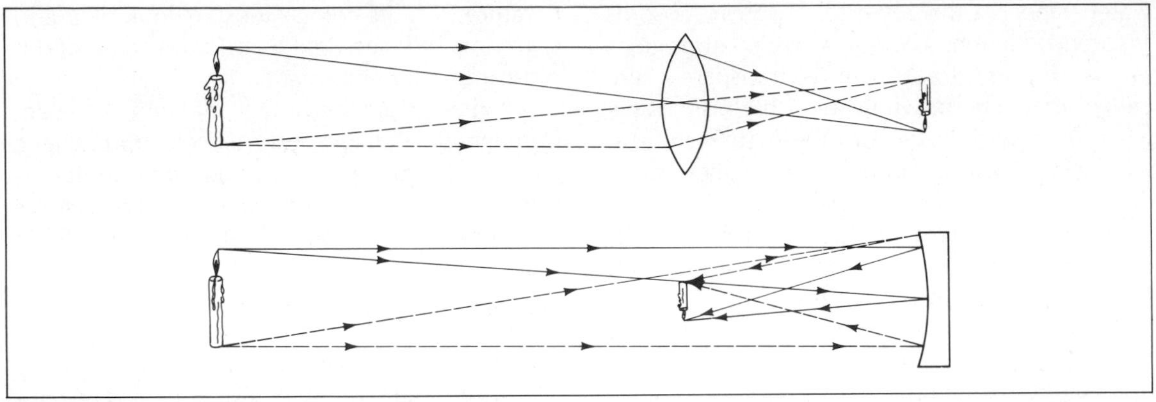

*Figure 3-4. A lens forms an image by refracting (bending) light. A curved mirror can form an image by reflecting light. In both cases, the image of distant objects is inverted.*

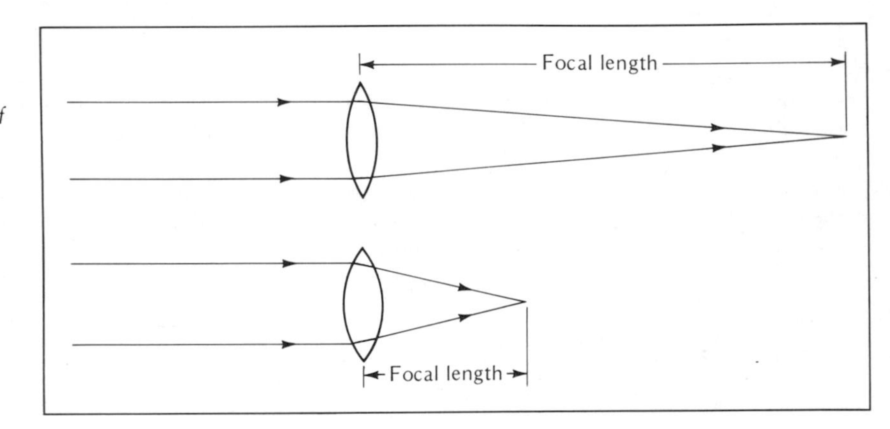


*Figure 3-5. The focal length of a lens is the distance from the lens to the point where parallel rays of light come to a focus. The lens above has a longer focal length than the lens below.*

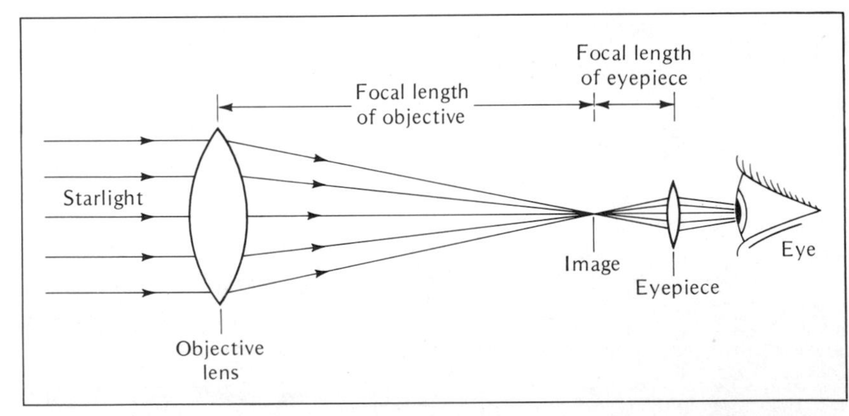


*Figure 3-6. In a refracting telescope, the objective lens forms an image that is magnified by the eyepiece.*

the easiest to understand and the most common. Later we will look at other telescopes that use other parts of the electromagnetic spectrum.

*Refracting Telescopes.* A **refracting telescope** works by refracting, or bending, light with a lens. Because of the shape of the lens, light

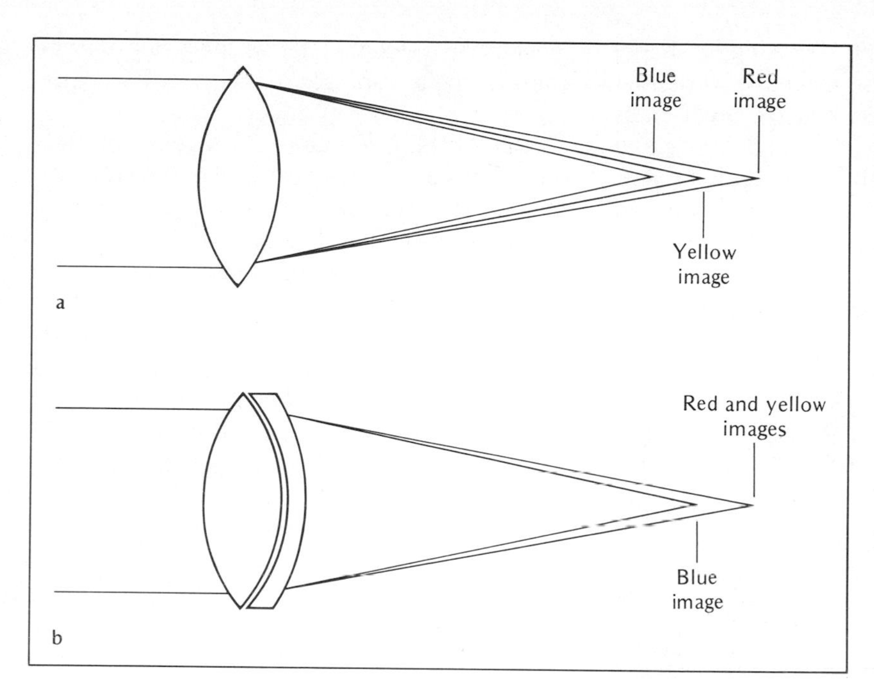


*Figure 3-7.* (a) *A normal lens suffers from chromatic aberration because short wavelengths bend more than long wavelengths.* (b) *An achromatic lens made in two parts can bring any two colors to the same focus, but other colors remain slightly out of focus.*

striking the edge bends more than light striking the central area, and the light that enters the lens from some distant object comes together to form a small, inverted image (Figure 3-4). The **focal length** of a lens is the distance from the lens to the point where it focuses parallel rays of light (Figure 3-5).

To build a refracting telescope we need a lens of relatively long focal length to form an image of the object we wish to view. This lens is often called the **objective lens** because it is closest to the object. To view the image we add a lens of short focal length called an **eyepiece** to enlarge the image and make it easy to see (Figure 3-6). Thus the eyepiece acts as a magnifier. By changing eyepieces we can easily change the magnification of the telescope.

Refracting telescopes suffer from a serious optical defect that limits what we can see through them. When light is refracted through glass, shorter wavelengths bend more than longer wavelengths, and blue light comes to a focus closer to the lens than does red light (Figure 3-7). If we focus the eyepiece on the blue image, the red light is out of focus, producing a red blur around the image. If we focus on the red image, the blue light blurs. This color separation is called **chromatic aberration.**

A telescope designer can partially correct for this by replacing the single objective lens with one made of two lenses, ground from different kinds of glass. Such lenses, called **achromatic lenses,** can be designed to bring any two colors to the same focus (Figure 3-7). Since our eyes are most sensitive to red and yellow light, we might bring these two colors to the same focus, but blue and violet would still be out of focus, producing a hazy blue fringe around bright objects.

Refracting telescopes were popular through the 19th century, but they are no longer economical. Achromatic lenses are expensive to make because the optician must grind four surfaces instead of two for a simple lens. Also, refractors cannot be made larger than about 1 m in diameter. Such large pieces of optically pure glass are hard to make and are not perfectly rigid. A large lens, supported only at its edge, sags under its own weight, distorting its shape. The largest refracting telescope in the world is at Yerkes Observatory in Wisconsin (Figure 3-8). Its achromatic objective lens, 40 in in diameter and weighing half a ton, was made in 1895. It seems unlikely that a larger refracting telescope will ever be built, not only because of the problems inherent in refracting telescopes, but also because of the advantages of reflecting telescopes.

*Figure 3-8. The 40-in telescope at Yerkes Observatory is the largest refracting telescope in the world. (Yerkes Observatory.)*

*Reflecting Telescopes.* A **reflecting telescope** uses a concave mirror, the **objective mirror,** to focus starlight into an image. Objective mirrors are usually made of glass or quartz covered with a thin layer of aluminum to act as a reflecting surface.

The objective mirror forms an image at the location called the **prime focus** at the upper end of the telescope tube (Figure 3-9). Since it is usually inconvenient to view the image there, a smaller **secondary mirror** reflects the light to a more accessible location. In one popular arrangement, the secondary mirror reflects the light back down the telescope tube through a hole in the center of the objective. This kind of telescope is called a **Cassegrain telescope** (Figure 3-10). Box 3-2 describes the most popular reflecting telescope designs.

Reflectors have no chromatic aberration because the light does not go through the glass. It reflects off of the aluminum surface of the mirror and all wavelengths come to the same focus. Other aberrations limit the field of view through reflecting telescopes, but usually the limitation is not serious.

Nearly all recently built telescopes are reflectors, not only because they avoid chromatic aberration, but also because mirrors are less ex-

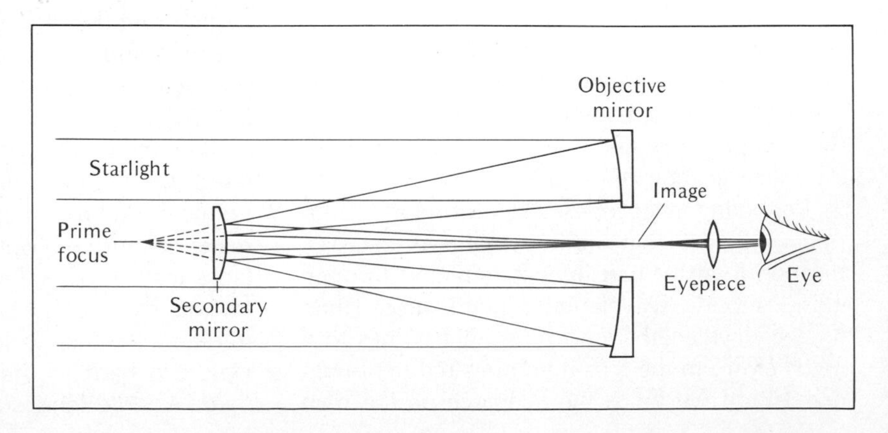


*Figure 3-9. In a Cassegrain reflecting telescope, the objective mirror forms an image that is magnified by the eyepiece. For convenient viewing a secondary mirror reflects the light through a hole in the objective.*

a

b

*Figure 3-10.* (a) Astronomer C. R. Lynds works in the Cassegrain observing cage beneath the objective mirror of the 4-m Mayall telescope at Kitt Peak National Observatory. (b) The 4-m (158-in) Mayall telescope. (Kitt Peak National Observatory.)

pensive than lenses of the same diameter. Since the light does not enter the mirror, the glass need not be perfectly clear and bubble free. Also, a mirror can be supported over its entire back surface, so sagging under its own weight is less serious a problem. Finally, a mirror is cheaper to make since it has only one optical surface to an achromatic lens's four.

Both refractors and reflectors need mountings to support them and to drive them slowly westward to compensate for the eastward rotation of the earth. However, because reflecting telescopes are free of chromatic aberration, they can be made with shorter focal lengths and thus use shorter tubes and smaller mountings than refracting telescopes of the same diameter.

The largest reflecting telescope in the world is the 6-m (236-in) telescope on Mount Pastukhov in the Soviet Union. Though observations have begun, the telescope is still under development, and Soviet astronomers plan to replace the present mirror to improve the quality of the images. The 5-m (200-in) Hale telescope on Mount Palomar is the largest fully functional

*Figure 3-11.* The six 1.8-m (72-in) mirrors of the multiple mirror telescope focus their light into a single image. This array creates the equivalent of a 4.5 m (176-in) telescope, making it the third largest in the world. (Courtesy of the Multiple Mirror Telescope Observatory, a joint facility of the University of Arizona and the Smithsonian Institution.)

telescope in the world. Other telescopes as large as 4 m (157 in) observe from both Northern and Southern Hemisphere observatories.

Smithsonian Astrophysical Observatory and University of Arizona astronomers have developed a new type of astronomical telescope that may be the model for future large telescopes. The multiple mirror telescope carries six mirrors each 1.8 m (72 in) in diameter on one telescope mounting (Figure 3-11). The light from the six

## BOX 3-2 FIVE FOCAL ARRANGEMENTS

A few telescopes are so large the observer can climb inside and observe in a prime focus cage, a small chamber about the size of a short telephone booth located at the top of the telescope tube (Figure 3-12). The observer, who is strapped in, can swing the telescope to any location in the sky and observe by looking downward toward the telescope mirror below. Though ideal for photographing faint objects, the prime-focus cage is small and cramped (Figure 3-13).

If the telescope is too small for a prime focus cage or if the observers need more room, they must use a different focus. A diagonal mirror can reflect the light out the side of the tube to the **Newtonian focus.** Although the observers have more room at the Newtonian focus, at a large telescope they are perched high above the observatory floor, which is both inconvenient and dangerous. A better focal arrangement keeps the observers close to the floor.

Using the Cassegrain focus, astronomers can observe standing securely on the floor. This is the most popular arrangement for large telescopes because it is convenient and relatively safe, and because it provides more room for large instruments that must be bolted to the telescope. In a few larger telescopes the observer rides inside a Cassegrain focus cage bolted to the back of the telescope (Figure 3-10a).

Sometimes an instrument needs even more room. In the larger telescopes, a series of mirrors guides the starlight through the observatory floor to the **coudé focus.** (*Coudé* comes from the French word for elbow and refers to the bent light path.) Here the observer can work in relative comfort with plenty of room for bulky equipment to analyze the starlight.

Unlike those listed above, a telescope called a **Schmidt camera** (Figure 3-14) cannot be used visually—we can't look through it. It is a photographic telescope that takes wide-angle, relatively distortion-free photographs of the sky. A Schmidt camera uses an objective mirror to form the image, while a thin correcting lens eliminates some of the distortion. Since the correcting lens is not strongly curved nor very thick, it does not introduce serious chromatic aberration. Such wide-angle telescopes are ideal for sky surveys to search out objects for study later with more conventional telescopes.

Astronomers usually design large telescopes so they can use different focal arrangements. The 5-m (200-in) Hale telescope at Mount Palomar, for example, can direct light to the prime focus cage, the Cassegrain focus, or the coudé focus. However, a conventional telescope cannot be converted into a Schmidt camera or vice versa. The Schmidt camera is a special-purpose telescope used only in a single focal arrangement.

*Figure 3-14. The 1.2-m (48-in) Schmidt camera is used solely for photography. The observer is looking through a guide telescope. (Hale Observatories.)*

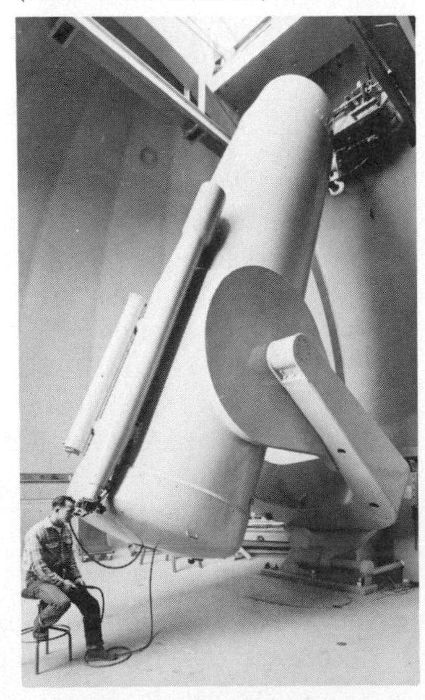

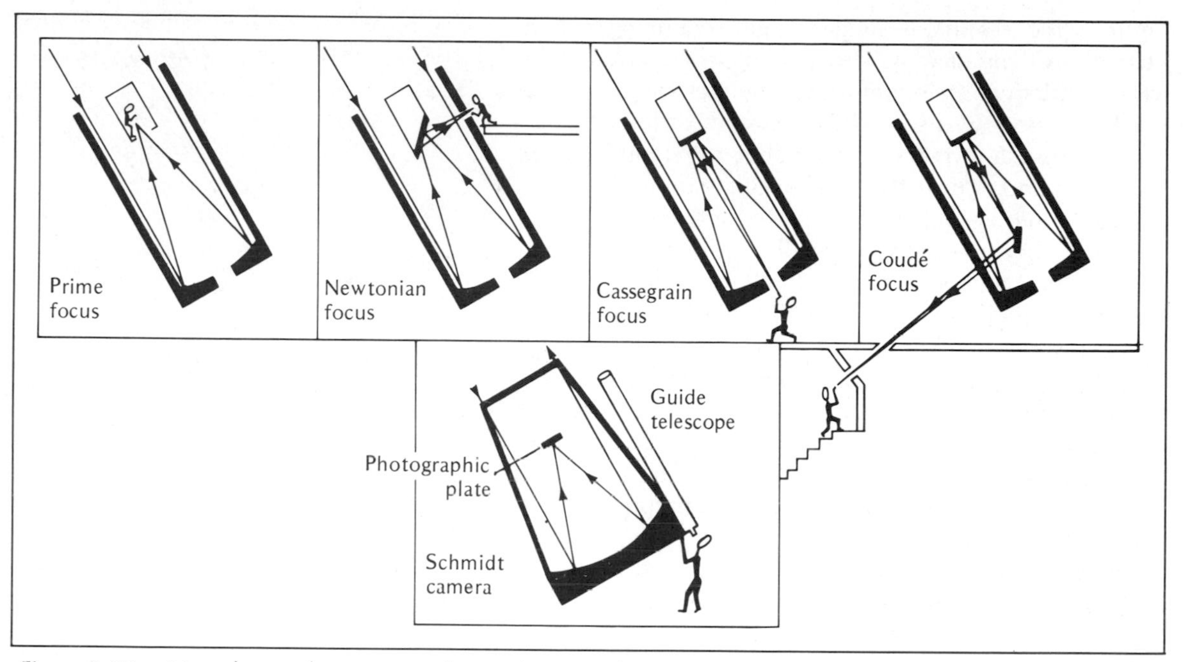


*Figure 3-12.* Many large telescopes can be used at a number of focal positions. Schmidt cameras, however, are used only for photography.

*Figure 3-13.* The prime focus cage of the 5-m (200-in) Hale Telescope on Mount Palomar. Note the objective mirror below. (Hale Observatories.)

mirrors focuses into a single image, producing the equivalent of a 4.5-m (176-in) telescope. The problem of maintaining the alignment of six heavy mirrors was solved by a system of laser beams and a small computer that constantly monitors and adjusts the position of the mirrors. Other multiple-mirror telescopes are being planned because several smaller mirrors are much easier and cheaper to build than one large mirror. Some plans call for multiple-mirror telescopes with effective diameters of 25 m (1000 in). Such a giant telescope could not be built with a single mirror.

*The Powers of a Telescope.* Refractor or reflector, the telescope can aid our eyes in three ways: light-gathering power, resolving power, and magnifying power.

The most important thing a telescope does is gather light. Most interesting celestial objects are faint sources of light, so we need a telescope that can gather large amounts of light to produce a bright image. **Light-gathering power** refers to the ability of a telescope to collect light. Catching light in a telescope is like catching rain in a bucket—the bigger the bucket, the more it catches (Figure 3-15). This is the main reason why astronomers use large telescopes.

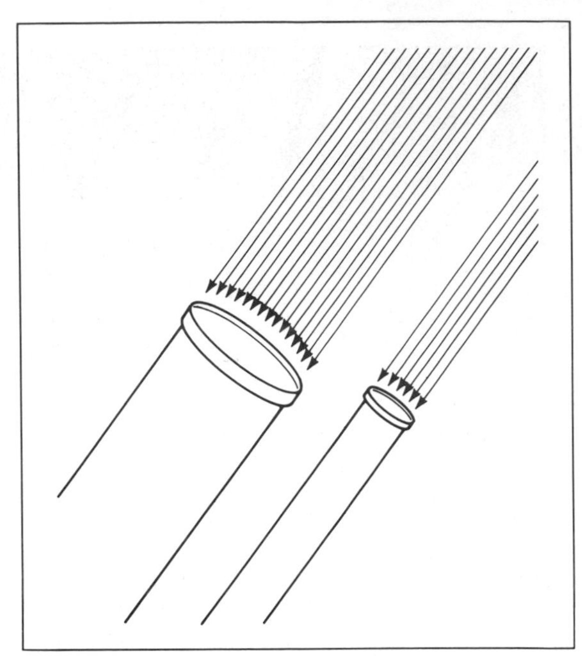

*Figure 3-15. Gathering light is like catching rain in a bucket. A large-diameter telescope gathers more light and has a brighter image than a smaller telescope of the same focal length.*

The second telescopic power, **resolving power,** refers to the ability of the telescope to reveal fine detail. Whenever light is focused to form an image, a small blurred fringe surrounds the image (Figure 3-16). Because this **diffraction fringe** surrounds every point of light in the image, we can never see any detail smaller than the fringe. There is nothing we can do to eliminate diffraction fringes; they are produced by the wave nature of light as it passes through the telescope. However, if we use a large-diameter telescope, the fringes are smaller and we can see smaller details. Thus the larger the telescope, the better its resolving power.

In addition to resolving power, two other factors—lens quality and atmospheric conditions—limit the detail we can see through a telescope. A telescope must contain high quality optics to achieve its full potential resolving power. Even a large telescope shows us little detail if its optics are marred with imperfections. Also, when we look through a telescope, we are looking through miles of turbulent air in earth's atmosphere, which makes the image dance and blur, a condition called **seeing.** On a night when the atmosphere is unsteady and the images are blurred, the seeing is bad. Even under good seeing conditions, the detail visible through a large telescope is limited, not by its diffraction fringes, but by the air through which the observer must look. A telescope performs better on a high mountain top where the air is thin and steady, but even there the earth's atmosphere limits the detail the telescope can reveal. (See Color Plate 4.)

The third and least important power of a telescope is **magnifying power,** the ability to make the image bigger. Since the amount of detail we can see is limited by the seeing conditions and

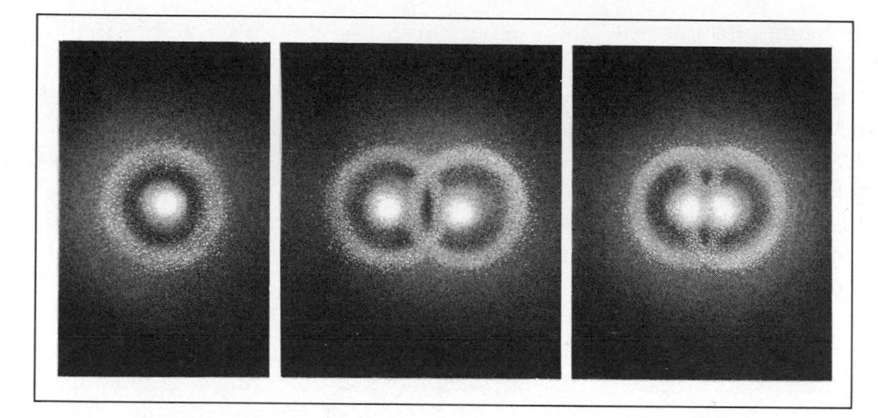

*Figure 3-16. (left) Diffraction fringes surround every star image. (center) Two stars close to each other have overlapping fringes. (right) Two stars very close to each other blend together.*

the resolving power, very high magnification does not necessarily show us more detail. Also, we can change the magnification by changing the eyepiece, but we cannot alter the telescope's light gathering power or resolving power.

If you visit a department store to shop for a telescope, you will probably find telescopes described according to magnification. One may be labeled an "80 power telescope" and another a "40 power telescope." However, the magnifying power really tells us little about the telescopes. Astronomers identify telescopes by diameter because that determines both light-gathering power and resolving power.

We have described the three powers of the telescope in general terms. Box 3-3 gives them mathematical form. Now we are ready to study the special instruments that analyze the light gathered by optical telescopes.

## SPECIAL INSTRUMENTS

Merely looking through a telescope tells us little about celestial objects. Our eyes are not accurate measuring instruments and our memories are not perfectly dependable. Also, most celestial objects are faint even when viewed through a large telescope. Consequently, astronomers have developed special instruments that, attached to telescopes, can dig out the clues hidden in starlight.

The photographic plate is a particularly valuable special instrument. A photograph, especially one taken on a glass photographic plate, can be measured precisely, and it can record a complex image and not "forget" the details as humans are apt to do. More important, time exposures can reveal features too faint to be visible to a human observer peering through an eyepiece. The photograph is also useful for recording information about the color of the light. Filters of colored glass placed in front of the photographic plate limit the range of wavelengths that reach the plate. Thus it is possible to photograph a cluster of stars "in red light" and again "in blue light" and by comparing the two photographs determine the colors of the stars. Ever since the first astronomical photograph was taken in 1840 (a daguerreotype of the moon) astronomers have made photography a part of their tool kit.

Photography is a familiar process in our world and you have probably taken photos yourself. However, most astronomical instruments are rather specialized and deserve some explanation. We will discuss two kinds—spectrographs, and photometers.

*The Spectrograph.* A **spectrograph** is a device that separates starlight according to wavelength to produce a spectrum. White light is a mixture of light of various wavelengths, a fact that we can demonstrate by shining a beam of white light through a glass prism. The angle through which the prism bends the light depends on wavelength—violet light bends most, and red least (Figure 3-17a). Thus a prism can separate a beam of light into a band of its component colors, a spectrum. The astronomer can build a spectrograph with a single prism or with a series of prisms to spread the wavelengths farther (Figure 3-18).

A spectrograph can be built with a grating in place of a prism. A **grating** is a piece of glass with thousands of microscopic parallel lines scribed into its surface. Different wavelengths of light reflect from the grating surface at slightly different angles, so white light striking a grating produces a spectrum (Figure 3-17b). Figure 3-19 shows a grating spectrograph.

Prism and grating spectrographs both work the same way. The telescope focuses starlight on a thin slit that admits light to the spectrograph. To keep the telescope properly pointed, the observer looks through a guiding eyepiece and keeps the image of the star on the slit. After passing through the slit, the light strikes a lens (or mirror) that guides it to the prism (or grating). A camera lens (or mirror) then focuses the various colors onto a photographic plate where the

---

BOX 3-3 THE POWERS OF THE TELESCOPE

Light-gathering power is proportional to the area of the telescope objective. A lens or mirror with a large area gathers a large amount of light. Since the area of a circular lens or mirror of diameter D is $\pi(D/2)^2$, we can compare the areas of two telescopes, and therefore their relative light-gathering powers, by comparing the square of their diameters. That is, the ratio of the light gathering powers of two telescopes A and B is equal to the ratio of their diameters squared:

$$\frac{LGP_A}{LGP_B} = \left(\frac{D_A}{D_B}\right)^2$$

For example, suppose we compare a 2-in telescope with a 10-in telescope. How much brighter would a star look with the 10-in telescope than with the 2-in?

$$\frac{LGP_{10}}{LGP_2} = \left(\frac{10}{2}\right)^2 = 5^2 = 25 \text{ times brighter}$$

Our eye acts like a telescope with a diameter of about ⅓ in, the diameter of the pupil. How much brighter will stars look if we use a 10-in telescope to aid our eyes?

$$\frac{LGP_{10}}{LGP_{eye}} = \left(\frac{10}{1/3}\right)^2 = (30)^2 = 900 \text{ times brighter}$$

The resolving power of a telescope is the angular distance between two stars that are just barely visible through the telescope as two separate images. For optical telescopes, the resolving power $\alpha$, in seconds of arc, equals 4.56 divided by the diameter of the telescope in inches:

$$\alpha = \frac{4.56}{D}$$

For example, a 10-in telescope has a resolving power of 0.456 seconds of arc. If the lenses are of good quality and if the seeing is good, we should be able to distinguish as separate points of light any pair of stars farther apart than 0.456 seconds of arc. If the stars are any closer together, diffraction fringes blur the stars together into a single image (Figure 3-16c). Obviously, we would like to use large telescopes to make $\alpha$ as small as possible.

The magnification of a telescope is the ratio of the focal length of the objective lens or mirror $F_o$ divided by the focal length of the eyepiece $F_e$:

$$M = \frac{F_o}{F_e}$$

For instance, if the focal length of a telescope is 80 in and we use an eyepiece with a focal length of 0.5 in, the magnification is 80/0.5 or 160 times.

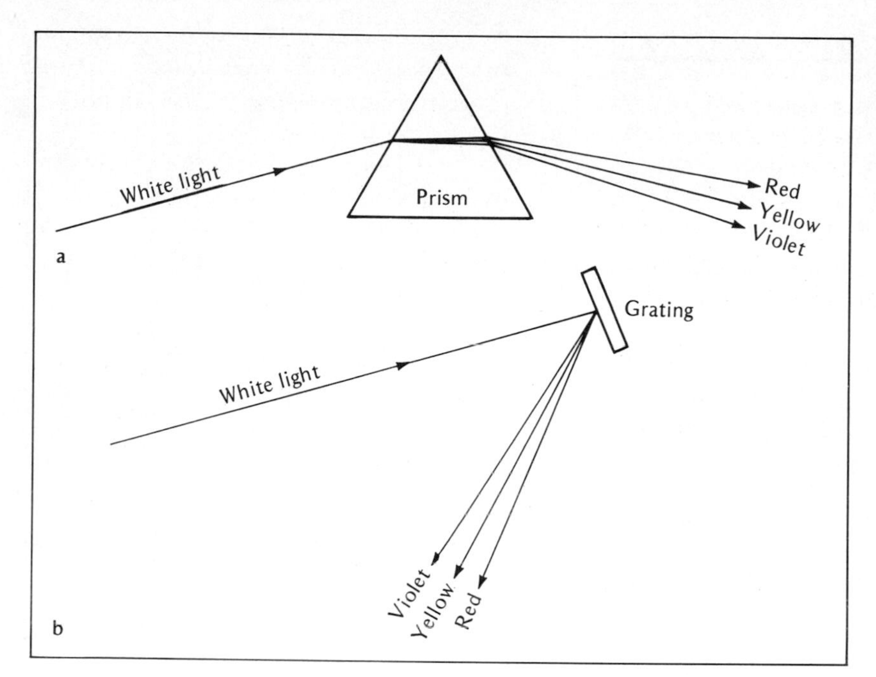


*Figure 3-17.* (a) A prism bends light by an angle that depends on the wavelength of the light. Short wavelengths bend most. (b) White light can be spread into a spectrum by reflection from a grating, producing the same effect as a prism.

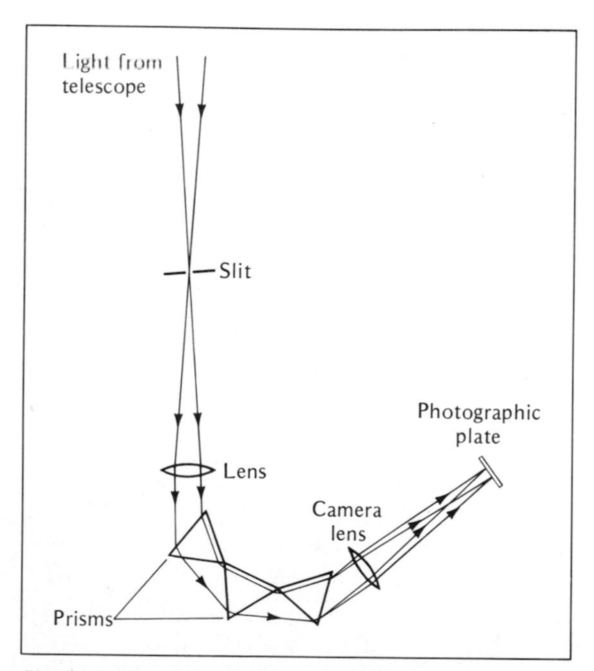


*Figure 3-18.* In a three-prism spectrograph, light from the telescope passes through the slit and through three prisms before being focused on the photographic plate.

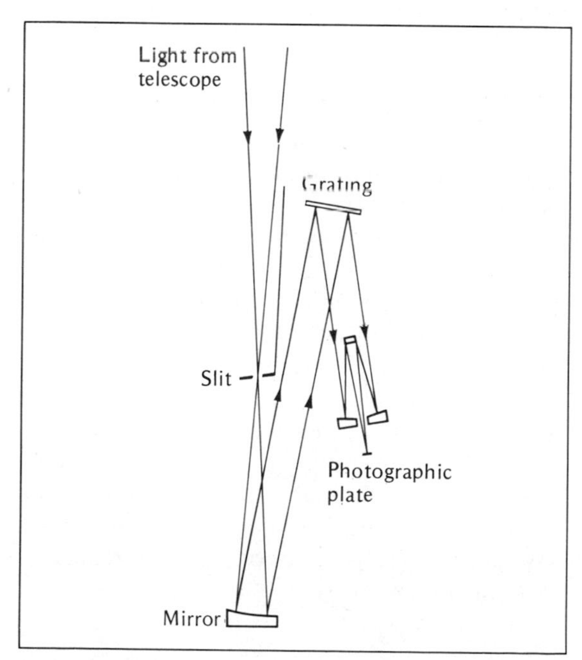


*Figure 3-19.* In this grating spectrograph, the grating disperses the light into a spectrum, and mirrors focus the spectrum on the photographic plate.

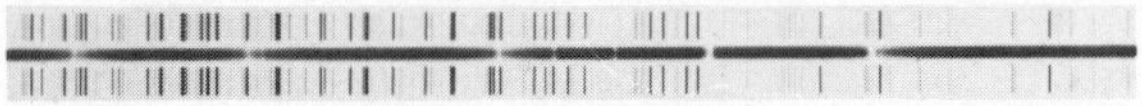

*Figure 3-20. The known wavelengths in the comparison spectrum (above and below) are guides to the wavelengths in the stellar spectrum (center). (Hale Observatories.)*

spectrum is recorded. Compare the spectrographs in Figures 3-18 and 3-19.

To help identify the lines in a stellar spectrum, the astronomer often adds a **comparison spectrum.** A mirror directs light from an iron arc or other source of light into the spectrograph, producing a spectrum above and below the star's spectrum. Since the wavelengths of the lines in the comparison spectrum are precisely known, the astronomer can use it as a road map to identify wavelengths in the stellar spectrum (Figure 3-20).

*The Photometer.* Another way to analyze starlight is to measure its intensity and color with a sensitive light meter called a **photometer.** A photometer contains a sensitive detector that produces an electric current when struck by light (Figure 3-21). The strength of this current is proportional to the intensity of the light, so a measurement of the current determines the brightness of the star.

In addition to brightness, a photometer can measure a star's color. A filter transmits only those wavelengths in a certain range, blue light between 4000 Å and 4800 Å for instance. The difference between the brightness of the star measured through a blue filter and a red filter gives a numerical index of the star's color. A blue star, for example, looks brighter through a blue filter than through a red filter. Such color measurements are important because, as we saw in Box 3-1, the color of the star is related to its temperature.

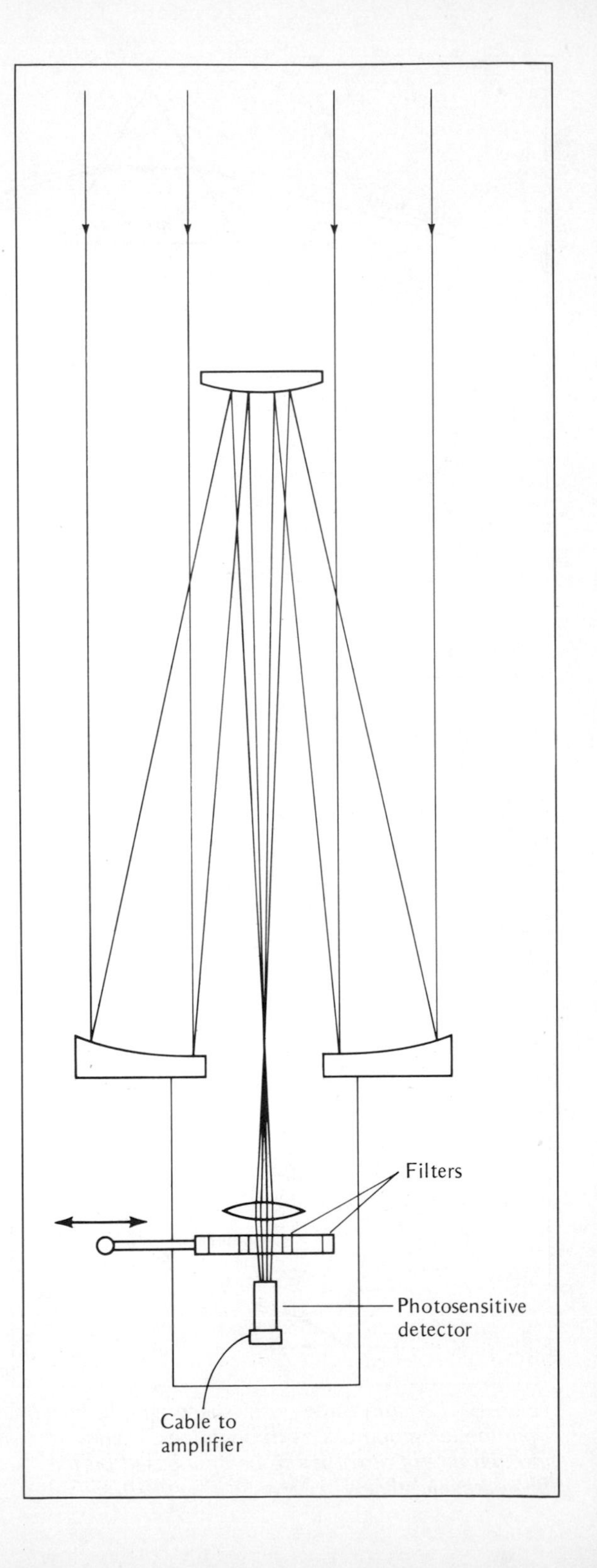


*Figure 3-21. A photometer contains a sensitive detector and a set of filters. The current from the detector is related to the brightness of the starlight passing through the filter.*

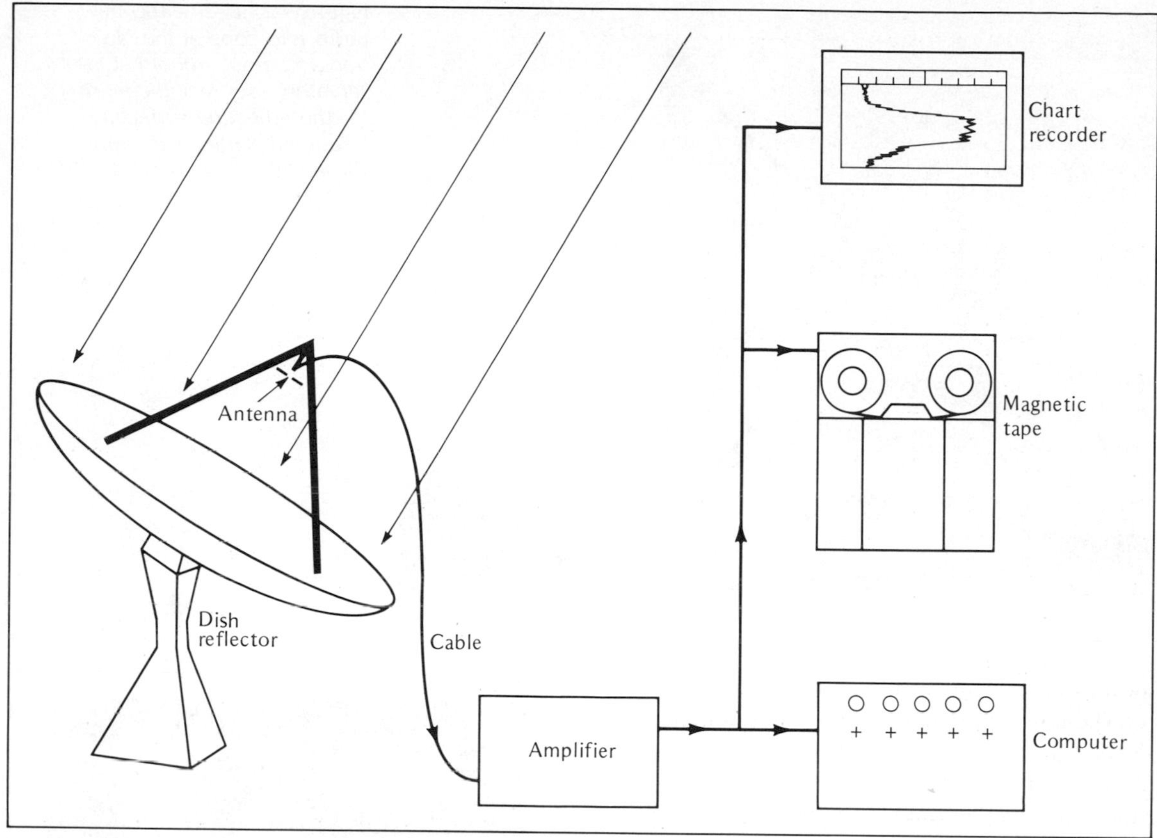


*Figure 3-22. In most radio telescopes, a reflector concentrates the radio signal on the antenna. The signal is then amplified and its intensity is recorded.*

The telescopes and special instruments discussed so far are important tools in the search for clues in starlight, but they tell only part of the story because they are limited to the visible spectrum. To get the information in the rest of the electromagnetic spectrum, we must use radio telescopes and space telescopes.

## RADIO TELESCOPES

*Operation of a Radio Telescope.* A radio telescope usually consists of four parts: a dish reflector, an antenna, an amplifier, and a recorder (Figure 3-22). The components, working together, make it possible for astronomers to detect radio radiation from celestial objects.

The dish reflector of a radio telescope, like the mirror of a reflecting telescope, collects and focuses radiation. Because radio waves are much longer than light waves, the dish need not be as smooth as a mirror. Wire mesh works well as a reflector of most radio waves (Figure 3-23). In some radio telescopes the reflector may not be dish shaped or the telescope may contain no reflector at all.

Though a radio telescope's dish may be hundreds of feet in diameter, the antenna may be as small as your hand. Like the antenna on a TV set, its only function is to absorb the radio energy and direct it along a cable to an amplifier. After amplification the signal goes to some kind of recording instrument. A paper chart recorder is

*Figure 3-23. The 100-m radio telescope at the National Radio Astronomy Observatory uses metallic mesh for the reflecting surface. (National Radio Astronomy Observatory.)*

common, but most radio observatories also record data on punched cards or magnetic tape or feed it directly to a computer. However it is recorded, an observation with a radio telescope measures the amount of radio energy coming from a specific point in the sky.

Because astronomers can't see radio waves, they must convert them into something perceptible. One way is to measure the strength of the signal at various places in the sky and draw a map on which contours mark areas of uniform radio intensity. Hundreds of measurements went into the contour map of the radio source Centaurus A shown in Figure 3-24. This radio object, associated with an unusual galaxy, stretches over 10° across the sky.

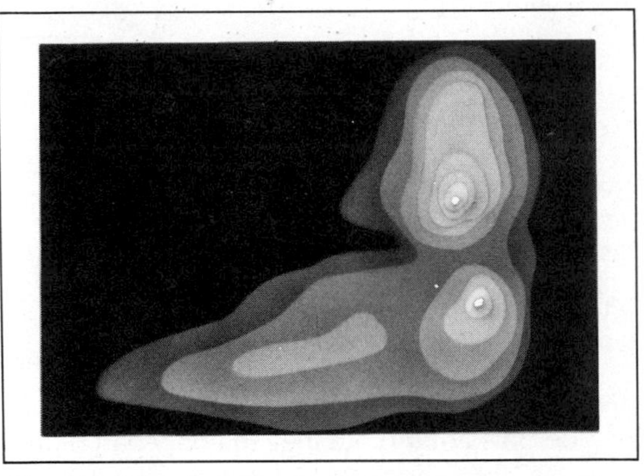

*Figure 3-24. A radio contour map shows the distribution of signal intensity over the sky. This source is associated with an unusual galaxy.*

*Limitations of the Radio Telescope.* A radio astronomer works under three handicaps: poor resolution, low intensity, and interference. We saw that the resolving power of an optical telescope depends on the diameter of the objective lens or mirror. It also depends on the wavelength of the radiation. At very long wavelengths, like those of radio waves, images become fuzzy because of the large diffraction fringes. As in the case of an optical telescope, the only way to improve the resolving power is to build a bigger telescope. Consequently radio telescopes must be quite large.

Even so, the resolving power of a radio telescope is not good. A dish 30 m in diameter receiving radiation with a wavelength of 21 cm has a resolving power of about 0.5°. Such a radio telescope would be unable to show us any details in the sky smaller than the moon. Fortunately, radio astronomers can combine two or more radio telescopes to improve the resolving power, as described in Box 3-4.

The second handicap radio astronomers face is the low intensity of the radio signals. We saw earlier that the energy of a photon depends on its wavelength. Photons of radio energy have such long wavelengths that their individual energies are quite low. In order to get strong signals focused on the antenna, the radio astronomer must build large collecting dishes.

The largest radio dish in the world is 1000 feet in diameter. So large a dish can't be supported in the usual way, so it is built into a mountain valley in Arecibo, Puerto Rico (Figure 3-25). A thin metallic surface supported above the valley floor by cables serves as the reflecting dish, and the antenna hangs on cables from three towers built on the three mountains that surround the valley. Although this telescope can look only overhead, the operators can change its aim slightly by moving the antenna and by waiting for the rotation of the earth to point the telescope in the proper direction. This may sound clumsy, but the telescope's ability to detect weak radio sources, together with its good resolution, makes it one of the most important radio observatories in the world.

*Figure 3-25. The largest radio telescope in the world is the 300-m (1000-ft) dish suspended in a valley in Arecibo, Puerto Rico. The antenna hangs above the dish on cables stretching from towers. The Arecibo Observatory is part of the National Astronomy and Ionosphere Center, which is operated by Cornell University under contract with the National Science Foundation.*

The third handicap the radio astronomer faces is interference. A radio telescope is an extremely sensitive radio receiver listening to radio signals thousands of times weaker than artificial radio and TV transmissions. Such weak signals are easily drowned out by interference. Sources of such interference include everything from poorly designed transmitters in earth satellites to automobiles with faulty ignition systems. To avoid this kind of interference, radio astronomers locate their telescopes as far from civilization as possible. Hidden deep in mountain valleys, they can listen to the sky protected from man-made radio noise. (See Color Plate 5.)

*Advantages of Radio Telescopes.* Building large radio telescopes in isolated locations is expensive, but two factors make it all worthwhile. First, and most important, a radio telescope can show us where clouds of cool hydrogen are located. Because 90 percent of the atoms in the universe are hydrogen, that is important information. Large clouds of cool hydrogen are completely invisible to normal telescopes, since they produce no light of their own and reflect too little to be detected on photographs. However, cool hydrogen emits a radio signal at the specific wavelength of 21 cm. (We will see how the hydrogen produces this radiation when we discuss atoms in the next chapter.) The only way we can detect these clouds of gas is with a radio telescope that receives the 21 cm signals. This alone is sufficient reason to build radio telescopes.

Nevertheless, there is a second reason. Because radio signals have relatively long wavelengths, they can penetrate the vast clouds of dust that obscure our view at visual wavelengths. Light waves are short and interact with tiny dust grains floating in space; thus the light is scattered and never gets through the clouds to optical telescopes on earth. However, radio signals from far across the galaxy pass unhindered through the dust, giving us an unobscured view.

## SPACE TELESCOPES

Ground-based telescopes can only operate at wavelengths in the optical and radio "windows." The rest of the electromagnetic radiation—infrared, ultraviolet, x-ray, and

---

### BOX 3-4 THE RADIO INTERFEROMETER

Because of the poor resolving power of radio telescopes, radio astronomers can see little detail. Consequently, they cannot pinpoint the location of radio objects. They can draw a box on a star chart and say that their radio source is somewhere in the box, but there may be hundreds of stars

*Figure 3-26. A radio interferometer consists of two or more radio telescopes whose signals are combined to give the resolving power of a much larger telescope.*

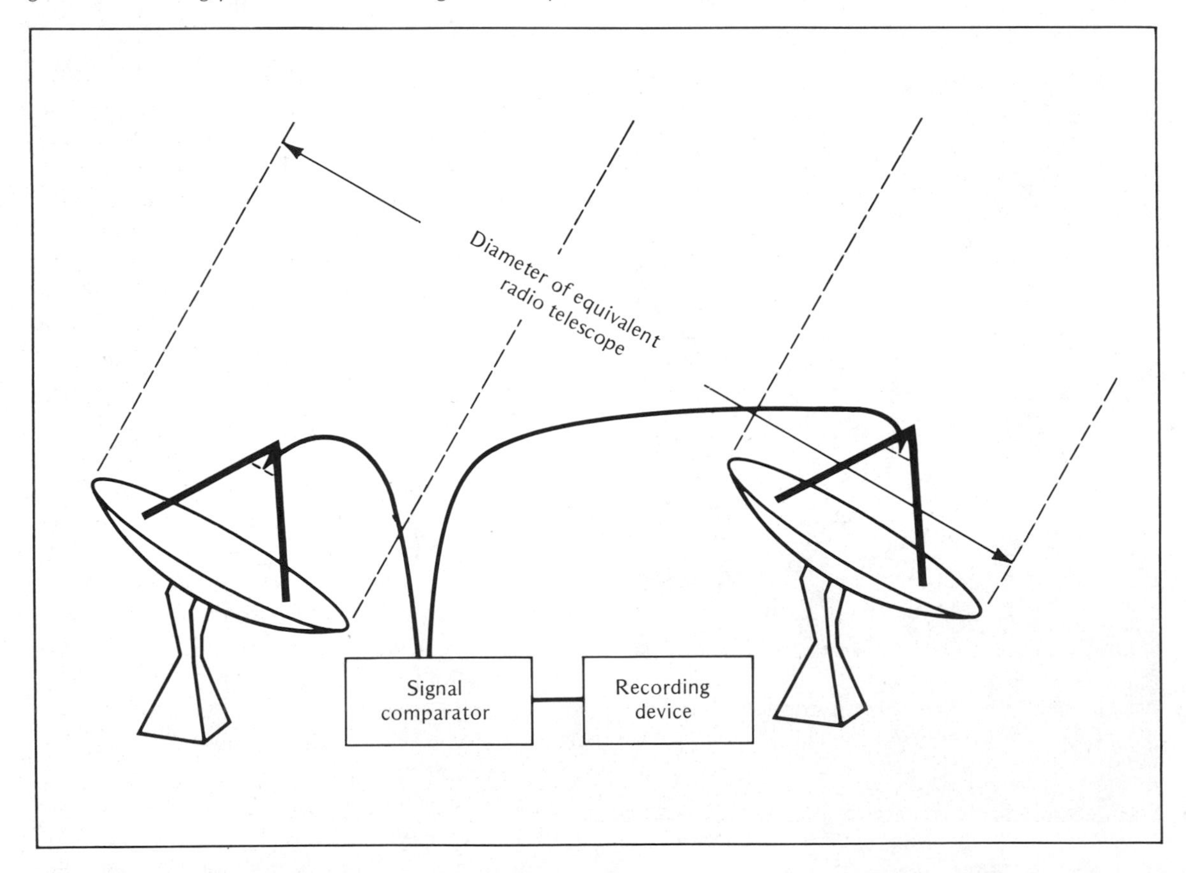

gamma ray—never reaches earth's surface. To observe at these wavelengths, telescopes must fly above the atmosphere in high-flying aircraft, rockets, balloons, and earth satellites.

*Infrared Astronomy.* Some infrared radiation does leak through our atmosphere. This radiation enters narrow, partially open atmospheric windows scattered from 12,000 Å to about 400,000 Å. In this range, called the near infrared, much of the radiation is absorbed by water vapor and carbon dioxide in earth's atmosphere, so it is an advantage to place telescopes on mountains where the air is thin and dry. The University of Hawaii operates one such observatory at 13,600 feet atop the volcano Mauna

---

and galaxies in the box. Which one is emitting radio signals?

To improve the resolving power of their telescopes, radio astronomers have devised the radio interferometer. A **radio interferometer** consists of two or more radio telescopes that combine their signals as if the two signals were coming from different parts of one big telescope (Figure 3-26 and 3-27). The system has the resolving power of a telescope whose diameter is equal to the separation between the two dishes. Since we could build radio telescopes miles apart, we could simulate a telescope much bigger than we could actually build.

The smallest radio sources are only a fraction of a second of arc in diameter. To study such sources, radio astronomers connected radio telescopes in Europe, the United States, Canada, and Australia into a planetwide radio interferometer. Because it was impractical to connect the telescopes with cables, they recorded the signals on magnetic tape together with time signals from atomic clocks. Tapes made simultaneously at different locations were later synchronized according to the time signals and then played together. The combined signal simulated a radio telescope nearly 8000 miles in diameter (the earth's diameter), giving very high resolution.

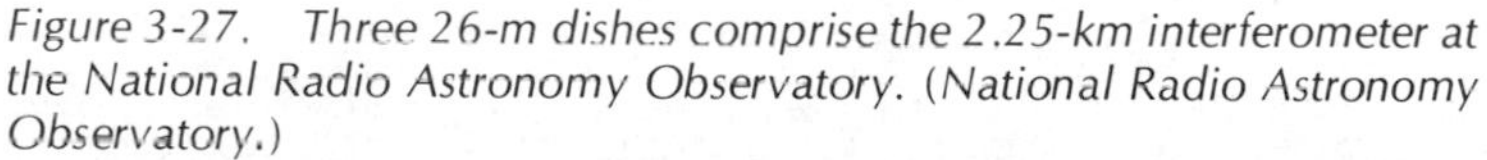

*Figure 3-27. Three 26-m dishes comprise the 2.25-km interferometer at the National Radio Astronomy Observatory. (National Radio Astronomy Observatory.)*

The National Radio Astronomy Observatory, with most of its telescopes located in Green Bank, West Virginia, has built a new radio observatory in the desert of New Mexico. The Very Large Array telescope (Figure 3-28), as it is known, consists of 28 radio dishes, each 25 m (82 ft) in diameter, located in a y-shaped pattern about 20 km (13 mi) on each leg. The signals from the dishes, combined by computer, simulate a radio telescope 27 km (17 mi) in diameter. When all of the dishes are in operation, this telescope will give us maps of radio sources as detailed as the best photographs taken from earth-based optical telescopes.

*Figure 3-28.* *The Very Large Array (VLA) radio interferometer uses up to 28 radio dishes located along the 20-km-long arms of a "y" to simulate a radio telescope 27 km in diameter. The VLA was developed by the National Radio Observatory with the support of the National Science Foundation. (National Radio Astronomy Observatory.)*

*Figure 3-29.* *At 13,600 feet, the top of Mauna Kea is home to (clockwise from the bottom) NASA-University of Hawaii 3-m infrared telescope, Canada-France-Hawaii 3.6-m telescope, Hawaii 0.6-m telescope, Hawaii 2.24-m telescope, United Kingdom 3.8-m Infrared Telescope, and a Hawaiian 0.6-m telescope. (Dale P. Cruikshank, Institute for Astronomy.)*

Kea (Figure 3-29). At this altitude, astronomers can use special spectrographs and photometers to observe in the windows of the near infrared.

The wavelength region longer than 400,000 Å, called the far infrared, is a rich source of data about the formation of planets and stars, but these longer infrared waves do not penetrate our atmosphere very far. To observe at these wavelengths, or at wavelengths between the narrow windows in the near infrared, telescopes must venture high in the atmosphere to get above the water vapor and carbon dioxide. Infrared telescopes suspended under balloons have reached altitudes as high as 41 km (25 miles), but these flights are unmanned and the telescopes must be operated automatically or by remote control.

Another solution is to modify a modern jet to carry both telescope and astronomer to high altitudes. Astronomers have flown to 50,000 feet to measure infrared radiation from distant galaxies. NASA now operates a flying astronomical observatory built into a Lockheed C-141 jet transport. The aircraft, called the Gerard P. Kuiper Airborne Observatory, can carry a 91-cm telescope, computer control systems, and a dozen astronomers to altitudes of 40,000 feet, where they can make infrared observations unaffected by 99 percent of the water vapor in the atmosphere (Figure 3-30).

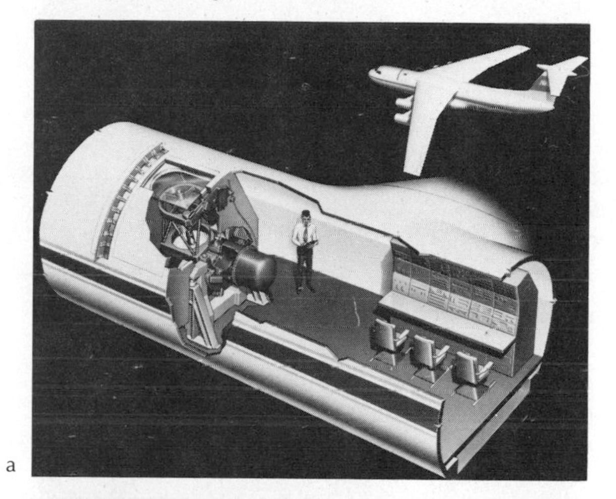

a

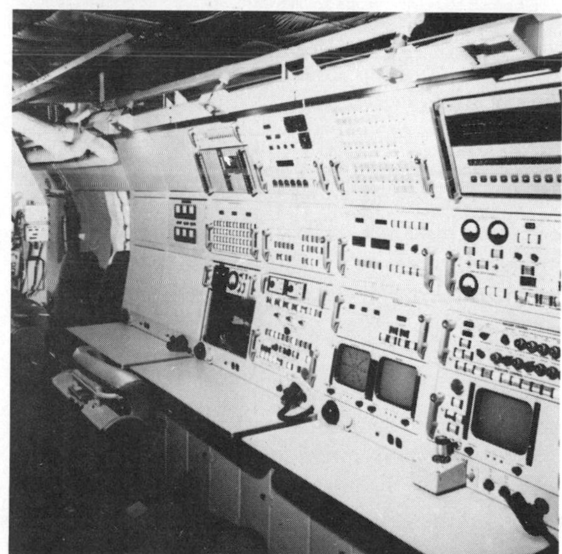

b

*Figure 3-30. (a) The Gerard P. Kuiper Airborn Observatory carries a 91-cm infrared telescope, astronomers, and operators to altitudes of 12 km (39,000 ft) to get above 99 percent of the water vapor in earth's atmosphere. (b) Control panel of the telescope, which is located behind the bulkhead in the background. (NASA.)*

Whether an infrared telescope is located in an airplane or on a mountain top, its operation remains the same. The observers can't see at infrared wavelengths, so they must have accurate control systems to point the telescope. Any radiation entering the telescope comes to a focus on a detector that, when illuminated by infrared radiation, conducts electricity. By measuring the current from the detector, the astronomers can measure the intensity of the infrared source. From measurements at a number of locations on the sky, they can draw a contour map of the infrared source just as radio astronomers map a radio source.

*Ultraviolet Astronomy.* The best clues are often the hardest to find. Earth-based telescopes can observe in the near ultraviolet, but radiation with wavelengths shorter than about 3000 Å is completely absorbed by the ozone layer high in our atmosphere. To get above this layer and seek data in the ultraviolet, astronomers must put telescopes into space.

NASA has launched a number of orbiting, ultraviolet telescopes to study the sun, planets, and stars. Also, the astronauts on Skylab made ultraviolet observations, and the astronauts of Apollo 16 photographed the sky with an ultraviolet camera from the airless surface of the moon. In addition, the International Ultraviolet Explorer (IUE) (Figure 3-31) was launched in

*Figure 3-31. The International Ultraviolet Explorer. (NASA.)*

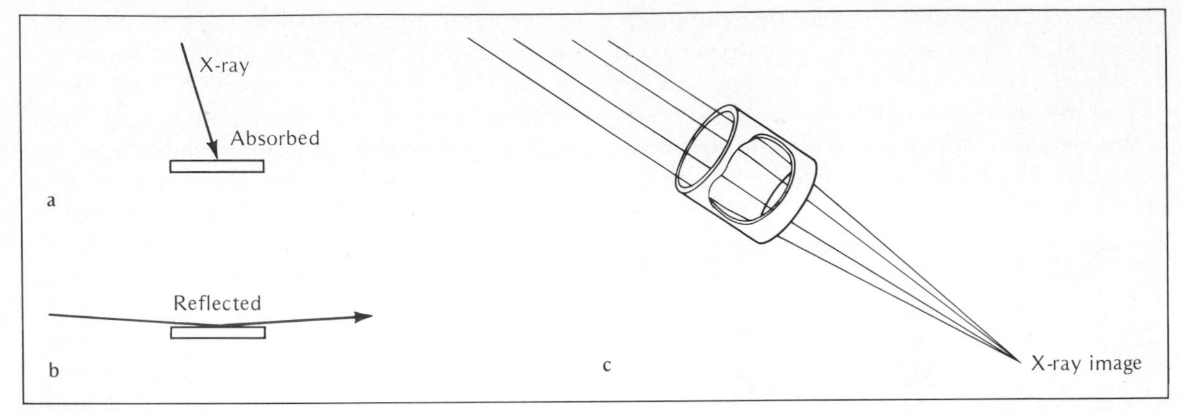


*Figure 3-32. X-rays penetrate if they strike at large angles (a), but reflect if they strike at small angles (b). Thus scientists can shape the inside of a cylinder to focus x-rays into an image (c).*

January 1978 for joint operation by American and European astronomers. It carries a 45-cm telescope with attached spectrographs using TV systems to record the spectra. The images are later transmitted to an earth-based computer system that controls the satellite and analyzes the data. The IUE can obtain spectra from 1150 Å to 3200 Å.

Because hot stars emit large amounts of ultraviolet radiation, the ultraviolet region is important. To understand these stars, astronomers must be able to study the spectral region containing the bulk of the stars' radiation. Also, as we will see in the next chapter, important spectral features lie in the ultraviolet region of stellar spectra.

*X-Ray Astronomy.* Of all the invisible clues sought by astronomers, some of the most exciting are the x-rays that tell us about distant violence. The x-ray region of the spectrum contains short wavelengths, ranging from 100 Å to about 0.1 Å. These very high energy photons can be produced only by violent, high energy events. We receive x-rays from exploding stars and galaxies and from matter smashing into the surface of a neutron star or falling into a black hole—phenomena we will be examining in greater detail in later chapters.

Because x-rays pass through or are absorbed by most materials, they can't be focused like light. Nevertheless, it is possible to build x-ray telescopes. Proper arrangements of baffles and counters can produce a detector that is sensitive only to x-rays coming from a single direction. Such a directional detector spinning on its axis while orbiting the earth sweeps the sky for x-rays. Scientists can construct an x-ray map of the sky by noting which way the detector was pointing when it detected x-rays.

A more sophisticated x-ray telescope can produce images instead of maps. To form an image, the x-rays must be focused. However, x-rays do not reflect from most material if they strike at large angles, as light does in a reflecting telescope. The solution is to arrange for the x-rays to graze the reflecting surfaces at very small angles (Figure 3-32). Under such circumstances, x-rays reflect from the surface like a stone skipping across a pond. Optical systems using this principle can focus x-rays to form images that can be recorded by film or by electronic camera systems.

An orbiting x-ray telescope called Uhuru (Swahili for "freedom") discovered nearly 170 x-ray sources when it was launched in 1970. The High Energy Astronomy Observatory (HEAO) satellites, carrying more sensitive and more sophisticated equipment, have pushed the total into many hundreds. The second HEAO

*Figure 3-33. The High Energy Astrophysical Observatory-2 (HEAO-2) can make x-ray and far-ultraviolet observations from orbit. (NASA.)*

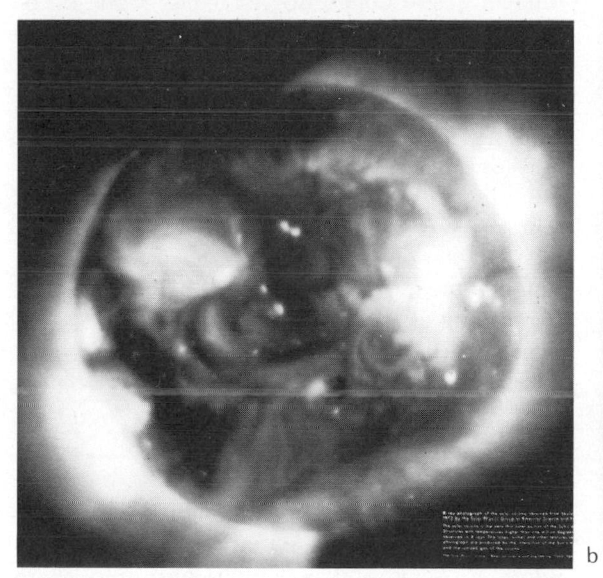

a

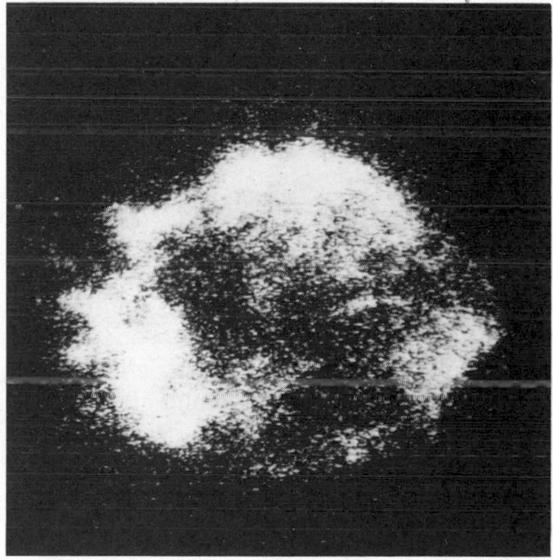

b

*Figure 3-34. (a) An x-ray image of the sun taken from Skylab reveals regions of x-ray emission in the corona. (NASA). (b) A photograph of the x-ray image of the Cassiopeia A supernova remnant obtained with the HEAO-2 orbiting observatory (Figure 2-36). The x-rays reveal a shell of gas expanding away from the site of a stellar explosion that occured about 310 years ago. (Reprinted courtesy of S. S. Murray, G. Fabbiano, A. Fabian, A. Epstein, and R. Giacconi, from* The Astrophysical Journal, *published by the University of Chicago Press; © 1979 The American Astronomical Society.)*

satellite (Figure 3-33), named the Einstein Observatory, uses grazing reflection optics to produce x-ray images. Similar x-ray optics aboard Skylab produced the x-ray image of the sun in Figure 3-34a.

The armada of space telescopes is growing rapidly. Satellites have been launched to monitor x-rays from the sun, and some satellites, such as HEAO-3, now detect gamma rays. Plans call for the space shuttle to launch a space telescope in the mid 1980s. That satellite will carry a large telescope, various electronic camera systems, spectrographs, and photometers, and will be the most sophisticated astronomical satellite ever launched. Other astronomical experiments are planned using the space shuttle, making the future of astronomy above the atmosphere an exciting prospect.

## SUMMARY

Electromagnetic radiation is an electric and magnetic disturbance that transports energy at the speed of light. The electromagnetic spectrum includes radio waves, infrared radiation, visible light, ultraviolet radiation, x-rays, and gamma rays.

We can think of "a particle of light," a photon, as a bundle of waves that sometimes acts as a particle and sometimes as a wave. The energy a photon carries depends on its wavelength. The wavelength of visible light, usually measured in Angstroms ($10^{-10}$ m), ranges from 4000 Å to 7000 Å. Infrared and radio photons have longer wavelengths and carry less energy. Ultraviolet, x-ray, and gamma ray photons have shorter wavelengths and carry more energy.

All objects emit some radiation, and the hotter the object, the more radiation it emits. This radiation contains all wavelengths, but it is most intense at the wavelength of maximum radiation, $\lambda_{max}$. This wavelength depends on the body's temperature. Hot objects emit mostly short wavelength radiation, while cool objects emit mostly long wavelength radiation. This effect gives us clues to the temperatures of stars—hot stars are blue and cool stars are red.

To obtain data, astronomers use telescopes to gather light, see fine detail, and magnify the image. The first two of these three powers of the telescope depend on the telescope's diameter; thus astronomical telescopes often have large diameters.

Astronomical telescopes are of two types, refractor and reflector. A refractor uses a lens to bend the light and focus it into an image. Because of chromatic aberration, refracting telescopes cannot bring all colors to the same focus, resulting in color fringes around the images. An achromatic lens partially corrects for this, but such lenses are expensive and cannot be made larger than about 1 m in diameter.

Reflecting telescopes use a mirror to focus the light, and are less expensive than refracting telescopes of the same diameter. In addition, reflecting telescopes do not suffer from chromatic aberration. Thus most recently built telescopes are reflectors.

To observe radio signals from celestial objects, we need a radio telescope, which usually consists of a dish reflector, an antenna, an amplifier, and a recorder. Such an instrument can measure the intensity of radio signals over the sky and construct radio maps. The poor resolution of the radio telescope can be improved by combining it with another radio telescope to make a radio interferometer. Radio telescopes have two important features—they can detect cool hydrogen, and they can see through dust clouds in space.

The earth's atmosphere admits radiation primarily through two wavelength intervals, or windows—the visual window and the radio window. At other wavelengths our atmosphere absorbs radiation. To observe in the far infared, astronomers must fly telescopes high in balloons or aircraft, though they can work at some wavelengths in the near infrared from high mountain tops. To observe in the ultraviolet and x-ray range, they must send their telescopes into space to get above our atmosphere.

## NEW TERMS

electromagnetic radiation
wavelength
photon
Angstrom (Å)
atmospheric window
black body radiation
wavelength of maximum ($\lambda_{max}$)
erg
refracting telescope
focal length
objective lens
eyepiece
chromatic aberration
achromatic lens
reflecting telescope
objective mirror
prime focus
secondary mirror
Cassegrain telescope
Newtonian focus
coudé focus
Schmidt camera
light-gathering power
resolving power
diffraction fringe
seeing
magnifying power
spectrograph
grating
comparison spectrum
photometer
radio interferometer

## QUESTIONS

1. What are the atmospheric "windows"?
2. Identify by name and approximate wavelength the different regions of the electromagnetic spectrum. How do astronomers observe in each region?
3. What are the three powers of a telescope?
4. Explain why one of the three powers of a telescope is less important than the other two.
5. How does a grating spectograph differ from a prism spectrograph?
6. Why are most modern astronomical telescopes reflectors?
7. Describe five focal arrangements that are possible for an astronomical telescope.
8. Explain the advantages and disadvantages of radio telescopes.
9. Why can we observe in the infrared from aircraft, but must go into space to observe in the ultraviolet?
10. Explain why only violent events produce x-rays.

## PROBLEMS

1. The wavelength of red light is about 6500 Å. How many of these waves would be needed to stretch 1 mm?
2. What is the approximate wavelength of blue light in Angstroms and in meters? (Hint: See Figure 3-2.)
3. Measure the actual wavelength of the wave in Figure 3-1. In what portion of the electromagnetic spectrum would it belong?
4. Human body temperature is about 310°K (98.6°F). At what wavelength do humans radiate the most energy? What kind of radiation do we emit?
5. Explain why a hot star looks blue and a cool star looks red.
6. Compare the light-gathering powers of the 200-in telescope and a 20-in telescope.

(*continued*)

7. What is the resolving power of a 10-in telescope? What do two stars, 1.5 seconds of arc apart, look like through this telescope?
8. If we build a telescope with a focal length of 50 in, what focal length should the eyepiece have to give a magnification of 100 times?
9. Use a ruler to measure the focal lengths of the objective and eyepiece of the telescope in Figure 3-6. What is the magnification?
10. Compare the radio-wave gathering power of a radio telescope 100 ft across with that of the Arecibo telescope.

## RECOMMENDED READING

Cameron, R. M. "NASA's 91-cm Airborne Telescope." *Sky and Telescope* 52 (Nov. 1976), p. 327.

Duncan, D. "The Schmidt Telescope and Modern Astronomy." *The Griffith Observer* 36 (January 1972), p. 2.

Hey, J. S. *The Evolution of Radio Astronomy.* New York: Neale Watson Academic Publishing, 1973.

Kellermann, K. I. "Intercontinental Radio Astronomy." *Scientific American* 226 (Feb. 1972), p. 72.

King, H. C. *The History of the Telescope.* Cambridge, Mass.: Sky Publishing, 1955.

Kraus, J. *Big Ear.* Powell, Ohio: Cygnus-Quasar Books, 1976.

Oberg, J. "Astronomers in Space." *Astronomy* 5 (April/May 1977), pp. 18 and 48.

Overbye, D. "The X-Ray Eyes of Einstein." *Sky and Telescope* 57 (June 1979), p. 527.

Pasachoff, J. M., Linsky, J. L., Haisch, B. M., and Boggess, A. "IUE and the Search for a Lukewarm Corona." *Sky and Telescope* 57 (May 1979), p. 438.

Sheaffer, R. "Radio Interferometry." *Astronomy* 4 (Oct. 1976), p. 6.

Spitzer, L. "The Space Telescope." *American Scientist* 66 (July/Aug. 1978), p. 426.

Van Heal, A. C. S. and Velzel, C. H. F. *What is Light?* New York: McGraw-Hill, 1968.

Verschuur, G. L. *The Invisible Universe.* New York: Springer-Verlag, 1974.

Verschuur, G. L. "Ultraviolet Astronomy." *Astronomy* 3 (Feb. 1975), p. 34.

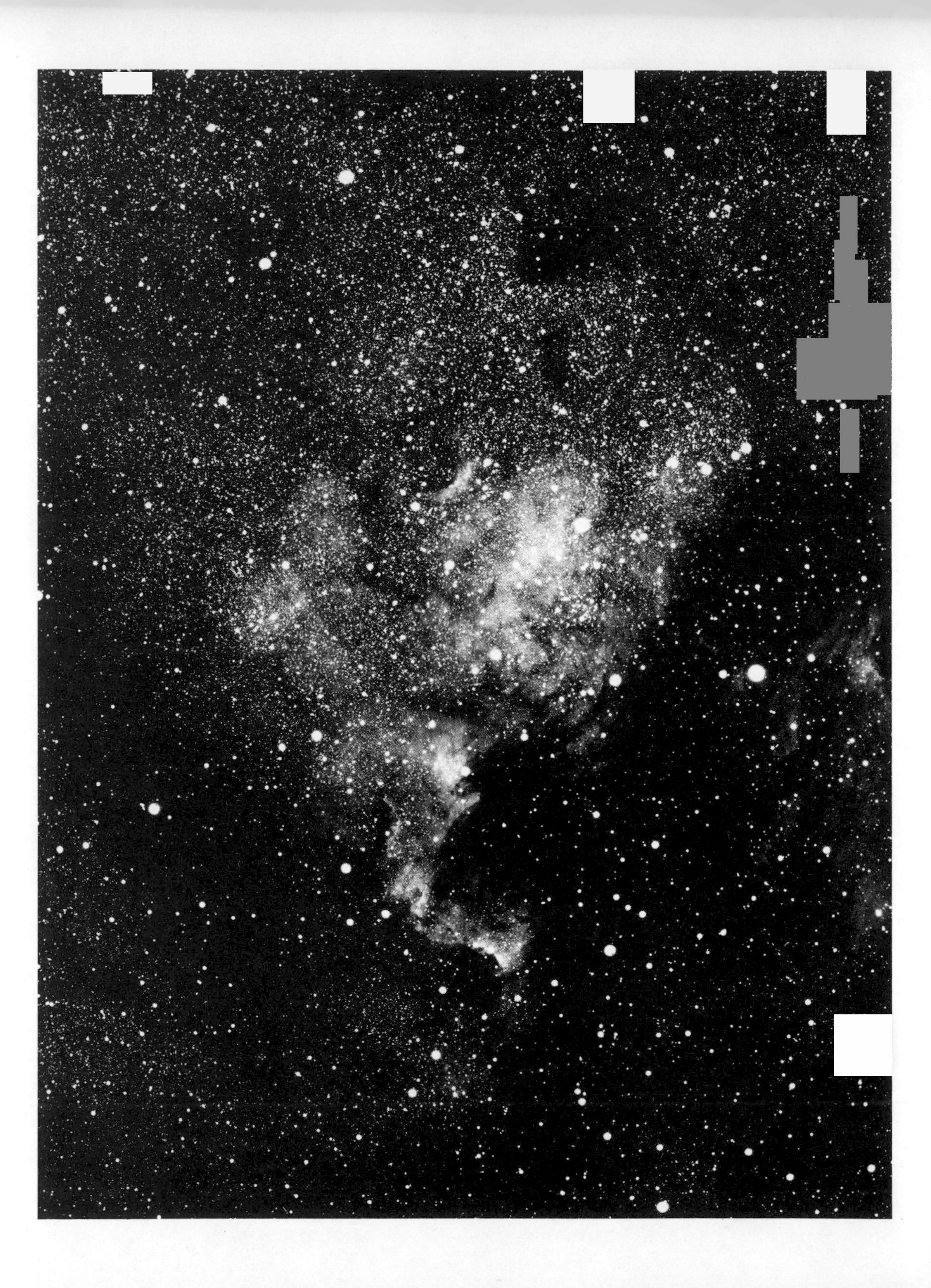

*The North American Nebula, so called because of its outline, is a glowing cloud of gas 3500 ly away. (Lick Observatory photograph.)*

# Chapter 4 ATOMS AND STARLIGHT

The stars are so far away that it would be unsurprising if earth-bound humans knew almost nothing about them. In fact, that was the case until the early 19th century, when the Munich optician Joseph Fraunhofer studied the solar spectrum and found it crossed by some 600 dark lines. As scientists realized that the lines were related to the various atoms in the sun and found that stellar spectra had similar patterns of lines, the door to an understanding of stars finally opened.

To go through that door we must consider how atoms interact with light to produce these spectral lines. We begin with the hydrogen atom because it is the most common atom in the universe and the simplest. Other atoms are larger and more complicated, but their properties resemble those of hydrogen. Another reason for studying hydrogen is to consider how the structure of its atoms can give rise to the 21-cm wavelength radiation that is so important to radio astronomers (see Chapter 3).

Once we understand how an atom's structure can interact with light to produce spectral lines, we will recognize certain patterns in stellar spectra. By classifying the spectra according to these patterns, we can arrange the stars in a sequence according to temperature. One of the most important pieces of information revealed in a star's spectrum is its temperature.

But, properly analyzed, a stellar spectrum can tell us much more. The strength of different features in the spectrum indicates the chemical composition of the star, and the wavelengths of such features can tell us how the star is moving with respect to the earth.

Finally we will study a special kind of spectrum that flashes into view for a few seconds during a total solar eclipse. Decoded, this spectrum tells us the conditions in the sun's atmosphere and gives hints to the nature of the atmospheres of other, more distant stars.

## ATOMS

*A Model Atom.* In Chapter 1 we devised a model of the sky, the celestial sphere, to help us think about the nature and motion of the heavens. In the case of the atom, we again need a model.

Our model of the atom consists of a small central **nucleus** surrounded by a cloud of whirling electrons. The nucleus has a diameter of about 0.000016 Å and the cloud of electrons has a diameter of about 1 to 5 Å. (Recall from Chapter 3 that one Angstrom is $10^{-10}$ m.) Household plastic wrap is about 100,000 atoms thick. This makes the atom seem very small, but the nucleus is 100,000 times smaller. Box 4-1 describes a scale model of an atom.

The nucleus of a typical atom consists of two different kinds of particles, protons and neutrons. **Protons** carry a positive electrical charge, and **neutrons** have no charge. Consequently, an atomic nucleus, made of protons and neutrons, has a net positive charge.

The **electrons** surrounding the nucleus are low-mass particles carrying a negative charge. In a normal atom, the number of electrons equals the number of protons. Thus the positive charge on each proton is balanced by the negative charge on an electron, and the atom is electrically neutral.

There are over a hundred kinds of atoms, called chemical elements. The kind of element an atom represents depends only on the number of protons in the nucleus. For example, carbon has six protons and six neutrons in its nucleus. Adding a proton produces nitrogen, and subtracting a proton produces boron.

However, we can change the number of neutrons in an atom's nucleus without changing the atom significantly. For instance, if we add a neutron to the carbon nucleus, we still have carbon, but it is slightly heavier than normal carbon. Atoms that have the same number of protons but a different number of neutrons are **isotopes.** Carbon has two stable isotopes. One form contains six protons and six neutrons, making a total of twelve particles, and is thus called carbon-12. Carbon-13 has six protons and seven neutrons in its nucleus. Figure 4-2 shows schematically the nuclei of a few isotopes.

Protons and neutrons are bound tightly into the nucleus, but the electrons are held loosely in the electron cloud. Running a comb through your hair creates a static charge by removing a few electrons from their atoms. This process is called **ionization,** and the atom that has lost one or more electrons is an **ion.** The neutral carbon atom, with six protons and six neutrons in its nucleus, has six electrons, which balance the positive charge of the nucleus. If we ionize the atom by removing one or more electrons, the atom is left with a net positive charge. Under some circumstances, an atom may capture an extra electron, giving it more negative charges than positive. Such a negatively charged atom is also considered an ion.

---

BOX 4-1 A SCALE MODEL ATOM

Suppose we could make a hydrogen atom bigger by a factor of $10^{12}$ (one million million). Only then would it be big enough to examine.

The nucleus of a hydrogen atom is a proton whose diameter is about 0.000016 Å, or $1.6 \times 10^{-13}$ cm. Multiplying by a factor of $10^{12}$ magnifies it to 0.16 cm, about the size of a grape seed. The electron cloud* has a diameter of about 5 Å, or $5 \times 10^{-8}$ cm. When we magnify the atom by $10^{12}$ this becomes 500 meters, or about 5½ football fields laid end to end (Figure 4-1). When you imagine a grape seed in the midst of 5½ football fields, orbited by one magnified electron still too small to be visible, you see that an atom is mostly empty space.

If we magnified the mass of the atom by $10^{12}$, we would find that it weighed less than $2 \times 10^{-12}$ gm. This is too small to be meaningful. We would have to multiply the mass by another factor of $10^{12}$ just to get a mass we could imagine. Individual atoms have such small masses that only by assembling vast numbers can nature build such massive objects as stars. The sun, for instance, contains about $10^{57}$ atoms.

Looking at our model atom, we find that the nucleus contains nearly all of the mass. Protons and neutrons have almost equal mass—about 1836 times the mass of an electron. Thus the electrons in an atom never represent more than about 0.05 percent of the atom's mass.

* For a representative diameter, we take the size of the atom's second orbit (Figure 4-3).

---

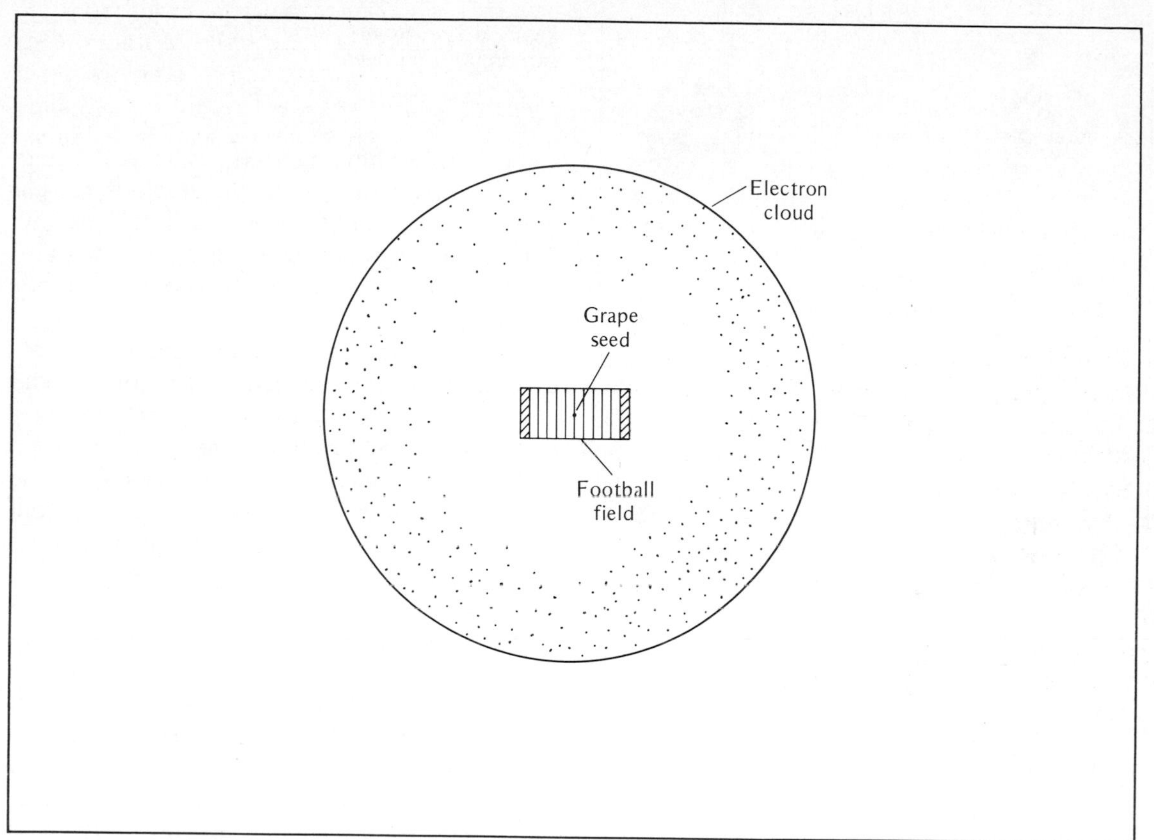


*Figure 4-1. Magnifying a hydrogen atom by $10^{12}$ makes the nucleus the size of a grape seed and the outer electron cloud about 5½ times bigger than a football field. The electron itself is still too small to see.*

Atoms form bonds with each other by exchanging or sharing electrons. Two or more atoms bonded together form a **molecule.** This chemical bonding of atoms is not usually important to astronomers since few atoms can form chemical bonds in stars. The high temperatures produce such violent collisions between atoms that most molecules would quickly break up. Only in the coolest stars are the collisions gentle enough to permit chemical bonds. We will see later that the presence of molecules such as titanium oxide (TiO) in a star is a clue that the star is cool. In later chapters we will see that molecules can form in cool gas clouds in space and in the atmospheres of planets.

*Electron Shells.* So far we have described the electron cloud only in a general way, but the specific way electrons behave within the cloud is very important in astronomy.

The electrons are bound to the atom by the attraction between their negative charge and the nucleus's positive charge. If we wish to ionize the atom, we need a certain amount of energy to pull an electron away from its nucleus. This energy is the electron's **binding energy,** the energy that holds it to the atom.

Within the electron cloud, an electron may orbit the nucleus at various distances. If the orbit is small, the electron is close to the nucleus, and we need a large amount of energy to pull it

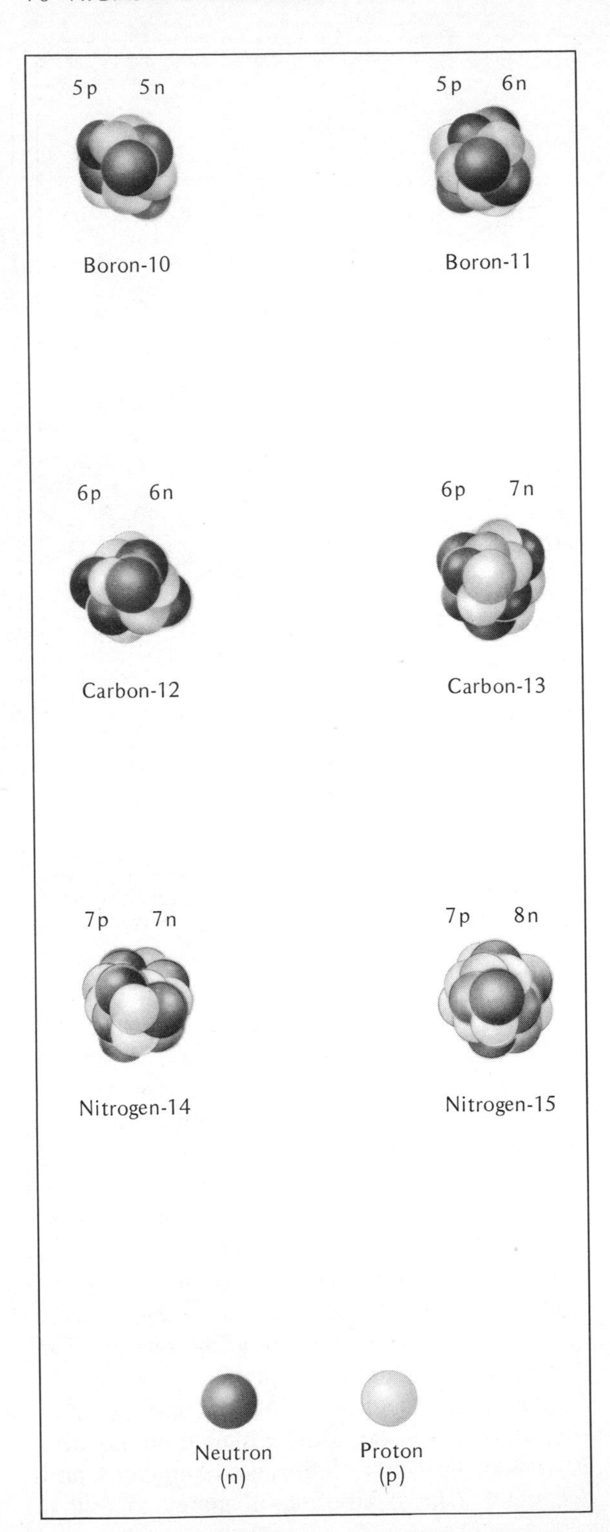


*Figure 4-2. Atomic nuclei of the isotopes of boron, carbon, and nitrogen.*

away. Thus, its binding energy is large. An electron orbiting farther from the nucleus is held more loosely, and less energy will pull it away. Thus it has less binding energy. The size of an electron's orbit is related to the energy that binds it to the atom.

Nature permits atoms only certain amounts of binding energy, which means our model atoms can have orbits of only certain sizes called **permitted orbits** (Figure 4-3). These are like steps in a staircase: you can stand on the number one step or the number two step, but not on the number one and one-quarter step. The electron can occupy any permitted orbit but not orbits in between.

The arrangement of permitted orbits depends primarily on the charge on the nucleus, which in turn depends on the number of protons. Thus each kind of element has its own pattern of permitted orbits (Figure 4-3). Isotopes of the same elements have nearly the same pattern because they have the same number of protons. However, ionized atoms have orbital patterns that differ from their un-ionized forms. Thus the arrangement of permitted orbits differs for every kind of atom and ion.

The properties of electron orbits are important to astronomy because the electrons can interact with light, our major clue from afar. By understanding this interaction we can interpret the lines in spectra and learn such things as the composition and temperature of stars, gas clouds in space, and atmospheres of planets.

## THE INTERACTION OF LIGHT AND MATTER

We begin our study of light and matter by considering the hydrogen atom. As we noted earlier, hydrogen is both simple and common. Roughly 90 percent of all atoms in the universe are hydrogen.

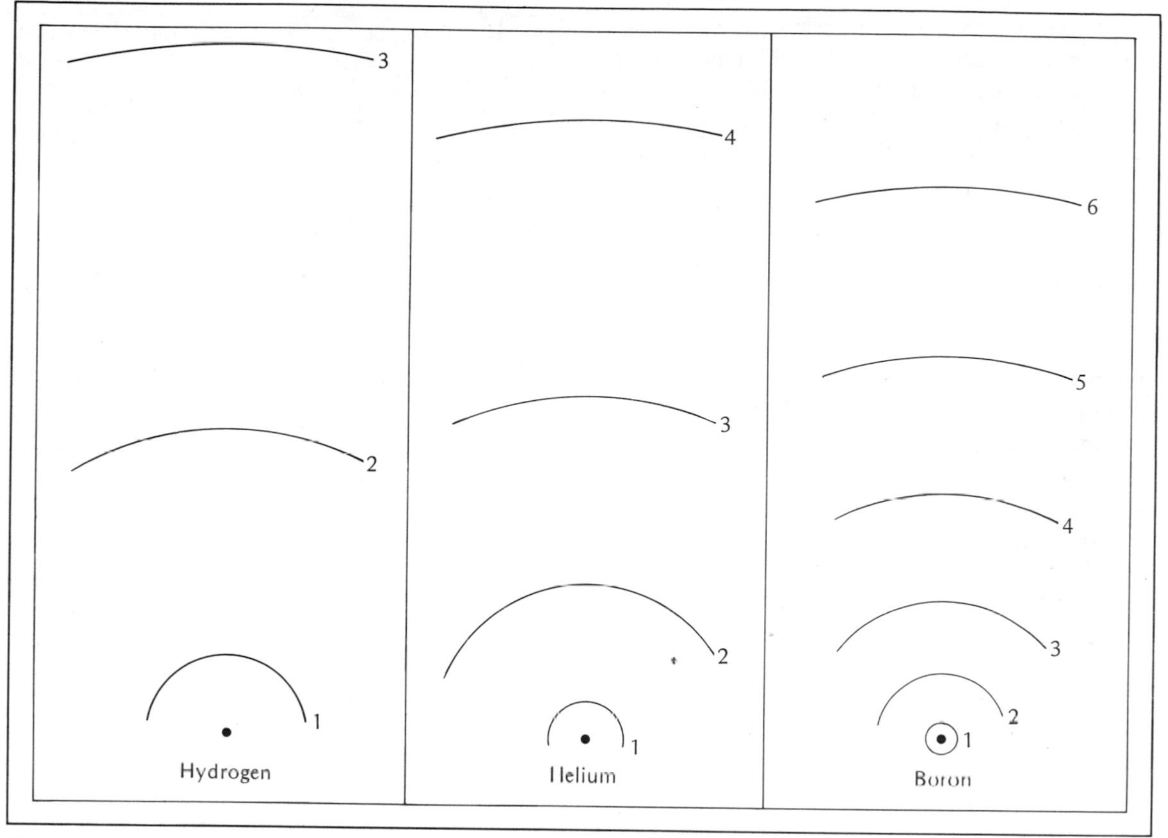


*Figure 4-3.* *The electrons in atoms may occupy only certain, permitted orbits. Since they have different charges on the nucleus, different elements have different patterns of permitted orbits.*

*The Excitation of Atoms.* The hydrogen atom in Figure 4-4 has its electron in the smallest permitted orbit, where it is tightly bound to the atom. We can move the electron to a higher orbit by supplying some energy. It is like moving a flower pot from the ground to a high shelf; the higher the shelf, the more energy we need to raise the pot. The amount of energy needed to move the electron is the energy difference between the two orbits.

If we move the electron from a low orbit to a higher orbit, we say the atom is **excited.** That is, we have added energy to the atom in moving its electron. If the electron falls back to the lower orbit, that energy is released.

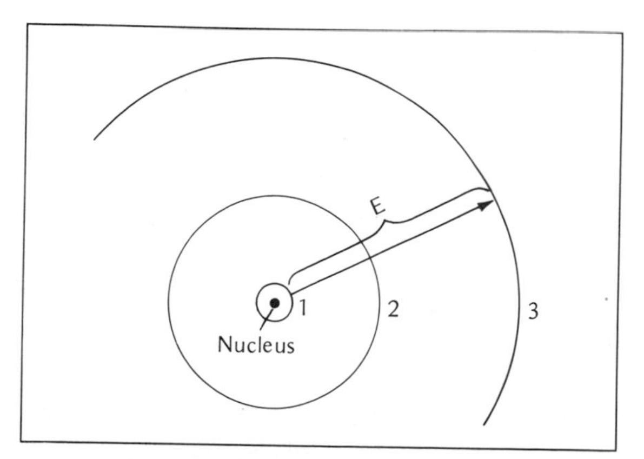


*Figure 4-4.* *A hydrogen atom with its electron in the first orbit can absorb energy E and move its electron to an upper orbit.*

An atom can become excited by collision. If two atoms collide, one or both may have its electron knocked into a higher orbit. This happens very commonly in hot gas where the atoms move rapidly and collide often.

Another way an atom can get the energy to move an electron to a higher orbit is to absorb a photon. Only a photon with exactly the right energy can move the electron from one orbit to another. If the photon has too much or too little energy, the atom cannot absorb it. Since the energy of a photon depends on its wavelength, only photons of certain wavelengths can be absorbed by a given kind of atom. The atom in Figure 4-5 can absorb any of three wavelengths, moving its electron up to any of three permitted orbits. Any other wavelength photon has too much or too little energy to be absorbed.

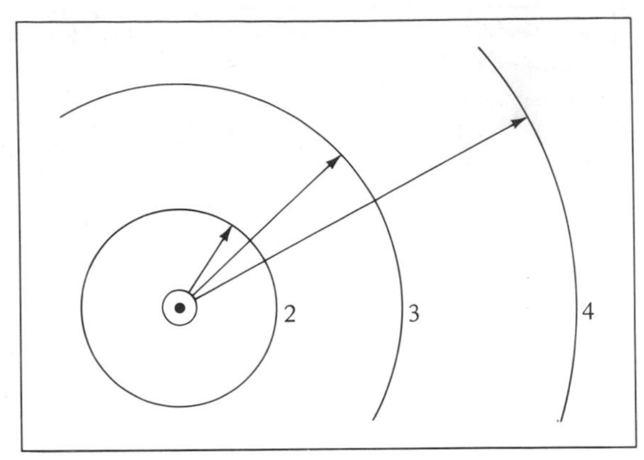


*Figure 4-5. A hydrogen atom can absorb one of a number of different wavelength photons and move the electron to one of its higher orbits.*

Atoms, like humans, cannot exist in an excited state forever. The excited atom is unstable and must eventually (usually within $10^{-6}$ to $10^{-9}$ seconds) give up the energy it has absorbed and return its electron to the lowest orbit. Since the electrons eventually tumble down to this bottom orbit, physicists call it the **ground state.**

When the electron drops from a higher to a lower orbit, it moves from a loosely bound orbit to one more tightly bound. The atom then has a surplus of energy—the energy difference between the orbits—which it can emit as a photon. Study the sequence of events in Figure 4-6 to see how an atom can absorb and emit photons.

Because only certain orbits are permitted in an atom, only certain energy differences can occur. Each type of atom or ion has its unique set of orbits, so each one absorbs and emits photons with a unique set of wavelengths. Thus we can identify the elements in a gas by studying the characteristic wavelengths of light absorbed or emitted.

This process of excitation and emission is a common sight. The gas in a neon sign glows

*Figure 4-6. An atom can only absorb a photon if it has the correct energy. The excited atom is unstable and within a fraction of a second returns to the lower orbit, reradiating the photon in a random direction.*

because a high voltage forces electrons to flow through the gas, exciting the atoms by collisions. Almost as soon as an atom is excited, its electron drops back to a lower orbit, emitting the surplus energy as a photon of a certain wavelength. Neon is a popular gas for signs because the pattern of its electron orbits makes it emit a rich reddish-orange light when it is excited. So-called neon signs of other colors contain other gases or mixtures of gases, but not pure neon.

*The Formation of a Spectrum.* To see how an astronomical object can produce a spectrum, imagine a cloud of hydrogen floating in space with an incandescent light bulb glowing behind it. The bulb glows because its filament is hot, producing black body radiation as described in Box 3-1. Thus it emits photons of all wavelengths, and a spectrum of its light would reveal an uninterrupted band of color called a **continuous spectrum.**

However, the light from this bulb must pass through the hydrogen gas before it can reach our telescope (Figure 4-7). Most of the photons will pass through the gas unaffected because they have wavelengths the hydrogen atoms cannot absorb, but a few photons will have the right wavelengths. These photons cannot pass through the gas because they are absorbed by the first atom they meet. The atom is excited for a fraction of a second, and the electron then drops back to a lower orbit and a new photon is emitted. The original photon was traveling through the gas toward our telescope, but the new photon is emitted in some random direction. Very few of these new photons leave the cloud in the direction of our telescope, so the light that finally enters the telescope has very few photons at the wavelengths the atoms can absorb. When we form a spectrum from this light, photons of these wavelengths are missing and the spectrum has dark lines at the positions these photons would have occupied. These dark lines, like those Fraunhofer saw in the solar spectrum in 1814 and 1815, are called **absorption lines** because the atoms absorbed the photons. A spectrum containing absorption lines is an **absorption spectrum** (also called a **dark line spectrum**).

What happens to the photons that were absorbed? They bounce from atom to atom, being absorbed and emitted over and over until they escape from the cloud. If, instead of aiming our telescope at the bulb, we swing it to one side so that no light from the bulb enters the telescope, we can take a spectrum of the light emitted by the gas atoms (Figure 4-8). In that case, the only photons entering the telescope are photons that were absorbed and re-emitted. A spectrum of this light is almost entirely dark except for the wavelengths corresponding to the photons the gas can absorb and re-emit. Thus we will see a spectrum containing only bright lines on a dark background. These bright lines are called **emis-**

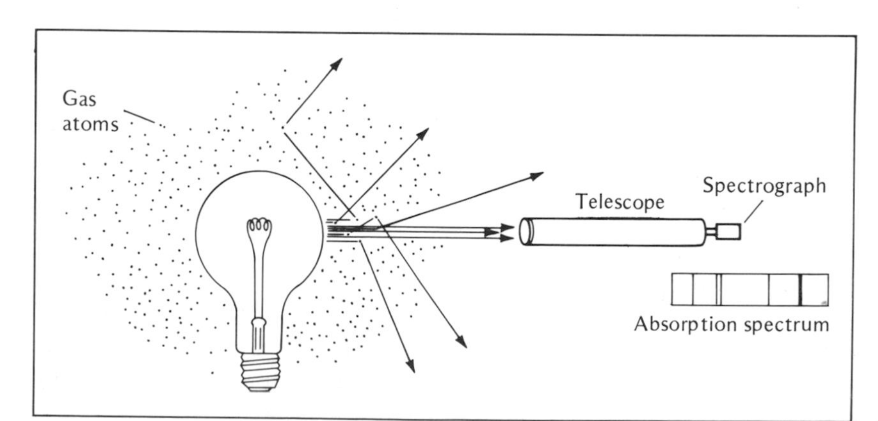


*Figure 4-7. Photons of the proper wavelengths can be absorbed by the gas atoms and reradiated at random directions. Since these photons do not reach the telescope, their wavelengths are dark, producing absorption lines in the spectrum.*

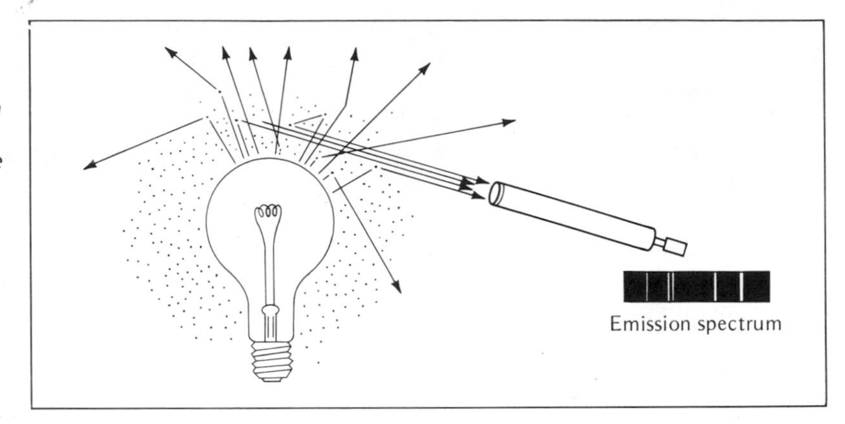


*Figure 4-8. Pointing the telescope away from the bulb, we can receive only those photons the atoms can absorb and reradiate, producing emission lines in the spectrum.*

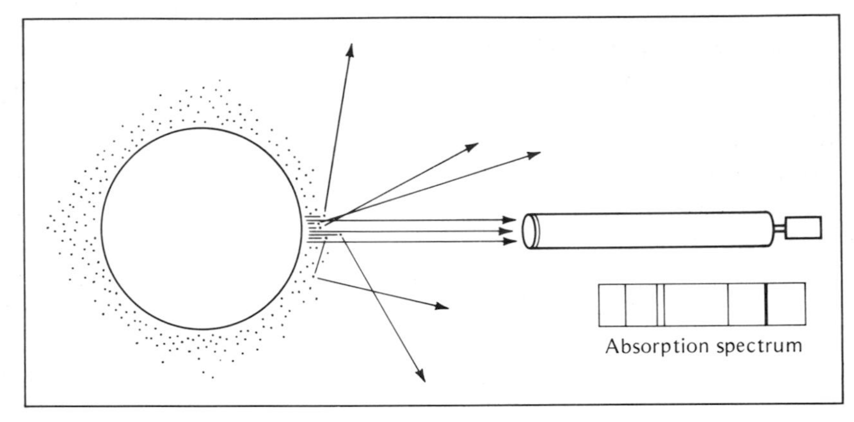


*Figure 4-9. A star produces an absorption spectrum because its atmosphere absorbs certain wavelengths in the spectrum.*

**sion lines,** and a spectrum with emission lines is an **emission spectrum** (also called a **bright line spectrum**). The spectrum of a neon sign, for instance, is an emission spectrum. (See Color Plate 8.)

The light bulb and hydrogen gas cloud produce a spectrum in the same way a star does. A star is all gas, however. Its outer layers act as a hot, bright surface emitting radiation at all wavelengths much like the filament in a light bulb. Above these layers lies the thinner gas of the star's atmosphere. As the light travels upward through the star's atmosphere, photons of certain wavelengths are absorbed by atoms and so never reach us. The spectrum of a star is an absorption spectrum whose dark lines indicate which wavelengths the atoms absorbed (Figure 4-9).

The absorption lines in stellar spectra provide a windfall of data about the star's surface layers. By studying the spectral lines we can identify the elements in the stellar atmosphere and find the temperature of the atoms. To see how to get all this information, we need to look carefully at the way the hydrogen atom produces lines in a star's spectrum.

*The Hydrogen Spectrum.* As you must have gathered by now, each element has its own spectrum, unique as a human fingerprint, and it can be recognized by its spectrum across trillions of miles. To see how hydrogen produces its unique spectrum we must draw a detailed diagram of its permitted orbits, making the size of the orbit proportional to its energy (Figure 4-10). Then we can examine the way such atoms interact with light.

A **transition** occurs in an atom when an elec-

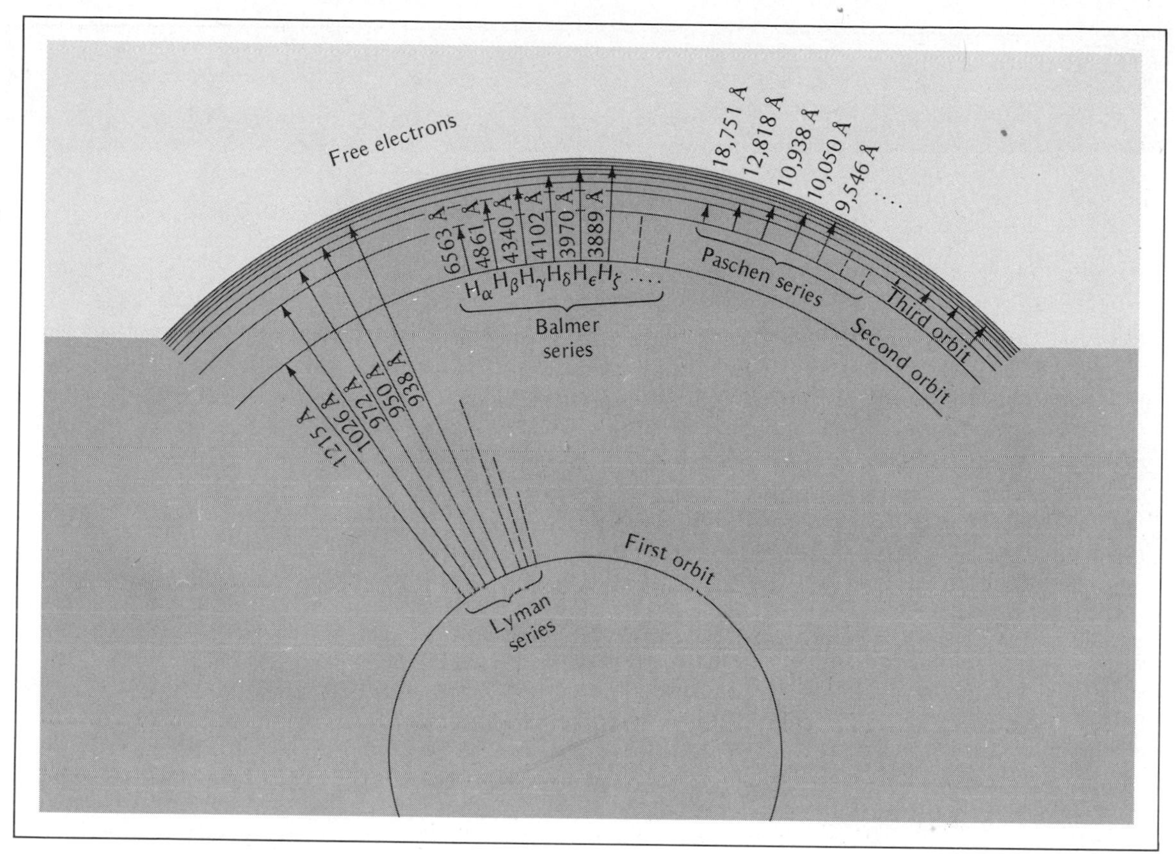


*Figure 4-10. The levels in this diagram are spaced to represent the energy the hydrogen atom's electron would have in each orbit. The transitions, drawn as arrows, can be grouped into series according to their lowest orbit. This drawing shows only a few of the infinity of transitions and series possible.*

tron changes orbits. In our diagram of a hydrogen atom, we can represent transitions by arrows pointing from one orbit to another. If the arrow points upward, the atom must absorb energy, and if the arrow points downward, the atom must emit a photon.

If the transition results in the absorption or emission of a photon, the length of the arrow tells us its energy. Long arrows represent large amounts of energy and thus short-wavelength photons. Short arrows represent smaller amounts of energy and longer-wavelength photons.

We can divide the possible transitions in a hydrogen atom into groups called series, according to their lowest orbit. Those arrows whose lower ends rest on the ground state represent the **Lyman series;** those resting on the second orbit, the **Balmer series;** and those resting on the third, the **Paschen series.** In principle, each series contains an infinite number of transitions, and there are an infinite number of series. Figure 4-10 shows only the first few transitions in the first few series.

The Lyman series transitions involve large energies, as shown by the long arrows in Figure 4-10. These energetic transitions produce lines in the ultraviolet part of the spectrum where they are invisible to the human eye. Nor do any Paschen lines lie in the visible part of the spectrum. These transitions involve small energies, and thus produce spectral lines in the infrared.

Balmer series transitions produce the only spectral lines of hydrogen in the visible part of

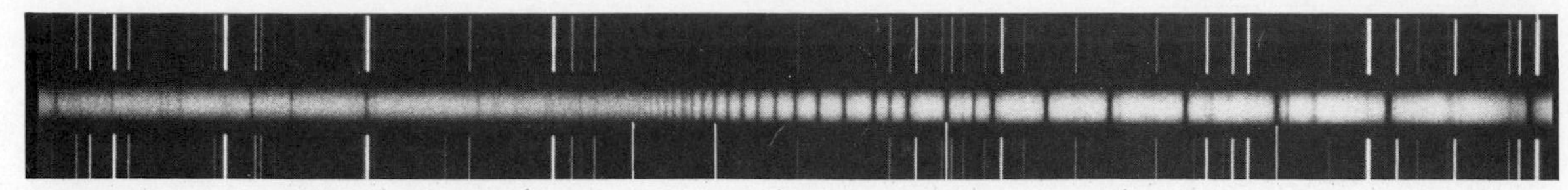

*Figure 4-11. The Balmer lines photographed in the near ultraviolet. (Hale Observatories.)*

the spectrum. Figure 4-10 shows that the first few Balmer series transitions are intermediate between the energetic Lyman transitions and the low energy Paschen transitions. These Balmer lines are labeled by Greek letters for easy identification. $H_\alpha$ is a red line, $H_\beta$ is blue, and $H_\gamma$ and $H_\delta$ are violet. The remaining Balmer lines have wavelengths too short to see, though they can be photographed easily (Figure 4-11). In fact, these four lines create the purple color charac-

## BOX 4-2 THE 21-CM RADIATION

As mentioned in Chapter 3, radio telescopes are important because they can detect the 21-cm wavelength radiation emitted by clouds of cool hydrogen in space. This radiation is emitted when a hydrogen atom's electron changes its energy by changing the direction of its spin.

Protons and electrons spin like tiny tops. This spin creates a magnetic field like that produced when electricity flows through a coil of wire in an electromagnet (Figure 4-12a). Because the proton and the electron have opposite charges their magnetic fields will be opposite when they spin in the same direction (Figure 4-12b and c).

The magnetic fields around the proton and the electron affect the energy of the ground state of hydrogen. If the two magnetic fields are opposite, north pole to south pole and south pole to north pole, they attract each other. If we try to pull the electron away from the atom, we need energy to overcome the binding energy of the

*Figure 4-12. (a) Electrons flowing through a coil of wire produce a magnetic field, as in an electromagnet. (b) The rotation of the positively charged proton produces a similar magnetic field. (c) The rotation of the negatively charged electron produces a magnetic field in the opposite direction from the field around the positive proton.*

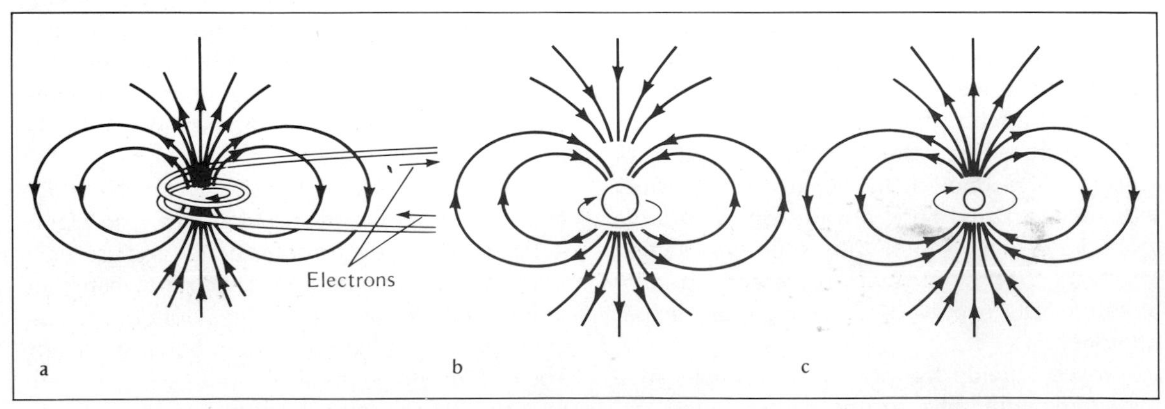

teristic of glowing clouds of hydrogen in space (see Color Plate 12).

Balmer series lines are important because, as we have seen, they are the only hydrogen lines in the visual part of the spectrum. In the next section, we will see how the Balmer series lines can tell us a star's temperature. But before we leave the subject of hydrogen and its electron orbits, we should note that the 21-cm radiation so important in radio astronomy comes from hydrogen atoms whose electrons are in their lowest possible orbit, the ground state. Thus only cool, unexcited hydrogen, such as that in cold, interstellar space, can emit 21-cm photons.

## STELLAR SPECTRA

By studying the interaction of light and matter, astronomers can get information from the spectra of celestial objects. A spectrum can tell us such things as temperature and composition, and even give us data about the motion of astronomical bodies. In later chapters, we will use spectra to study galaxies and planets, but we begin by studying stellar spectra. They are the

---

electron, plus a little extra energy to pull the magnetic fields apart. This means the electron is more tightly bound when the fields are opposite to each other.

However, if the electron spins the other way, its magnetic field is the same as the proton's, north pole to north pole. Oriented this way, the magnets try to repel each other, and less energy pulls the electron away. In other words, the electron is less tightly bound to the proton when it spins in the opposite direction.

Because of these magnetic fields, the ground state of the hydrogen atom is really two orbits of different energy (Figure 4-13). The upper energy orbit is for an electron spinning opposite to its proton, and the lower energy orbit is for an electron spinning in the same direction as its proton. If the electron is in the higher orbit, it can spontaneously flip over and spin the other way, dropping to the lower orbit and emitting a photon. The two energy levels are so close together that the photon emitted in the transition must have a very low energy—corresponding, in fact, to a wavelength of 21 cm.

Only cold, low-density clouds of hydrogen produce 21 cm radiation. If the gas is warm and dense, the atoms collide so often that the electrons are never in the ground state long enough to flip their direction of spin and emit a 21-cm photon.

*Figure 4-13. The magnetic fields of the proton and electron split the ground state of the hydrogen atom into two very close energy levels. A transition from the upper to the lower level emits a 21-cm photon.*

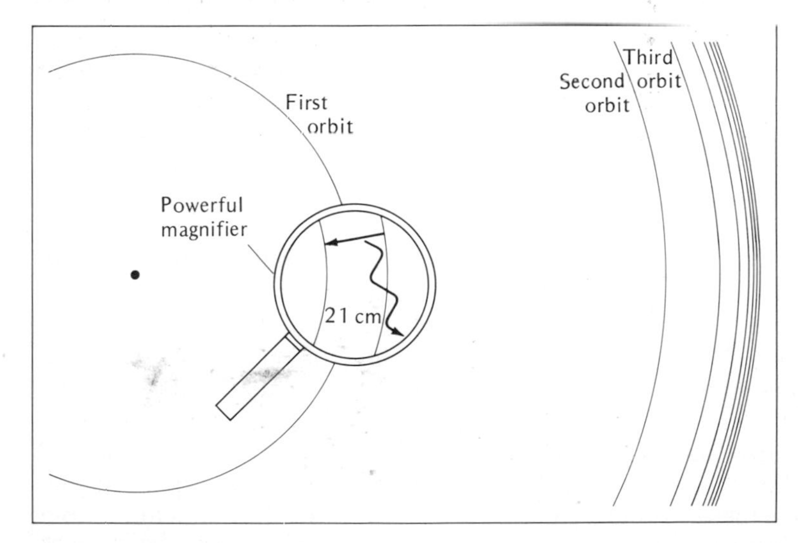

easiest to understand and the nature of stars is central to our study of all celestial objects.

*The Balmer Thermometer.* We can use the Balmer absorption lines as a thermometer to find the temperatures of stars. Chapter 3 showed how to estimate temperature from color, but the strengths of the Balmer lines in a star's spectrum give a much more accurate estimate of the star's temperature.

The Balmer thermometer works because the Balmer absorption lines are only produced by atoms whose electrons are in the second orbit (Figure 4-10). If the star is cool, there are few violent collisions between atoms to excite the electrons, and most atoms have their electron in the ground state. If most electrons are in the ground state, they can't absorb photons in the Balmer series. As a result, we should expect to find weak Balmer absorption lines in the spectra of cool stars.

In hot stars, on the other hand, there are many violent collisions between atoms, exciting electrons to high orbits or knocking the electron clear out of some atoms. That is, some atoms are ionized. Thus, few atoms have electrons in the second orbit to form Balmer absorption lines, and we should expect hot stars, like cool stars, to have weak Balmer absorption lines.

At some intermediate temperature the collisions are just right to excite large numbers of electrons into the second orbit. With many atoms excited to the second orbit, the gas absorbs Balmer-wavelength photons well and thus produces strong Balmer lines.

To summarize, the strength of the Balmer lines depends on the temperature of the star's surface layers. Both hot and cool stars have weak Balmer lines, but medium-temperature stars have strong Balmer lines.

Theoretical calculation can predict just how strong the Balmer lines should be for stars of various temperatures. The details of these calculations are not important to us, but the results are. Figure 4-14 shows the strength of the Balmer lines for various stellar temperatures. We could use this as a temperature indicator except that the curve gives us two answers. A star with Balmer lines of a certain strength might have either of two temperatures, one high and one low. We must examine other spectral lines to choose the correct temperature.

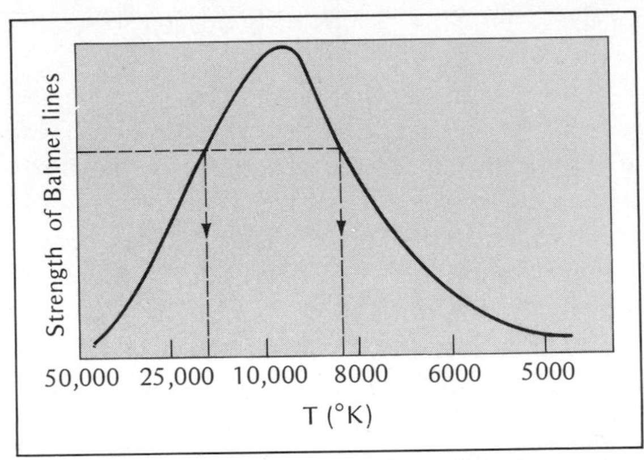


*Figure 4-14. The strength of the Balmer lines in a stellar spectrum gives us two possible temperatures for the star.*

We have seen how the strength of the Balmer lines depends on temperature. The same process affects the spectral lines of other elements, but the temperature at which they reach maximum strength differs for each element. If we add these elements to our graph, we get a handy tool for taking the stars' temperatures (Figure 4-15).

We can determine a star's temperature by comparing the strengths of its spectral lines with our graph. For instance, if we photographed a spectrum of a star and found medium-strength Balmer lines and strong helium lines, we could conclude it had a temperature of about 20,000°K. But if the star had weak hydrogen lines and strong lines of ionized iron, we would assign it a temperature of about 5800°K, similar to the sun.

The spectra of the coolest stars contain dark bands produced by molecules such as titanium oxide (TiO). Because of their structure, molecules can absorb photons at many wavelengths, producing numerous, closely spaced spectral

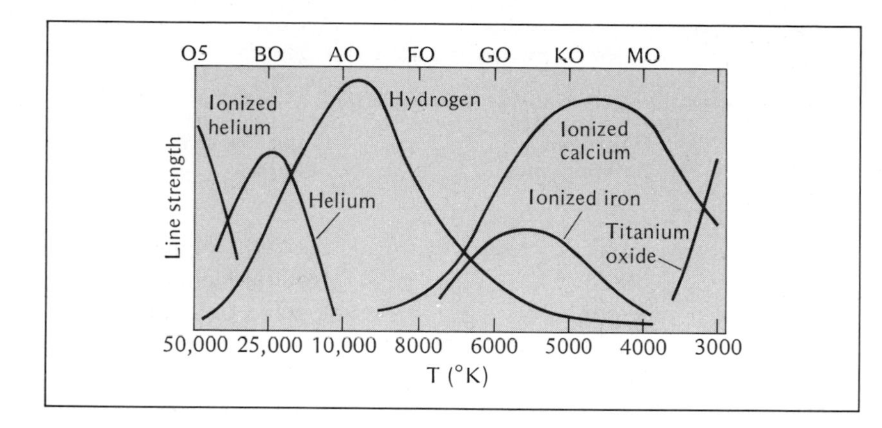


*Figure 4-15. The strengths of spectral lines can tell us the temperature of a star. This relationship is the basis for the spectral classification system.*

lines that blend together to form bands. These molecular bands appear only in the spectra of the coolest stars because, as mentioned before, only there can molecules avoid the collisions that would break them up in hotter stars. Thus the presence of dark bands in a star's spectrum indicates that the star is very cool.

By analyzing stellar spectra, astronomers have found that the hottest stars have surface temperatures above 40,000°K. The coolest have temperatures of about 2000°K. Compare these with the surface temperature of the sun, about 5800°K.

*Spectral Classification.* We have seen that the strengths of spectral lines depend on the surface temperature of the star. From this we can predict that all stars of a given temperature should have similar spectra. If we learn to recognize the pattern of spectral lines produced by a 6000°K star, for instance, we need not use Figure 4-15 every time we see that kind of spectrum. In other words, we can save time by classifying stellar spectra rather than analyzing each one individually.

The first useful classification system was made by Annie J. Cannon, an astronomer at the Harvard College Observatory during the early 1900s. She studied spectra of 200,000 stars and classified them according to the appearance of their spectral lines. She first labeled the classes alphabetically but later rearranged their order and discarded unnecessary classes, arriving at the seven **spectral classes,** or **types,** still used today: O, B, A, F, G, K, M.

This sequence of spectral types, called the **spectral sequence,** is important because it is a temperature sequence. The O stars are the hottest, the B stars next hottest, and so on. The temperature continues to decrease down to the M stars, the coolest of all.

We can classify a star by examining features in its spectrum, as described in Table 4-1. For example, if it has weak Balmer lines and lines of ionized helium, it must be an O star. This table is based on the same information we used in Figure 4-15.

The spectra shown in Figure 4-16 illustrate how spectral features change from class to class. Note that the Balmer lines are strongest in A stars where the temperature is moderate but still high enough to excite the electrons in hydrogen atoms to the second orbit, where they can absorb Balmer-wavelength photons. In the hotter stars (O and B) the Balmer lines are weak because the higher temperature excites the electrons to orbits above the second. The Balmer lines in cooler stars (F through M) are also weak but for a different reason. The lower temperature cannot excite many electrons to the second orbit, so few hydrogen atoms are capable of absorbing Balmer-wavelength photons.

**Table 4-1 Spectral Classes**

| Spectral Classes | T (°K) | Hydrogen Balmer Lines | Other Spectral Features |
|---|---|---|---|
| O | 40,000 | Weak | Ionized Helium |
| B | 20,000 | Medium | Neutral Helium |
| A | 10,000 | Strong | Ionized Calcium Weak |
| F | 7,500 | Medium | Ionized Calcium Weak |
| G | 5,500 | Weak | Ionized Calcium Medium |
| K | 4,500 | Very Weak | Ionized Calcium Strong |
| M | 3,000 | Very Weak | TiO strong |

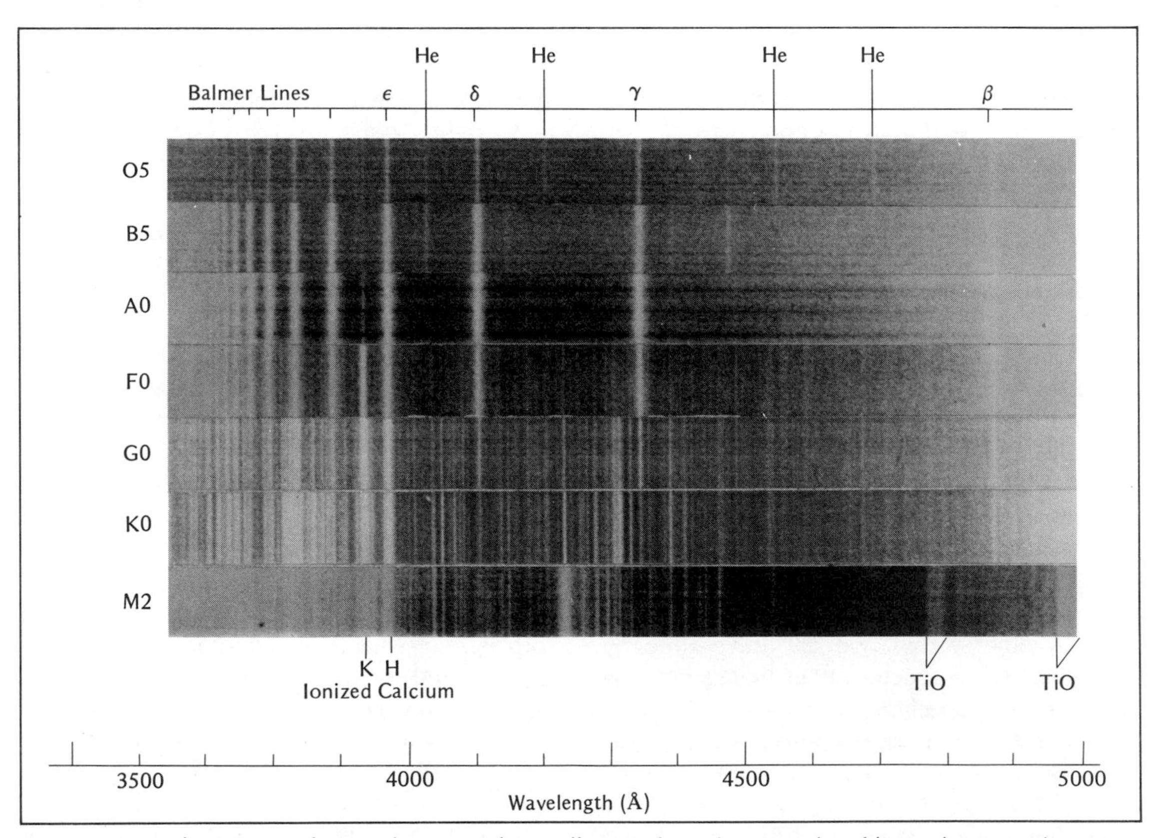


*Figure 4-16. The spectra of stars of various classes illustrate how the strengths of lines change with temperature. (Adapted from H. A. Abt, A. B. Meinel, W. W. Morgan, and J. W. Tapscott,* An Atlas of Low-Dispersion Grating Stellar Spectra, *Kitt Peak National Observatory, 1968.)*

The spectral lines of other atoms also change from class to class. Helium is visible only in the spectra of the hottest classes, and the titanium oxide bands only in the coolest. The two lines of ionized calcium, labeled **H** and **K,** increase in strength from A to K and then decrease from K to M. Because the strength of these spectral features depends on temperature, it requires only a few minutes to compare a star's spectrum with Table 4-1 or Figure 4-16 and determine its temperature.

We can be more precise if we divide each spectral class into 10 subclasses. For example, spectral class A consists of the subclasses A0, A1, A2 . . . A8, A9. Next comes F0, F1, F2, and so on. This finer division, of course, demands that we look carefully at a spectrum, but it is worth the effort, for the subclasses give us a star's temperature with an accuracy of about 5 percent. The sun, for example, is not just a G star, but a G2 star, with a temperature of about 5800°K.

A stellar spectrum can tell us many things besides temperature. Box 4-3 describes how a spectrum can reveal a star's motion. Lines in the spectrum of a star coming toward earth are shifted slightly toward the blue end of the spectrum, and lines in the spectrum of a star moving away from earth are shifted slightly toward the red. This effect, called the **Doppler effect,** can tell us the star's velocity with respect to the earth.

*Composition.* On earth many branches of science and industry use spectrographs to analyze samples to determine their chemical composition. Astronomers too can use spectra in this way to tell which chemical elements are present in a celestial body in what proportion.

Identifying the elements in a star by identifying the lines in the star's spectrum is relatively straightforward. For example, two dark absorption lines appear in the yellow region of the solar spectrum at the wavelengths 5890 Å and 5896 Å. The only atom that can produce this pair of lines is sodium, so we must conclude that the sun contains sodium. Over 90 elements have been found this way in the sun.

However, just because the spectral lines characteristic of an element are missing, we cannot conclude that the element itself is absent. For example, the hydrogen Balmer lines are weak in the sun's spectrum, yet the sun is almost 80 percent hydrogen by mass. The reason for this apparent paradox is that the sun is too cool to produce strong Balmer lines. Similarly, an element's spectral lines may be absent from a star's spectrum because the star is too hot or too cool to excite those atoms to the orbits that produce visible spectral lines.

Detailed spectral analysis taking the star's temperature into consideration can reveal the abundance of the chemical elements in the star. The results of such studies show that nearly all stars have compositions similar to the sun's—about 73 percent hydrogen, 25 percent helium, and small traces of heavier elements (see Table 4-2).

**Table 4-2 The Most Abundant Elements in the Sun**

| Element | Percentage by Number of Atoms | Percentage by Mass |
|---|---|---|
| Hydrogen | 92.0 | 73.4 |
| Helium | 7.8 | 25.0 |
| Carbon | 0.03 | 0.3 |
| Nitrogen | 0.008 | 0.1 |
| Oxygen | 0.06 | 0.8 |
| Neon | 0.008 | 0.1 |
| Magnesium | 0.002 | 0.05 |
| Silicon | 0.003 | 0.07 |
| Sulfur | 0.002 | 0.04 |
| Iron | 0.004 | 0.2 |

Source: Adapted from C. W. Allen, *Astrophysical Quantities,* London: The Athlone Press, 1976.

This chapter has assembled a set of powerful tools for the analysis of starlight. In the next chapter we will begin using them to determine the properties of stars. However, the sun is an average sort of star, so we can begin our study of stars with the sun. We have already discussed the sun's composition. Now we can consider the nature of its atmosphere.

## PERSPECTIVE: THE FLASH SPECTRUM

The **flash spectrum** occurs during a total eclipse. Just as the moon covers the sun completely, an astronomer looking through a spectroscope sees the normal solar absorption spectrum suddenly flash into a beautifully colored emission spectrum. The flash spectrum remains

### BOX 4-3 THE DOPPLER EFFECT

The **Doppler effect** is the change in wavelength of radiation due to the relative motion of the source and observer. To see how this works, imagine standing on a railroad track as a train approaches with the engine bell ringing once each second (Figure 4-17). When the bell rings, the sound travels ahead of the engine to reach your ears. One second later the bell rings again, but not at the same place. During that one second the engine moved closer to you, so the bell is closer at its second ringing. Now the sound has a shorter distance to travel and reaches your ears a little sooner than it would have if the engine had not moved. The third time the bell rings it is even closer. By timing the ringing of the bell, you would observe that the bell seemed to be ringing more often than once each second, all because the engine was approaching.

Standing behind the engine would give the opposite effect. You would find that each successive ring takes place farther away from you and the rings would sound more than one second apart. These apparent changes in the rate of the ringing bell is an example of the Doppler effect.

We can think of the peaks of the electromagnetic waves leaving a star as a series of clangs from a bell. If a star is moving toward us, we see the peaks of the light waves closer together than expected, making the wavelengths slightly shorter than they would have been if the star were not moving. If the star is going away from us, the peaks of the light waves are slightly farther apart and the wavelengths are longer. Thus the lines in a star's spectrum are shifted slightly toward the blue if the star is approaching, and toward the red if it is receding.

For convenience, we have assumed that the earth is standing still and the star is moving, but the Doppler effect depends only on relative motion. Thus we cannot say that either earth or star is stationary—only that there is relative motion between them. In addition, the Doppler effect depends only on the **radial velocity,** that part of the velocity directed away from or toward earth. The Doppler effect cannot reveal relative motion to right or left.

How much the spectral lines change depends on the radial velocity. This can be expressed as a simple ratio relating the radial velocity $\mathbf{V_r}$, divided by the speed of light $\mathbf{c}$, to the change in wavelength $\mathbf{\Delta\lambda}$ divided by the unshifted wavelength $\boldsymbol{\lambda_0}$.

$$\frac{V_r}{c} = \frac{\Delta\lambda}{\lambda_0}$$

This expression is quite accurate for the low velocities of stars, but we will need a better version later when we discuss objects moving with very high velocities.

Suppose we observe a line in a star's spectrum with a wavelength of 6001 Å. Laboratory measurements show that the line should have a wavelength of 6000 Å. That is, its unshifted wavelength is 6000 Å. What is the star's radial velocity? First we note that the change in wavelength is 1 Å.

$$\frac{V_r}{c} = \frac{1}{6000} = 0.000167$$

visible for only a few seconds, during which time the lines fade and change.

The flash spectrum is actually the spectrum of the chromosphere, the thin layer of the sun's atmosphere about 10,000 km thick located between the bright photosphere and the corona (Figure 4-18a). (Compare the chromosphere's thickness with the earth's diameter of 12,600 km.) This layer is responsible for the brilliantly colored emission lines in the flash spectrum. In fact, chromosphere comes from the Greek word *chroma,* meaning color.

*Observing the Flash Spectrum.* If astronomers observe the spectrum of the sun during a total solar eclipse, they see the usual ab-

Multiplying by the speed of light, $3 \times 10^5$ km/sec, gives the radial velocity, 50 km/sec. Since the wavelength is shifted to the red (lengthened), the star must be receding from us.

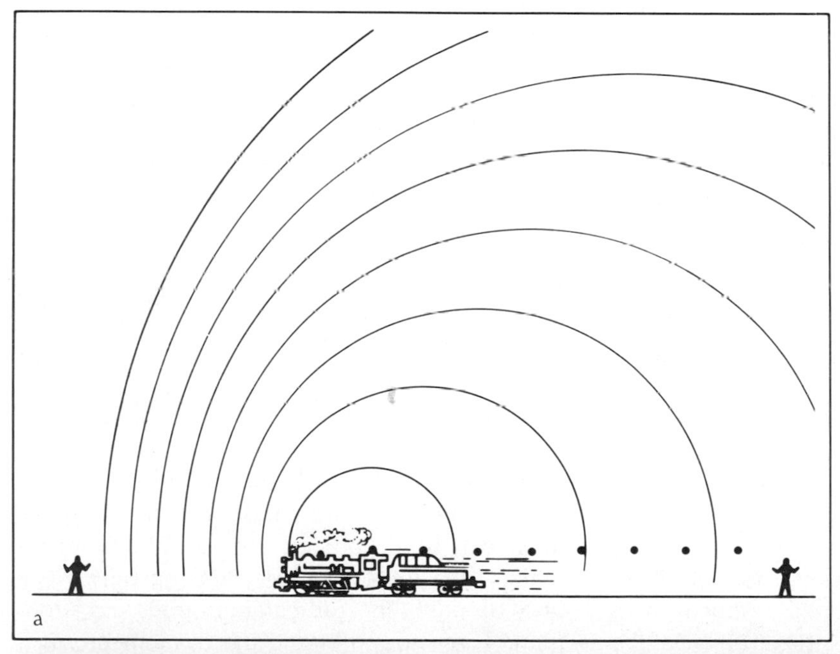


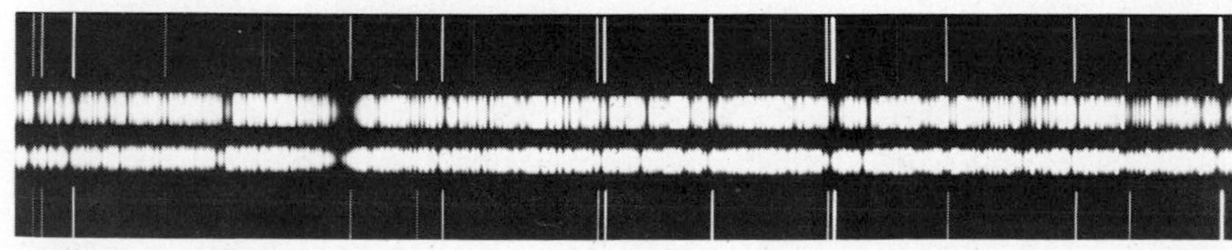


*Figure 4-17.* (a) *Successive clangs of the engine bell (marked by dots) occur closer to the observer ahead, decreasing the distance the sound must travel. Thus the observer hears the bell ring more often than it really does. The observer behind the train hears the bell ring less often. This is an example of the Doppler effect.* (b) *The upper spectrum of Arcturus was taken when earth's orbital motion carried it toward the star. The lower spectrum was taken six months later when earth was receding from Arcturus. The difference in the wavelengths of the lines is due to the Doppler shift. (Hale Observatories.)*

*Figure 4-18.* (a) *The chromosphere is a thin layer (exaggerated here) between the photosphere and corona.* (b) *A thin crescent of chromosphere is visible for a few seconds as the moon covers the photosphere during a solar eclipse.*

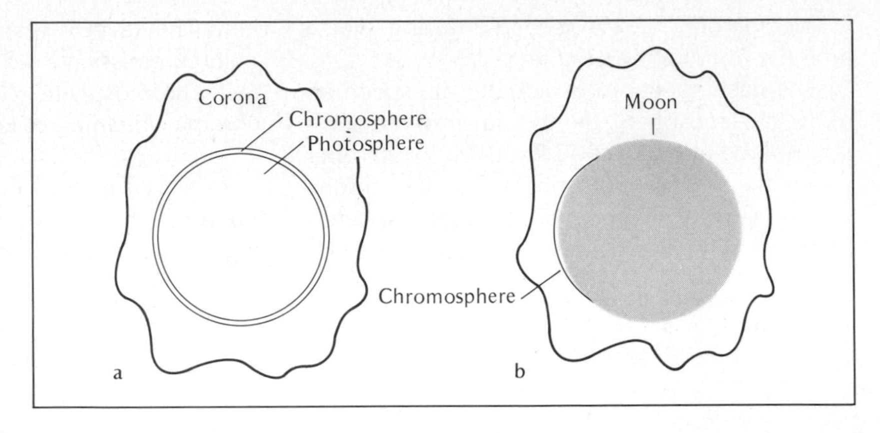


sorption spectrum so long as any part of the photosphere is visible. This photosphere, only a few hundred kilometers thick, is the layer from which most of the sunlight we see originates. Deeper layers are hotter and denser and the atoms emit a great many photons, but most are reabsorbed and do not escape from the sun. Higher in the sun's atmosphere, above the photosphere, the gas is less dense and emits fewer photons than the photosphere. Thus the photosphere is the layer in the sun that is hot enough and dense enough to emit many photons, yet transparent enough for the photons to escape.

Within the photosphere the photons interact with the particles of the gas so often the radiation is in equilibrium with the gas. This means the wavelength distribution of the radiation depends, not on the composition of the gas, but only on its temperature. Instead of producing a spectrum of the gas, it produces a continuous spectrum like that emitted by a black body (Box 3-1). This is similar to the hot filament of a light bulb, which does not produce a spectrum of tungsten, but a continuous spectrum. As the black body radiation passes through the gas of the photosphere and higher layers, gas atoms absorb photons with specific wavelengths, producing dark absorption lines. Thus the spectrum of the photosphere as seen from the earth is an absorption spectrum.

When the moon covers the photosphere during a total solar eclipse, the chromosphere is visible extending beyond the edge of the moon in a thin crescent (Figure 4-18b). This hot, low-density gas produces a faint emission spectrum, so at the moment when the moon conceals the last of the brilliant photosphere, the solar absorption spectrum suddenly flashes into an emission spectrum (Figure 4-19). The lines of this flash spectrum curve because the spectrograph does not contain a slit to isolate the light, but merely uses the narrow crescent of chromosphere.

The lines in the emission spectrum are the same as those in the solar absorption spectrum, with the addition of the lines of un-ionized helium, which is normally invisible in the sun's spectrum.

As the moon covers more and more of the chromosphere, the lines change. Balmer and neutral helium lines disappear, as do lines of sodium and calcium, and only the lines of ionized helium, ionized iron, and titanium remain. Then these too fade, leaving only weak lines of highly ionized calcium, iron, and strontium. These too fade and the flash spectrum is gone only seconds after it appeared.

*Analysis of the Flash Spectrum.* The flash spectrum is filled with information about the structure of the sun's atmosphere. The appearance of the emission lines, their strengths, and the way they fade away as the chromo-

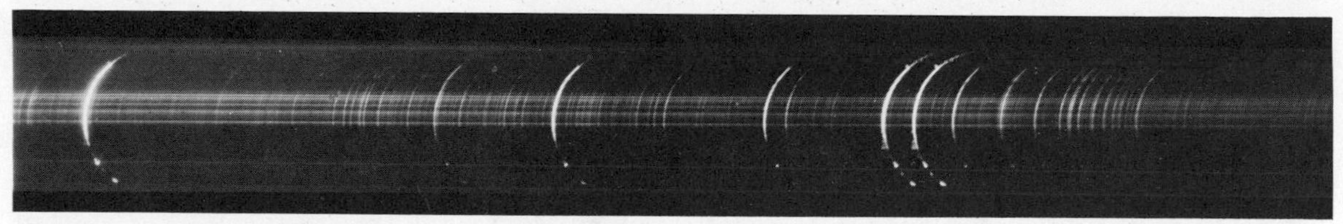

*Figure 4-19. The flash spectrum of the chromosphere shows emission lines of high ionization. The lines are curved because the thin crescent of chromosphere acts as the slit of the spectrograph. (Hale Observatories.)*

sphere is covered can tell us many things if we can untangle the clues.

The presence of emission lines in the flash spectrum is itself important. Recall from Figure 4-7 our example of a gas cloud surrounding a light bulb. There we saw an absorption spectrum when we looked at the bulb through the gas, and an emission spectrum when we looked at the gas alone. In the case of the sun, the photosphere plays the role of the glowing filament, and the chromosphere plays the part of the gas cloud. The atoms in the chromosphere absorb photons as they leave the photosphere below. This forms the absorption lines we see when we look at the photosphere. But these same atoms emit photons in random directions. When the moon blots out the brighter photosphere, we can see the fainter light emitted by the atoms of the chromosphere. Thus we see emission lines where we saw absorption lines before. This accounts for the beautiful color of the chromosphere—it is glowing like a giant neon sign. (See Color Plate 1.)

When the flash spectrum first appears, we receive light from the top, middle, and bottom of the chromosphere. Since the bottom is brightest, it dominates the spectrum. In the spectrum of this lowest layer, we see the Balmer lines plus lines of neutral helium. In order for neutral helium to radiate, the temperature must be at least 10,000°K. But if it is much higher, the Balmer lines would be weak since the hydrogen atoms would be excited to higher electron orbits. So the temperature of the lowest layers of the chromosphere must be about 10,000°K.

When the moon conceals these lower layers, we can see the emission of the middle layers dominating the flash spectrum. The Balmer lines start to fade and lines of ionized helium appear. To ionize helium the temperature must be at least 20,000°K, so this tells us the temperature at this layer. To confirm our estimate of the temperature, we also find lines of ionized iron and titanium which require high temperatures.

When the moon moves on and covers all but the top of the chromosphere, we see weak lines of very highly ionized atoms such as calcium, iron, and strontium. One line, for example, is produced by iron atoms that have lost 13 electrons. The temperature must be very high indeed to produce such extreme ionization.

*The Sun's Atmosphere.* Working from these data, solar astronomers can determine the temperature at each layer in the sun's atmosphere (Figure 4-20). At the photosphere the temperature is about 5800°K. It decreases slightly as we go up through the chromosphere, reaching a minimum of about 4000°K only a few hundred kilometers above the photosphere. Above that the temperature increases rapidly to 1,000,000°K at a height of 10,000 km, the beginning of the corona.

The flash spectrum can even tell us how dense the sun's atmosphere is. The emission lines fade away not only because the temperature increases with height, but also because the density of the gas decreases. Near the photosphere the gas is rather dense, only about $10^4$ times thinner than the air we breathe, but at the top of the chromosphere it is nearly a vacuum, about $10^{13}$ times thinner than air.

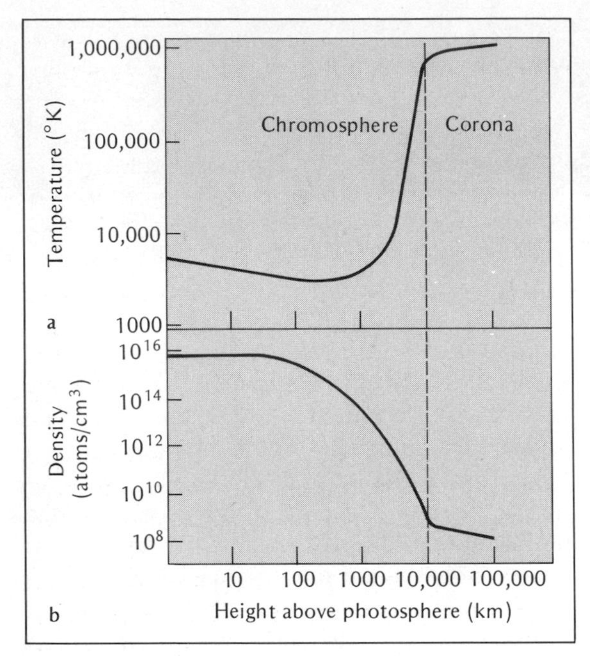


*Figure 4-20. The temperature increases with height in the chromosphere, and the density of the gas decreases.*

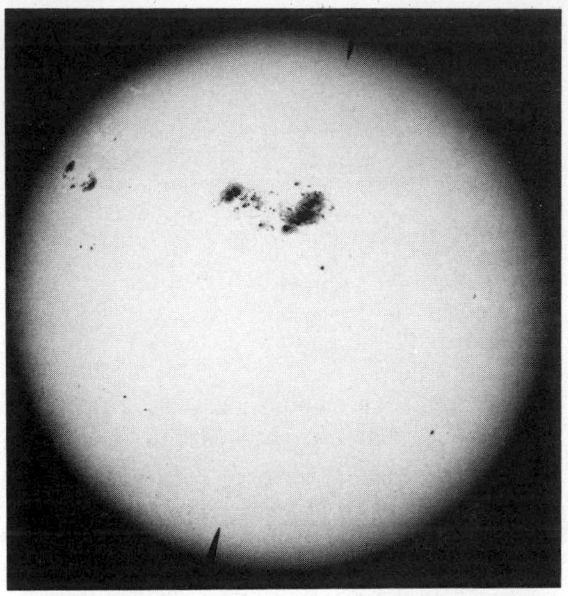

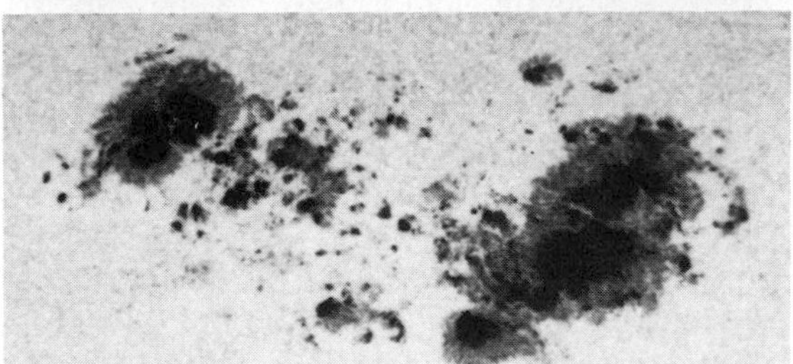

*Figure 4-21. Sunspots look like dark areas on the solar disk. They are known to be intense magnetic storms related to the sun's surface activity. This unusually large group of spots appeared in 1947. (Hale Observatories.)*

Although the flash spectrum tells us how the temperature and density change in the chromosphere, it doesn't explain why the chromosphere and corona are so hot. The best theory supposes that sound waves, produced by gas motions below the photosphere, rise up through the chromosphere. As the sound waves reach the lower-density gas, they become sonic booms that agitate the atoms and thus heat the chromosphere and corona.

In the very hot corona, some atoms and ions move so rapidly they escape the sun's gravity and stream away in a flow of thin gas called the **solar wind.** This breeze blows past the earth at a speed of 500 km/sec and continues on into space. If we consider the solar wind an extension of the corona, then the earth orbits inside the sun's atmosphere.

Scientists are now searching for relationships between the solar wind and climatic changes on earth. They have known for two centuries that the sun is not perfectly stable. **Sunspots,** dark blemishes that appear and disappear from the solar surface, are now known to be magnetic storms reflecting the sun's activity (Figure 4-21). The number of sunspots varies with a period of 11 years, indicating that the sun is subject to an 11-year cycle of activity. Earth satellites have shown that the solar wind is also affected by the 11-year cycle, and that may have important effects on our planet.

Unfortunately, we can't directly observe the chromosphere and corona of other stars. However, a few stars have spectral lines that are very broad, showing that the atoms are moving so fast that some must escape from the star. This produces a large, hot corona flowing outward as a stellar wind much stronger than the solar wind. Observations made by orbiting telescopes show that many stars are sources of ultraviolet emission lines and that some are emitting x-rays—phenomena that could only originate in hot chromospheres and coronas. These data suggest that the structure of the sun's atmosphere is not unusual, and that many stars have similar chromospheres, coronas, and stellar winds.

This chapter has discussed the kinds of information we can get from spectra of the sun and stars. In the next chapter, we will consider other sources of data about stars, and begin assembling the information we need to understand the life history of a star.

## SUMMARY

An atom consists of a nucleus surrounded by a cloud of electrons. The nucleus is made up of two kinds of particles: positively charged protons and uncharged neutrons. The number of protons in an atom determines which element it is. Atoms of the same element (i.e., having the same number of protons) with different numbers of neutrons are called isotopes.

The negatively charged electrons surrounding an atomic nucleus may occupy various permitted orbits. An electron may be excited to a higher orbit during a collision between atoms, or it may move from one orbit to another by absorbing or emitting a photon of the proper energy. If the energy of an absorbed photon is too large, the atom may lose an electron and become ionized.

Because only certain orbits are permitted, only photons of certain wavelengths can be absorbed or emitted. Each kind of atom has its own characteristic set of spectral lines. The hydrogen atom has the Lyman series in the ultraviolet, the Balmer series in the visual, and the Paschen series (and others) in the infrared.

In cool stars the Balmer lines are weak because atoms are not excited out of the ground state. In hot stars the Balmer lines are weak because atoms are excited to higher orbits or ionized. Only at medium temperatures are the Balmer lines strong. We can use this effect as a thermometer for determining the temperature of a star. In its simplest form this amounts to classifying the stars' spectra in the spectral sequence; O, B, A, F, G, K, M.

When a source of radiation is approaching us, we observe shorter wavelengths, and when it is receding, we observe longer wavelengths. This Doppler effect makes it possible for the astronomer to measure a star's radial velocity, that part of its velocity directed toward or away from earth.

The flash spectrum gives us a chance to study the outer atmosphere of the sun. From the way the chromospheric emission lines change and fade during a total eclipse, we can find how the temperature and density vary with height in the chromosphere. Though the photosphere has a temperature of only a few thousand degrees, the corona has a temperature of about 1,000,000°K. Astronomers may have detected the chromosphere and corona of some other stars.

## NEW TERMS

nucleus
proton
neutron
electron
isotope
ionization
ion
molecule
binding energy
permitted orbit
excited atom
ground state
continuous spectrum
absorption line
absorption spectrum (dark line spectrum)
emission line
emission spectrum (bright line spectrum)
transition
Lyman series
Balmer series
Paschen series
spectral class or type
spectral sequence
Doppler effect
radial velocity ($V_r$)
flash spectrum
solar wind
sunspot

## QUESTIONS

1. Describe the structure of a simple atom.
2. Why is the binding energy of an electron related to the size of its orbit?
3. Describe two ways an atom can become excited.
4. Draw a diagram of a light source and a cloud of gas, and explain how they produce a continuous spectrum, an absorption spectrum, and an emission spectrum.
5. Why do different atoms have different lines in their spectra?
6. Explain why the strengths of the Balmer lines depend on temperature.
7. How do hydrogen atoms produce the 21-cm radio signal?
8. Explain the similarities between Figure 4-15 and Table 4-1.
9. Why do we expect TiO bands to be weak in the spectra of all but the coolest stars?
10. Describe the chemical composition of the sun and stars.

## PROBLEMS

1. If the average atom is 5 Å in diameter, how many atoms are needed to reach 1 cm?
2. Transition A produces light with a wavelength of 5000 Å. Transition B involves twice as much energy as A. What wavelength light does it produce?
3. Where would the arrow for the 21-cm transition be located in Figure 4-10?
4. Determine the temperatures of the following stars based on their spectra. Use Figure 4-15.
   (a) Medium strength Balmer lines, strong helium lines
   (b) Medium strength Balmer lines, weak ionized calcium lines
   (c) TiO bands strong
   (d) Very weak Balmer lines, strong ionized calcium lines
5. To which spectral classes do the stars in problem 4 belong?
6. In a laboratory the Balmer alpha line has a wavelength of 6563 Å. If the line appears in a star's spectrum at 6565 Å, what is the star's radial velocity? Is it approaching or receding?
7. If the diameter of the sun is $1.4 \times 10^6$ km, what percent of the diameter is taken up by the chromosphere?

## RECOMMENDED READING

Aller, L. H. *Atoms, Stars, and Nebulae*. Cambridge, Mass.: Harvard University Press, 1971.

Boorse, H. and Motz, L. *The World of the Atom*. New York: Basic Books, 1966.

Booth, V. *The Structure of Atoms*. New York: Macmillan, 1964.

Feinberg, G. *What is the World Made of?* Garden City, N. Y.: Anchor Press/Doubleday, 1977.

Gibson, E. G. *The Quiet Sun*. Washington, D. C.: NASA SP-303, 1973.

Goldberg, L. "Ultraviolet Astronomy." *Scientific American* 220 (June 1969), p. 92.

Holzinger, J. R., and Seeds, M. A. *Laboratory Exercises in Astronomy*. Ex. 25, 28, 29. New York: Macmillan, 1976.

Marschall, L. A. "A Tale of Two Eclipses." *Sky and Telescope* 57 (Feb. 1979), p. 116.

Parker, E. N. "The Sun." *Scientific American* 233 (Sept. 1975), p. 42.

Pasachoff, J. M., Linsky, J. L., Haisch, B. M., and Boggess, A. "IUE and the Search for a Lukewarm Corona." *Sky and Telescope* 57 (May 1979), p. 438.

Pasachoff, J. M. "Our Sun." *Astronomy* 6 (Jan. 1978), p. 6. Reprinted in *Astronomy: Selected Readings,* ed. M. A. Seeds. Menlo Park, Calif.: Benjamin/Cummings, 1980.

Snow, T. P. "Ultraviolet Spectroscopy with Copernicus." *Sky and Telescope* 54 (Nov. 1977), p. 371.

Thackeray, A. D. *Astronomical Spectroscopy*. New York: Macmillan, 1961.

*The Eta Carinae Nebula, so called after the bright star that lies at its center, is in the southern constellation Carina. It is a complex region of bright and dark clouds apparently in the process of forming stars.* (*Cerro Tololo Inter-American Observatory.*)

# Chapter 5 INTRODUCING STARS

Our task is to try to understand the universe, and since the universe is filled with stars, we must discover how stars are born, live, and die. We begin our study in this chapter by gathering data about the intrinsic properties of stars—those properties inherent in the nature of the stars. In the next two chapters, we will use these data to deduce the life stories of different kinds of stars.

Unfortunately, determining a star's intrinsic properties is quite difficult. When we look at a star through a telescope, we see only a point of light that tells us nothing about the star's energy production, temperature, diameter, or mass. Because we cannot visit stars, we can only observe from earth and unravel the properties of stars through the analysis of starlight. One of the reasons astronomy is interesting is that it contains so many such puzzles, each demanding a different method of solution.

To simplify our task in this chapter, we will concentrate on three intrinsic stellar properties. Our goals will be to find out how much energy stars emit, how large stars are, and how much mass they contain. These three parameters, combined with stellar temperatures—an intrinsic stellar property discussed in the preceding chapter—will give us an overview of the nature of stars and provide us with the data we need to consider the lives of stars.

Although we begin with three goals firmly in mind, we immediately meet a short detour. To find out how much energy a star emits, we must know how far away it is. If at night we see bright lights approaching on the highway, we cannot tell whether the lights are the intrinsically bright headlights of a distant truck or the intrinsically faint lights on a pair of nearby bicycles. Only when we know the distance to the lights can we judge their intrinsic brightness. In the same way, to find the intrinsic brightness of a star, and thus the amount of energy it emits, we must know its distance. Our short detour will provide us with a method of measuring stellar distances.

Having reached our three goals, we will pause to consider the densities of stars, an intrinsic property that is easily determined once we know a star's size and mass. The densities of different kinds of stars will be helpful when we consider the internal structure of stars in the next two chapters.

We will conclude this chapter by considering the frequency of stellar types—that is, which kinds of stars are common and which are rare. That, like all of the data in this chapter, is aimed at helping us understand what stars are.

## MEASURING THE DISTANCES TO STARS

Determining the distance to a star is difficult because astronomers cannot journey to the star. They must, instead, measure the distance indirectly, much as surveyors measure the distance across a river they cannot cross. We will begin by reviewing this method and then apply it to stars.

*The Surveyor's Method.* To measure the distance across a river, a team of surveyors begins by driving two stakes into the ground. The distance between the stakes is the baseline of the measurement. The surveyors then choose a landmark on the opposite side of the river, a tree perhaps, thus establishing a large triangle marked by the two stakes and the tree. Using their surveyor's instruments they sight the tree from the two ends of the baseline and measure the two angles on their side of the river (Figure 5-1).

Knowing two angles and the length of the side between them, the surveyors can find the distance across the river by trigonometry or by constructing a scale drawing. For example, if the baseline was 50 m and the angles were 66° and 71°, they could draw a line 50 mm long to represent the baseline. Using a protractor, they could construct angles of 66° and 71° at each end of the baseline, and then extend the two sides until they met at C, the location of the tree. Measuring the height of the triangle in the drawing, they would find it was 64 mm high, and thus conclude that the distance across the river to the tree was 64 m.

*The Astronomer's Method.* To find the distance to a star we must use a very long baseline, the diameter of earth's orbit. If we took a photograph of a nearby star and then waited six months, the earth would have moved halfway around its orbit. We could then take another photograph of the star at a point in space 2 AU (astronomical units) from the point where the first photograph was taken. Thus our baseline would equal the diameter of earth's orbit, or 2 AU.

We would then have two photographs of the same part of the sky taken from slightly different locations in space. If we examined the photographs, we would discover that the star was not in exactly the same place in the two photographs. This apparent shift in the position of the star is called parallax (Figure 5-2).

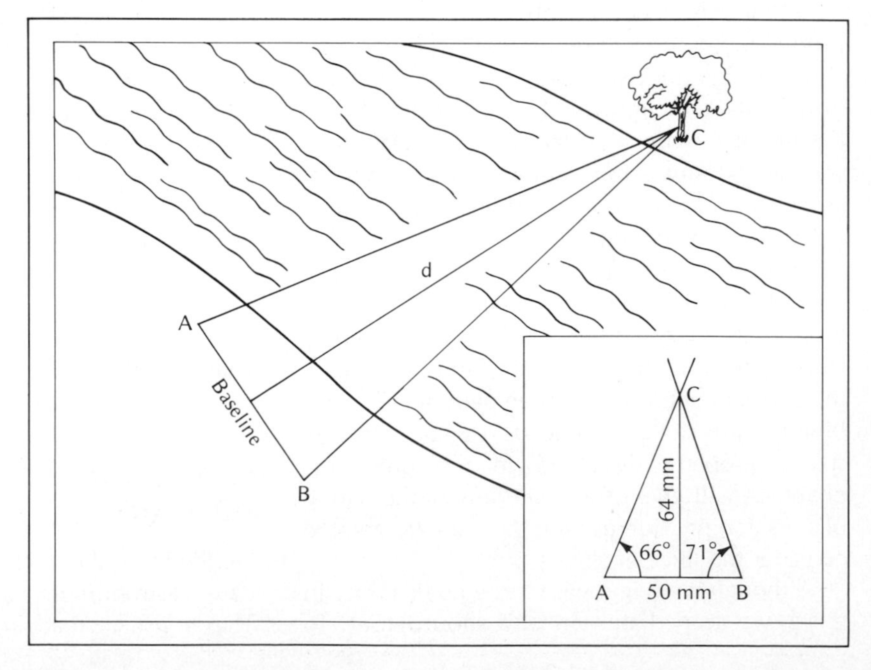


*Figure 5-1. Surveyors can find the distance* **d** *across the river by measuring the baseline and the angles* **A** *and* **B** *and then constructing a scale drawing of the triangle.*

**Parallax** is the apparent change in the position of an object due to a change in the location of the observer. It is actually an old friend, though you may not have known its name, for you use parallax to judge the distance to things. To see how this works, close your right eye and use your thumb, held at arm's length, to cover some distant object, a building perhaps. Now look with your right eye. Your thumb seems to move to the left, uncovering the building (Figure 5-3). This apparent shift is parallax, and your brain uses it to estimate distances to objects around you.

The size of the shift depends on the distance. If you hold your thumb close to your eyes, the shift is large; the farther away you hold your thumb, the smaller the shift. By measuring the parallax of an object we can find its distance.

Because the parallax of a star is such a small angle, we express it in seconds of arc. The quantity that astronomers call stellar parallax **p** is half the total shift of the star, as shown in Figure 5-2. Astronomers measure the parallax and surveyors measure the angles at the ends of the baseline, but both measurements tell us the same thing—the shape of the triangle and thus the distance to the object in question.

Measuring the small angle **p** is very difficult. The nearest star, $\alpha$ Centauri, has a parallax of only 0.76 seconds of arc, and the more distant stars have even smaller parallaxes. To see how small these angles are, hold a piece of paper edgewise at arm's length. The thickness of the

*Figure 5-3. To demonstrate parallax, close one eye and cover a distant object with your thumb held at arm's length. Look with the other eye and your thumb appears to have shifted position.*

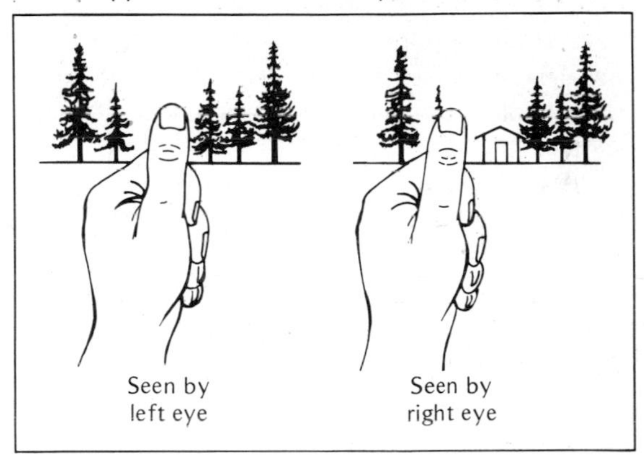


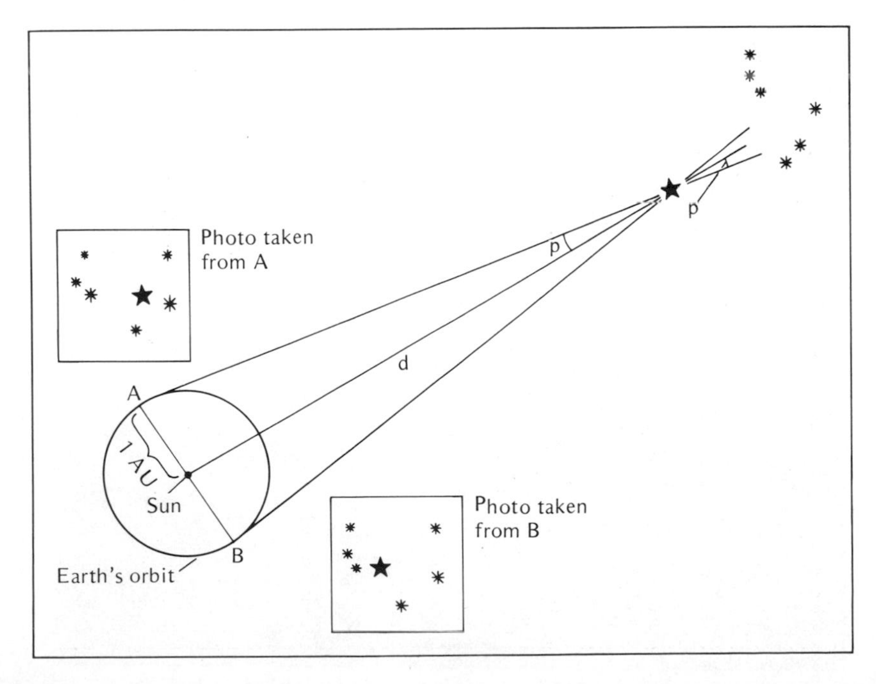


*Figure 5-2. We can measure a nearby star's parallax by photographing it from two points along the earth's orbit, **A** and **B.** Half of the star's total change in position from one photograph to the other is its parallax **p.***

paper covers an angle of about 30 seconds of arc.

We cannot use a scale drawing to find the distance to a star because the distance is so large and the angle is so small. Even for the nearest star, the triangle would have to be 300,000 times longer than it was wide. If the baseline in our drawing was 1 in, the triangle would have to be about 5 mi long. Box 5-1 describes how we could find the distance from the parallax without drawing scale triangles.

The distances to the stars are so large it is not convenient to use astronomical units. As Box 5-1 explains, when we measure distance via parallax, it is convenient to use the unit of distance called a **parsec** (abbreviated pc). One parsec is 206,265 AU, roughly 3.26 light-years.*

Since we can't measure **p** when it is smaller than about 0.01 seconds of arc, the farthest star whose distance we can measure by parallax is about 100 pc away. Less than 1000 stars have accurately measured parallaxes. After we know more about the different types of stars, we will find other ways to determine stellar distances.

Having found a way to determine the distances to some of the nearer stars, we are ready to discuss the first of the three stellar parameters—brightness. Our goal is to find out how much energy stars emit.

* The parsec is used throughout astronomy because it simplifies the calculation of distance. However, there are instances where the light-year is also convenient. Consequently, the chapters that follow use either parsecs or light-years as convenience and custom dictate.

## INTRINSIC BRIGHTNESS

If we view a street light from nearby, it may seem quite bright, but if we view it from a hilltop miles away, it appears faint. Its apparent brightness depends on its distance, but its intrinsic brightness, the amount of light it emits, is independent of distance. When we look at stars, we face the same problem we might face trying to judge the brightness of city lights viewed from a distant hilltop (Figure 5-4). We can judge apparent brightness easily, but unless we know the distances to individual points of light we cannot determine their intrinsic brightnesses. We could not, for instance, tell distant street lights from dimmer, but nearer light bulbs. Once an astronomer determines the distance to a star, however, it is simple to calculate its intrinsic brightness from its apparent brightness and distance.

We will use two terms to refer to a star's intrinsic brightness. One, related to the magnitude system, is common in astronomy because its use makes calculations involving distance easier. A second term refers directly to the amount of energy the star emits in one second.

*Figure 5-4. The valley below Mount Wilson sparkles with the lights of Los Angeles, Pasadena, Hollywood, and over forty other communities. Without knowing the distance to a light on the valley floor, we cannot estimate its intrinsic brightness. Similarly, we must know a star's distance before we can find its intrinsic brightness. (Hale Observatories.)*

## BOX 5-1 PARALLAX AND DISTANCE

We wish to find the distance to a star from its measured parallax. To see how this is done, imagine that we observe the earth from the star. Figure 5-5 shows that the angular distance from the sun to the earth would equal the star's parallax **p.** To find the distance, we recall that the small-angle formula in Box 1-3 relates an object's angular diameter, its linear diameter, and its distance. In this case, the angular diameter is **p** and the linear diameter is 1 AU. Then the small-angle formula, rearranged slightly, tells us that the distance to the star in AU is equal to 206,265 divided by the parallax in seconds of arc.

$$d = \frac{206{,}265}{p}$$

Because the parallaxes of even the nearest stars are less than 1 second of arc, the distances in AU are inconveniently large numbers. To keep the numbers manageable, astronomers have defined the **parsec** as their unit of distance in a way that simplifies the arithmetic. One parsec equals 206,265 AU, so the equation becomes

$$d = \frac{1}{p}$$

Thus a parsec is the distance to an imaginary star whose parallax is one second of arc.

For example, the star Altair has a parallax of 0.20 seconds of arc. Then its distance is

$$d = \frac{1}{0.2} = 5 \text{ pc}$$

Since one parsec equals about 3.26 light-years, Altair is about 16.3 light-years away.

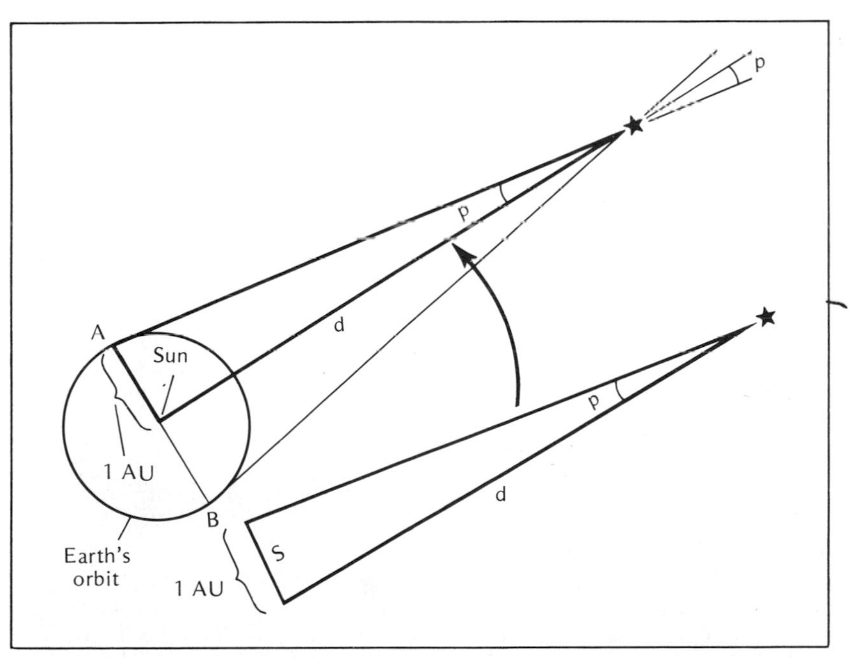


*Figure 5-5. If we were located at the star and looked back at the earth, the angular distance from the sun to the earth would equal the star's parallax. This means we can use the small angle formula (Box 1-3) to find the star's distance from its parallax.*

*Absolute Visual Magnitude.* Judging the intrinsic brightness of stars would be easier if they were all at the same distance. Though astronomers can't move stars about and line them up at some standard distance, they can calculate how bright a star of known distance would appear at any other distance. In this way, they refer to the intrinsic brightness of a star as its **absolute visual magnitude** $\mathbf{M_v}$—the apparent visual magnitude the star would have if it were 10 pc away (Box 5-2).

The symbol for absolute visual magnitude is a capital **M** with a subscript **v.** The subscript tells us it is a visual magnitude based only on the wavelengths of light we can see. Other magnitude systems are based on other parts of the electromagnetic spectrum such as the infrared, ultraviolet, and so on. Yet another magnitude system refers to the total energy emitted at all wavelengths. We will limit our discussions to visual magnitudes.

The intrinsically brightest stars known have absolute magnitudes of about −8, and the faintest about +15 or fainter. The nearest star to the sun, $\alpha$ Centauri, is only 1.4 pc away, and its apparent magnitude is 0.0, indicating that it looks bright in the sky. However, its absolute magnitude is 4.39, telling us it is not intrinsically very bright. Because we know the distance to the sun and can measure its apparent magnitude, we can find its absolute magnitude: about 4.78. If the sun were only 10 pc away from us, it would look no brighter than the faintest star in the handle of the Little Dipper.

---

BOX 5-2 ABSOLUTE MAGNITUDE

The absolute visual magnitude $\mathbf{M_v}$ of a star is the apparent visual magnitude of the star if it were 10 pc away. If we know a star's apparent magnitude and its distance, we can calculate its absolute magnitude. The equation that allows this calculation relates apparent magnitude **m,** distance in parsecs **d,** and absolute magnitude $\mathbf{M_v}$:

$$m - M_v = -5 + 5 \log_{10} (d)$$

This equation shows that the difference between apparent and absolute magnitude depends only on the distance to the star.

The quantity $\mathbf{m - M_v}$ is called the **distance modulus,** a measure of how far away the star is. If the star is very far away, the distance modulus is large; if the star is close, the distance modulus is small. We could use the equation given above to make a table of distance moduli (Table 5-1).

If we know the distance to a star, we can find its distance modulus from the table. If we subtract the distance modulus from the apparent magnitude, we get the star's absolute magnitude. For example, Deneb is 490 pc away and has an apparent magnitude of 1.26. From the table we find that its distance modulus is about 8.5. Then its absolute magnitude is about 1.26 − 8.5 or about −7.2. Deneb is intrinsically a very bright star. If it were only 10 pc away it would dominate the night sky, shining over 200 times brighter than Sirius.

**Table 5-1 Distance Moduli**

| $m - M_v$ | d (in pc) |
|---|---|
| 0 | 10 |
| 1 | 16 |
| 2 | 25 |
| 3 | 40 |
| 4 | 63 |
| 5 | 100 |
| 6 | 160 |
| 7 | 250 |
| 8 | 400 |
| 9 | 630 |
| 10 | 1000 |
| . | . |
| . | . |
| . | . |
| 15 | 10,000 |
| . | . |
| . | . |
| . | . |
| 20 | 100,000 |
| . | . |
| . | . |

*Luminosity.* The second method of sorting out the brightnesses of stars is to measure their brightnesses directly. A photometer attached to a telescope measures the amount of light received on earth from a star, and then, if the star's distance is known, we calculate the total amount of energy the star is producing. This is called the star's luminosity.

The **luminosity (L)** is the total amount of energy a star radiates in one second—not just visible light, but all wavelengths. However, the actual number of ergs per second is such a large number it has little meaning for us. Thus it is often more useful to give a star's luminosity in terms of the sun's. For example, Canopus emits 100 times more energy per second than the sun, and we say its luminosity is 100 $L_{\odot}$. The symbol $L_{\odot}$ represents the luminosity of the sun, about $4 \times 10^{33}$ ergs per second.

The range of luminosities is very large. The most luminous stars emit more than $10^5$ $L_{\odot}$, and the least luminous stars roughly $10^{-4}$ $L_{\odot}$.

Both absolute magnitude and luminosity are measures of the intrinsic brightness of a star. Thus we have reached our first goal, determining the energy output of stars. Our second goal, finding the diameters of stars, is related to stellar luminosity.

## THE DIAMETERS OF STARS

Two factors influence an object's luminosity: temperature and surface area. For example, you can eat dinner by candlelight because the candle flame has a small surface area, and consequently a small luminosity. However, if the flame were 12 ft tall, it would have a very large surface area from which to radiate, and, even though it was no hotter than a normal candle flame, its luminosity would drive you away from the table.

In a similar way, a hot star may not be very luminous if it has a small surface area, although it could be highly luminous if it were larger. Even a cool star could be luminous if it had a large surface area. Because of this dependence on both temperature and surface area, we can use stellar luminosities to determine the diameters of stars if we can separate the effects of temperature and surface area.

*The H-R Diagram.* The **Hertzsprung-Russell diagram,** named after its discoverers Ejnar Hertzsprung and Henry Norris Russell, is a graph that separates the effects of temperature and surface area on stellar luminosities and enables us to sort the stars according to their diameters. Before we discuss the details of the **H-R diagram** (as it is often called), let us look at a similar diagram we might use to sort automobiles.

We could plot a diagram such as Figure 5-6 to show horsepower versus weight for various makes of cars. We would find that in general the more a car weighs, the more horsepower it has. Most cars would fall somewhere along the sequence of cars running from heavy, high-powered cars to lightweight, low-powered models. We could call this the main sequence of cars. But some would have much more horsepower than normal for their weight—the sport or racing models—and the economy models would have less power than normal for cars of the same weight. Just as this diagram would help us understand the different kinds of autos, the H-R diagram helps us understand the different kinds of stars.

The H-R diagram relates the intrinsic brightness of stars to their surface temperatures (Figure 5-7). We may plot either absolute magnitude or luminosity on the vertical axis of the graph, since both refer to intrinsic brightness. As you will remember from Chapter 4, spectral type is related to temperature, so we may plot either spectral type or temperature on the horizontal axis. Technically only graphs of absolute magnitude versus spectral type are H-R diagrams. However, we will refer to plots of luminosity versus either spectral type or surface temperature by the generic term *H-R diagram.*

A point on an H-R diagram shows a star's luminosity and surface temperature. Points near the top of the diagram represent very luminous stars, and points near the bottom represent very

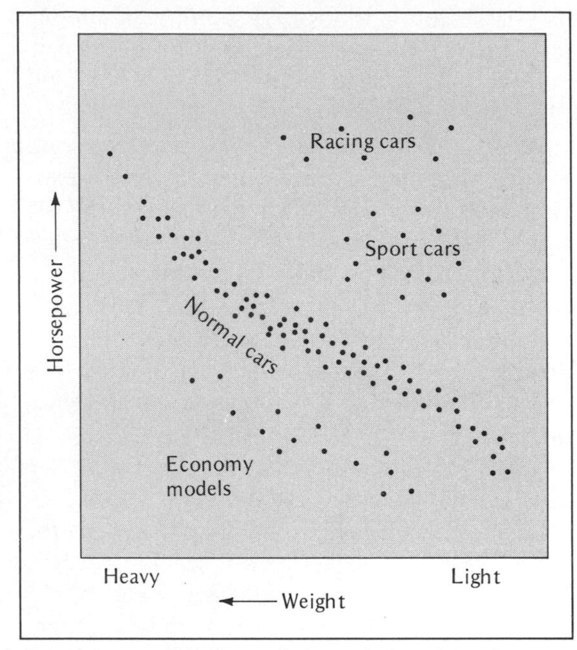


*Figure 5-6. We could analyze automobiles by plotting their horsepower versus their weight, and thus reveal relationships between various models. Most would lie somewhere along the main sequence of "normal" cars.*

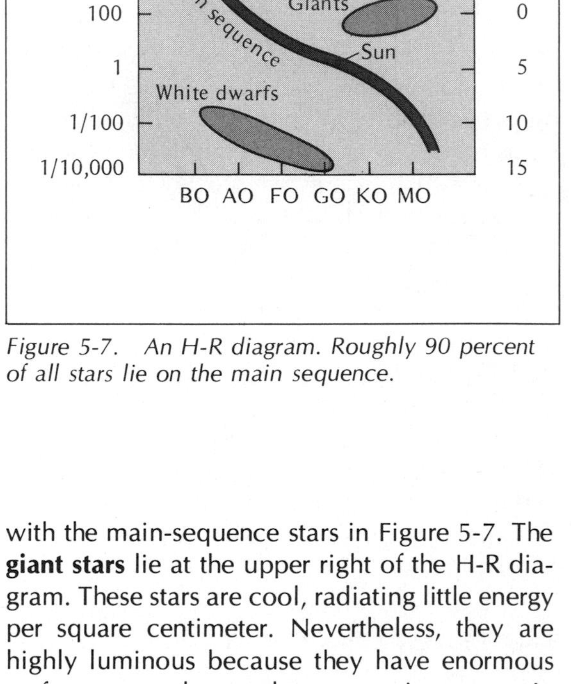


*Figure 5-7. An H-R diagram. Roughly 90 percent of all stars lie on the main sequence.*

faint stars. Points on the left represent hot stars, and points on the right represent cool stars. Notice that the location of a star in the H-R diagram has nothing to do with its location in space. Also, as a star ages, its luminosity and surface temperature change and it moves in the H-R diagram, but this has nothing to do with the star's actual motion through space.

*Dwarfs, Giants, and Supergiants.* The **main sequence** is the region of the H-R diagram running from upper left to lower right (shaded area of Figure 5-7), which includes roughly 90 percent of all stars. These are the "ordinary" stars. As we might expect, the hot main-sequence stars are brighter than the cool main-sequence stars. The sun is a medium-temperature main-sequence star.

Just as sports cars do not fit in with the normal cars in Figure 5-6, some stars do not fit in with the main-sequence stars in Figure 5-7. The **giant stars** lie at the upper right of the H-R diagram. These stars are cool, radiating little energy per square centimeter. Nevertheless, they are highly luminous because they have enormous surface areas; hence the name *giant stars.* In fact, we can estimate the size of these giants with a simple calculation. Notice from the H-R diagram that they are about 100 times more luminous than the sun even though they have about the same surface temperature. Thus they must have about 100 times more surface area than the sun, indicating that their diameters must be about 10 times larger than the sun's (Figure 5-8).

Near the top of the H-R diagram we find a few stars called **supergiants.** These exceptionally luminous stars are 10 to 1000 times the sun's diameter. The stars Betelgeuse and Rigel ($\alpha$ and $\beta$ Orionis) are both supergiants. The star $\epsilon$

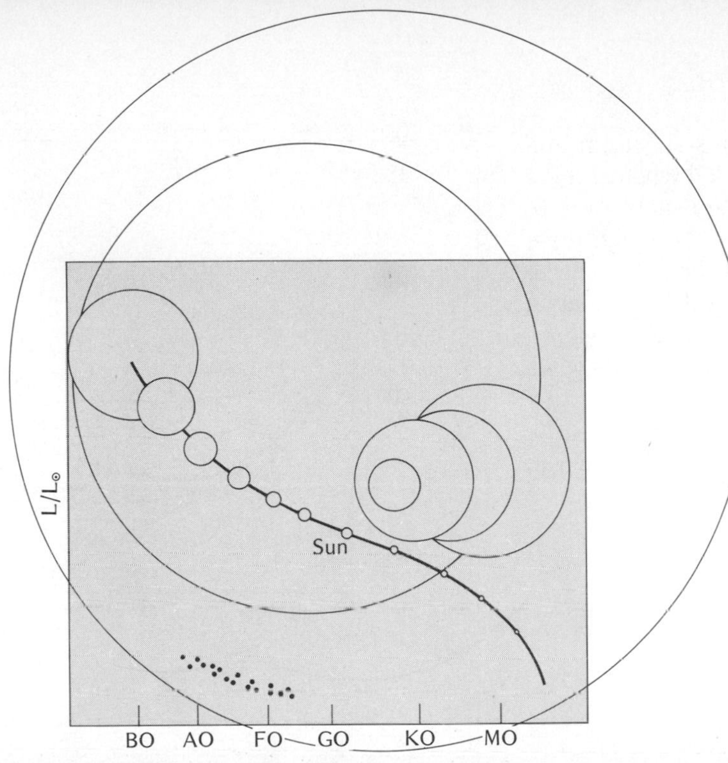


*Figure 5-8.* *An H-R diagram showing the relative sizes of stars. The giant and supergiant stars are much larger than the sun. The white dwarf stars are about the size of the earth. On this scale, the largest supergiants would be over 1 m in diameter.*

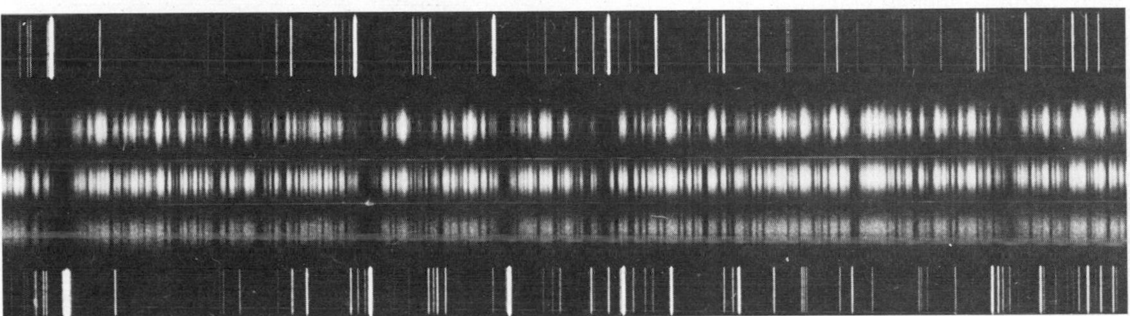

*Figure 5-9.* *Slight differences in the widths and strengths of spectral lines distinguish the spectra of supergiants, giants, and main-sequence stars, thus making the luminosity classification possible. (Lick Observatory photograph.)*

Aurigae consists of a pair of stars orbiting each other, one of which may be as large as 2800 times the sun's diameter. If such a supergiant magically replaced the sun at the center of our solar system, it would swallow up Mercury, Venus, Earth, Mars, Jupiter, and Saturn.

In the lower left of the H-R diagram are the economy models, stars that are very faint even though they are hot. Clearly such stars must be small. These are the **white dwarf stars,** dying stars that have collapsed to about the size of the earth and are slowly cooling off.

*Luminosity Classification.* We can tell from a star's spectrum what kind of star it is. Main-sequence stars are relatively small and have dense atmospheres. The gas atoms collide with each other often and the electron energy levels become distorted, smearing the spectral lines. On the other hand, giant stars are larger, their atmospheres are less dense, and the atoms disturb each other relatively little (Figure 5-9). The lines in the spectra of giant stars are sharp, and the lines in the spectra of supergiants are even sharper.

Thus we can look at a star's spectrum and classify its luminosity. We can tell whether it is a supergiant, a bright or ordinary giant, a subgiant, or a main-sequence star. Although these are the **luminosity classes,** the names refer to the sizes of the stars because size is the dominating factor in determining luminosity. Supergiants, for example, are very bright because they are very large.

These luminosity classes are represented by the roman numerals I through V, as shown in this list.

**Luminosity Classes**

| | |
|---|---|
| Ia | Bright Supergiant |
| Ib | Supergiant |
| II | Bright Giant |
| III | Giant |
| IV | Subgiant |
| V | Main-Sequence Star |

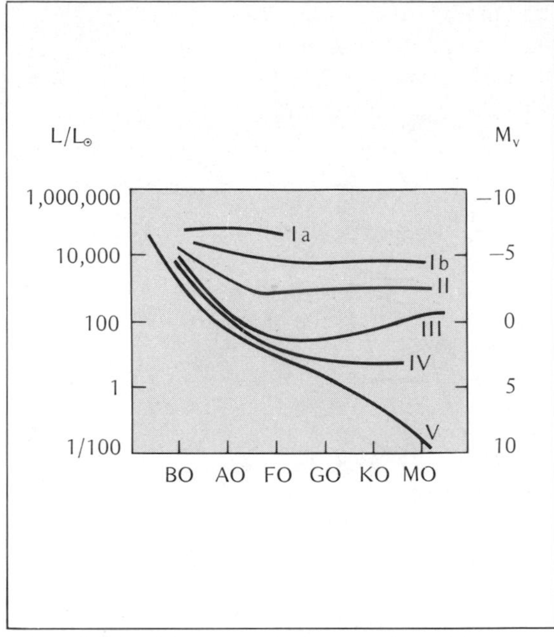


*Figure 5-10. The approximate location of the luminosity classes on the H-R diagram.*

Using letters for subclasses, we can distinguish between the bright supergiants (Ia) such as Rigel ($\beta$ Orionis) and the regular supergiants (Ib) such as Polaris, the North Star. The star Adhara ($\epsilon$ Canis Majoris) is a bright giant (II), Capella ($\alpha$ Aurigae) is a giant (III), and Altair ($\alpha$ Aquilae) is a subgiant (IV). The sun is a main-sequence star (V). The luminosity class usually appears after the spectral type, as in G2 V for the sun. White dwarf stars don't enter into this classification because their spectra are quite unlike those of other stars.

We can, as in Figure 5-10, plot the locations of the luminosity classes on the H-R diagram. Remember that these are rather broad classifications. A star of luminosity class III may lie slightly above or below the line labeled III. The lines are only approximate.

The luminosity classes are an important tool because the luminosity of a star can give us a clue to its distance. This method of finding distance is called **spectroscopic parallax** (Box 5-3), and we can use it to estimate the distance to stars that are too far away to have measurable parallaxes. As long as we can photograph a star's spectrum, we can estimate its distance.

In reaching our second goal, finding the diameters of stars, we discovered four different kinds of stars: main-sequence stars, giants, supergiants, and white dwarfs. This raises the question, Are giant stars large because they contain more mass, or do all stars have about the same mass? Considering this question leads to our third goal, finding the masses of stars.

## THE MASSES OF STARS

A star's mass is difficult to determine. Observing a single star through a telescope, we see nothing but a small, fuzzy point of light that tells us nothing about the star's mass. To measure the masses of stars, we must discuss the motions of **binary stars,** pairs of stars that orbit each other.

*Binary Stars in General.* The key to finding the mass of a binary star is understanding orbital motion—which provides clues to mass.

To illustrate the principle of orbital motion, imagine that we construct a large cannon at the top of a mountain, point the cannon horizontal, and fire it (Figure 5-11). The cannon ball falls to earth some distance from the foot of the mountain. The more gunpowder we use, the faster the ball travels, and the farther from the foot of the mountain it falls. If we use enough powder, the ball travels so fast it never strikes the ground. Earth's gravity pulls it toward the earth's center, but the earth's surface curves away from it at the same rate at which it falls. Thus we say it is in orbit. If the cannon ball is high above the atmosphere, where there is no friction, it will fall around the earth forever.

Simple physics says that an object in motion tends to stay in motion in a straight line unless acted upon by some force. Thus the cannon ball in the example above travels in a curve around the earth only because earth's gravity acts to pull it away from its straight line motion.

Similarly, the two stars in Figure 5-12 should move at constant speed in the straight line paths shown by the dashed lines unless forces act on them. In a binary star system, the stars move around each other in orbits because their mutual

*Figure 5-11. A cannon on a high mountain could put its projectile into orbit if it could achieve a high enough velocity.*

## BOX 5-3 SPECTROSCOPIC PARALLAX

Driving along the highway at night, we often see lights dotting the countryside. We cannot immediately tell how far away these lights are unless we know how bright they really are. If we know that one of the lights is an airport searchlight and another is the headlight on a bicycle, we can judge their distances. The same is true of stars. If we can discover the luminosity of a star, we can use its apparent brightness to estimate its distance.

The method of **spectroscopic parallax** lets us find the distance to a star by classifying its spectrum according to spectral type and luminosity class. We can then look it up on an H-R diagram such as Figure 5-10 and read off its absolute magnitude $\mathbf{M_v}$. Once we measure its apparent magnitude **m**, we can calculate the distance using either the equation

$$m - M_v = -5 + 5 \log_{10}(d)$$

or a distance modulus table such as Table 5-1.

For example, Spica is classified as B1 V, and its apparent magnitude is +1. From Figure 5-10 we can estimate that a B1 V star should have an absolute magnitude of about −3. Therefore, its distance modulus is 4, and the distance (taken from the table) is about 63 pc.

This method is not very accurate because there is some uncertainty in Figure 5-10 due to individual differences between stars. Consequently when we classify a star's spectrum, we can't be sure of its exact absolute magnitude. It might be a little brighter or fainter than the diagram predicts. If the star is just one magnitude fainter than we expect, the distance we calculate is 37 percent too small. Although this method is not very accurate, spectroscopic parallax is often the only method available to measure distance.

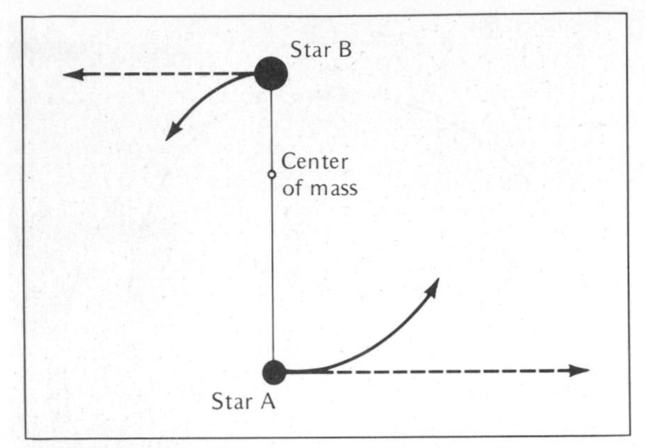


*Figure 5-12. Without gravity the two stars would follow the straight-line paths shown by dashed lines. Gravity pulls the stars into curved orbits about their common center of mass.*

gravitational attraction pulls them away from straight line motion.

Each star in a binary system moves in its own orbit around the system's center of mass, the balance point of the system. If the stars were connected by a massless rod and placed in a uniform gravitational field such as that near the earth's surface, the system would balance at its center of mass like a child's seesaw (Figure 5-13). If one star is more massive than its companion, then the massive star is closer to the center of mass and moves in a smaller orbit, while the lower-mass star whips around in a larger orbit. By observing the relative sizes of the two orbits, we could determine the ratio of the masses. That is, we could say which was more massive and by how much. Unfortunately this does not tell us the individual masses of the stars, which is what we really want to know. That requires further analysis.

To find the mass of a binary star system we must know the size of the orbits and the orbital period—the length of time the stars take to complete one orbit. The smaller the orbits are and the shorter the orbital period is, the stronger the stars' gravity must be to hold each other in orbit. However, the strength of a star's gravity depends only on its mass, so if we knew the gravitational

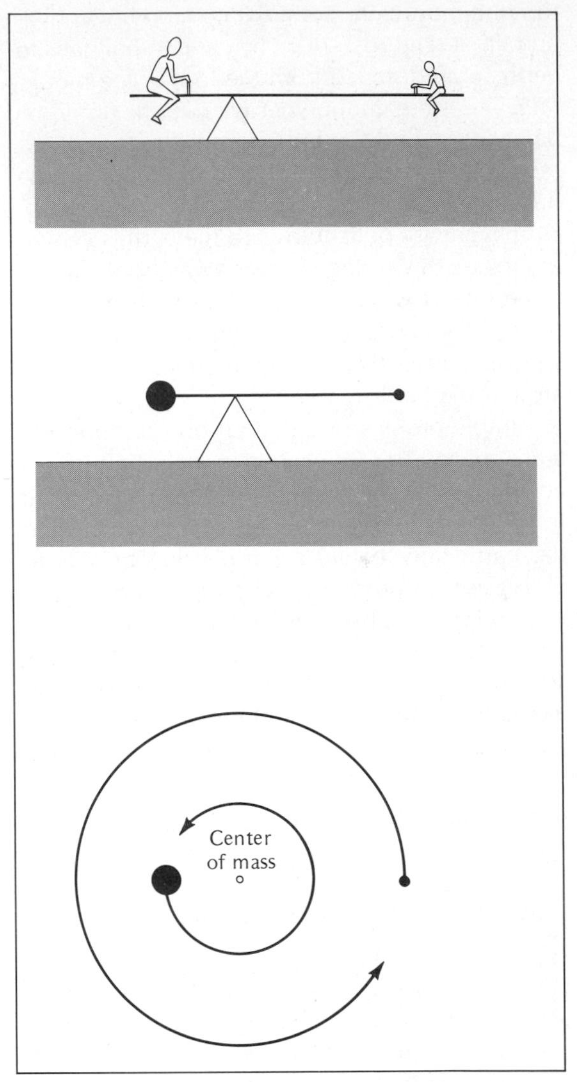


*Figure 5-13. If the two stars in a binary system were connected by a massless rod and placed in a uniform gravitational field like that at the earth's surface, it would balance like a seesaw at its center of mass. The stars orbit around their center of mass, and thus the more massive star follows a smaller orbit than its lower mass companion.*

force between the stars, we could figure out how much mass they contain. The result of such a calculation is the total mass, which, combined with the ratio of the masses found from the relative sizes of the orbits, can tell us the individual masses of the stars.

Actually, figuring out the mass of a binary star system is not so easy as solving our example. The orbits of the stars are often elliptical and tipped at an unknown angle to our line of sight. For some kinds of binary star systems astronomers can overcome these problems, analyze the system, and find the masses of the stars.

Although there are many different kinds of binary stars, three types are especially important for determining stellar masses. We will discuss these separately in the next three sections.

*Visual Binaries.* In a **visual binary,** the two stars are separately visible in the telescope. Only a pair with large orbits can be separated visually, for if the orbits are small, the star images blend together in the telescope and we see only a single point of light (Figure 5-14a). Since visual binaries have such large orbits, they also have long orbital periods. Some take hundreds or even thousands of years to complete a single revolution.

Astronomers study visual binaries by measuring the position of the two stars directly at the telescope or on photographic plates. In either case, they need measurements over many years to map the orbits. Figure 5-14b shows the orbits of the visual binary Sirius and its white dwarf companion. The dates show the observed location of the stars over the last 30 years and the predicted locations in the future. The orbital period of this system is about 50 years.

Cygnus (the Swan) is an interesting constellation in terms of binary stars. Albireo ($\beta$ Cygni) is a beautiful sight through a small telescope, appearing as a golden yellow star of spectral type K3 and a sapphire blue B8 star. Albireo is probably a binary system, but no orbital motion has been detected because the stars are at least 4400 AU apart and thus would have a very long period, perhaps as long as 100,000 years. Another visual binary in Cygnus is the star labeled

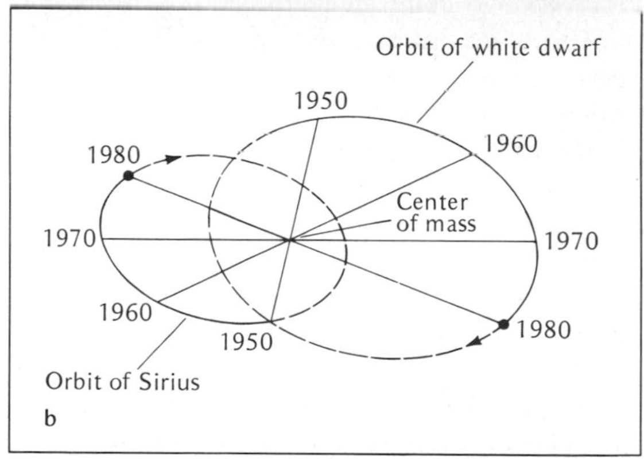


*Figure 5-14. (a) Sirius and its white-dwarf companion, a visual binary system. The size of the images is related to the brightness of the stars and not to their diameters. (Lick Observatory photograph.) (b) Sirius and its white-dwarf companion move in elliptical orbits around their common center of mass.*

*Figure 5-15. The visual binary 61 Cygni. (Lick Observatory photograph.)*

61 Cygni (Figure 5-15). It has a period of only 653 years, which is still too long for convenient analysis, but it is of special interest for a number of reasons. It is the fourth closest star to the sun, and in 1838 it became the first star to have its parallax measured. In addition, spectroscopic observations suggest one of its stars may have a planetlike companion about eight times the mass of Jupiter.

Although many stars are binaries, few can be analyzed completely. Many are so far apart that their periods are much too long for practical mapping of their orbits. Others are so close together they are not visible as separate stars.

*Spectroscopic Binaries.* If the stars of a binary system are too close together to be visible separately, the telescope shows us a single point of light. Only by taking a spectrum, which is formed by light from both stars and contains spectral lines from both, can we tell that there are two stars present and not one. Such a system is called a **spectroscopic binary.**

Because the stars in a spectroscopic binary orbit each other, they alternately approach and recede from us, as in Figure 5-16. As one star comes toward us, its spectral lines are Doppler-shifted toward the blue. The other star is moving

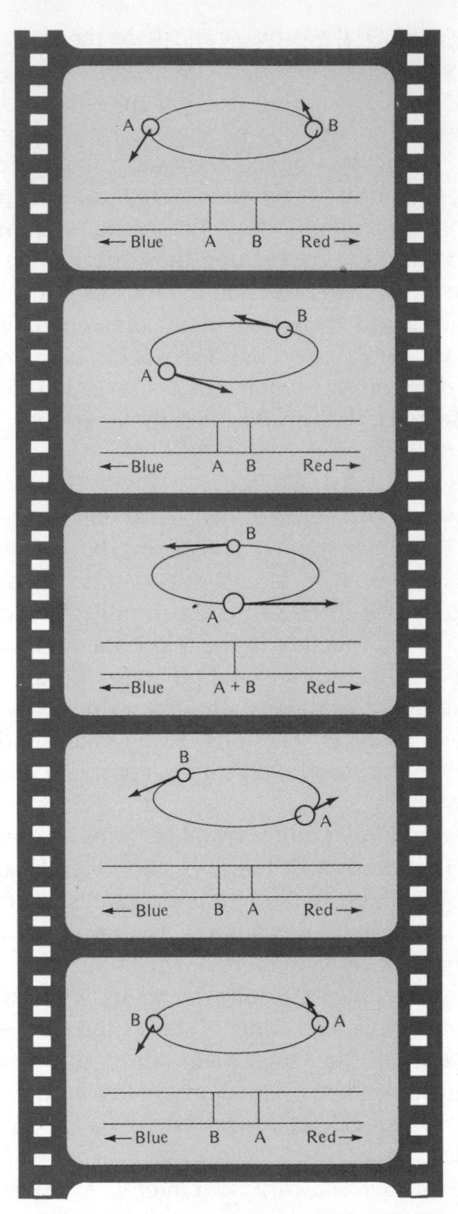


*Figure 5-16. As the stars of a spectroscopic binary revolve around their common center of mass, they alternately approach and recede from the earth. The Doppler shifts cause their spectral lines to move back and forth across each other.*

away from us and its spectral lines are shifted toward the red. Half an orbit later the star that was approaching is receding. As we watch the spectrum of the binary system, we see the spec-

*Figure 5-17.* *Mizar, at the bend of the handle of the Big Dipper, is a spectroscopic binary, as these two spectra demonstrate. The upper spectrum was taken when the stars were moving perpendicular to the line of sight and the lines were single. The lower spectrum was taken when one star was approaching and the other receding; thus the lines are double. (Hale Observatories.)*

tral lines split into two parts that move apart and then move together as the stars follow their orbits (Figures 5-16 and 5-17).

We can find the orbital period of a spectroscopic binary by observing its changing radial velocities. We can also get the velocity of each star. If we know the velocity and the length of time it takes for the star to complete one orbit, we can multiply to find the circumference of the orbit. From that we can find the radius of the orbit. Thus we would know most of what we need to find the masses of the stars—the size and period of their orbits. However, one important detail is still missing. We don't know how the orbits are inclined to our line of sight.

We can find the inclination of a visual binary because we can see the two stars moving along their orbits. In a spectroscopic binary, however, we cannot see the individual stars nor find the inclination nor untip the orbits. The velocities we observe are not the true orbital velocities but only the part of that velocity directed radially toward or away from earth. Because we cannot find the inclination, we cannot correct these radial velocities to their true orbital velocities. Therefore we cannot find the true masses. All we can find from a spectroscopic binary is a lower limit to the masses.

Spectroscopic binaries are common. Capella (α Aurigae), for instance, is a spectroscopic binary with a period of 104 days. A small telescope shows that Mizar, the star at the bend of the handle of the Big Dipper, is a visual binary (Figure 5-18). Spectroscopic observations show that both of the stars in the visual binary are themselves spectroscopic binaries, making Mizar a "double double star." Near Mizar is

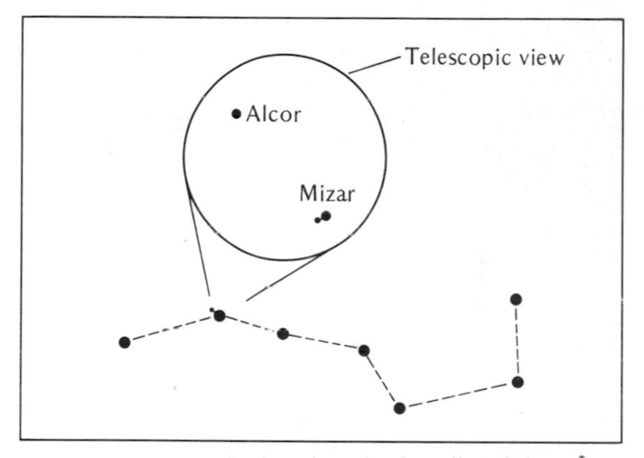


*Figure 5-18.* *At the bend in the handle of the Big Dipper lies Mizar, a visual binary. It and its companion, visible through small telescopes, plus the nearby star Alcor, are all spectroscopic binaries. (See Figure 5-17.)*

Alcor, a fainter star just visible to the naked eye. It too is a spectroscopic binary.

So far we have discussed an ideal spectroscopic binary in whose spectrum lines of both stars appear. However, in many spectroscopic binaries, one of the stars is significantly fainter than its companion, and the lines of the fainter star are invisible in the spectrum. These systems are even more difficult to analyze and provide even less information about stellar masses.

*Eclipsing Binaries.* Rare among binary stars are those with orbits tipped so the stars cross in front of each other as seen from earth. Imagine a model of a binary star system in which a cardboard disk represents the orbital plane as in Figure 5-19. If the orbits are seen edge on from earth, then the two stars cross in front of each other. The small star crosses in

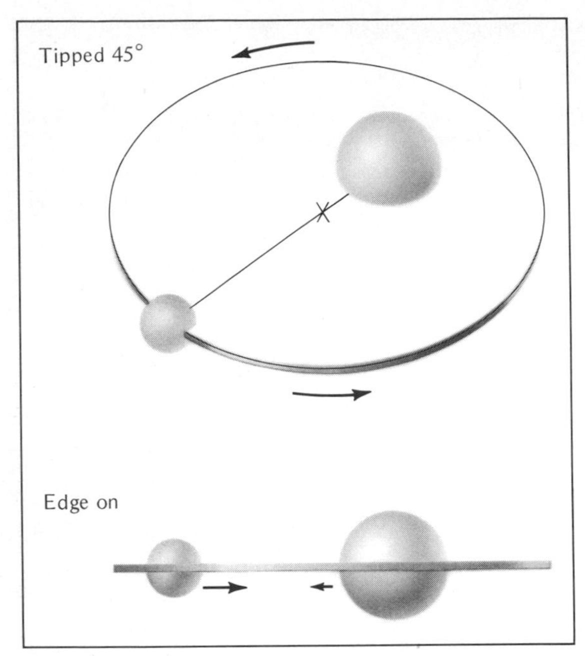


*Figure 5-19. Imagine a model of a binary system with balls for stars and a disk of cardboard for the plane of the orbits. Only if we view the system edge on do we see the stars cross in front of each other. These eclipsing binary systems are rare.*

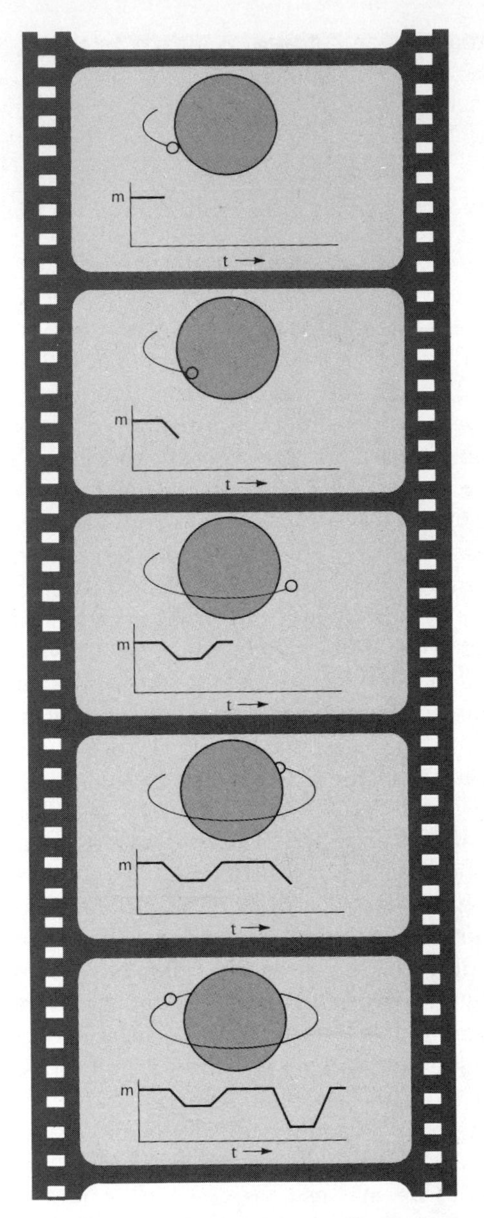


*Figure 5-20. As the stars in an eclipsing binary cross in front of each other, the total brightness of the system declines.*

front of the large star and then, half an orbit later, the larger star crosses in front of the small star. We say the stars are eclipsing each other, and we call the system an **eclipsing binary.**

Seen from the earth, the two stars are not resolvable; that is, they are not visible separately. The system looks like a single point of light. But when one star moves in front of the other star, part of the light is blocked, and the total brightness of the system decreases. Figure 5-20 shows a smaller star moving in an orbit around a larger star, first eclipsing the larger star as it crosses in front and then being eclipsed as it moves behind. The resulting variation in brightness is recorded in a graph called the **light curve** of the system.

It is often possible to find the masses of eclipsing binaries. We could find the orbital period easily, and if we could get spectra showing the Doppler shifts of the two stars, we could find the orbital velocity. Then we could find the size of the orbits and the masses of the stars. The inclination of the orbits poses no problem because we know that we must be observing the orbits nearly edge on. Otherwise we would not see the stars eclipse each other.

Eclipsing binaries are especially important because they enable us to measure the diameters of the stars. From the light curve we can tell how long it took for the small star to cross the large star. Multiplying this time interval by the orbital velocity of the small star gives us the diameter of the larger star. We could also determine the diameter of the small star by noting how long it took to disappear behind the edge of the large star. For example, if it took 300 seconds for the small star to disappear while traveling 500 km/sec relative to the large star, then it must be 150,000 km in diameter.

Of course there are complications due to the inclination and eccentricity of orbits, but often these effects can be taken into account, and the system can tell us not only the masses of its stars but also their diameters.

Algol ($\beta$ Persei) is one of the best known eclipsing binaries because its eclipses are visible to the naked eye. Normally about 2.15 magnitude, its brightness drops to 3.4 in eclipses that occur every 68.8 hours. Although the nature of the star was not recognized until 1783, its periodic dimming was probably known to the ancients. Algol comes from the Arabic for "the demon's head," and it is associated in constellation mythology with the severed head of Medusa, the sight of whose serpentine locks turned mortals to stone (Figure 5-21). Indeed, in some accounts, Algol is the winking eye of the demon.

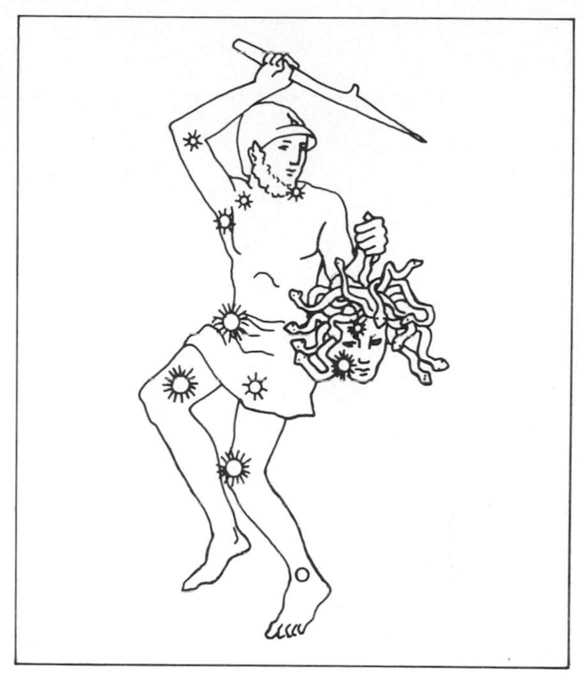

*Figure 5-21. The eclipsing binary Algol is the star on the demon's forehead in this drawing showing Perseus and the head of the gorgon Medusa. Algol comes from the Arabic for "the demon's head." (Adapted from Duncan Bradford,* The Wonders of the Heavens. *Boston: John B. Russell, 1837.)*

*Mass, Luminosity, and Density.* Although binary systems are common, masses can be accurately derived for fewer than 100. Nevertheless, this handful of stars reveals important relationships between a star's mass, luminosity, and density.

The most massive stars known are about 55 times the mass of the sun, and the least massive are slightly less than 0.1 solar masses. Among main-sequence stars the most massive are the O and B stars, the so called upper-main-sequence stars. As we run our eye down the main sequence, we find lower-mass stars. The lowest mass are the K and M stars, the lower-main-sequence stars. Thus a star's location along the main sequence depends on its mass.

Stars that do not lie on the main sequence do not appear to be arranged in any particular pattern according to mass. Some giants and supergiants are quite massive, while others are no more massive than our sun. White dwarfs are about the mass of the sun or slightly less.

Because of the systematic ordering of mass along the main sequence, these stars obey a **mass-luminosity relation**—the more massive a star is, the more luminous it is (Figure 5-22). In fact, the mass-luminosity relation can be expressed as a simple formula, as in Box 5-4. Giants and supergiants do not follow the mass-luminosity relation very closely, and white

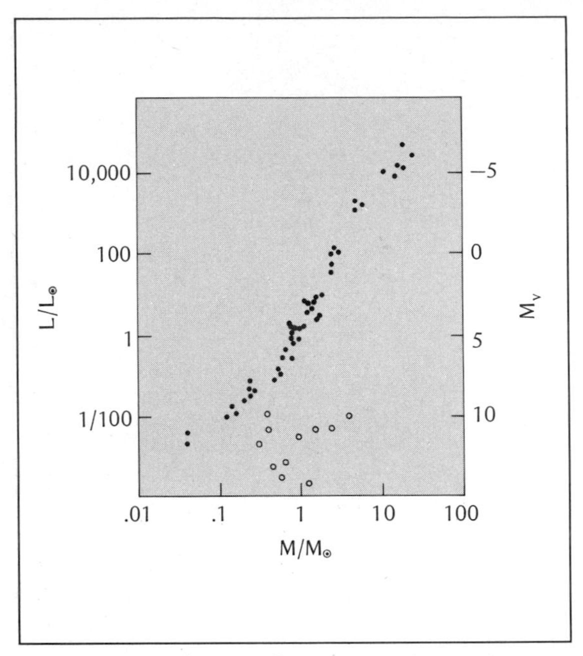


*Figure 5-22. The mass-luminosity relation shows that the more massive a star is, the more luminous it is. The open circles represent white dwarfs, which do not obey the relation.*

---

BOX 5-4 THE MASS-LUMINOSITY RELATION

We can calculate the approximate luminosity of a star from a simple equation. A star's luminosity in terms of the sun's luminosity equals its mass in solar masses raised to the 3.5 power:

$$L = M^{3.5}$$

This is the mathematical form of the mass-luminosity relation.

We can do simple calculations with this equation if we remember that raising a number to the 3.5 power is the same as cubing it and then multiplying by its square root. For example, suppose we observe a star whose mass is four times the mass of the sun. Thus:

$$L = M^{3.5} = 4^{3.5} = 4 \cdot 4 \cdot 4\sqrt{4} = 64 \cdot 2 = 128$$

The star emits 128 times as much energy as the sun.

---

dwarfs not at all. In the next chapters the mass luminosity relation will help us understand how stars generate their energy.

Though mass alone does not reveal any pattern among giants, supergiants, and white dwarfs, density does. Once we know a star's mass and diameter, we can calculate its average density by dividing its mass by its volume. Stars are not uniform in density, but are most dense at their centers and least dense near their surface. The center of the sun, for instance, is about 100 times as dense as water; its density near the visible surface is about 1000 times less dense than the earth's atmosphere at sea level. A star's average density is intermediate between its central and surface densities. The sun's average density is about 1 gm/cm³—about the density of water.

Main-sequence stars have average densities similar to the sun's, but giant stars, being large, have low average densities ranging from 0.1 to 0.01 gm/cm³. The enormous supergiants have still lower densities, ranging from 0.001 to 0.000001 gm/cm³. This is thinner than the air we breathe, and, if we could insulate ourselves from the heat, we could fly an airplane through these stars. Only near the center would we be in any danger, for there the material is very dense—about 3,000,000 gm/cm³.

The white dwarfs have masses of about 1 $M_\odot$ but are very small, only about the size of the earth. Thus the matter is compressed to densities of 10,000,000 gm/cm³ or more. On earth, 1 cm³ of this material would weigh about 20 tons.

Density divides stars into three groups. Most stars are main-sequence stars with densities like the sun's. Giants and supergiants are very low density stars, and white dwarfs are high-density objects. We will see in later chapters that these densities reflect different stages in the evolution of stars.

In this chapter we have tried to describe the properties of stars. We have found ways to find their luminosity (or absolute magnitude), their diameter, and their mass. Along the way we have discovered relationships between mass and luminosity and between density and position in the H-R diagram. Thus it might appear that we have sufficient data to begin the study of the birth, evolution, and death of stars. However, before we begin, we must answer one last question: Among the different kinds of stars, which are common and which are rare? The only way we can find the answer is by taking a survey.

## PERSPECTIVE: A NEIGHBORHOOD SURVEY

*Surveying a Representative Sample.* Suppose you took a survey in your neighborhood to find how many people had gray eyes. If you knew the area of your neighborhood in square miles, you could then say, "In my neighborhood, x people per square mile have gray eyes." Next you might think of extending your conclusion to the country as a whole: "In America x people per square mile have gray eyes." Of course if you did not live in an average neighborhood, your result would be wrong, but if you had sampled a truly representative neighborhood, your survey would give valid results about the entire population.

We can do the same thing with the stars. We can ask how many stars of each spectral type are whirling through each million cubic parsecs of space. We can't count every star in our galaxy, but we can take a survey in the region of space near the sun. Because we think the solar neighborhood is a fairly average sort of place, we can use this local survey to reach conclusions about the entire population of stars.

*Units of Stellar Density.* The result of such a survey of stars is called a **stellar density function,** a description of the abundance of different types of stars in space. A simple form of the stellar density function appears in Figure 5-23, giving the abundance of the stars of each spectral type in terms of the number of each type we would expect to find in 1,000,000 pc$^3$.

The stellar density function does not tell us

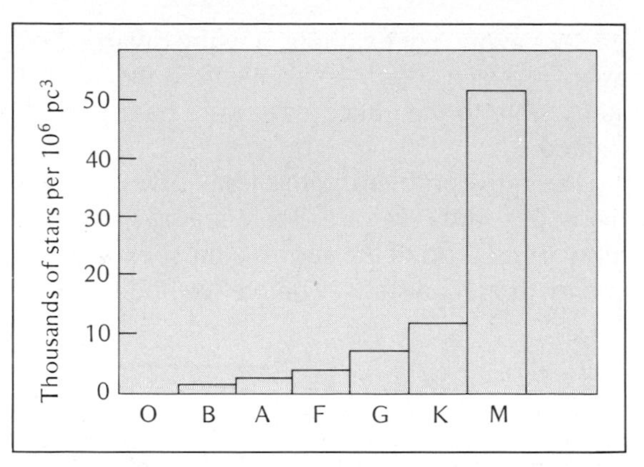


*Figure 5-23. The stellar density function shows that M stars are the most common stars in space and that O stars are very rare.*

how many stars are giants and how many are main-sequence stars. To distinguish among the luminosity classes, we would have to take a survey to find the number of stars of a given luminosity per million cubic parsecs.

*Three Problems in Counting Stars.* The astronomer could make these surveys in the neighborhood of the sun by counting all the stars within a given distance. A sphere of radius 62 pc contains 1,000,000 pc$^3$. Thus if we could count all of the M stars, for example, within 62 pc of the sun, we would have the frequency of M stars per million cubic parsecs.

But this survey of the stars poses three problems. First, to determine which stars are within 62 pc of the sun, we must measure their distances. However, stars near the outer edge of a sphere 62 pc in radius have small parallaxes that are difficult to measure accurately, and the method of spectroscopic parallax is not accurate enough for this purpose. We could count stars in a smaller sphere, but some stars are so rare that we might not find any in such a small volume of space.

A second problem for the stellar surveyor is the intrinsic faintness of stars such as M stars and white dwarfs. These are so faint that they are very hard to see if they are only a few dozen parsecs away. For example, a white dwarf 62 pc away is over 1500 times fainter than the faintest star visible to the naked eye—very hard to find, indeed.

The third problem with these surveys is that the hottest stars are rare. There are no O stars at all within 62 pc of the sun. We must extend our survey to great distances before we find many of these hot, luminous stars. At such distances the parallaxes are too small to be measured, and we have to find distances in other, less accurate ways, such as spectroscopic parallax.

*The Frequency of Stellar Types.* In spite of these difficulties, astronomers have discovered the stellar density function of the stars in

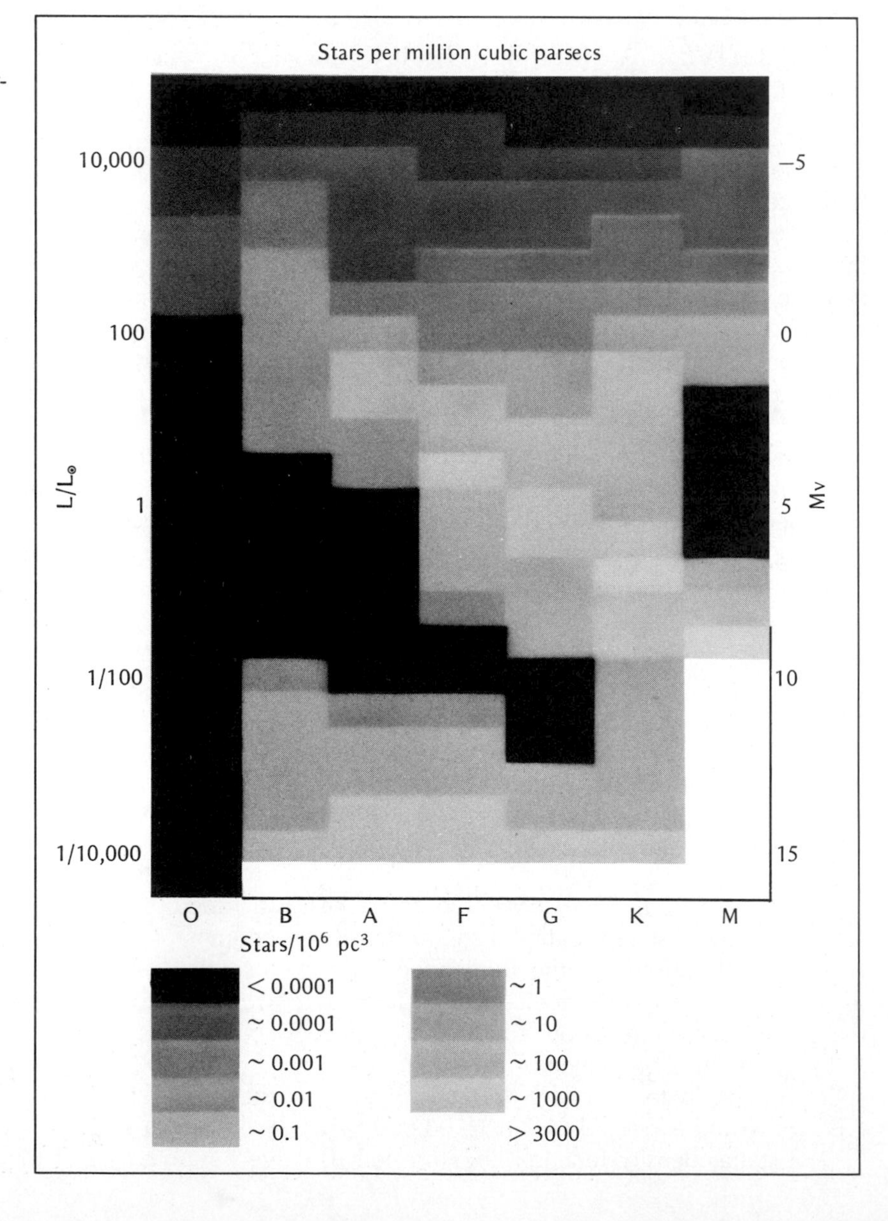


*Figure 5-24. The shading in this H-R diagram represents the frequency of different kinds of stars in space. White areas represent very common stars such as M stars and white dwarfs. Giants, supergiants, and O stars are rare.*

our galaxy. Using statistical methods, they can now tell us not just the abundance of stars of a given spectral type, but also how many stars of that type are likely to be giants, how many supergiants, and how many white dwarfs. This detailed version of the stellar density function is shown in Figure 5-24, an H-R diagram in which the shading represents the frequency of stars in space. A region of the diagram that is nearly white represents very common stars, while regions that are nearly black represent very rare stars. Compare this figure with the H-R diagram in Figure 5-7 and locate the main sequence, giants, white dwarfs, and supergiants.

Notice how common the main-sequence M stars are. There are about 50,000 in every million cubic parsecs. Notice also how few M giants there are, only about 20 per million cubic parsecs. The white dwarfs and the main-sequence M stars are the most common kinds of star. Fortunately for us, these stars are very faint, for if they were as bright as supergiants, the sky would be filled with their glare and we would hardly see anything else. As it is, these most common stars are so faint they are hard to find even with large telescopes.

The most luminous stars are very rare. On the average there are only 0.03 main-sequence O stars per million cubic parsecs. That is, we would have to search through about 30,000,000 $pc^3$ to find an O star. Put yet another way, only one star in 4 million is an O star. The giants are slightly more abundant than this. There are a few hundred giants and subgiants in every million cubic parsecs, but the supergiants are very rare. There are about 0.07 supergiants per million cubic parsecs. Luckily these very rare kinds of stars are also very luminous. We can see them from great distances. If they were as faint as the main-sequence M stars, we might never know they existed.

## SUMMARY

We can measure the distance to the nearer stars by observing their parallaxes. The more distant stars are so far away that their parallaxes are unmeasurably small. To find the distances to these stars, we must use spectroscopic parallax. Stellar distances are commonly expressed in parsecs. One parsec is 206,265 AU—the distance to an imaginary star whose parallax is one second of arc.

Once we know the distance to a star, we can find its intrinsic brightness, expressed as its absolute magnitude or its luminosity. A star's absolute magnitude is the apparent magnitude we would see if the star were only 10 pc away. The luminosity of the star is the total energy radiated in one second, usually expressed in terms of the luminosity of the sun.

The H-R diagram plots stars according to their intrinsic brightness and their surface temperature. In the diagram, roughly 90 percent of all stars fall on the main sequence, the more massive being hotter, larger, and more luminous. However, the giants and supergiants are much larger and lie above the main sequence, more luminous than main-sequence stars of the same temperature. The white dwarfs are hot stars, but they fall below the main sequence because they are so small.

The large size of the giants and supergiants means their atmospheres have low densities and their spectra have sharper spectral lines than the spectra of main-sequence stars. In fact, it is possible to assign stars to luminosity classes by the widths of their spectral lines. Class V stars are main-sequence stars with broad spectral lines. Giant stars (III) have sharper lines and supergiants (I) have extremely sharp spectral lines.

The only direct way we can find the mass of a star is by studying binary stars. When two stars orbit a common center of mass, we can find their masses by observing the period and sizes of their orbits.

Given the mass and diameter of a star, we can find its average density. On the main sequence, the stars are about as dense as the sun, but the giants and supergiants are very low density stars. Some are much thinner than air. The white dwarfs, lying below the main sequence, are tremendously dense.

The mass-luminosity relation says that the more massive a star is, the more luminous it is. Main-sequence stars follow this rule closely, the most massive being the upper-main-sequence stars and the least massive the lower-main-sequence stars. Giants and supergiants do not follow the relation precisely, and white dwarfs not at all.

A survey in the neighborhood of the sun shows us that the most common kind of stars are the lower-main-sequence stars. The hot stars of the upper-main sequence are very rare. Giants and supergiants are also rare, but white dwarfs are quite common, although they are faint and hard to find.

## NEW TERMS

parallax (**p**)
parsec (pc)
absolute visual magnitude ($\mathbf{M_v}$)
distance modulus ($\mathbf{m - M_v}$)
luminosity (**L**)
H-R diagram
main sequence
giant stars
supergiant stars
white dwarf stars
luminosity class
spectroscopic parallax
binary stars
visual binary
spectroscopic binary
eclipsing binary
light curve
mass-luminosity relation
stellar density function

## QUESTIONS

1. Why are parallax measurements limited to the nearest stars?
2. Describe two ways of specifying a star's intrinsic brightness.
3. How can a cool star be more luminous than a hot star? Give some examples.
4. Draw and label an H-R diagram. Show the proper locations of the main sequence, giants, white dwarfs, and the sun.
5. How can we be certain the giant stars are actually larger than the sun?
6. Describe the steps in using the method of spectroscopic parallax. Do we really measure a parallax?
7. Give the approximate radii in terms of the sun of stars in the following classes: G2 V, G2 III, G2 Ia. Which of these is most dense, and which least dense?
8. Why is it impossible to determine the masses of binary stars that are very far apart?
9. What is the most common type of star?

## PROBLEMS

1. If a star has a parallax of 0.050 seconds of arc, what is its distance in parsecs? In light-years? In AU?
2. If a star has a parallax of 0.016 seconds of arc and has an apparent magnitude of 6, how far away is it and what is its absolute magnitude?

3. Complete the following table.

| m | $M_V$ | d (pc) | p (sec of arc) |
|---|---|---|---|
| — | 7 | 10 | — |
| 11 | — | 1000 | — |
| — | −2 | — | 0.025 |
| 4 | — | — | 0.040 |

4. If a main-sequence star has a luminosity of 400 $L_\odot$, what is its spectral type? (Hint: See Figure 5-7.)
5. If a star has an apparent magnitude equal to its absolute magnitude, how far away is it in parsecs? In light-years?
6. An O8 V star has an apparent magnitude of +1. Use the method of spectroscopic parallaxes to find the distance to the star. Why might this distance be inaccurate?
7. Find the luminosity and spectral type of a 5 $M_\odot$ main-sequence star.
8. In the following table, which star is brightest in apparent magnitude? Most luminous in absolute magnitude? Largest? Least dense? Farthest away?

| Star | Spectral Type | m |
|---|---|---|
| a | G2 V | 5 |
| b | B1 V | 8 |
| c | G2 Ia | 15 |
| d | M5 III | 12 |
| e | white dwarf | 15 |

## RECOMMENDED READING

DeVorkin, D. H. "Steps Toward the Hertzsprung-Russell Diagram." *Physics Today* 31 (March 1978), p. 32.

Evans, D. S., Barnes, T. G., and Lacy, C. H. "Measuring Diameters of Stars." *Sky and Telescope* 58 (August 1979), p. 130.

Gingerich, O., ed. *New Frontiers in Astronomy.* San Francisco: W. H. Freeman, 1976.

Hack, M. "The Hertzsprung-Russell Diagram Today." *Sky and Telescope* 31 (May/June 1966), pp. 260 and 333.

Holzinger, J. R., and Seeds, M. A. *Laboratory Exercises in Astronomy.* Ex. 23, 24, 30, 31. New York: Macmillan, 1976.

Irwin, J. B. "The Case of the Degenerate Dwarf." *Mercury* 7 (Nov./Dec. 1978), p. 125.

Philip, A. G. D., and Green, L. C. "Henry Norris Russell and the H-R Diagram." *Sky and Telescope* 55 (April 1978), p. 306.

Philip, A. G. D., and Green, L. C. "The H-R Diagram as an Astronomical Tool." *Sky and Telescope* 55 (May 1978), p. 395.

Struve, O., and Zebergs, V. *Astronomy of the 20th Century.* New York: Crowell, Collier and Macmillan, 1962.

Verschuur, G. L. "Measuring Star Diameters." *Astronomy* 2 (Dec. 1974), p. 36.

Wilson, R. E. "Binary Stars: A Look at Some Interesting Developments." *Mercury* 3 (Sept./Oct. 1974), p. 4.

Zeilik, M., Feldman, P. A., and Walter, F. "The Strange RS Canum Venaticorum Binary Stars." *Sky and Telescope* 57 (Feb. 1979), p. 132.

*The Lagoon nebula in Sagittarius is a complex gas cloud surrounding the very young star cluster NGC 6530. The small dark dust clouds are called Bok globules and are believed to be related to star formation.* (Kitt Peak National Observatory.)

# Chapter 6 THE FORMATION OF STARS

Stars exist because of gravity. They form because gravity makes clouds of gas contract, and they generate nuclear energy because gravity squeezes them to unearthly densities and temperatures. In the end, stars die because they exhaust their fuel supply and can no longer withstand the force of their own gravity.

A star can remain stable only by maintaining great pressure in its interior. Gravity tries to make it contract, but if the internal temperature is high enough, pressure pushes outward just enough to balance gravity. Thus a star is a battlefield where pressure and gravity struggle for dominance. If gravity wins, the star must contract, but if pressure wins, the star must expand.

Only by generating tremendous amounts of nuclear energy can a star keep its interior hot enough to maintain the gravity-pressure balance. The sun, for example, generates $6 \times 10^{13}$ times more energy per second than all of the coal, oil, natural gas, and nuclear power plants on earth. Most stars generate this energy by reactions that combine four hydrogen atoms to make one helium atom plus energy. These reactions, called fusion reactions because they fuse atoms together, convert a small fraction of the atom's original mass into energy. Later in their lives stars may generate energy by fusing helium, carbon, and other atoms.

The overall anatomy of a stable star is elegant in its simplicity—four laws of stellar structure suffice to describe how mass and energy behave inside a star. Using these laws, astrophysicists can create mathematical models of stars that tell them what internal conditions are like and how the star will change over billions of years. Nearly everything we know about the internal structure and evolution of stars comes from such models.

In this chapter we will see how the simple laws of stellar structure rule the formation of stars, the ignition of their nuclear fuels, and their long, stable lives on the main sequence. In the next chapter, we will follow the life story of stars from the main sequence to their deaths.

## THE BIRTH OF STARS

Stars have been forming continuously since our galaxy took shape over 10 billion years ago. We know this for two reasons. First, the sun is only about 5 billion years old, a relative newcomer compared to the older stars in our galaxy. Second, we can see hot, blue stars such as Spica ($\alpha$ Virginis), a B1 main-sequence star. As we will see in Box 6-2, such massive stars have very short lives. In fact, a star like Spica can last only 10 million years and thus must have formed recently.

The key to understanding star formation is the correlation between young stars and clouds of gas. Where we find the youngest groups of stars

*Figure 6-1. Extremely young groups of stars are often located in regions filled with clouds of gas and dense globules of dust, suggesting that stars condense from such nebulae. Note the young star cluster, NGC 6611, near the center of the nebula. (Kitt Peak National Observatory.)*

we also find large clouds of gas illuminated by the hottest and brightest of the new stars (Figure 6-1). This leads us to suspect that stars form from such clouds, just as raindrops condense from the water vapor in a thunder cloud. To study the formation of stars, we must examine these disorganized clouds of gas that float between the stars. (See Color Plate 9.)

*The Interstellar Medium.* The gas and dust distributed between the stars is called the **interstellar medium.** The gas contains about 75 percent hydrogen and 25 percent helium plus traces of carbon, nitrogen, oxygen, calcium, sodium, and heavier atoms. The dust grains, about the same size as the particles in cigarette smoke, are carbon, iron, and silicates (rocklike minerals). Although this dust makes up only 1 to 2 percent of the mass of the interstellar medium, it plays an important role in star formation.

The interstellar medium is not uniformly distributed through space; it is concentrated in cool, dense clouds that float through a hot, low-density medium like lumps of ice floating in water. Although these clouds contain only 10 to 1000 atoms/cm$^3$ (nearly a vacuum), we refer to them as dense in contrast with the intercloud medium, which contains only 0.1 atoms/cm$^3$.

These clouds make their presence known by affecting starlight passing through them. If a cloud is unusually dense, it totally obscures our view of the stars beyond, and we see it as a dark cloud. Photographs of such dark clouds (Figures 6-1 and Color Plate 11) reveal that they are generally not spherical, but torn and twisted like wisps of smoke distorted by a breeze. If a cloud is less dense, starlight may be able to penetrate, but the stars look dimmer because the cloud absorbs some of the light.

Besides dimming the star, the dust in the cloud makes the star look redder. Because the dust grains have diameters comparable to the wavelength of light, they can interact with photons and deflect them. Red photons, due to their longer wavelengths, are less likely to be deflected than shorter-wavelength blue photons (Figure 6-2). Thus the light that reaches us is relatively rich in long-wavelength photons, making it seem redder. This **interstellar reddening** is an important source of information about the

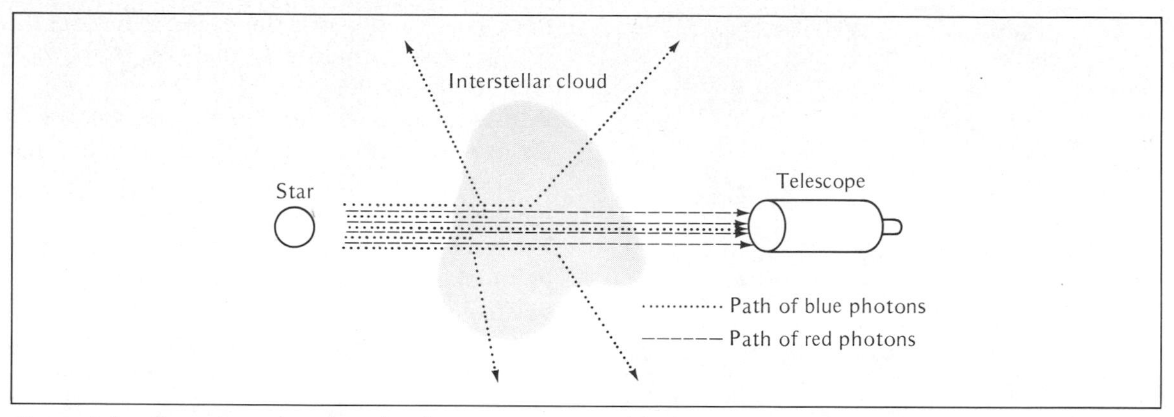


*Figure 6-2. Seen through a dust cloud, a star appears redder because the blue photons, having shorter wavelengths, are more likely to be scattered by the dust grains.*

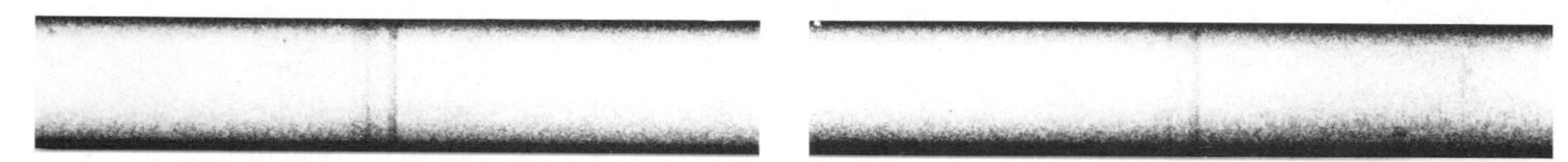

*Figure 6-3. Interstellar lines of calcium in the spectra of hot stars show that the light has passed through interstellar gas clouds. Hot stars do not produce calcium lines in their own spectra. Multiple lines indicate multiple clouds with different velocities. (Hale Observatories.)*

nature and distribution of the clouds and about the size and composition of the grains.

The clouds also reveal their presence by forming interstellar absorption lines in the spectra of distant stars (Figure 6-3). As starlight passes through a cloud, gas atoms of elements such as calcium and sodium absorb certain wavelength photons, producing narrow interstellar absorption lines. We can be sure these lines originate in the interstellar medium because they appear in the spectra of O and B stars—stars that are too hot to form their own calcium and sodium absorption lines. The narrowness of the lines tells us the clouds are very cool, about 10 to 50°K. If they were warmer, the lines would be broadened by the motions of the individual atoms. It is not unusual to find interstellar absorption lines split into several components, indicating that the starlight passed through more than one cloud. Small differences in the velocities of the clouds Doppler-shift the components and make them appear at slightly different wavelengths as in Figure 6-3.

The most intriguing parts of the interstellar medium are the **Bok globules,*** small clouds only about one light-year in diameter that contain 10 to 1000 solar masses of gas and dust. Since these clouds produce no light of their own, they are only seen silhouetted against bright nebulae, where they look like dark specks on photographs (Figure 6-4). Though some astronomers believe the globules are collapsing clouds that will become stars, others suggest they are fragments of larger clouds that were shredded by the violence of the star formation process. In either case, Bok globules are believed to be associated with the birth of new stars.

Evidence shows that interstellar matter is the source of new stars, but how can such cold, low-density clouds turn into hot, dense stars? The answer lies in the irresistible force of the clouds' gravity.

* Named after the astronomer Bart Bok.

*Figure 6-4. Bok globules are small, black clouds of interstellar matter. They may be collapsing clouds about to form stars or the fragments of clouds that have already formed stars. The smallest globules visible in this photograph are roughly 100 times the diameter of our solar system. (Lick Observatory photograph.)*

*Contraction and Heating.* The combined gravitational attraction of the atoms in a cloud of gas squeezes the cloud, pulling every atom toward the center. Thus we might expect every cloud eventually to collapse and become a star, but two factors oppose gravity and resist collapse. First, the atoms of the gas move quickly, even at temperatures of 10°K. The average hydrogen atom in such a cloud moves about 0.5 km/sec (1100 mph). This thermal motion would make the cloud drift apart if gravity were too weak to hold it together. The second factor that resists collapse is turbulence. Gas in the cloud may be churned by the heating effects of nearby stars, collisions with other gas clouds, or currents pushing through the interstellar medium from the explosion of distant stars. Whatever the cause, these turbulent motions resist the collapse of the cloud.

If we measured the diameter and density of an interstellar cloud, we could find its mass, and thus its gravity. Comparing its gravity with its temperature and estimating the effects of turbulence would tell us whether the cloud was dominated by gravity and would collapse, or whether it was dominated by thermal and turbulent motions and would drift apart. The disappointing result of such measurements is that few if any clouds are dominated by gravity. That is, few are collapsing.

Until recently, this finding left astronomers with no mechanism to explain the new stars we see all around us, but recent mathematical models of gas clouds colliding with **shock waves** (the astronomical equivalent of sonic booms) show that the clouds are compressed and disrupted into fragments (Figure 6-5). Some of these fragments probably become dense enough to collapse and form stars.

Thus interstellar clouds may not collapse and form stars until they are triggered by the compression of a shock wave. Happily for this hypothesis, space is filled with shock waves. Supernova explosions (exploding stars described in the next chapter) produce shock waves that compress the interstellar medium, and recent observations show young stars forming at the edges of such shock waves. Another source of shock waves may be the ignition of very hot stars. If one part of a cloud forms a massive star, the sudden flare of radiation as it turns on could push against other parts of the cloud, compress the gas, and trigger more star formation. Radio and infrared observations have revealed a number of nebulae where this process seems to be making new stars.

Although these are important sources of shock waves, the dominant trigger of star formation in our galaxy may be the spiral pattern itself. One account of the spiral arms supposes that they are shock waves traveling through the interstellar medium (Chapter 8). If they are, then

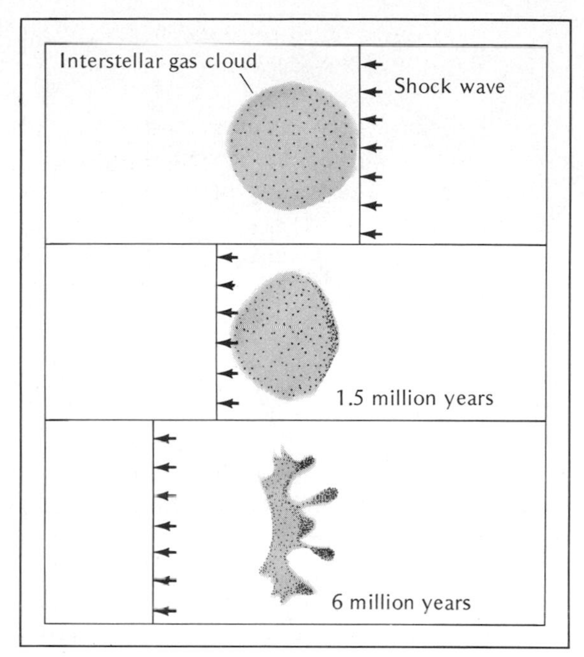


*Figure 6-5. An idealized interstellar gas cloud is struck by a passing shock wave and compressed to densities that may trigger star formation. (Adapted from computer models by Paul R. Woodward.)*

the collision of an interstellar cloud with a spiral arm could trigger the formation of a cluster of new stars.

Once a cloud is triggered into collapse, its gravity draws each atom toward the center. The infalling atoms pick up speed as they fall until, by the time the gas becomes dense enough for the atoms to collide often, they are traveling at high velocities. Since temperature is related to the velocity of the atoms, the temperature of the gas increases. Thus the contraction of an interstellar cloud heats the gas by converting gravitational energy into thermal energy.

Such a collapsing cloud of gas does not form a single object; because of instabilities it fragments, producing a cluster of 10 to 1000 stars. As the cluster grows older, its stars wander away and the cluster gradually disappears within a few hundred million years. Probably the sun formed in such a cluster about 5 billion years ago.

*Protostars.* To follow the story of star formation further we must concentrate on a single fragment of a collapsing cloud as it forms a star. The initial collapse of the material is almost unopposed by any internal pressure. The material falls inward, picking up speed and increasing the density and temperature. This free fall contraction transforms the dark, cool gas into a glowing red **protostar,** an object that will eventually become a star.

We must extend the H-R diagram to very low temperatures to follow this process. Figure 6-6 shows the initial collapse of a 1 $M_\odot$ cloud fragment, raising its temperature from 50°K to more than 1000°K. The exact process is poorly understood because of the complex effects of the dust in the cloud, but eventually the object must enter the red giant region of the H-R diagram. At this stage the protostar is a cool, red object a few thousand times larger than the sun. Its original free fall contraction is slowed by the increasing density and internal pressure, but since the protostar is not yet hot enough to begin nuclear reactions, it must contract further as it radiates energy from its surface. This contraction is much slower than the free fall contraction.

Throughout its contraction, the protostar converts its gravitational energy into thermal energy. Half of this energy radiates into space, but the remaining half raises the internal temperature closer to the ignition point for nuclear reactions.

The rising central temperature ignites nuclear reactions burning hydrogen, but, because of the low temperature, the reactions generate too little energy to stop the contraction. However, as the temperature increases, the reactions generate more energy, making the core even hotter. The higher temperature raises the internal pressure until it balances gravity, and the star's contraction slows to a stop. The star settles onto the main sequence where it remains through most of its life.

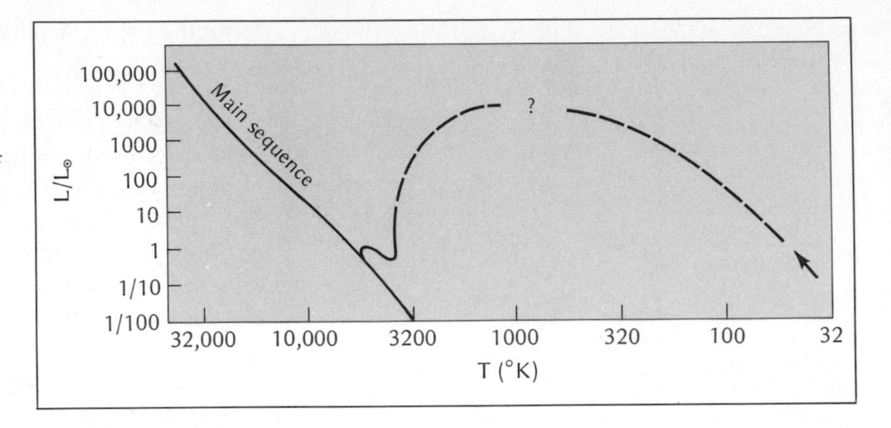


*Figure 6-6. To follow the contraction of a collapsing gas cloud (dashed line), we must extend the H-R diagram to very low temperatures. The exact behavior of such a cloud is uncertain because of the dust.*

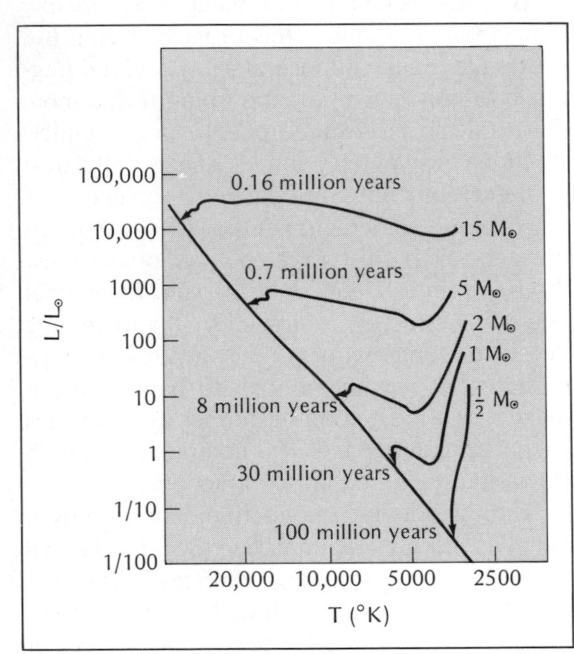


*Figure 6-7. The more massive a protostar is, the faster it contracts. A 1 M$_\odot$ star requires 30 million years to reach the main sequence. (Adapted from Michael A. Seeds, "Stellar Evolution," Astronomy, February 1979.)*

The time a protostar takes to contract from a cool interstellar gas cloud to a main-sequence star depends on its mass. The more massive the star, the stronger its gravity and the faster it contracts (Figure 6-7). The sun took about 30 million years to reach the main sequence, but a 30 M$_\odot$ star can contract in less than 30,000. Conversely, a star of 0.2 M$_\odot$ takes 1 billion years to reach the main sequence.

*Star Formation Confirmed.* The preceding scenario of star formation is based on theory. To test the theory and its supporting calculations, we should look for protostars and study their behavior.

Unfortunately a protostar is not easy to find. The protostar stage is very short, less than 0.1 percent of the star's total lifetime, and thus we are unlikely to find many protostars in the sky. Also when it forms, the protostar is surrounded by a cloud of gas and dust called a **cocoon** within which the protostar gradually becomes a star. The thick dust in the cocoon absorbs all of the light radiated by the protostar, so, although we plot the protostar in the red giant region of the H-R diagram, we cannot see it within its cocoon.

In spite of these problems we can detect protostars. Although the dust blocks our view of the protostar, it reveals star formation by radiating away the energy it absorbs as infrared radiation. Infrared observations reveal many bright sources that are probably cocoon stars.

The dust cocoon around the protostar slowly disappears as the protostar grows hotter, absorbs the inner cocoon, and blows away the rest. **Herbig-Haro objects,** small nebulae that vary irregularly in brightness, may be protostars be-

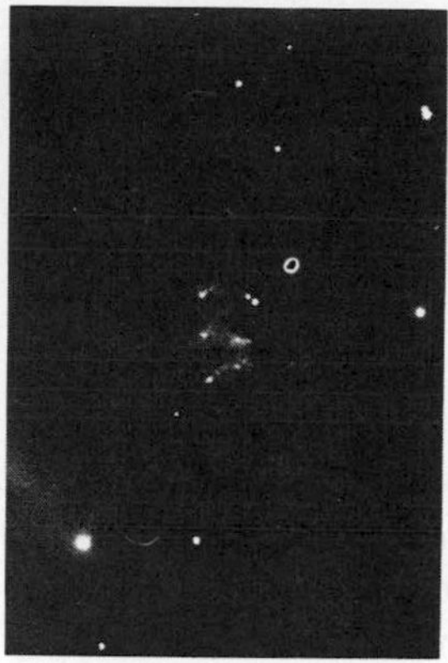
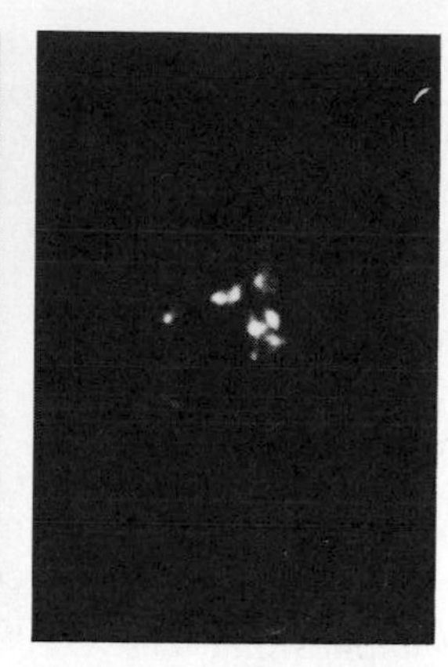

*Figure 6-8. Herbig-Haro objects are small nebulous condensations found within dense clouds of gas and dust. They are believed to be stars in very early stages of evolution. (Lick Observatory photographs.)*

ginning to clear their cocoons. At least one such nebula shows changes over recent decades that suggest the emergence of new stars (Figure 6-8).

Other objects associated with cocoon stars are the **T Tauri stars,** named after variable star T in Taurus. These objects may be lower-mass versions of the Herbig-Haro objects. In the typical T Tauri star, the cocoon is nearly gone and the star, though still highly variable and not yet on the main sequence, is clearly visible. The remnants of the cocoon are detectable only in the spectrum, where Doppler-shifted lines show that the star is surrounded by expanding clouds of gas. Since some T Tauri stars are sources of infrared radiation, dust must also be present, evidently the last shreds of the cocoon.

We know that many infrared objects, Herbig-Haro objects, and T Tauri stars are young because they are located in regions where hot blue stars are found. Since these stars have such short lives, they too must be young, a fact that implies that the entire region is a site of active star formation.

In addition, T Tauri stars are found in some very young clusters such as NGC 2264. This cluster of stars formed so recently that the less massive stars, contracting slower than the massive stars, have not yet reached the main sequence (Figure 6-9). Scattered throughout the cluster are a number of T Tauri stars, whose temperature and luminosity place them to the right of the main sequence, just where we would expect to find contracting protostars.

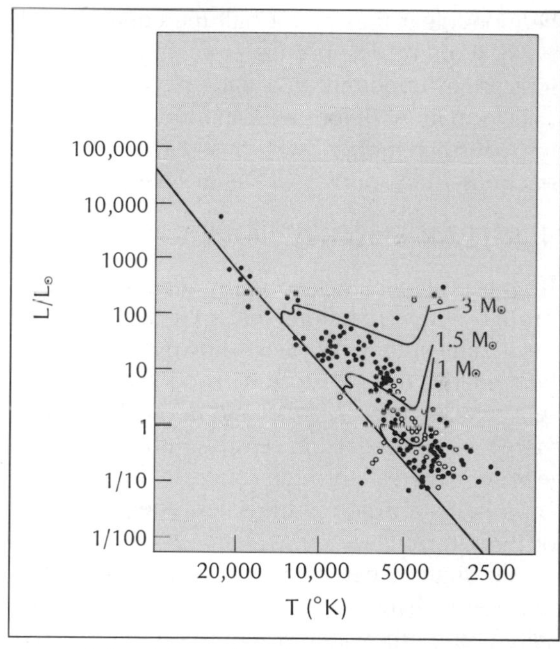


*Figure 6-9. Many of the lower-mass stars in NGC 2264 have not yet reached the main sequence. The cluster is only a few million years old and contains many T Tauri stars (open circles). (Adapted from data by M. Walker.)*

All of these observations confirm our theoretical model of contracting protostars. Although star formation still holds many mysteries, the general process seems clear. In at least some cases, interstellar gas clouds are compressed by passing shock waves, and the clouds' gravity, acting unopposed, draws the matter inward to form protostars.

Most astronomers now believe that the planets of our solar system formed at the same time the sun formed. At some point during the sun's contraction, when it was surrounded by swirling clouds of gas and dust, solid bodies condensed and grew into planets. We will discuss this process in Chapter 12, but it is important to note here because it implies that many stars may form planets as they contract.

Only when the star is hot enough to ignite its thermonuclear fires can it halt the collapse and reach stability. Clearly, internal nuclear reactions are as important in a star's life as gravity is in its formation. Before we can follow the evolution of stars further, we must examine these reactions—the sources of stellar energy.

## NUCLEAR ENERGY SOURCES

Atomic power plants on earth generate energy through fission reactions that split heavy atoms into lighter fragments. Stars, however, generate their energy through fusion reactions that fuse light atoms into heavier atoms and liberate energy. At present no earthly power plants generate energy through nuclear fusion, but some believe fusion is the safest energy source for the future.

Although 90 percent of all stars, including the sun, burn hydrogen, some older stars burn helium and other heavier atoms. (In this use, the word *burn* refers not to chemical burning as in a candle flame, but to thermonuclear reactions.) Whatever the fuel, the energy generated by these fusion reactions keeps the star hot and the pressure high, balancing the compressive force of gravity. In a sense, the matter in a star is supported against its own gravity by the nuclear reactions.

*Hydrogen Burning.* The nuclear reactions that burn hydrogen in stars join four hydrogen atoms to make one helium atom. Since one helium atom has 0.7 percent less mass than four hydrogen atoms, it seems that some mass vanishes in the process.

$$\begin{array}{rl} \text{4 hydrogen atoms} &= 6.693 \times 10^{-24}\ \text{gm} \\ \text{1 helium atom} &= 6.645 \times 10^{-24}\ \text{gm} \\ \hline \text{difference in mass} &= 0.048 \times 10^{-24}\ \text{gm} \end{array}$$

However, this mass does not actually vanish; it merely changes form. As Einstein stated in his equation $E = mc^2$, mass and energy are different forms of the same thing, and under certain circumstances, mass may become energy and vice versa. Thus the $0.048 \times 10^{-24}$ gm do not vanish, but merely become energy. To see how much, we use Einstein's equation:

$$\begin{aligned} E &= mc^2 \\ &= (0.048 \times 10^{-24}\ \text{gm})(3 \times 10^{10}\ \text{cm/sec})^2 \\ &= 0.000043\ \text{ergs} \end{aligned}$$

This is a very small amount of energy, hardly enough to raise a housefly one-thousandth of an inch. Since one reaction produces such a small amount of energy, it is obvious that many reactions are necessary to support the voracious energy appetite of a star. The sun, for example, needs $10^{38}$ reactions per second, transforming 5,000,000 tons of mass into energy every second, just to resist its own gravity.

These nuclear reactions can only occur when the nuclei of two atoms approach close to each other. However, since atomic nuclei carry positive charges, they repel each other with an electrostatic force called the **Coulomb barrier.** To overcome this barrier, atomic nuclei must collide violently. Violent collisions are rare unless the gas is very hot, in which case the atoms move at high speeds and collide violently. (Remember, an object's temperature is just a measure of the speed with which its atoms or molecules move.)

Thus nuclear reactions can only occur near the centers of stars, where they are hot and dense. High temperature ensures that collisions between nuclei are violent enough to overcome the Coulomb barrier, and high density assures that there are enough collisions, and thus enough reactions, to meet the star's energy needs.

We can symbolize this process with a simple nuclear reaction:

$$4\ {}^{1}H \rightarrow {}^{4}He + \text{energy}$$

In this equation, ${}^{1}H$ represents a proton, the nucleus of the hydrogen atom, and ${}^{4}He$ represents the nucleus of a helium atom. The superscripts indicate the approximate weight of the atoms. The actual steps in the process are more complicated than this convenient summary suggests. Instead of waiting for four hydrogen atoms to collide simultaneously, a highly unlikely event, the process can proceed step by step in a chain of reactions that can occur in either of two ways—the proton-proton chain, or the CNO cycle.

The **proton-proton chain** is a series of three nuclear reactions that builds a helium atom by adding together protons. This process is efficient at temperatures above 10,000,000°K. The sun, for example, manufactures over 90 percent of its energy in this way.

The three steps in the proton-proton chain entail these reactions:

$$ {}^{1}H + {}^{1}H \rightarrow {}^{2}H + e^{+} + \nu $$
$$ {}^{2}H + {}^{1}H \rightarrow {}^{3}He + \gamma $$
$$ {}^{3}He + {}^{3}He \rightarrow {}^{4}He + {}^{1}H + {}^{1}H $$

In the first step, two hydrogen nuclei (two protons) combine to form a heavy hydrogen nucleus, emitting a particle called a positron (a positively charged electron) and another called a neutrino. (See Box 6-1 for more about neutrinos.) In the second reaction, the heavy hydrogen nucleus absorbs another proton, and, with the emission of a gamma ray, becomes a lightweight helium nucleus. Finally, two light helium nuclei combine to form a normal helium nucleus and two hydrogen nuclei. Since the last reaction needs two ${}^{3}He$ nuclei, the first and second reactions must occur twice (Figure 6-10). The net result of this chain reaction is the transformation of four hydrogen nuclei into one helium nucleus plus energy.

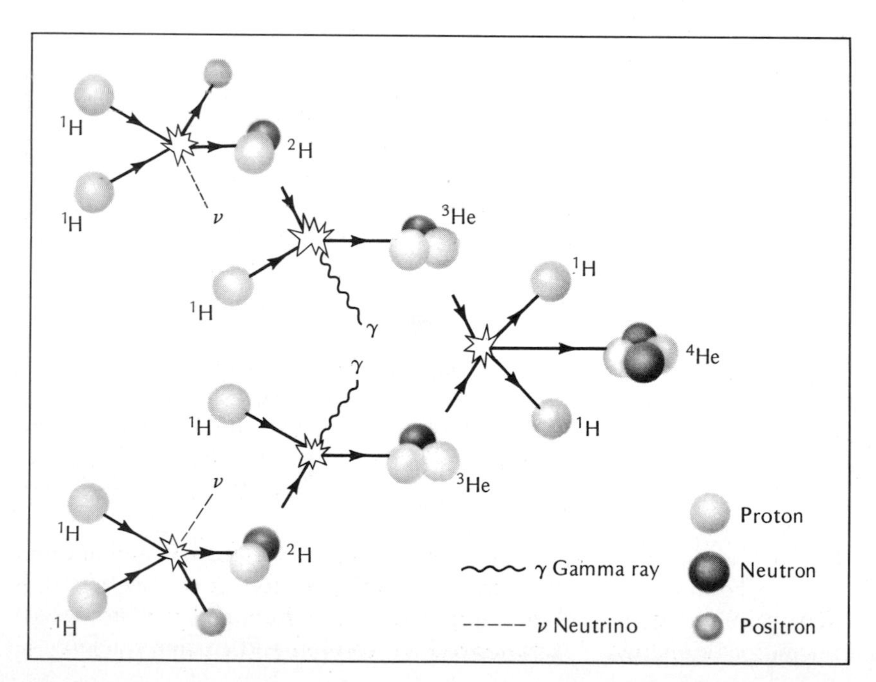


*Figure 6-10. The proton-proton chain combines four protons (at left) to produce one helium nucleus (at right) plus energy.*

The energy appears in the form of gamma rays, positrons, and neutrinos. The gamma rays are photons that are absorbed by the surrounding gas before they can travel more than a few centimeters. This heats the gas and helps maintain the pressure. The positrons produced in the first reaction combine with free electrons and both particles vanish, converting their mass into gamma rays. Thus the positrons also help keep the center of the star hot. The neutrinos, however, are massless particles that resemble photons except that they almost never interact with other particles. The average neutrino could pass unhindered through a lead wall 1 ly thick. Thus the neutrinos do not help heat the gas, but race out of the star at the speed of light, carrying away roughly 2 percent of the energy produced.

Like the proton-proton chain, the **CNO cycle** is a series of nuclear reactions that combines four hydrogen atoms to make one helium atom plus energy. However, the CNO cycle needs carbon as a catalyst. This process is most efficient at temperatures above 16,000,000°K. Since the sun's central temperature is only about 15,000,000°K, the CNO cycle creates less than 10 percent of the sun's energy.

The steps in the CNO cycle begin with carbon-12 ($^{12}C$), pass through stages involving isotopes of carbon, nitrogen (N), and oxygen (O), and finish with the reappearance of a carbon-12 nucleus just like the one that started the process. Thus we say the carbon is a catalyst; it makes the reactions possible but is not altered in the end.

$$^{12}C + {}^{1}H \rightarrow {}^{13}N + \gamma$$
$$^{13}N \rightarrow {}^{13}C + e^{+} + \nu$$
$$^{13}C + {}^{1}H \rightarrow {}^{14}N + \gamma$$
$$^{14}N + {}^{1}H \rightarrow {}^{15}O + \gamma$$
$$^{15}O \rightarrow {}^{15}N + e^{+} + \nu$$
$$^{15}N + {}^{1}H \rightarrow {}^{4}He + {}^{12}C$$

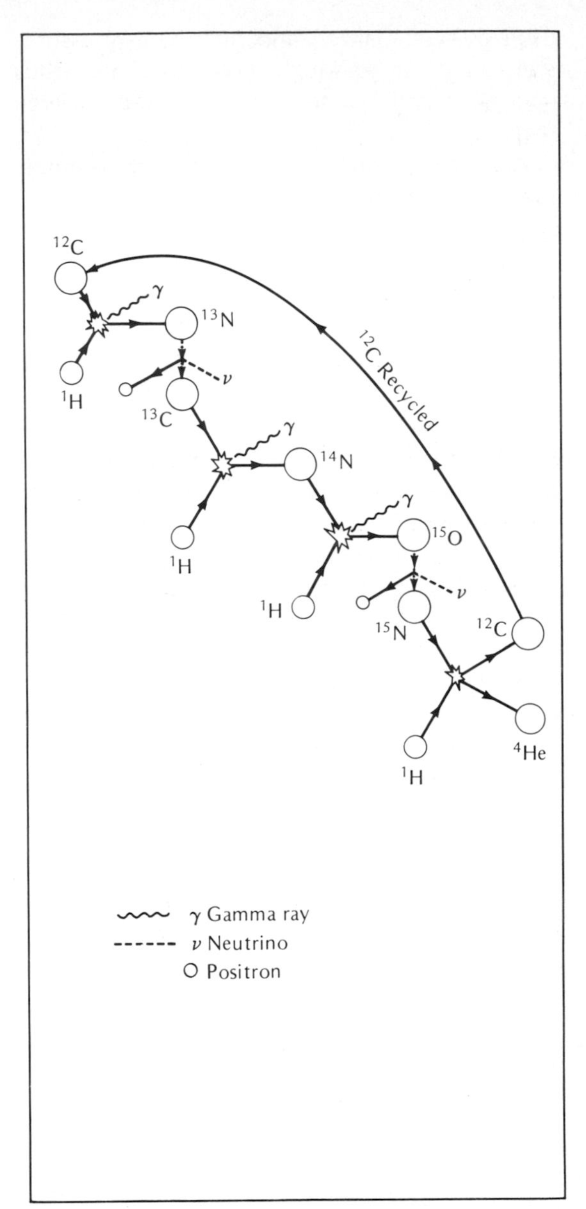


*Figure 6-11. The CNO cycle uses $^{12}C$ as a catalyst to combine four hydrogen atoms ($^{1}H$) to make one helium atom ($^{4}He$) plus energy. The carbon atom reappears at the end of the process ready to start the cycle over.*

Counting protons in the CNO cycle reactions above or in Figure 6-11 shows that the net result of the CNO cycle is four hydrogen nuclei combined to make one helium nucleus plus energy.

Because the carbon nucleus has a charge six times that of hydrogen, the Coulomb barrier is high, and much hotter temperatures are necessary to force the proton and carbon nucleus to-

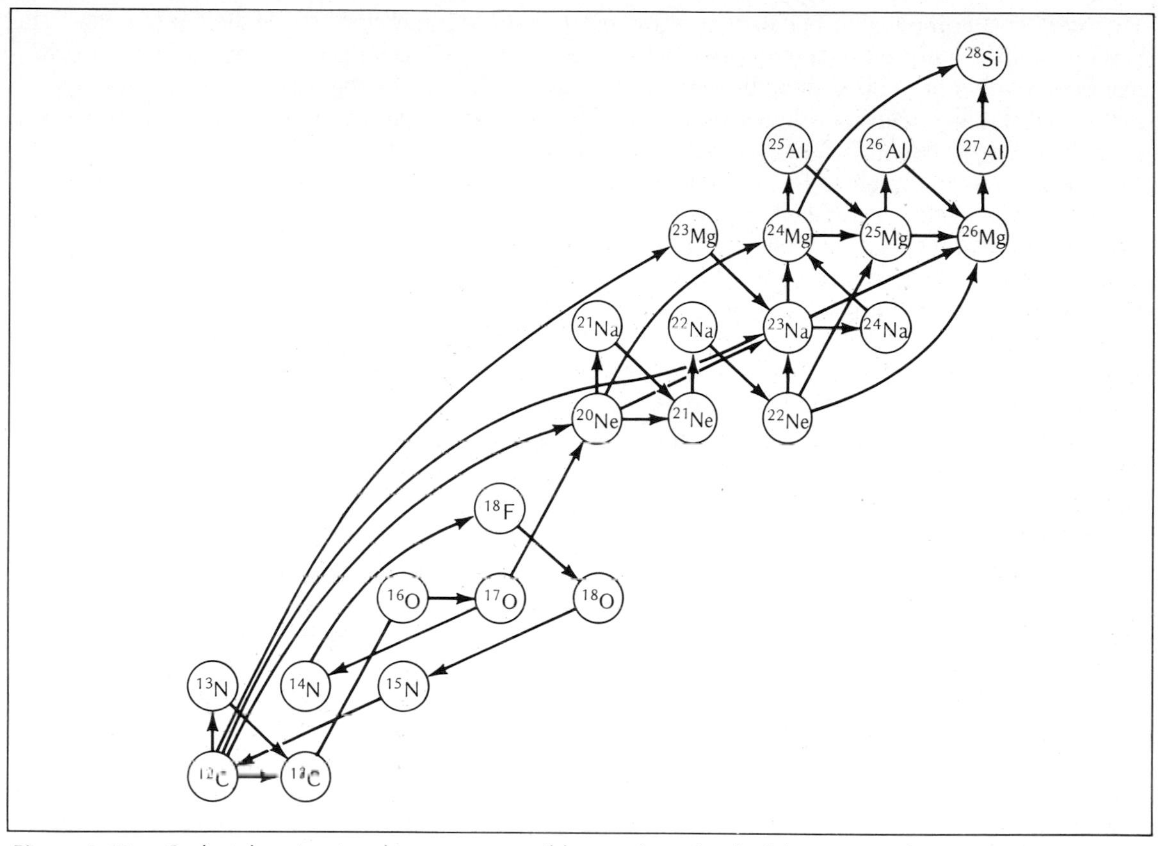


*Figure 6-12. Carbon burning involves many possible reactions that build numerous heavy atoms.*

gether. Thus the CNO cycle is important only in stars more massive than about 1.1 $M_\odot$. These stars have central temperatures hotter than about 16,000,000°K. Stars less massive than 1.1 $M_\odot$, such as the sun, are not hot enough at their centers to burn hydrogen on the CNO cycle.

*Heavy Element Burning.* At later stages in the life of a star, when it has exhausted its hydrogen fuel, it may burn other nuclear fuels. As we saw above, the ignition of these fuels requires high temperatures. Helium burning occurs at temperatures above 100,000,000°K, and carbon burning does not begin until the temperature exceeds 600,000,000°K.

We can summarize the helium-burning process in two steps:

$$^{4}\text{He} + {}^{4}\text{He} \rightarrow {}^{8}\text{Be} + \gamma$$
$$^{8}\text{Be} + {}^{4}\text{He} \rightarrow {}^{12}\text{C} + \gamma$$

This process is complicated by the fact that beryllium-8 is very unstable and may break up into two helium nuclei before it can absorb another helium nucleus. Three helium nuclei can also form carbon directly, but such a triple collision is unlikely.

At temperatures above 600,000,000°K, carbon burns rapidly in a complex network of reactions illustrated in Figure 6-12, where each arrow represents a different nuclear reaction.

The process is complicated because nuclei can react by adding a proton, a neutron, or a helium nucleus or by combining directly with other nuclei. Unstable nuclei can decay by ejecting an electron, a positron, or a helium nucleus or by splitting into fragments. The complexity of this process makes it difficult to determine exactly how much energy will be generated and how many heavy atoms will be produced.

Reactions at still higher temperatures can convert magnesium and silicon into yet heavier atoms. These reactions involving heavy elements will be important in the study of the deaths of massive stars in the next chapter.

*The Pressure-Temperature Thermostat.* Nuclear reactions in stars manufacture energy and heavy atoms under the supervision of a built-in thermostat that keeps the reactions from erupting out of control. That thermostat is the relation between pressure and temperature discussed at the beginning of this chapter.

In a star, the nuclear reactions generate just enough energy to balance the inward pull of gravity. Consider what would happen if the reactions begin to produce too much energy. The extra energy would raise the internal temperature of the star, and, since the pressure of the gas depends on its temperature, the pressure would also rise. The increased pressure would make the star expand. Expansion of the gas would cool the star slightly, slowing the nuclear reactions. Thus the star has a built-in regulator that keeps the nuclear reactions from going too rapidly.

The same thermostat also keeps the reactions from dying down. Suppose the nuclear reactions began to produce too little energy. Then the inner temperature would decrease, lowering the pressure and allowing gravity to compress the star slightly. As the gas was compressed, it would heat up, increasing the nuclear energy generation until the star regained stability.

The stability of a star depends on this relation between pressure and temperature. If an increase or decrease in temperature produces a corresponding change in pressure, then the thermostat functions correctly and the star is stable. We will see in this chapter how the thermostat accounts for the mass-luminosity relation. In the next chapter we will see what happens to a star when the thermostat breaks down completely and the nuclear fires burn unregulated.

## STELLAR STRUCTURE

To understand how stars work, we must look more closely at how the generation of energy at a star's core can balance gravity. That is, we must study the internal structure of stars. By structure, we mean the variation in temperature, density, pressure, and so on from the surface of the star to its center. A star's structure depends on how it generates its energy and on four laws of stellar structure that describe how matter and energy behave inside the star.

It will be easier to think about stellar structure if we imagine that the star is divided into concentric shells like those in an onion (Figure 6-13). We can then discuss the temperature, density, pressure, and so on in each shell. Keep in mind, however, that these helpful shells do not really exist, for stars have no separable layers.

*The Laws of Mass and Energy.* The first two laws of stellar structure have something in common. They are both laws of continuity. They tell us that the distribution of matter and energy inside the star must vary smoothly from surface to center. No gaps are allowed.

The **continuity of mass** law says that the total mass of the star must equal the sum of the masses of its shells and that the mass must be distributed smoothly throughout the star. In a sense, the continuity of mass law is the law of conservation of mass with the added proviso that the mass be distributed smoothly.

The **continuity of energy** law says that the amount of energy flowing out the top of a shell must equal the amount coming in at the bottom plus whatever energy is generated within the shell. Further, it says that the energy leaving the surface of the star—the luminosity—must equal

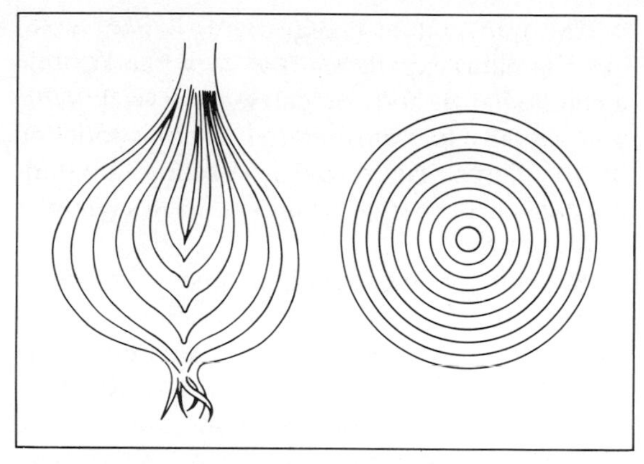

*Figure 6-13. The layers of the model star resemble the layers of the common onion.*

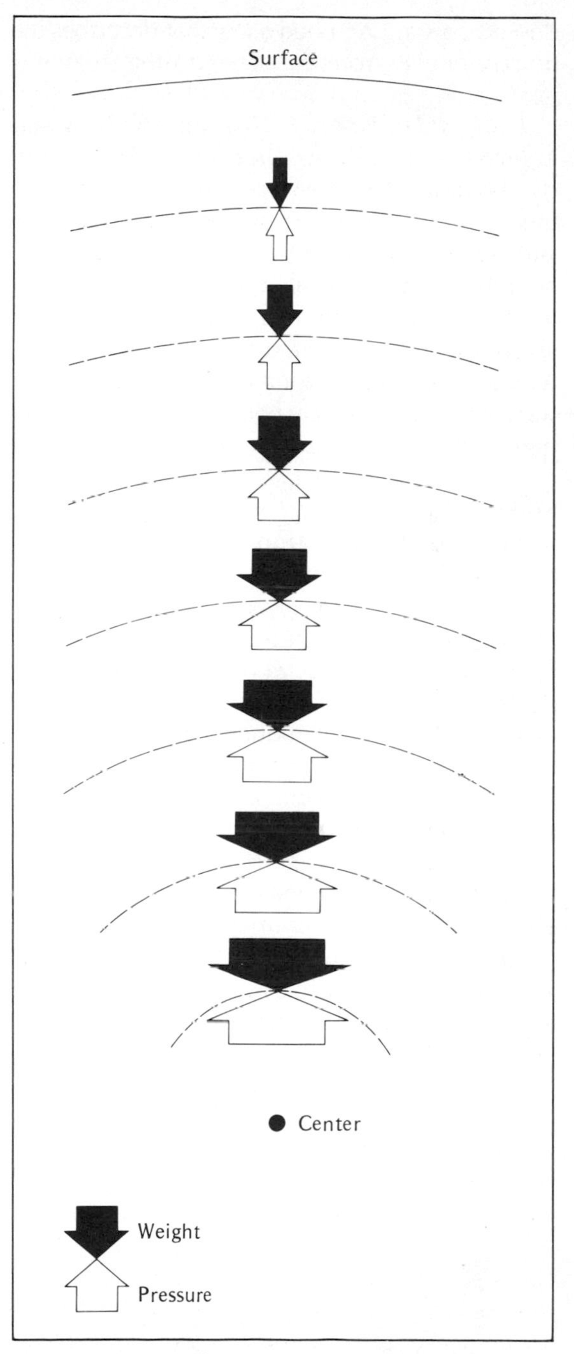


*Figure 6-14. Hydrostatic equilibrium says the pressure in each layer must balance the weight on that layer. As a result, pressure and temperature must increase from the surface of a star to its center.*

the sum of the energies generated in the shells. Thus the energy law is really a version of the law of conservation of energy.

In fact, both laws may seem familiar since we have all heard of the conservation of mass and energy. The third law of stellar structure is also familiar because we have been using it under a different name in the previous sections.

*Hydrostatic Equilibrium.* The law of **hydrostatic equilibrium** says that, in a stable star, the weight of the material pressing downward on a layer must be balanced by the pressure of the gas in that layer. Hydrostatic equilibrium is just a new name for the gravity-pressure balance discussed earlier. *Hydro* implies we are discussing a fluid, the gases of a star, and *static* implies that the fluid is stable, neither expanding nor contracting.

The law of hydrostatic equilibrium can prove to us that the temperature must increase as we descend into a star. Near the surface, there is little weight pressing down on the gas, so the pressure must be low, implying a low temperature. But as we go deeper into the star, the weight becomes larger, so the pressure, and therefore the temperature, must also increase (Figure 6-14).

Although the law of hydrostatic equilibrium can tell us some things about the inner structure of stars, we need one more law to completely describe a star. We need a law that describes the flow of energy from the center to the surface.

*Energy Transport.* The surface of a star radiates light and heat into space and would quickly cool if that energy were not replaced. Since the inside of the star is hotter than the surface, energy must flow outward from the core, where it is generated, to the surface, where it radiates away. This flow of energy through the shells determines their temperature, which, as we saw previously, determines how much weight each shell can balance. To understand the structure of a star, we must understand how energy moves from the center through the shells to the surface.

The law of **energy transport** says that energy must flow from hot regions to cooler regions by conduction, convection, or radiation.

Conduction is the most familiar form of heat flow. If you hold the bowl of a spoon in a candle flame, the handle of the spoon grows warmer. Heat, in the form of motion among the molecules of the spoon, is conducted from molecule to molecule up the handle, until the molecules of metal under your fingers begin to move faster and you sense heat (Figure 6-15). Thus conduction requires close contact between the molecules. Since matter in most stars is gaseous, conduction is unimportant. Conduction is significant only in white dwarfs, which have tremendous internal densities.

The transport of energy by radiation is another familiar experience. Put your hand beside a candle flame, and you can feel the heat. What you actually feel are infrared photons radiated by the flame (Figure 6-15). Because photons are packets of energy, your hand grows warm as it absorbs them.

Radiation is the principal means of energy transport in the sun's interior. Photons are absorbed and reemitted in random directions over and over as they work their way outward. It takes about 1 million years for the average photon to travel from the sun's center to its surface.

The flow of energy by radiation depends on how difficult it is for the photons to move through the gas. If the gas is cool and dense, the photons are more likely to be absorbed or scattered and thus the radiation does not get through easily. We would call such a gas opaque. In a hot, thin gas, the photons can get through more easily; such a gas is less opaque. The **opacity** of the gas, its resistance to the flow of radiation, depends strongly on its temperature.

If the opacity is high, radiation cannot flow through the gas easily and it backs up like water behind a dam. When enough heat builds up, the gas begins to churn as hot gas rises upward and cool gas sinks downward. This is convection, the third way energy can move in a star. Convection is a common experience; the wisp of smoke rising above a candle flame travels up-

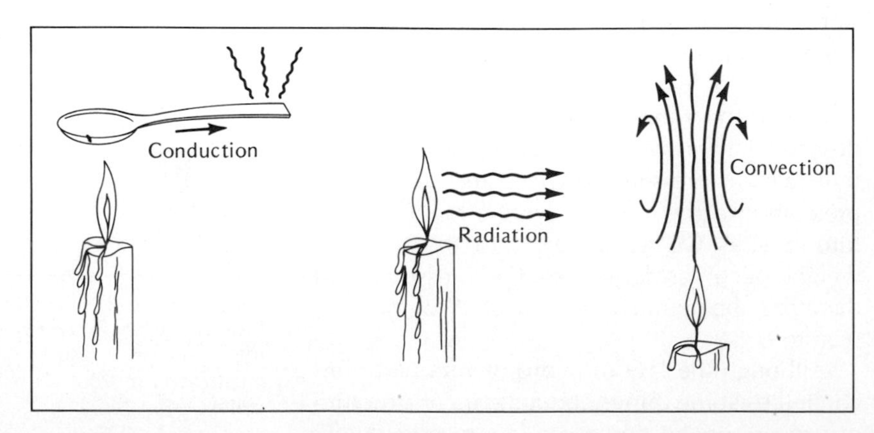


*Figure 6-15. Conduction, radiation, and convection are the only modes of energy transport within a star.*

ward in a small convection current (Figure 6-15). If you hold your hand above the flame, you can feel the rising current of hot gas. In stars, energy may be carried upward by rising currents of hot gas hundreds or thousands of miles in diameter.

Convection is important in stars because it carries energy and because it mixes the gas. Convection currents flowing through the layers of a star tend to homogenize the gas, giving it a uniform composition throughout the convective zone. As you might expect, this mixing affects the fuel supply of the nuclear reactions, just as the stirring of a campfire makes it burn more efficiently.

These four laws of stellar structure give us a deep insight into how stars work. Each of them is a basic law of physics applied to conditions inside a star.

*Stellar Models.* The four laws of stellar structure are much more precise than the four statements in Table 6-1. In fact, the four laws can be expressed as four equations. What these equations look like is not important to us, but applying them to build mathematical models of stars can tell us how stars work, how they are born, and how they die.

**Table 6-1 The Four Laws of Stellar Structure**

| | | |
|---|---|---|
| 1. | Continuity of mass | Total mass equals sum of shell masses. |
| 2. | Continuity of energy | Total luminosity equals sum of energies generated in each shell. |
| 3. | Hydrostatic equilibrium | The weight on each layer is balanced by the pressure in that layer. |
| 4. | Energy transport | Energy moves from hot to cool by conduction, radiation, or convection. |

If we wanted to build a model of a star, we would have to divide the star into about 100 concentric shells, and then write down the four equations of stellar structure for each shell. We would then have 400 equations that would have 400 unknowns; namely, the temperature, density, mass, and energy flow in each shell. Solving 400 equations simultaneously is not easy, and the first such solutions, done by hand before the invention of the electronic computer, took months of work. Now a properly programmed computer can solve the equations in a few seconds and print a table of numbers that represents the conditions in each shell of the star. Such a table, shown in Figure 6-16, is a **stellar model.**

The model shown in Figure 6-16 represents our sun. As we scan the table from top to bottom, we descend from the surface of the sun to its center. The temperature increases rapidly as we move downward, reaching a maximum of about 15,000,000°K at the center. At this temperature, the gas is not very opaque and the energy can flow outward as radiation. In the cooler outer layers, the gas is more opaque and the outward flowing energy forces these layers to churn in convection. This model of the sun, like all stellar models, lets us study the otherwise inaccessible layers inside a star.

The sun offers us an opportunity to check the results of the stellar models by direct observation. We cannot see the inside of the sun, but we can see the top of the convection zone, which lies just below the observable surface. High quality photographs of the solar surface show a patchwork pattern of bright areas with dark borders called **solar granulation** (Figure 6-17). Each granule is about 700 km in diameter, an area slightly smaller than Texas, and lasts for only about four minutes. Solar astronomers have identified the bright granules as the tops of rising currents of hot gas in the convective zone. The dark boundaries of the granules are regions of cooler gas sinking downward (Figure 6-18).

Stellar models also let us look into the star's past and future. In fact, we can use models as

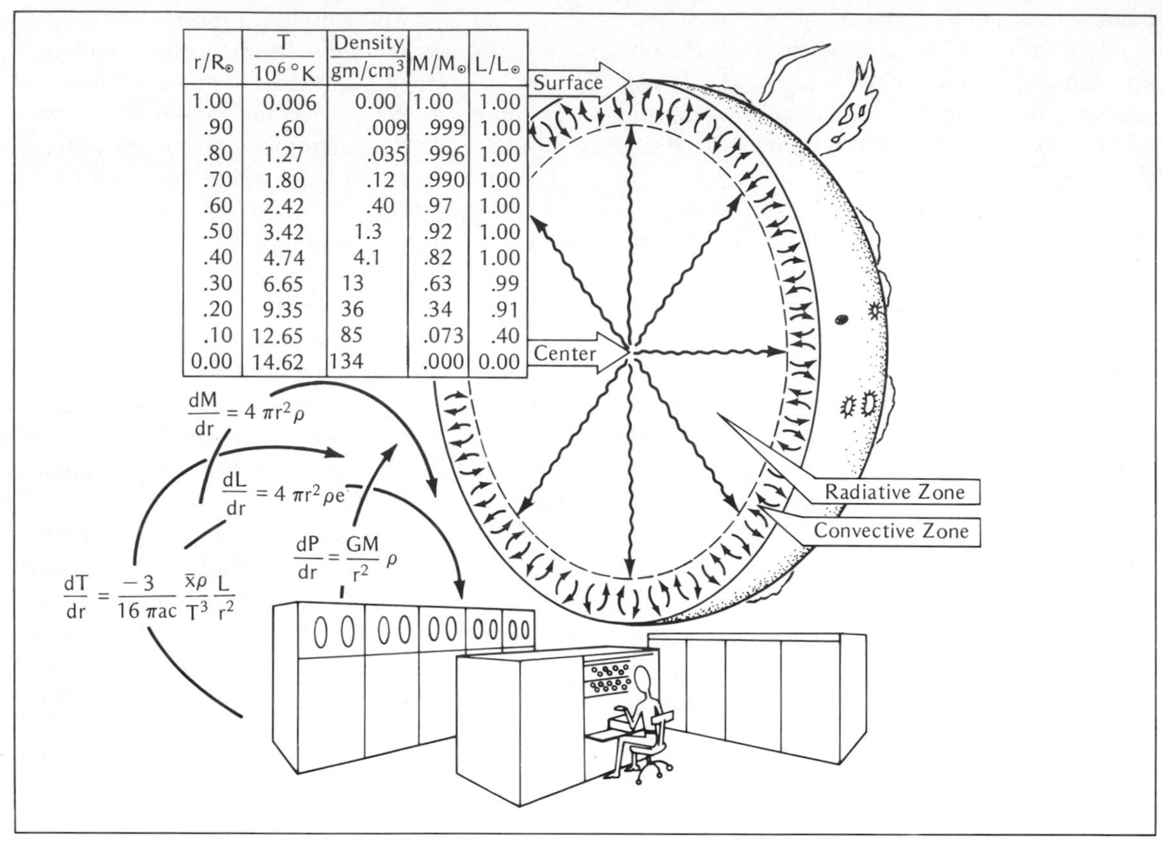

| $r/R_\odot$ | $T$ ($10^6$ °K) | Density ($gm/cm^3$) | $M/M_\odot$ | $L/L_\odot$ |
|---|---|---|---|---|
| 1.00 | 0.006 | 0.00 | 1.00 | 1.00 |
| .90 | .60 | .009 | .999 | 1.00 |
| .80 | 1.27 | .035 | .996 | 1.00 |
| .70 | 1.80 | .12 | .990 | 1.00 |
| .60 | 2.42 | .40 | .97 | 1.00 |
| .50 | 3.42 | 1.3 | .92 | 1.00 |
| .40 | 4.74 | 4.1 | .82 | 1.00 |
| .30 | 6.65 | 13 | .63 | .99 |
| .20 | 9.35 | 36 | .34 | .91 |
| .10 | 12.65 | 85 | .073 | .40 |
| 0.00 | 14.62 | 134 | .000 | 0.00 |



*Figure 6-16. A stellar model is a table of numbers that represents conditions inside a star. This table describes the interior of the sun. (Adapted from Michael A. Seeds, "Stellar Evolution,"* Astronomy, *February 1979.)*

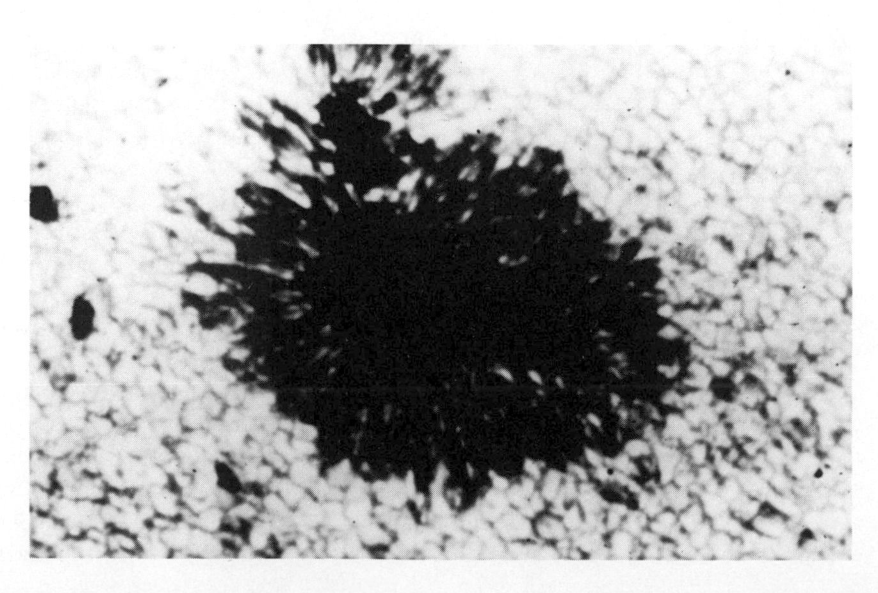

*Figure 6-17. Granulation on the solar surface is direct evidence of convection cells just below the visible surface. The sunspot is roughly the size of the earth. (NASA.)*

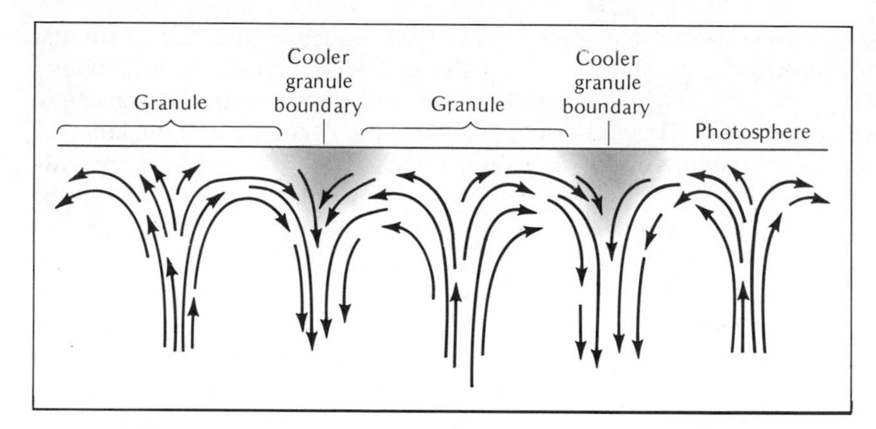


*Figure 6-18. Rising convection currents in the sun heat areas of the surface producing bright granules. Cooler material sinks at the darker edges of the granule.*

time machines to follow the evolution of stars over billions of years. To look into a star's future, for instance, we use a stellar model to determine how fast the star burns its fuel in each shell. As the fuel is consumed, the chemical composition of the gas changes and the amount of energy generated declines. By calculating the rate of these changes, we can predict what the star will look like at any point in the future.

Although this sounds simple, it is actually a highly challenging problem involving nuclear and atomic physics, thermodynamics, and sophisticated computational methods. Only since the 1950s have electronic computers made the rapid calculation of stellar models possible, and the advance of astronomy since then has been heavily influenced by the use of such models to study the structure and evolution of stars. Our summary of star formation in this chapter is based on thousands of stellar models. We will continue to rely on theoretical models as we study the main-sequence lives of stars in the next section and the deaths of stars in the next chapter.

## MAIN-SEQUENCE STARS

When a contracting protostar becomes hot enough to burn hydrogen, it becomes stable and stops contracting. The location of these stable stars in the H-R diagram is marked by the main sequence. Since 90 percent of all stars lie on the main sequence, it is important that we understand their structure.

*The Mass-Luminosity Relation.* In the previous chapter, we saw that the more massive stars are also more luminous. This mass-luminosity relation can be explained by the theory of stellar structure, giving us direct observational confirmation of the theory.

The key to the mass-luminosity relation is the law of hydrostatic equilibrium, which says that pressure must balance weight, and the pressure-temperature thermostat, which regulates energy production. We have seen that a star's internal pressure stays high because generation of thermonuclear energy keeps its interior hot. Since more massive stars have more weight pressing down on the inner layers, their interiors must have high pressures and thus must be hot. For example, the temperature at the center of a 15 $M_\odot$ star is about 34,000,000°K, more than twice the central temperature of the sun.

Since massive stars have hotter cores, their nuclear reactions burn more fiercely. That is, their pressure-temperature thermostat is set higher. One gram of material at the center of a 15 $M_\odot$ star burns over 3000 times more rapidly than 1 gm of material at the center of the sun. The rapidly burning reactions in massive stars make them more luminous than the lower-mass stars. Thus the mass-luminosity relation results from the requirement that a star support its weight by generating nuclear energy.

*The Life of a Main-Sequence Star.* Since a main-sequence star maintains its stability by burning hydrogen, it is inevitable that the star change when its hydrogen fuel is exhausted. However, long before the star runs out of hydrogen, its changing composition forces small alterations in its structure. When these small changes occur in the sun, life on earth will vanish.

Hydrogen burning combines four atoms into one. Thus, as a main-sequence star burns its hydrogen, the total number of atoms in its interior decreases. Each newly made helium atom can exert the same pressure as a hydrogen atom, but since the gas has fewer atoms, its total pressure is less. This unbalances the gravity-pressure stability, and gravity squeezes the core of the star more tightly. As the core contracts, its temperature increases and the nuclear reactions burn faster, releasing more energy and making the star more luminous. This additional energy flowing outward through the envelope forces the outer layers to expand and cool, so the star becomes slightly larger, brighter, and cooler.

As a result of these gradual changes in main-sequence stars, the main sequence is not a sharp line across the H-R diagram but rather a band (shaded in Figure 6-20). Stars begin their stable lives burning hydrogen on the lower edge of this band, the **zero age main sequence,** but gradual changes in luminosity and surface temperature move the stars upward and to the right. By the time they reach the upper edge of the main sequence (the dashed line in Figure 6-20), they have exhausted nearly all of the hydrogen in

## BOX 6-1 THE SOLAR NEUTRINO PROBLEM

Because the study of stellar structure is highly theoretical, astronomers are ever alert for ways to confirm their theories. One method measures the rate at which the sun produces energy by trapping **neutrinos,** massless atomic particles that travel at the speed of light.

The nuclear reactions in the sun's core produce a flood of neutrinos that rush outward into space. During the moment it takes to read this sentence, approximately $10^{12}$ solar neutrinos pass through your body, through the earth, and eventually on through the solar system. If we could catch and study these particles, we could learn about conditions at the sun's center. But neutrinos almost never react with other particles, so they are very difficult to detect.

In 1970, an experiment to trap solar neutrinos began. Because a few of the many solar neutrinos do interact with chlorine atoms, Raymond Davis, Jr., a research chemist from the Brookhaven National Laboratory, filled a 100,000-gallon tank with a type of cleaning fluid that contains a large percentage of chlorine atoms. This neutrino trap had to be buried nearly a mile underground in a South Dakota gold mine to shield it from cosmic rays from space, which could react with the chlorine and imitate neutrinos (Figure 6-19).

Of the trillions of solar neutrinos that pass through the tank each second, one should occasionally interact with a chlorine atom, transforming it into a radioactive argon atom. The argon

*Figure 6-19. The solar neutrino experiment consists of 100,000 gallons of cleaning fluid held in a tank nearly a mile underground. Solar neutrinos trapped in the cleaning fluid convert chlorine atoms into argon atoms that can be counted by their radioactivity. (Brookhaven National Laboratory.)*

their centers. Thus we find main-sequence stars scattered throughout this band at various stages of their main-sequence lives.

These gradual changes in the sun will spell trouble for earth. When the sun began its main-sequence life about 5 billion years ago, it was only about 60 percent as luminous as it is now, and by the time it leaves the main sequence in 5 billion years the sun will have twice its present luminosity. This will raise the average temperature on earth by at least 19°C (34°F). As this happens over the next few billion years, the polar caps will melt, the oceans will evaporate, and much of the atmosphere will vanish into space. Clearly the future of the earth as the home of life is limited by the future evolution of the sun.

Once a star leaves the main sequence, it evolves rapidly and dies. The average star spends 90 percent of its life burning hydrogen on the main sequence. This explains why 90 percent of all stars are main-sequence stars. We are most likely to see a star during that long, stable period while it is on the main sequence.

The number of years a star spends on the main sequence depends on its mass (Table 6-2). Massive stars burn fuel rapidly and live short lives, but low-mass stars conserve their fuel and shine for billions of years. For example, a 25 $M_{\odot}$ star will exhaust its hydrogen and die in only about seven million years. Very low mass stars, the red dwarfs, burn their fuel so slowly they last for 200 to 300 billion years. Since the universe seems to be only 10 to 20 billion years old, red dwarfs must still be in their infancy. Box 6-2 explains how we can quickly estimate the life expectancies of stars from their masses.

Nature makes more low-mass stars than mas-

---

atoms can be counted via their radioactivity. If we know the probability that a neutrino will interact with a chlorine atom, the number of argon atoms that appear in the tank can tell us how many neutrinos the sun is producing, and that should tell us how rapidly the nuclear reactions are burning at the sun's core.

As the experiment began, the best models of the sun predicted that the tank should catch about five neutrinos per week. After the tank accumulated argon atoms for months, the experimenters found that only one neutrino was caught per week. The experiment has been repeated many times and its accuracy has been improved, but it still detects too few neutrinos coming from the sun.

The missing neutrinos are a serious problem because they seem to refute the theoretical models of the sun and stars on which much of modern astronomy is based. Can there be something wrong with the theory, or is there something wrong with the experiment? One possibility is the uncertainty in the theoretically predicted rates of neutrino production in the sun. Since these nuclear reactions occur at temperatures that cannot be duplicated on earth, their rates must be extrapolated from measurements made at lower temperatures. Other theories suggest that the sun pulsates with a period of a few hundred years, and we happen to be observing it while the nuclear reactions are not operating at peak power. Another possible explanation lies in the elusive nature of the neutrinos themselves. Recently, physicists have suggested that the neutrino may oscillate among three different states. The cleaning fluid neutrino trap is capable of catching only one type, so if the neutrinos emitted by the sun oscillate to other types during the eight minutes they take to reach the earth, the experiment would detect fewer than expected.

Whatever the explanation, modern astronomers are not ready to abandon existing theories of stellar structure on the basis of one experiment. At the same time, the results of the neutrino experiment cannot be ignored. At present, an uneasy truce reigns as experimenters try to design new equipment to search for the missing neutrinos, and theorists try to adjust their models to account for the deficiency.

**Table 6-2 Main-Sequence Stars**

| Spectral Type | Mass (Sun = 1) | Luminosity (Sun = 1) | Approximate Years on Main Sequence |
|---|---|---|---|
| O5 | 40 | 405,000 | $1 \times 10^6$ |
| B0 | 15 | 13,000 | $11 \times 10^6$ |
| A0 | 3.5 | 80 | $440 \times 10^6$ |
| F0 | 1.7 | 6.4 | $3 \times 10^9$ |
| G0 | 1.1 | 1.4 | $8 \times 10^9$ |
| K0 | 0.8 | 0.46 | $17 \times 10^9$ |
| M0 | 0.5 | 0.08 | $56 \times 10^9$ |

sive stars, but this fact is not sufficient to explain the vast numbers of low-mass stars that fill the sky. An additional factor is the stellar lifetimes. Because low-mass stars live long lives, there are more of them in the sky than massive stars. Look at Figure 5-24 and notice how much more common the lower-main-sequence stars are than the massive O and B stars. The main-sequence K and M stars are so faint they are difficult to locate but they are very common. The O and B stars are luminous and easy to locate, but, because of their fleeting lives, there are never more than a few on the main sequence at any one time.

When a star finally exhausts its hydrogen fuel, it can no longer resist the pull of its own

BOX 6-2 THE LIFE EXPECTANCIES OF THE STARS

We can estimate the amount of time a star spends on the main sequence—its life expectancy T—by estimating the amount of fuel it has and the rate at which it consumes that fuel.

$$T = \frac{\text{fuel}}{\text{rate of consumption}}$$

The amount of fuel a star has is proportional to its mass, and the rate at which it burns its fuel is proportional to its luminosity. Thus its life expectancy must be proportional to M/L. But we can simplify this further because, as we saw in the last chapter, the luminosity of a star depends on its mass raised to the 3.5 power ($L = M^{3.5}$). Thus the life expectancy is

$$T = \frac{M}{M^{3.5}}$$

or

$$T = \frac{1}{M^{2.5}}.$$

If we express the mass in solar masses, the lifetime will be in solar lifetimes. For example, a 4 solar mass star will last for

$$T = \frac{1}{4^{2.5}} = \frac{1}{4 \cdot 4\sqrt{4}} = \frac{1}{32} \text{ solar lifetimes.}$$

Studies of solar models show that the sun, presently 5 billion years old, can last another 5 billion years. Thus a solar lifetime is approximately 10 billion years, and a 4 solar mass star will last for about

$$T = \frac{1}{32} \times (10 \times 10^9 \text{ yr}) = 310 \times 10^6 \text{ years.}$$

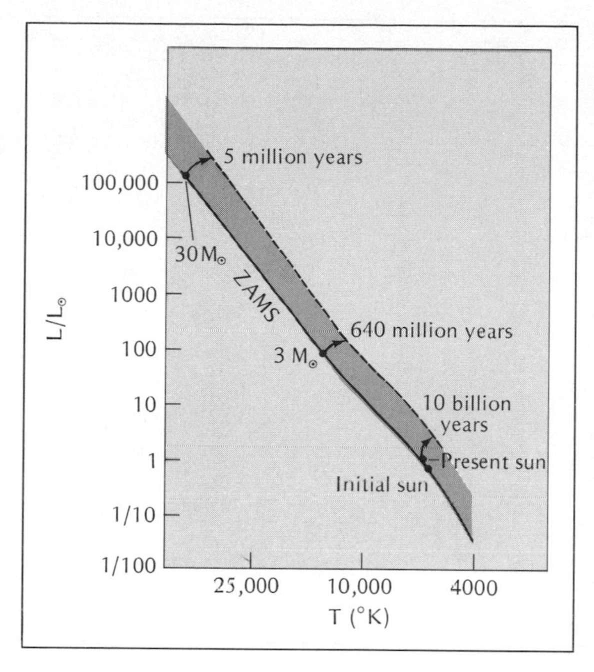


*Figure 6-20. The main sequence is not a line but a band (shaded). Stars begin their main-sequence lives on the lower edge, which is called the zero age main sequence (ZAMS). As hydrogen burning changes their composition, the stars slowly move across the band.*

*Figure 6-21. The Great Nebula in Orion is a glowing cloud of hot gas over 8 pc in diameter. It is excited by a cluster of hot stars, the Trapezium, that lies at its center. (Lick Observatory photograph.)*

gravity. Contraction resumes and the star collapses. As the star dies, it can delay its end by burning other fuels, but as we will discover in the next chapter, nothing can steal gravity's final victory.

Before we examine the deaths of stars, we should look once more at regions where stars are forming. Recent observations in the ultraviolet, infrared, and radio parts of the spectrum have revealed that a nebula familiar to generations of astronomers is actually a region of intense star formation.

## PERSPECTIVE: THE ORION NEBULA

The fuzzy wisp of nebula visible to the naked eye in Orion's sword has attracted the attention of astronomers and casual stargazers throughout history. Commonly refered to as the Great Nebula in Orion, it is a striking sight through binoculars or a small telescope, and through a large telescope it is breathtaking. (See Color Plate 12.) At the center lie four brilliant blue-white stars known as the Trapezium, and surrounding them are the glowing filaments of a nebula over 8 pc across (Figure 6-21). Like a great thunder cloud illuminated from within, the churning currents of gas and dust testify to the violence of the mechanisms that created them. However, a deeper significance lies hidden, figuratively and literally, behind the Great Nebula, for in the last decade radio and infrared astronomers have discovered a vast dark cloud lying just beyond the visible nebula—a cloud in which stars are now being created.

*Star Formation in Orion.* It should not surprise us that Orion is a site of star formation. The stars that make up the constellation are mostly hot, bright, main-sequence stars. These massive stars have short lifetimes and therefore must have formed recently. In addition, the region contains associations of T Tauri stars, which are probably stars in the later stages of contraction to the main sequence. These new stars most likely formed from the large gas and dust clouds that fill the Orion region.

The Great Nebula represents a late stage of star formation. The stars in the Trapezium reached the main sequence only a few million years ago. The most massive, a star of about 40 solar masses, must burn its fuel at a tremendous rate to support its large mass. Thus it is hot and luminous. Its surface temperature is about 30,000°K, and it is about 300,000 times more luminous than the sun. The other stars are too cool to affect the surrounding gas very much, but the 40 solar mass star is so hot it radiates large amounts of ultraviolet radiation (Figure 6-22). These ultraviolet photons have enough energy to ionize the hydrogen gas in the region near the Trapezium, creating the glowing clouds we see as the Great Nebula.

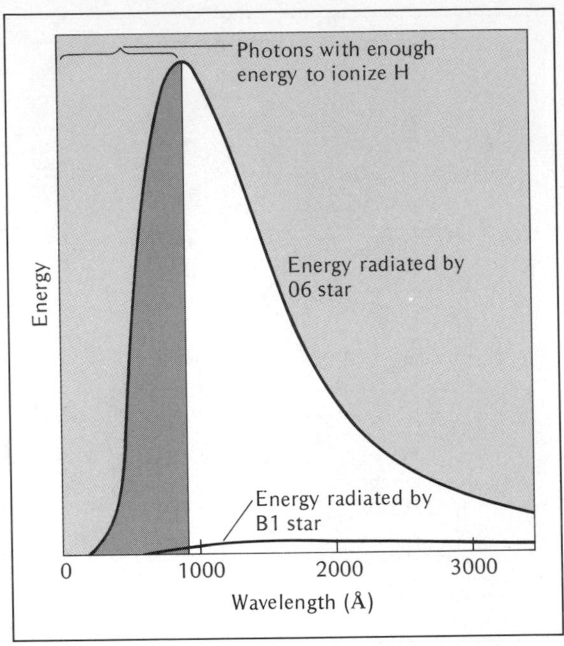


*Figure 6-22. Photons with wavelengths shorter than 912 Å have enough energy to ionize hydrogen. The O6 star is the only star in the Trapezium hot enough to produce appreciable ionization.*

Although the nebula looks impressive, it is nearly a vacuum, containing a mere 600 atoms/cm$^3$. For comparison, the interstellar medium has an average density of about 1 atom/cm$^3$, and the density of air at sea level is about $10^{19}$ atoms/cm$^3$. Infrared observations show that the thin, ionized gas is mixed with sparsely scattered dust, heated by the central stars to a temperature of about 70°K.

*The Molecular Cloud.* The importance of the Orion region was established when observations at wavelengths longer than visible light revealed a dense cloud of gas lying just beyond the Great Nebula. These observations spanned the region of the spectrum that includes infrared and short-wavelength radio waves. At these wavelengths, hot stars and ionized gas are invisible, but cool dense gas is detectable. The gas cloud beyond the Great Nebula is so dense that molecules such as carbon monoxide have formed where the dense dust protects them from ultraviolet radiation. These molecules emit the radio signals that make the cloud detectable. As shown in Figure 6-23, the cloud contains two dense regions, named the Orion Molecular Clouds 1 and 2 (OMC1 and OMC2).

These clouds are significantly different from the Great Nebula. The molecular clouds contain at least $10^6$ atoms/cm$^3$ and large amounts of dust. In addition, observations at different wavelengths show that the clouds are warmest near their centers. This heating cannot be due to massive stars because the gas is not being ionized. Instead, astronomers conclude that the clouds are growing hotter at their centers because they are contracting.

*The Infrared Clusters.* Infrared observations reveal that clusters of warm objects lie at the centers of both OMC1 and OMC2. Invisible at optical wavelengths, these objects are evidently stars in pre-main-sequence stages,

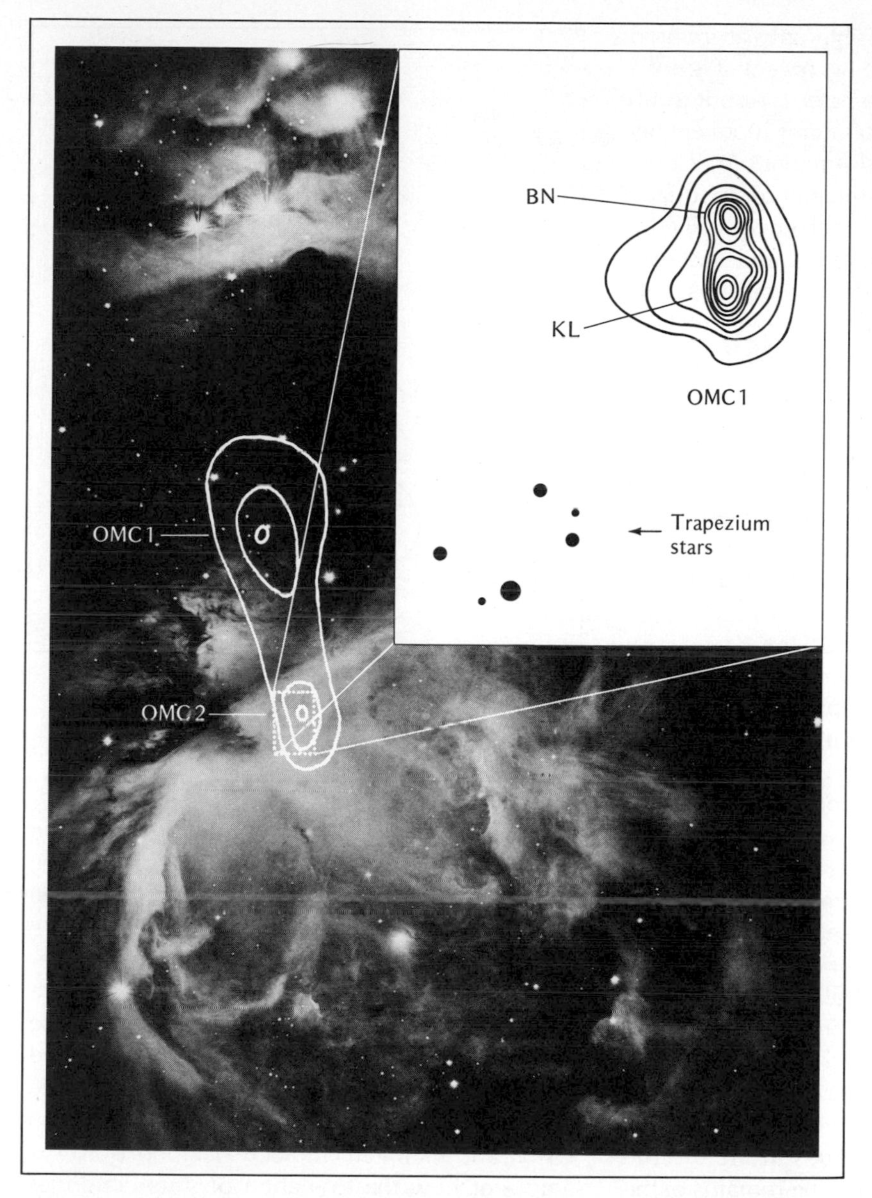


*Figure 6-23. Molecular radio emissions outline the Orion molecular clouds (OMC1 and OMC2). Invisible behind the Great Nebula, they are sites of star formation. An infrared map (inset) shows the location of the BN object and KL nebula imbedded in OMC1, both of which are discussed in the text. Nearby are the Trapezium stars. (Map adapted from Werner, Becklin, and Neugebauer; Lick Observatory photograph.)*

wrapped deep in dust cocoons. Two especially interesting objects lie near the center of OMC1.

The Becklin-Neugebauer object (BN object), named after its discoverers, is probably a single star nearing the main sequence. Although the star may have a surface temperature exceeding 5000°K, it is invisible within a cloud of dust no more than 200 AU in diameter. The dust absorbs the energy radiated by the star and reradiates it as infrared radiation. Radio astronomers strengthen this hypothesis by reporting that the gas near the object is not ionized. Thus the object is not a hot main-sequence star like that found in the Trapezium.

The second infrared object in OMC1 is the Kleinmann-Low nebula (KL nebula). Like the BN object, the KL nebula is visible only at infrared wavelengths, but unlike the BN object, it is not a single object. The total diameter is about 2000 AU, and it emits about 2000 times more energy than the sun, although its temperature is only 100°K. With such a low temperature its surface must be large to radiate so much energy. The KL nebula is probably a cloud of gas and dust collapsing to form a small cluster of protostars at its center. The lack of ionized gas tells us that no massive stars have yet reached the main sequence.

*The Future of the Orion Cloud.* All evidence points to star formation within the molecular cloud just behind the Great Nebula in Orion. To predict the future of this cloud, we have only to look at the ionized nebula around the Trapezium stars.

About 6 million years ago the Trapezium stars were contracting protostars buried within a molecular cloud. As they approached the main sequence, their temperatures increased and their radiation drove away their cocoons. But not until the 40 $M_{\odot}$ star turned on was there sufficient ultraviolet radiation to ionize the gas. The ionization transformed the cloud from an opaque shield into a transparent gas and forced it to expand, an expansion that continues to this day. We can see evidence of this expansion in the Doppler shifts of spectral lines and in the twisted filaments of gas within the nebula.

Once the hot star ionized the gas, star formation in the Trapezium region stopped. In more distant parts of the cloud, the contraction of the gas into protostars continues. Indeed, the ionized gases pushing against the remains of the cloud may have triggered the collapse of more protostars. We now see the front of the cloud torn by the expanding ionized gases around the Trapezium cluster, while deeper within the cloud more protostars are forming (Figure 6-24).

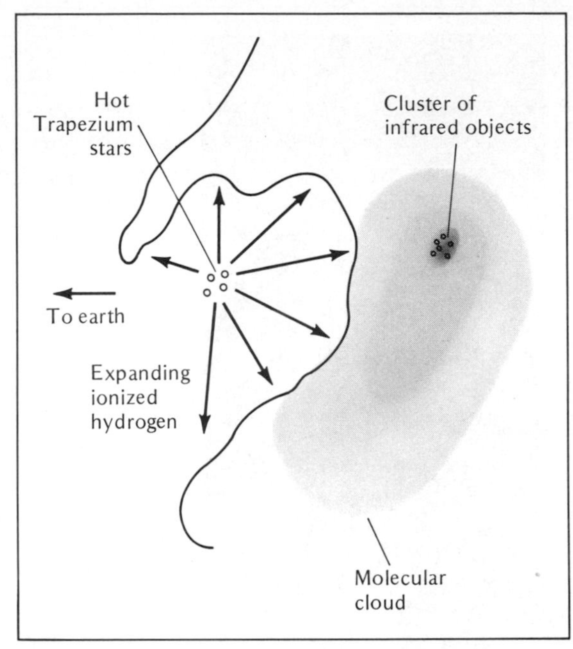


*Figure 6-24. A side view of the Great Nebula would show the newly formed Trapezium stars heating and driving away the surrounding gas as new stars form in OMC1.*

In the next few thousand years, the familiar outline of the Great Nebula will change, and a new nebula may form as the protostars in the molecular clouds reach the main sequence and the most massive become hot enough to ionize the surrounding gas. Thus the Great Nebula in Orion and its parent molecular cloud is an example of how the formation of stars continues even today.

## SUMMARY

The interstellar medium contains large, cool clouds of gas, most of which are not dense enough to collapse spontaneously to form stars. Apparently they must be triggered into collapse by collision with a shock wave. Such a collision fragments and compresses the gas, producing a cluster of protostars.

A contracting protostar is large and cool and would appear in the red giant region of the H-R diagram were it not surrounded by a cocoon of dust and gas. The dust in the cocoon absorbs the protostar's light and reradiates it as infrared radiation. Many infrared sources are probably protostars. Herbig-Haro objects and T Tauri stars may be cocoon stars in later stages as the cocoon is absorbed by the protostar or driven into space. Very young star clusters, like NGC 2264, contain large numbers of T Tauri stars.

As a protostar's center grows hot enough to burn hydrogen, it settles onto the main sequence to begin its long hydrogen-burning phase. If it is a low-mass star, it cannot become very hot and must burn by the proton-proton chain. But if it is a massive star, it will become hot enough to burn by the CNO cycle. In either case, the nuclear reactions are regulated by the pressure-temperature thermostat in the star's core.

Almost everything we know about the internal structure of stars comes from mathematical stellar models. The models are based on four simple laws of stellar structure. The first two laws say that mass and energy must be conserved and spread smoothly through the star. The third, hydrostatic equilibrium, says the star must balance the weight of its layers by its internal pressure. The fourth says energy can only flow outward by conduction, convection, or radiation.

The mass-luminosity relation is explained by the requirement that a star support the weight of its layers by its internal pressure. The more massive a star is, the more weight it must support, and the higher its internal pressure must be. To keep its pressure high, it must be hot and generate large amounts of energy. Thus the mass of a star determines its luminosity. The massive stars are very luminous and lie along the upper main sequence. The less massive stars are fainter and lie lower on the main sequence.

How long a star can stay on the main sequence depends on its mass. The more massive a star is, the faster it uses up its hydrogen fuel. A 25 $M_\odot$ star will exhaust its hydrogen and die in only about 7 million years, but the sun can last for 10 billion years.

## NEW TERMS

interstellar medium
interstellar reddening
Bok globule
shock wave
protostar
cocoon
Herbig-Haro object
T Tauri star
Coulomb barrier
proton-proton chain
CNO cycle
continuity of mass law
continuity of energy law
hydrostatic equilibrium
energy transport
opacity
stellar model
neutrino
solar granulation
zero age main sequence

## QUESTIONS

1. How can we detect and study the interstellar medium?
2. How does a protostar convert gravitational energy into thermal energy?
3. What observational evidence do we have that (a) star formation is a continuous process? (b) protostars really exist? (c) the Orion region is actively forming stars?
4. How do the proton-proton chain and the CNO cycle resemble each other? How do they differ?
5. How does the pressure-temperature thermostat control the nuclear reactions inside stars?
6. What is the solar neutrino problem? How might it be resolved?
7. What is solar granulation?
8. Why is there a mass-luminosity relation?
9. Why does a star's life expectancy depend on its mass?
10. Describe the BN object and the KL nebula. What is the significance of the lack of ionization?

## PROBLEMS

1. Sketch a series of diagrams to show how a gas cloud behaves when it encounters a shock wave.
2. On an H-R diagram, draw the path taken by a protostar as it contracts to the main sequence.
3. In the H-R diagram from problem 2, indicate the location of the stars in a young cluster containing T Tauri stars.
4. How much energy is produced when the sun converts one gram of mass into energy?
5. How much energy is produced when the sun converts 1 gram of hydrogen into helium?
6. Circle all $^1$H and $^4$He nuclei in Figures 6-10 and 6-11. Explain how both the proton-proton chain and the CNO cycle can be summarized by $4^1\text{H} \rightarrow {}^4\text{He} + \text{energy}$.
7. In the model shown in Figure 6-16, how much of the sun's mass is hotter than 12,000,000°K?
8. What is the life expectancy of a 16 $M_\odot$ star?
9. The hottest star in the Orion nebula has a surface temperature of 30,000°K. At what wavelength does it radiate the most energy? (See Box 3-1.)

## RECOMMENDED READING

Bok, B. J. "The Birth of Stars." *Scientific American* 222 (Aug. 1972), p. 49.

Cohen, M. "Star Formation and Early Evolution." *Mercury* 4 (Sept./Oct. 1975), p. 10.

Cohen, M. "Stellar Formation." *Astronomy* 7 (Sept. 1979), p. 66.

Dickinson, D. "Cosmic Masers." *Scientific American* 238 (June 1978), p. 90.

Heiles, C. "The Structure of the Interstellar Medium." *Scientific American* 238 (Jan. 1978), p. 74.

Knacke, R. F. "Solid Particles in Space." *Sky and Telescope* 57 (April 1979), p. 347.

Loren, R. B., and Vrba, F. J. "Starmaking with Colliding Molecular Clouds." *Sky and Telescope* 57 (June 1979), p. 521.

Seeds, M. "Stellar Evolution." *Astronomy* 7 (February 1979), p. 6. Reprinted in *Astronomy: Selected Readings,* ed. M. A. Seeds. Menlo Park, Calif.: Benjamin/Cummings, 1980.

Strom, S., and Strom, K. "The Early Evolution of Stars." *Sky and Telescope* 45 (May/June 1973), pp. 279 and 359.

Turner, B. "Interstellar Molecules." *Scientific American* 228 (March 1973), p. 50.

Ulrich, R. K. "Solar Neutrinos and Variations in the Solar Luminosity." *Science* 190 (14 Nov. 1975), p. 619.

Zeilik, M. "Birth Places of Stars." *Astronomy* 4 (Oct. 1976), p. 31.

Zeilik, M. "The Birth of Massive Stars." *Scientific American* 238 (April 1978), p. 110.

*The Veil nebula is a portion of the expanding shell of gas ejected from a supernova explosion that occurred roughly 50,000 years ago. (See Figure 7-12.) (Lick Observatory photograph.)*

# Chapter 7 THE DEATHS OF STARS

The previous chapter explored the formation of stars and their long stable lives on the main sequence. In this chapter, we will complete the picture by studying the way stars exhaust their fuels and die. Once we have the stellar life cycle worked out, we will be prepared to understand astronomical problems as diverse as the origin of galaxies, the structure and evolution of the universe, and the origin of planetary systems.

The stellar death process begins when the star runs out of hydrogen at its core and is no longer able to resist its own gravity. As it delays its end by burning helium and other elements, it swells into a giant star over 10 times its original diameter. Soon the star exhausts all of its available fuels and dies.

Perhaps the most important principle of stellar evolution is that the properties of stars depend mainly on their mass. For one thing, massive stars evolve faster and die sooner than low-mass stars. The most massive stars may live only a few million years, while the lowest mass stars live hundreds of billions. In addition, stars of different mass die in different ways. Low-mass stars die quietly, but massive stars die in tremendous explosions called supernovae. To follow the evolution of stars through the giant stage to their final collapse, we must consider stars of various masses.

The deaths of stars lead invariably to one of three final states. If the star is a medium- to low-mass star, it becomes a white dwarf, a star about the size of the earth with no usable fuels left. Massive stars explode and may leave behind either a neutron star or a black hole. Which of these objects remains after the explosion depends on how much mass is left.

Neutron stars and black holes were originally predicted theoretically, but observers have now located certain celestial objects that seem to confirm the predictions. Radio astronomers have located sources of radio pulses that are apparently neutron stars spinning faster than once a second. So far, x-ray astronomers have discovered at least two x-ray objects that may be black holes.

The final states of stellar evolution, white dwarfs, neutron stars, and black holes, represent the final victory of gravity. Understanding stellar evolution and death means we must first understand how the star delays its inevitable collapse. We begin this chapter at the beginning of the end, the evolution of giant stars.

## GIANT STARS

A star spends over 90 percent of its life on the main sequence, and only a small percentage of its life as a giant star. Thus by the time a star begins to swell into a giant, its life is nearly over and the death process has begun.

*The Expansion into a Giant.* The nuclear reactions in a star's core burn hydrogen and produce helium, which accumulates at the star's center like ashes in a fireplace. Because the helium is cooler than 100,000,000°K, it cannot burn in nuclear reactions. Initially this helium ash has little effect on the star, but as hydrogen is exhausted and the stellar core becomes almost pure helium, the star loses the ability to generate nuclear energy. Since it is the energy generated at the center that opposes gravity and supports the star, the core begins to contract as soon as the energy generation starts to die down.

Although the contracting helium core cannot generate nuclear energy, it does grow hotter because it converts gravitational energy into thermal energy (see Chapter 6). The rising temperature heats the unprocessed hydrogen just outside the core, hydrogen that was never before hot enough to burn. When the temperature of the surrounding hydrogen becomes high enough, it ignites in a burning hydrogen shell. Like a grass fire burning outward from an exhausted campfire, the hydrogen-burning shell burns outward, leaving helium ash behind and increasing the mass of the helium core.

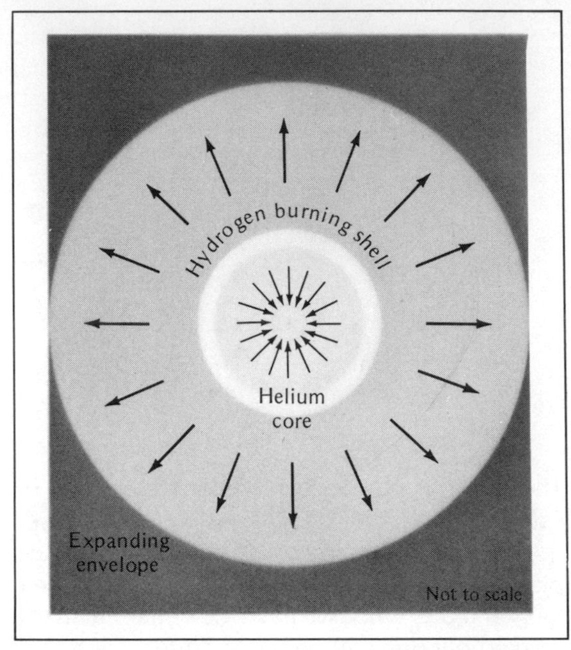


*Figure 7-1. When a star runs out of hydrogen at its center, it ignites a hydrogen-burning shell. The helium core contracts and heats, while the envelope expands and cools. (For a scale drawing see Figure 7-3.)*

The flood of energy produced by the hydrogen-burning shell pushes toward the surface, heating the layers of the envelope, forcing them to expand, and swelling the star into a giant (Figure 7-1). This accounts for the large diameters and low densities of the giant stars. In Chapter 5, we found that giant and supergiant stars were 10 to 1000 times larger in diameter than the sun and from 10 to $10^6$ times less dense. Thus giant and supergiant stars are normal main-sequence stars that expanded to large size and low density when hydrogen-shell burning began.

The expansion of the envelope dramatically changes the star's location in the H-R diagram. As the outer layers expand outward, they cool, and the star moves quickly to the right into the red giant region. As the radius continues to increase, the enlarging surface area makes the star more luminous, moving it upward in the H-R diagram (Figure 7-2). Aldebaran ($\alpha$ Tauri), the glowing red eye of Taurus the bull, is such a red giant, having a diameter 25 times that of the sun, but with a surface temperature only half that of the sun.

*Helium Burning.* Although the hydrogen-burning shell can force the envelope of the star to expand, it cannot stop the contraction of the helium core. Since the core has no energy source, gravity squeezes it tighter and the temperature continues to rise. When the helium reaches 100,000,000°K, it begins to burn in nuclear reactions that convert it to carbon.

Some stars begin helium burning gradually, but others begin with an explosion called the **helium flash.** This explosion is caused by the density of the helium, which can reach 15,000 gm/cm³. On earth, a teaspoon of this material

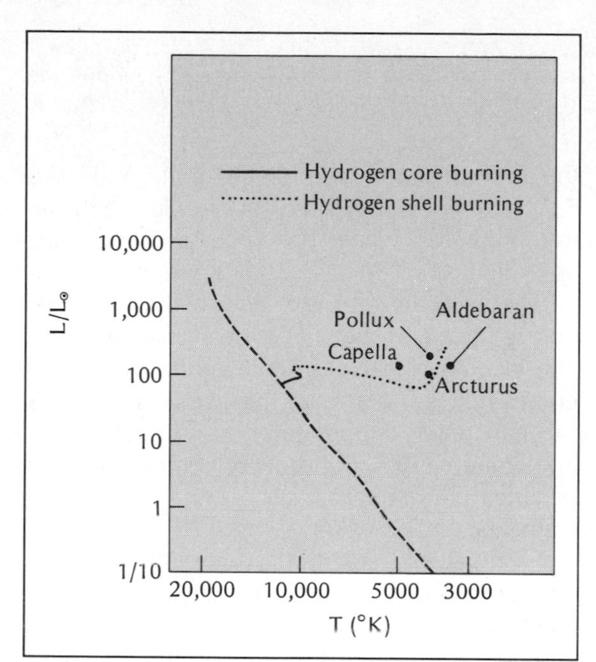


*Figure 7-2.* *Many of the bright, red stars in the night sky are giants. Not long ago they were main-sequence stars, but they expanded when their cores ran out of hydrogen. (Adapted from Michael A. Seeds, "Stellar Evolution"* Astronomy, February *1979.)*

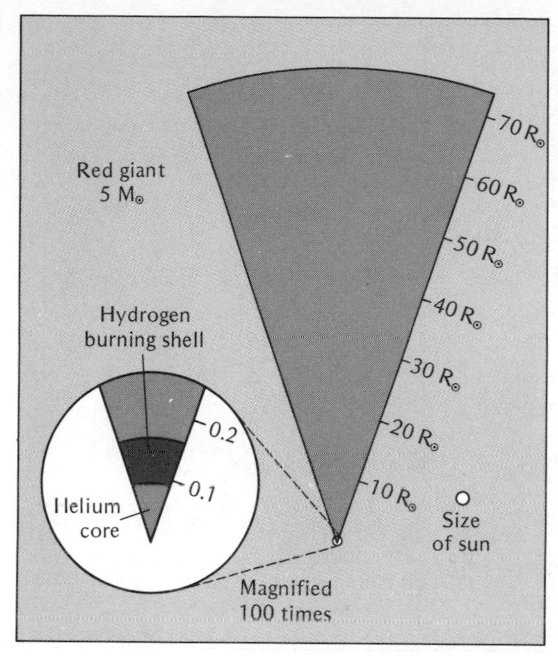


*Figure 7-3.* *A plug cut from a 5 $M_\odot$ red giant would be almost all low-density envelope. The magnified image shows the hydrogen-burning shell and the helium core. The core contains roughly 12 percent of the star's mass. (Adapted from Michael A. Seeds, "Stellar Evolution," * Astronomy, February *1979.)*

would weigh as much as a small automobile. At these densities, the gas becomes **degenerate** (Box 7-1) and its pressure no longer depends on its temperature. Thus the pressure-temperature thermostat that automatically controls the nuclear reactions no longer works, and when the helium ignites, it explodes. The explosion is so violent that for a few moments the helium core may generate more energy per second than an entire galaxy.

Although the helium flash is sudden and powerful, it does not destroy the star. In fact, if you were observing a giant star as it experienced the helium flash, you would see no outward evidence of the eruption. The helium core is quite small (Figure 7-3) and all of the energy of the explosion is absorbed by the distended envelope. In addition, the helium flash is a very shortlived event in the life of the star. In a matter of hours, the core of the star becomes so hot it is no longer degenerate, the pressure-temperature thermostat brings the helium burning under control, and the star proceeds to burn helium steadily in its core (Figure 7-4).

Not all stars experience a helium flash. Stars less massive than about 0.4 $M_\odot$ can never get hot enough to ignite helium, and stars more massive than about 3 $M_\odot$ ignite helium before their cores become degenerate. In such stars, pressure depends on temperature, so the pressure-temperature thermostat keeps the helium burning under control.

Helium burning produces carbon and oxygen, atoms that do not burn at the low temperatures of the helium core. Thus when the helium in the core is used up, the core contracts, grows hotter, and ignites a helium-burning shell. The star expands and moves back toward the right in

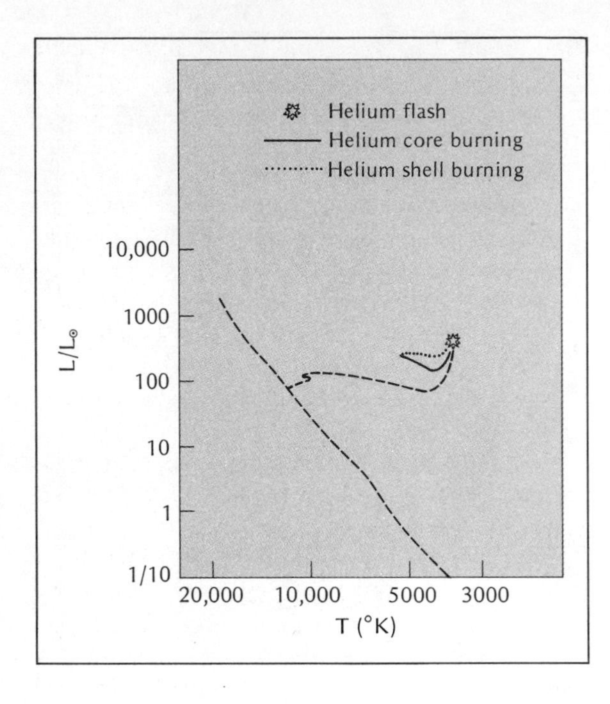


*Figure 7-4. After the helium flash at the tip of the red giant track, the star moves through a loop in the diagram as it burns helium in its core and then in a shell.*

the H-R diagram, completing a loop (Figure 7-4). This loop may carry the star through the instability strip, where it becomes a variable star, as explained in Box 7-2 and Figure 7-6.

The evolution of a star after helium exhaustion is uncertain, but the general plan is clear. The inert carbon-oxygen core contracts and becomes hotter. Stars more massive than about 3 $M_{\odot}$ can reach temperatures of 600,000,000°K and ignite carbon. Subsequent contraction may be able to burn oxygen, silicon, and other heavy elements.

Eventually the star must exhaust all of its fuels and die in a final collapse. Exactly how the star collapses depends on which fuels it has burnt and on the strength of its gravity. Both of these

## BOX 7-1 DEGENERATE MATTER

Normally the pressure in a gas depends on its temperature, but under certain circumstances nature may break that rule, often with catastrophic consequences for the star. To see how this works, we must consider the energy of the free electrons in an ionized gas.

The gas inside a star is completely ionized. That is, it consists of the bare nuclei of the atoms mixed with free electrons. At high densities and temperatures the pressure of the gas is due almost entirely to these electrons.

If the density is very high, the particles of the gas are forced close together, and two laws of quantum mechanics become important. First, quantum mechanics says that the moving electrons confined in the star's core can have only certain amounts of energy just as the electron in an atom can occupy only certain energy levels (see Chapter 4). We can think of these permitted energies as the rungs in a ladder. An electron can occupy any rung, but not the spaces between. The second quantum-mechanical law (the Pauli exclusion principle) says that no two electrons can occupy the same energy level. That is, two electrons cannot occupy the same rung.

In a low-density gas, there are few electrons per cubic centimeter, so there are plenty of empty energy levels (Figure 7-5). However, if a gas becomes very dense, nearly all of the lower energy levels may be occupied, and the gas is termed **degenerate.** In such matter, an electron cannot slow down because there are no open energy levels for it to drop down to. It can speed up only if it can absorb enough energy to leap to the top of the energy ladder, where there are empty energy levels.

This has two important effects in stars. First, the pressure of degenerate gas resists compression. To compress the gas we must push against the moving electrons, and changing their motion means

factors depend on the star's mass, so we must divide the stars into two groups, low-mass stars and high-mass stars.

## THE DEATHS OF LOW-MASS STARS

Since contracting stars heat up by converting gravitational energy into thermal energy, low-mass stars cannot get very hot. This limits the fuels they can ignite.

Structural differences divide the low-mass stars into two subgroups—very low mass red dwarfs, and medium-mass stars such as the sun. The critical difference between the two groups is the extent of interior convection. If the star is convective, fuel is constantly mixed, and the resulting evolution is drastically altered.

*Red Dwarfs.* Stars less massive than about 0.4 $M_\odot$ are totally convective. The gas is constantly mixed, so hydrogen is consumed and helium accumulates uniformly throughout the star. Since the star cannot develop an inert helium core surrounded by a shell of unprocessed hydrogen, it never ignites a hydrogen shell and cannot become a giant.

As nuclear reactions convert hydrogen to helium, the star slowly contracts and heats up. Because it has a low mass, it cannot get hot enough to ignite its helium. Contraction continues until the gas becomes degenerate and, like solid steel, resists further compression.

Thus red dwarfs contract, heat up, and move to the left side of the H-R diagram to become white dwarfs. As we will see, white dwarfs are small, degenerate stars, unable to burn their remaining nuclear fuels.

*Medium-Mass Stars.* Stars with masses between 0.4 $M_\odot$ and about 3 $M_\odot$* exhaust hy-

* This mass limit is uncertain, as are many of the masses quoted here. The evolution of stars is highly complex, and such parameters are not well known.

---

changing their energy. That requires tremendous effort because we must boost them to the top of the energy ladder. Thus degenerate matter, though still a gas, takes on the consistency of hardened steel.

In addition, the pressure of degenerate gas does not depend on temperature. The pressure depends on the speed of the electrons, which cannot be changed without tremendous effort. The temperature, however, depends on the motion of all of the particles in the gas, both electrons and nuclei. If we add heat, most of it goes to speed up the motions of the nuclei, and only a few electrons can absorb enough energy to reach the empty energy levels at the top of the energy ladder. Thus changing the temperature of the gas has almost no effect on the pressure.

These two properties of degenerate matter become important when stars die. The hardened steel effect supports white dwarfs, while the independence of pressure and temperature causes the helium flash, the carbon detonation, and the supernova explosion.

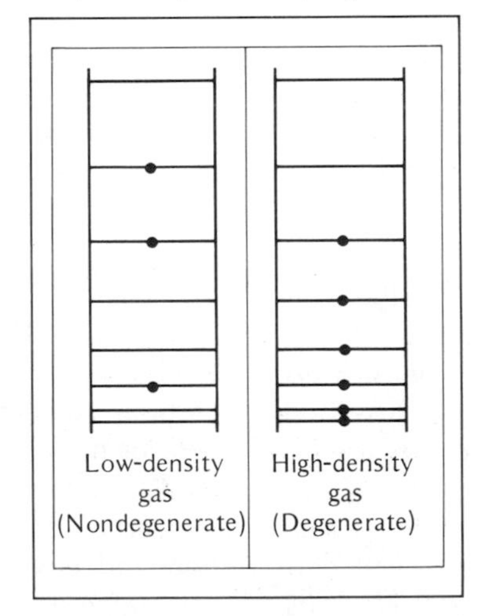


*Figure 7-5. Electron energy levels are arranged like rungs on a ladder. In a low-density gas, many levels are unoccupied, but in a degenerate gas all lower energy levels are filled.*

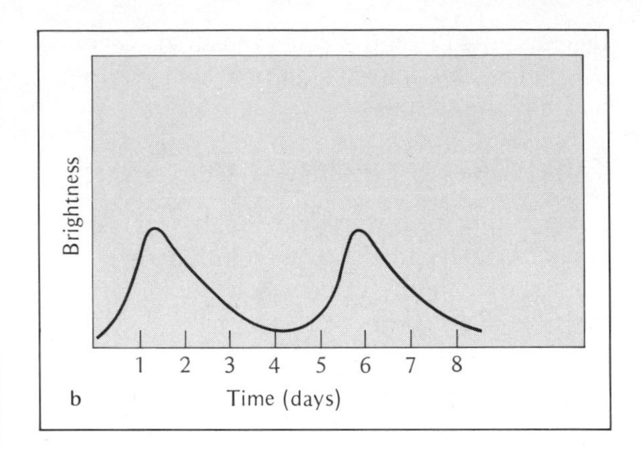


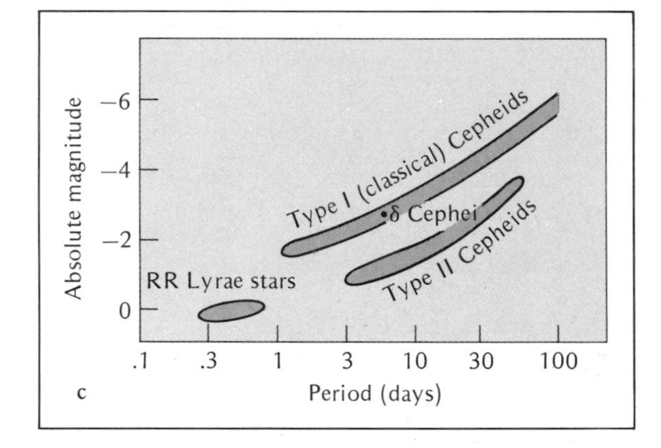


*Figure 7-6. (a) When a star enters the instability strip (shaded), it becomes a variable star. Massive stars cross the strip at higher luminosities than lower-mass stars, and, because of their higher mass and larger radius, they pulsate with longer periods. (b) A Cepheid variable can be identified by its brightness variation. (c) A period-luminosity diagram shows the different types of variables.*

drogen in their cores, burn hydrogen shells, become giant stars, and ignite helium, but they cannot get hot enough to burn carbon, the next fuel in the sequence. When they reach that impasse, they collapse and become white dwarfs. There are two keys to the evolution of these stars: the lack of complete mixing and the cooling of the surface as the star expands.

The interiors of medium-mass stars are not completely mixed. Stars of 1.1 $M_\odot$ or less have no convection near their centers so they are not mixed at all. Stars more massive than 1.1 $M_\odot$ have small zones of convection at their centers but this mixes no more than about 12 percent of the star's mass. Thus medium-mass stars, whether they have convective cores or not, are not mixed and the helium accumulates in an inert helium core surrounded by unprocessed hydrogen. When this core contracts, the unprocessed hydrogen ignites in a shell and swells the star into a giant.

As a giant, the star burns helium in its core and then in a shell surrounding a core of carbon and oxygen. This core contracts and grows hotter, but cannot become hot enough to ignite the carbon. Thus the carbon-oxygen core is a dead end for these medium-mass stars.

Because no nuclear reactions can begin in the carbon core, nothing stops the core contraction and envelope expansion. The star grows very large and cool, until the runaway expansion blows the stellar surface into space, forming a **planetary nebula,** an expanding shell of gas around a dying star (Figure 7-7 and Color Plate 14).

These objects are called planetary nebulae because, through a small telescope, they look like the greenish disks of distant planets such as Neptune and Uranus. Larger telescopes and time-exposure photographs show that they are spherical shells of gas. The Doppler shifts in their spectra show that they are expanding with typical velocities of 30 km/sec. Within about 100,000 years this expansion disperses the nebula, which mixes into the interstellar medium. Thus the planetary nebulae we see in the sky are transient stages in the evolution of medium-mass stars.

Although the ejected gas of a planetary nebula is only a small fraction of the star's total mass, it is an important part of a stable star—namely, the insulating blanket that confines the star's internal heat. When the surface puffs away into space, the white-hot core is exposed and the star emits intense ultraviolet radiation. In the H-R diagram, the evolutionary path of the star turns sharply left, toward the hot side.

The stellar remains consist of a small carbon-oxygen core covered by a thin layer of hydrogen and helium. With the insulating surface layers gone, the star loses heat rapidly and continues to contract until it becomes degenerate and enters the region of the white dwarfs in the lower left corner of the H-R diagram (Figure 7-8).

*White Dwarfs.* Both low-mass red dwarfs and medium-mass stars eventually become white dwarfs. Our survey of neighboring stars (Chapter 5) showed that most stars have masses less than that of the sun, but these are very longlived stars. Our galaxy is probably not old enough for many of these red dwarfs to have become white dwarfs. However, white dwarfs are very numerous—our galaxy probably contains billions—so they must be the remains of stars with masses similar to the sun's.

The first white dwarf discovered was the faint companion to Sirius. In that visual binary system, the bright star is Sirius A. The white dwarf, Sirius B, is 10,000 times fainter than Sirius A. The orbital motions of the stars (shown in Figure 5-14b) tell us that the white dwarf's mass is about 1 $M_{\odot}$, and its blue-white color tells us that its surface is hot, about 32,500°K. Since its luminosity is low, it must have a small surface area—in fact, it is about 85 percent earth's diameter. The mass and size imply that its average density is about $1.8 \times 10^6$ gm/cm$^3$. On earth, a teaspoonful of Sirius B material would weigh over 100 tons.

A normal star is supported by energy flowing outward from its core, but a white dwarf has no internal energy source, so there is nothing to oppose gravity and the gas becomes degenerate (Box 7-1). Thus a white dwarf is supported not by energy flowing outward, but by the refusal of its electrons to pack into a smaller volume.

Not all of the white dwarf is degenerate. Computer models predict that a crust about 50 km (30 mi) thick forms at the surface. The bottom of this crust consists of atoms locked in a rigid crystalline lattice, while the upper layers blend gradually into an intensely hot atmosphere of hydrogen and helium gas. Because the surface gravity of a white dwarf is over 200,000 times that on earth, the white dwarf's atmosphere is pulled down into a very shallow layer. If the earth's atmosphere were equally shallow, people on the top floors of skyscrapers would have to wear oxygen masks.

Clearly, a white dwarf is not a normal star. It generates no nuclear energy, is almost totally degenerate, and, except for a thin layer at its surface, contains no gas. Instead of calling a white dwarf a "star," we can call it a "compact object." Later in this chapter we will discuss two other kinds of compact objects—neutron stars and black holes.

A white dwarf's future is bleak. As it radiates energy into space, its temperature gradually falls, but it cannot shrink any smaller because its degenerate electrons cannot get closer together. This degenerate matter is a very good thermal

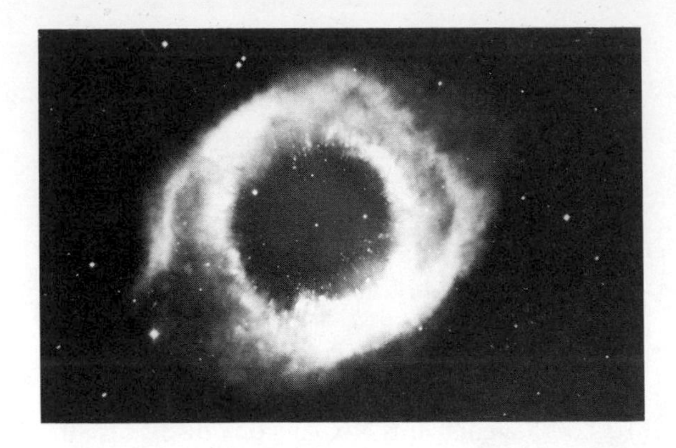

*Figure 7-7. A planetary nebula is the ejected surface of a medium-mass star. The remains of the star collapse into a hot, faint object (visible at the center of the nebula) that eventually becomes a white dwarf. (Hale Observatories.)*

## BOX 7-2 THE INSTABILITY STRIP

The evolution of a star after it leaves the main sequence moves it back and forth in the giant region of the H-R diagram. If a star passes through a region called the **instability strip** (Figure 7-6a), it becomes unstable and pulsates as a variable star.

Variable stars in the instability strip pulsate like beating hearts because of an energy-absorbing layer in their outer envelopes. They not only pulsate in size but also in temperature, causing their total brightness to vary (Figure 7-6b). These stars vary with periods from a few hours to hundreds of days.

The evolutionary track of a star may carry it through the instability strip a number of times, resulting in episodes of variability that may last millions of years. As the star moves across the strip, its changing radius and density cause its period to change. Such slow changes can be detected by careful observation and confirm that the variable stars we find in the instability strip are evolving and will someday move out of the strip and stop pulsating. Of course, other stars will evolve into the strip and become variable.

Though there are many types of variable stars, two kinds are especially important. The **RR Lyrae stars,** named after variable star RR in the constellation Lyra, have periods from 12 to 24 hours and are common in some star clusters (see Box 7-3). All RR Lyrae stars have about the same absolute magnitude, +0.5.

Another type of variable star is the **Cepheid,** named after δ Cephei, the first star of this type to be found. Cepheids have periods of from one to 60 days, but they do not all have the same absolute magnitude. More massive stars cross the instability strip higher in the diagram and are thus more luminous. Because of their greater mass and larger radius, they pulsate more slowly than the lower-mass stars. Consequently, there is a relation between the period and luminosity of Cepheid variables: The longer the period, the greater the luminosity.

First discovered by Harvard astronomer Henrietta Leavitt in 1912, this **period-luminosity relation** has become a powerful tool for the determination of distance. The period-luminosity diagram in Figure 7-6c shows the RR Lyrae stars and the two kinds of Cepheids. Type I Cepheids (also known as classical Cepheids) tend to be younger than type II. An astronomer who identifies one of these stars in a star cluster or galaxy and determines the star's period of pulsation, can find its absolute magnitude from the diagram. Comparing this with the apparent magnitude yields the distance to the star cluster or galaxy (see Box 5-2).

In the next chapter, we will see how such measurements of the distances to star clusters led to the determination of the size of our galaxy, and in Box 9-1 we will see how Cepheids can tell us the distance to other galaxies.

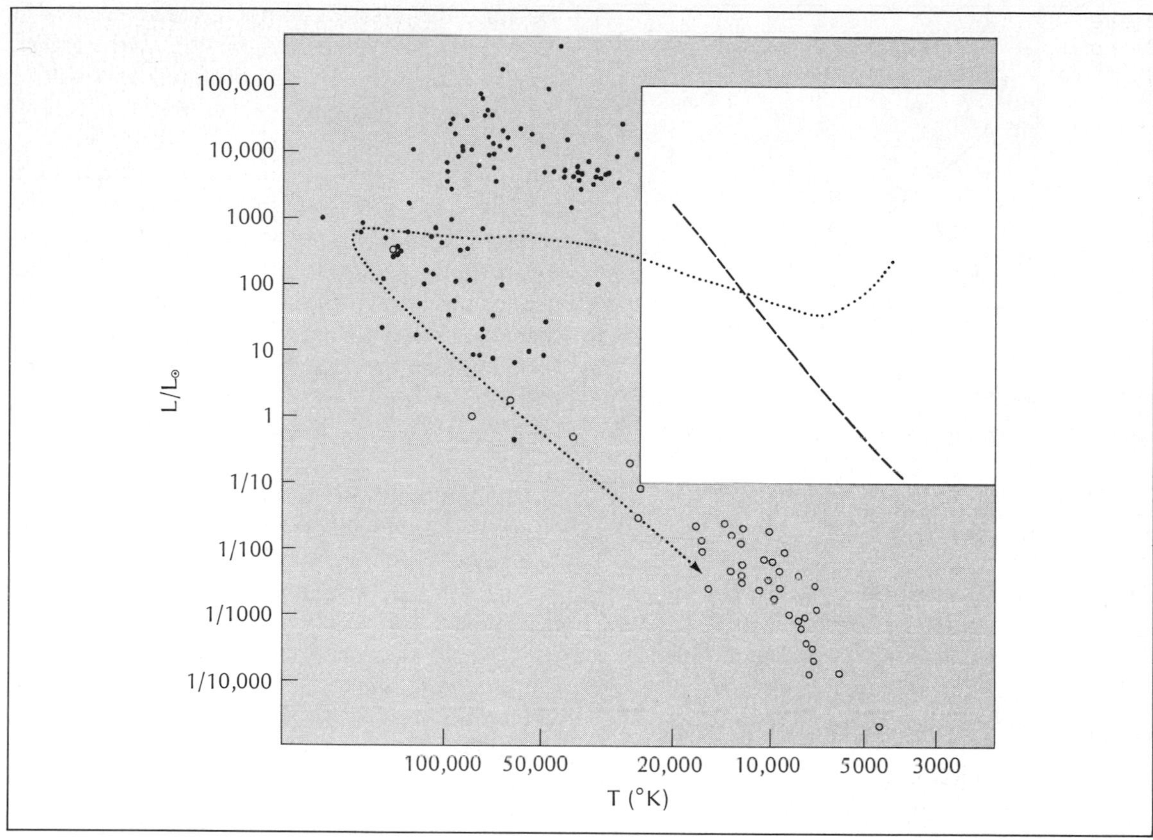


*Figure 7-8. Our customary H-R diagram must be expanded (shaded) to show the evolution of a star after it ejects a planetary nebula. Central stars of planetary nebulae (filled circles) are hotter and more luminous than most white dwarfs (open circles). The dotted line shows the evolution of a 0.8 $M_\odot$ star as it collapses toward the white dwarf region. (Adapted from diagrams by C. R. O'Dell and S. C. Villa.)*

conductor, so heat flows to the surface and escapes into space and the white dwarf gets fainter and cooler, moving downward and to the right in the H-R diagram. Because the white dwarf contains a tremendous amount of heat, it needs billions of years to radiate that heat through its small surface area. Eventually, such objects may become cold and dark, so-called **black dwarfs.** Our galaxy is probably not old enough to contain many black dwarfs.

Perhaps the most interesting thing about white dwarfs appeared in mathematical models. The equations predict that if we added mass to a white dwarf, its radius would *shrink* because added mass would increase its gravity and squeeze it tighter. If we added enough to raise its total mass to about 1.4 $M_\odot$, its radius would shrink to zero (Figure 7-9). This is called the **Chandrasekhar limit** after the astronomer who discovered it. It seems to imply that a star more massive than 1.4 $M_\odot$ could not become a white dwarf unless it shed mass in some way. Since spectra of many stars showed evidence of mass ejection, astronomers assumed that all stars lost mass to get under the Chandrasekhar limit. But the unsettling question remained—what would happen to a star that collapsed with a mass greater than 1.4 $M_\odot$? As we will see in the next

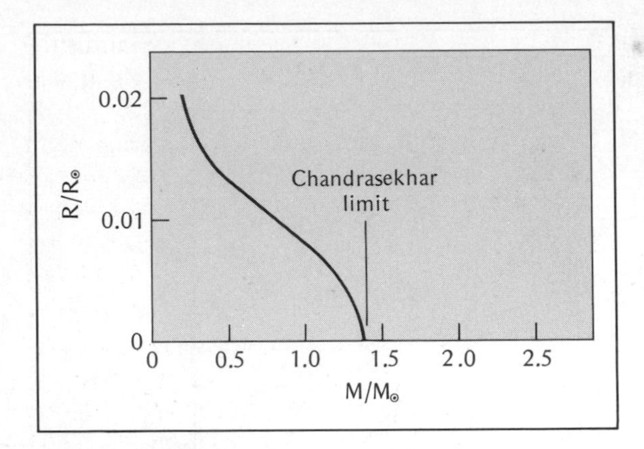


*Figure 7-9. The more massive a white dwarf is, the smaller its radius. Stars more massive than the Chandrasekhar limit of 1.4 $M_\odot$ cannot be white dwarfs.*

section, that question leads to one of the most exciting advances in modern astronomy, the discovery of neutron stars and black holes.

## THE DEATHS OF MASSIVE STARS

We have seen that low- and medium-mass stars die relatively quietly as they exhaust their nuclear fuels. The most violent event in the lives of these stars is the ejection of their surface layers to form planetary nebulae. However, massive stars live much more violent lives, and they die in spectacular supernova explosions that destroy the star.

*Hydrogen, Helium, and Carbon.* The evolution of stars more massive than about 3 $M_\odot$ begins as does the evolution of medium-mass stars. The star exhausts the hydrogen in its core, and, when the core contracts, a hydrogen-burning shell ignites, swelling the star into a giant. As a giant, it burns helium in its core and then in a shell, leaving behind a carbon-oxygen core that contracts and grows hotter.

In stars of approximately 3 to 9 $M_\odot$, the contracting carbon-oxygen core becomes degenerate before it can get hot enough to ignite the carbon. This is a lethal condition for the star, because the pressure-temperature thermostat is turned off in degenerate matter. The carbon-oxygen core is a bomb, destined to explode when the temperature reaches 600,000,000°K, the ignition temperature for carbon burning. This explosion is called the **carbon detonation.**

The carbon detonation may be powerful enough to blow some stars apart, and it could be responsible for some of the violent stellar explosions known as supernovae. As we will see later, supernovae are rare and poorly understood. Other stars may be able to survive carbon detonation since the rising temperature eventually forces the core to expand and the gas stops being degenerate.

Stars more massive than about 9 $M_\odot$ do not face the carbon detonation because their cores are so hot carbon ignites before the gas becomes degenerate. Thus the pressure-temperature thermostat is in working order and carbon burning turns on gradually. Stars that survive carbon ignition continue their lives for a short while as they burn heavier elements (Table 7-1), but they too eventually face an explosive end. The development of an iron core spells their absolute and final collapse.

**Table 7-1 Heavy Element Burning in a 25 $M_\odot$ Star**

| Fuel | Time | Percentage of lifetime |
|---|---|---|
| H | 7,000,000 years | 93.3 |
| He | 500,000 years | 6.7 |
| C | 600 years | 0.00008 |
| O | 0.5 years | 0.00000007 |
| Si | 1 day | 0.0000000004 |

*The Iron Core.* The evolution of stars that survive carbon ignition is uncertain because of the complexity of the nuclear reactions involving carbon and heavier elements (see Figure

6-12). Nevertheless, the general outline is clear. As the star burns heavy elements, each new fuel ignites first in the core and then in a shell, building layer after layer of heavy elements. This heavy element burning must stop when the star develops an iron core because the nucleus of the iron atom is exceptionally stable and cannot participate in energy-producing reactions. Reactions that convert iron into other atoms absorb rather than release energy.

The star, of course, is unaware of nuclear physics, and, when the iron core develops, it contracts and heats up, trying to ignite the iron. Any nuclear reactions involving iron that do occur absorb energy and make the core contract even faster. In addition, some reactions may produce large numbers of neutrinos that rush out of the star without interacting with the gas, and thus carry away energy needed to support the core. The result is a catastrophic collapse of the core that ends in a supernova explosion.

Exactly why the collapsing star explodes is not known, but we can examine a few possibilities. First, as the core collapses, its density grows rapidly and it becomes degenerate. Such a degenerate center would resist further compression, and the outer layers might bounce off of the core just as a rubber ball bounces off a brick wall. If the bounce ejects large amounts of matter into space, it could explain the supernova explosion. Another factor is the large mass of unburnt fuel that falls inward when the star collapses. The compression and heating could ignite that fuel in a violent explosion. In addition, the outer layers of the immense star do not take part in the collapse, but are puffed gently into space. If a bounce occurs in the collapsing core, it could drive a shock wave outward that would overtake the ejected gas, and the collision might be the violent explosion we see as a supernova. Whatever the mechanism, it must generate tremendous energy to account for the appearance of supernova explosions.

*Supernovae.* In 1054 AD, Chinese astronomers observed the appearance of a "guest star" in the constellation we know as Taurus the Bull. The star rapidly became so bright it was visible in broad daylight. Afterward it slowly faded, though it remained visible in the night sky for nearly two years. When modern astronomers turned their telescopes to the location of the "guest star," they found a cloud of gas about 1.35 pc in radius expanding at 1400 km/sec. Projecting the expansion back in time, they found that the cloud must have begun its expansion about 900 years ago, just about the time the "guest star" made its visit. Thus we think the nebula, now called the Crab nebula because of its shape (Figure 7-10), marks the site of the 1054 AD supernova.

Supernovae are very rare astronomical events; our galaxy has produced only four visible from the earth in recorded history. Arab astronomers saw one in 1006 AD and the Chinese saw the one in 1054 AD. European astronomers

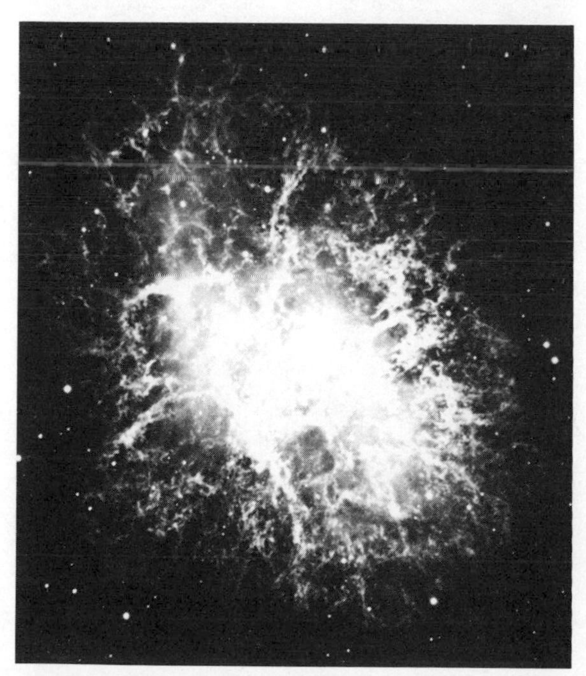

*Figure 7-10. The Crab nebula is a supernova remnant, the remains of a supernova observed by the Chinese in 1054 AD. (Hale Observatories.)*

observed two, one in 1572 AD (called Tycho's supernova) and one in 1604 AD (called Kepler's supernova). The only supernovae that have been observed since the invention of the astronomical telescope in 1609 AD have been in distant galaxies, and are thus more difficult to study (Figure 7-11).

Modern studies of supernovae paint a violent picture of the eruption. As the explosion begins, the star's luminosity rises rapidly, reaching a peak that can exceed 100 million times the sun's luminosity. Following maximum, the supernova's luminosity declines gradually. Spectra show emission lines characteristic of hot gas, and Doppler shifts indicate ejected matter.

Although the supernova fades to obscurity in a year or two, the expanding shell of gas marks the site of the explosion. These shells, called **supernova remnants,** last for tens of thousands of years before they gradually mix with the interstellar medium and vanish. The Crab nebula is a young remnant, only 900 years old. The Veil nebula (Figure 7-12 and Color Plate 15) in Cygnus is larger and more diffuse, having originated in a supernova about 50,000 years ago. Some supernova remnants are visible only at radio and x-ray wavelengths. They have become too tenuous to emit detectable light, but the collision of the expanding gas shell with the interstellar medium can generate radio and x-ray radiation. We saw in Chapter 6 that the compression of the interstellar medium by supernova remnants can also trigger star formation.

*Figure 7-11. Supernovae are so rare the only ones observed since the invention of the astronomical telescope in 1609 AD have been in other galaxies. These two views of galaxy NGC 7331 show the appearance of a supernova near the galaxy's outer edge. (Lick Observatory photograph.)*

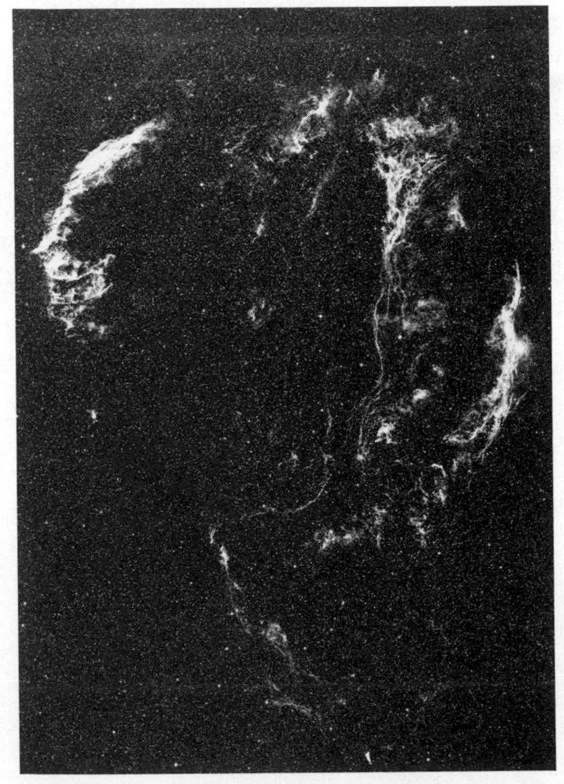

*Figure 7-12. The Veil nebula, a supernova remnant about 50,000 years old, is about 40 pc in diameter. (Hale Observatories.)*

The supernova death of massive stars leaves us with an interesting speculation. Suppose such a star ejects most of its mass in the supernova explosion, but retains a core more massive than the Chandrasekhar limit of 1.4 $M_\odot$. Such a collapsing object could not become a white dwarf. Then what could it become? The answer is a neutron star or a black hole, objects we will discuss in the next two sections.

Let us pause and review stellar evolution as summarized in Figure 7-13. The life cycle of a star depends on its mass. The least massive stars, red dwarfs, evolve directly to white dwarfs, but stars with masses between approximately 0.4 $M_\odot$ and 3 $M_\odot$ become giants, and eject planetary nebulae, before becoming white dwarfs. The most massive stars become giants and supergiants and burn heavier elements until they form an iron core. When the iron core collapses, the star explodes as a supernova and leaves behind a neutron star or a black hole. Notice that a star may lose mass as it evolves and move to the right in Figure 7-13.

Stellar evolution can give us new insights into the evolution of stellar groups, as described in Box 7-3. In addition, the H-R diagrams of such star clusters confirm the theory of stellar evolution. However, our theory is incomplete until we study the extreme states of matter in collapsed massive stars—neutron stars and black holes.

## NEUTRON STARS

A **neutron star** is the core of a star that has collapsed to a radius of only 10 km and to a density so high only neutrons can exist. The proof that these theoretical neutron stars actually exist has combined theoretical astrophysics, radio astronomy, and optical astronomy.

*Predicting the Properties of Neutron Stars.* Basic physics describes what happens to the core of a massive star after the supernova explosion. If the remaining mass is greater than 1.4 $M_\odot$, gravity is so strong that not even the

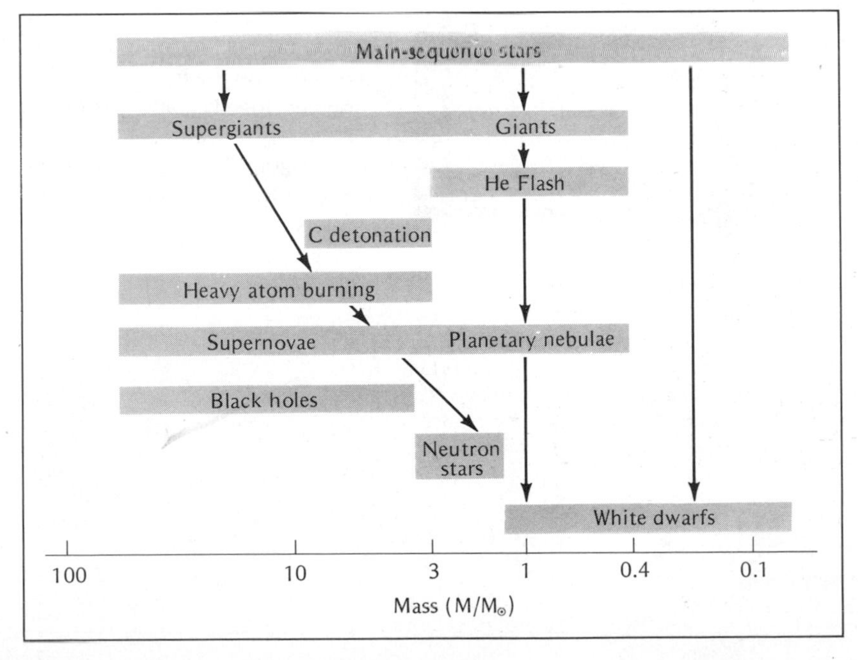


*Figure 7-13. How a star evolves depends on its mass. The lowest-mass stars become white dwarfs, while stars like the sun become giants and then white dwarfs. Mass loss can move a star to the right in this diagram, as shown for the massive star. (Adapted from Michael A. Seeds, "Stellar Evolution," Astronomy, February 1979.)*

## BOX 7-3 CLUSTER H-R DIAGRAMS

The theory of stellar evolution is so complex and involves so many assumptions that astronomers would have little confidence in it were it not for H-R diagrams of star clusters. Because the stars in a cluster begin their lives at about the same time, an H-R diagram of the cluster makes the slow evolution of the stars visible.

Suppose we follow the evolution of a star cluster by making H-R diagrams like frames in a film (Figure 7-14). Our first frame shows the cluster only $10^6$ years after it began forming, and already the most massive stars have reached the main sequence, consumed their fuel, and moved off to become supergiants. However, the medium- to low-mass stars have not yet reached the main sequence.

Because evolution is such a slow process, we cannot make the time step between frames equal or we would fill over 1000 pages with nearly identical diagrams. Instead, we increase the time step by a factor of 10 with each frame. Thus the second frame shows the cluster after $10^7$ years and the third after $10^8$ years.

By the third frame, all massive stars have died, and stars slightly more massive than the sun are beginning to leave the main sequence. Notice that the lowest-mass stars have finally begun to burn hydrogen. Only after $10^{10}$ years does the sun begin to swell into a giant.

These five frames were made from theoretical models of stellar evolution, but they compare very well with H-R diagrams of real star clusters. NGC 2264 is only a few million years old and still has many of its lower-mass stars contracting toward the main sequence (Figure 7-15a). The Pleiades, a cluster visible to the naked eye in Taurus, is older than NGC 2264, dating back about 100 million years (Figure 7-15b). Compare these younger clusters with M 67, a faint cluster of stars about 5 billion years old (Figure 7-15c).

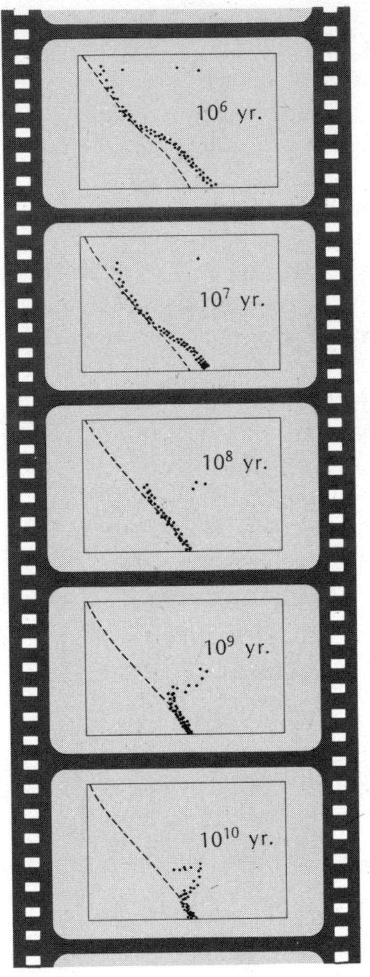


*Figure 7-14. A series of H-R diagrams, like frames in a film, illustrate the evolution of a cluster of stars. Massive stars approach the main sequence faster, live shorter lives, and die sooner than lower-mass stars. Compare with Figure 7-15.*

We can estimate the age of a star cluster by noting the point where its stars turn off the main sequence and move toward the red giant region. The masses of the stars at this **turn-off point** will tell us the age of the cluster, because those stars are on the verge of exhausting their hydrogen burning cores. Thus the life expectancy (Box 6-2) of the stars at the turn-off point equals the age of the cluster.

The clusters we have just discussed are called **open star clusters** because their stars are not crowded together and the cluster has an open, transparent appearance. Figure 7-16 shows the combined H-R diagrams of nine open clusters of different ages.

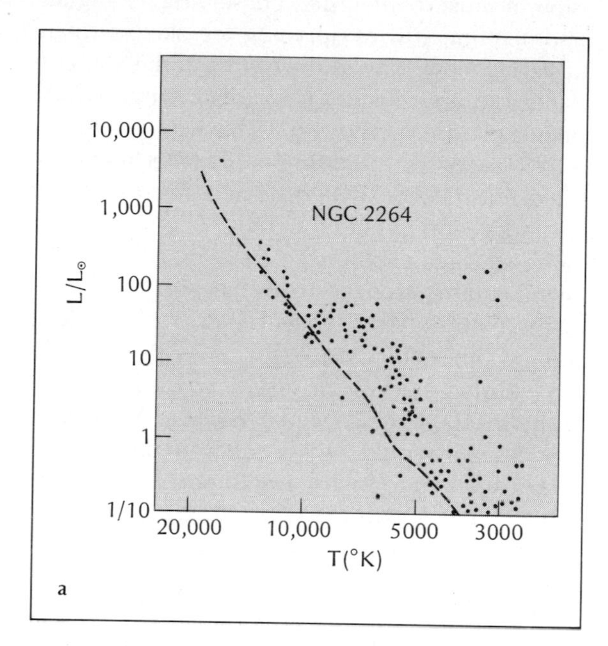


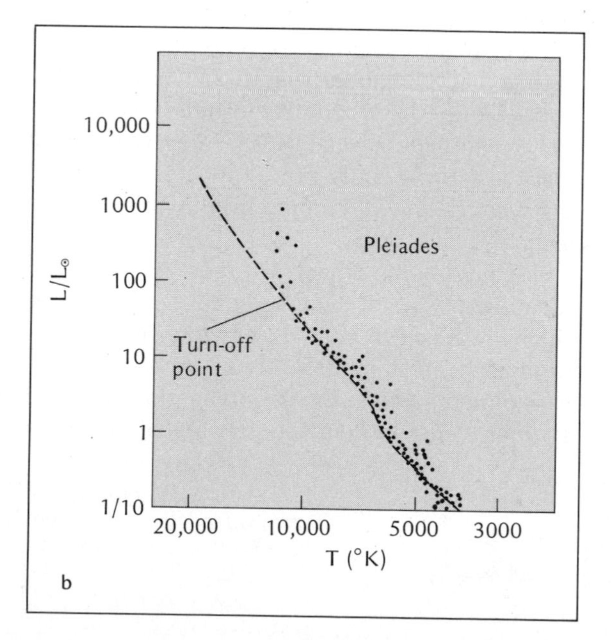


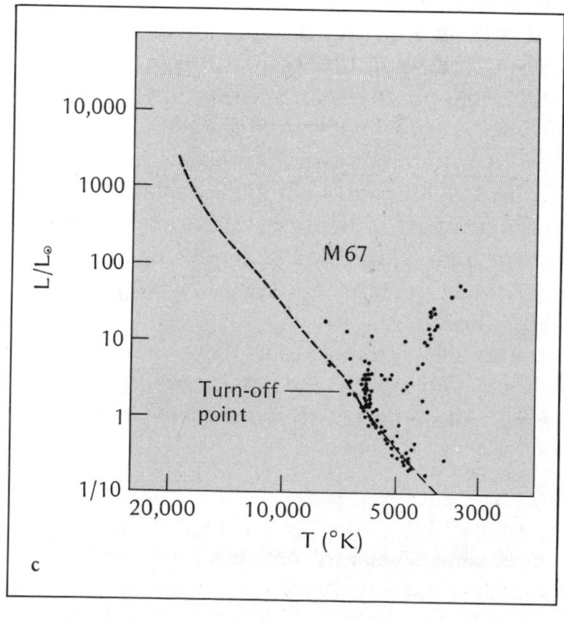


*Figure 7-15. (a) NGC 2264 is a cluster only a few million years old. Its lower-mass stars are still approaching the main sequence (b) The Pleiades is an older star cluster, but most of its stars are still on the main sequence. (c) M 67 is about 5 billion years old, and all of its more massive stars have died.*

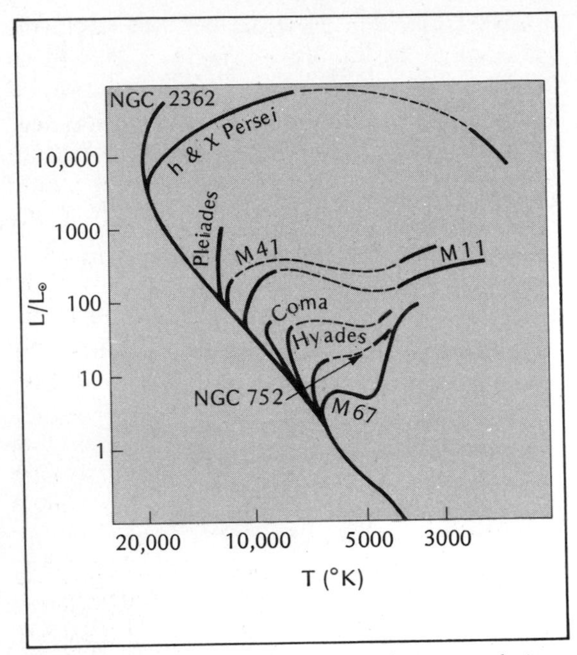


*Figure 7-16. The combined H-R diagrams of nine open star clusters illustrate how clusters of different ages have different turn-off points.*

degenerate electrons can stop the collapse. The pressure forces electrons into the atomic nuclei, where they combine with protons, transform them into neutrons, and convert the matter into a gas of pure neutrons. Contraction continues until the neutrons become degenerate, refusing to pack tighter and halting the collapse. The resulting object, a neutron star, is only about 10 km in radius and has a density of about $10^{15}$ gm/cm$^3$. On earth, a teaspoonful would weigh 50 billion tons.

We can predict that these neutron stars should spin rapidly and have powerful magnetic fields. All stars rotate to some extent because they condensed from the swirling interstellar matter. As such a star collapses after a supernova explosion, it must rotate faster, conserving its angular momentum, just as spinning ice skaters rotate faster as they draw in their arms. If the sun collapsed to a radius of 10 km, its rotation would increase from once every 25 days to once every 0.001 seconds. In addition, whatever magnetic field the star had would be compressed by the collapse and would become 1 billion times stronger. Since some stars have magnetic fields significantly stronger than the sun's, neutron stars might have magnetic fields 1 billion to 1 trillion times solar strength.

Neutron stars were first described theoretically in 1934 by R. Minkowski and F. Zwicky, but no one could be sure they really existed. There seemed to be no way to observe a distant object only 10 km in radius. Progress came not from theory, but from observation.

*The Discovery of Pulsars.* In November 1967, Jocelyn Bell, a graduate student at the University of Cambridge, England, found a peculiar pattern on a paper chart from a radio telescope. Unlike other radio signals from celestial objects, this was a series of pulses (Figure 7-17) with a highly regular period of 1.33733 seconds. Bell and Anthony Hewish, the director of the experiment, investigated further and found that the signals could not be local interference, but were definitely coming from a celestial object.

Identifying the object was more difficult than locating it. One of the first theories was that the signals were from intelligent beings on a planet orbiting a distant star. That theory quickly evaporated when three more pulsed radio sources were found in various parts of the sky. It

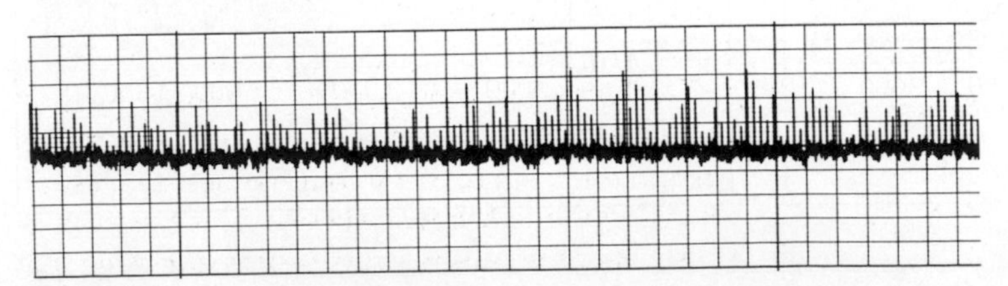

*Figure 7-17. A typical pulsar produces pulses shown here by the vertical spikes. This pulsar, PSR 2021 + 51, has a period of 0.529 seconds. (National Radio Astronomy Observatory.)*

seemed unlikely that four civilizations in four parts of the galaxy would transmit similar signals. In addition, the signals were not limited to one frequency, as are radio and TV transmissions, but were spread over the entire radio spectrum. Only nature could afford to spread its energy over so many wavelengths, so the tentative name LGM (standing for "Little Green Men") was replaced by **pulsar**—an astronomical source of radio pulses.

Radio astronomers found more pulsars, each with its own period of pulsation, ranging from 0.033 seconds to 3.75 seconds. The pulsations were highly precise, almost as regular as atomic clocks. But when observers measured the periods regularly over months, they found a gradual increase. Many of the pulsars were slowing by a few billionths of a second per day.

Another surprising observation uncovered sudden changes, called **glitches,** in the periods of some pulsars. The pulsar would pulse regularly, gradually slowing down, and then suddenly increase its rate of pulsation, and immediately resume its steady decline. For example, the pulsar in Vela increased its pulse rate by 0.2 millionths of a second in early March 1969 (Figure 7-18).

The pulses themselves last for only about 0.001 second, which is too short for them to have originated in pulsating stars like Cepheids, or in white dwarfs. If a white dwarf blinked on and then off in that interval of time, we would not see a pulse 0.001 seconds long. The near side of the white dwarf would be about 6000 km closer to us, and light from the near side would arrive 0.022 second before the light from the bulk of the white dwarf. Thus its short blink would be smeared out into a longer pulse. This is an important principle in astronomy—an object cannot change its brightness appreciably in an interval shorter than the time light takes to cross its diameter. That the pulses from pulsars were no longer than 0.001 seconds meant that the pulsating object could not be larger than about 300 km (190 mi) in diameter, much smaller than any star or white dwarf.

*Figure 7-18. The Vela pulsar was slowing gradually when a glitch occurred in late February or early March 1969, decreasing its period by 0.2 millionths of a second.*

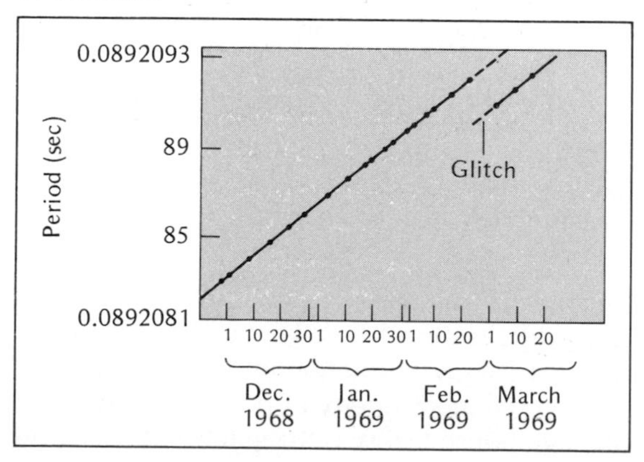


Another suggestion, that pulsars were rapidly spinning white dwarfs with hot spots on their surfaces, had to be abandoned because a white dwarf spinning that rapidly would fly apart. The only thing left was a neutron star. Although a neutron star could pulsate, its small size and strong gravity would make it pulsate too rapidly to be a pulsar, but it could spin at about the right speed. Thus astronomers began to suspect that the pulsars were spinning neutron stars.

The missing link between the theoretical neutron stars and the observational pulsars appeared when radio astronomers found a pulsar in the heart of the Crab nebula (Figure 7-19). The Crab nebula is the remnant of the supernova of 1054 AD, and theory predicts that some exploding stars may leave behind neutron stars. The Crab nebula pulsar is evidently such an object.

If we reconsider the theoretical properties of neutron stars and combine them with the observed properties of pulsars, we can devise a model of a pulsar.

*A Model of a Pulsar.* The exact method by which a rotating neutron star produces pulses of radiation is a mystery, but the best suggestion is called the **lighthouse theory.** This theory supposes that the pulsar emits beams of energy that

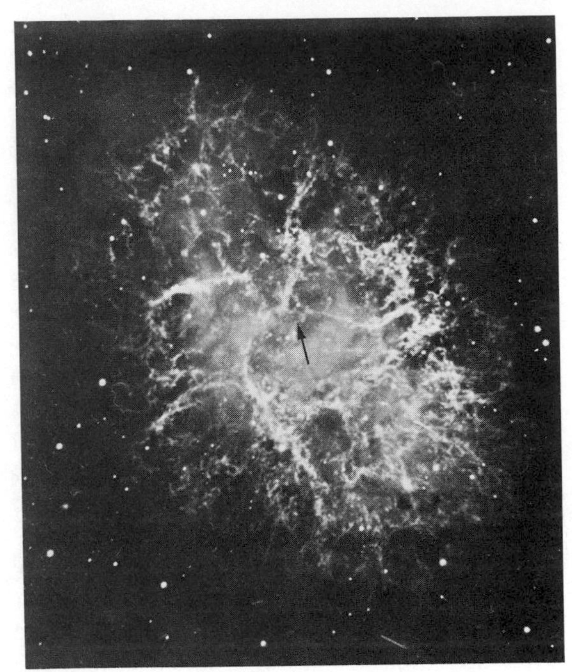

*Figure 7-19.* *The discovery that a faint star at the center of the Crab nebula was a pulsar linked neutron stars, the theoretical remains of supernovae, with the observational pulsars. (Lick Observatory photograph.)*

sweep around the sky like the beams of light from a lighthouse. When the beams sweep over us, we detect a pulse of radio radiation.

The powerful magnetic field of the neutron star may explain how it can emit beams. Energy emitted by the neutron star could be guided by the magnetic field into beams emerging from the magnetic poles. If the magnetic axis were inclined with respect to the axis of rotation, then the spinning neutron star would sweep the beams around the sky (Figure 7-20).

Two properties of pulsars suggest that the lighthouse theory is correct. First, many pulsars are slowing down by a few billionths of a second per day. The amount of energy a pulsar radiates into space per day, about $10^5$ times more than the sun, is approximately equal to the amount of energy a spinning neutron star would loose by slowing down a few billionths of a

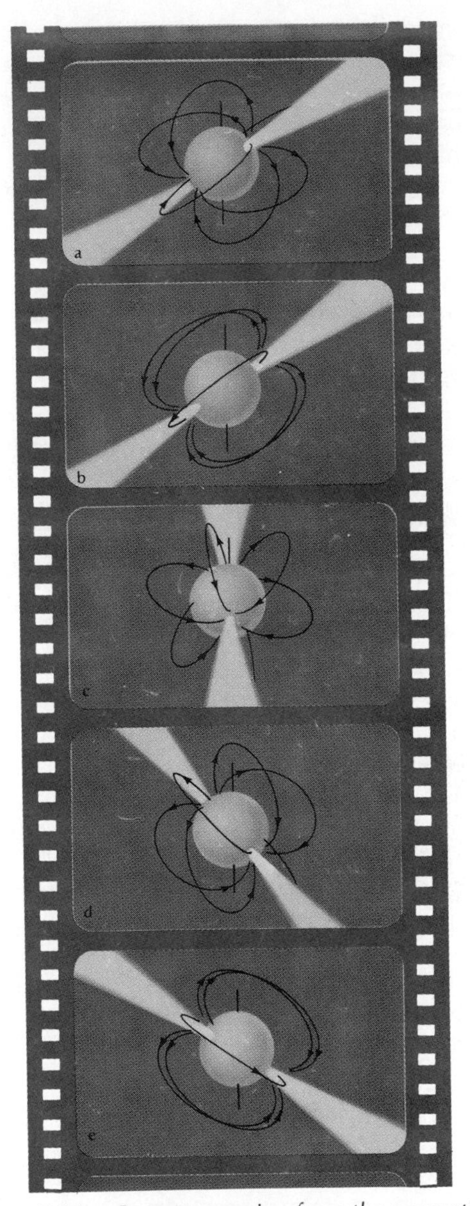


*Figure 7-20.* *Beams emerging from the magnetic poles of a neutron star sweep around the sky like beams of light from a lighthouse.*

second. Thus it seems that the spinning neutron star and its intense magnetic field somehow convert energy of rotation into radiation.

The second property of pulsars that supports the lighthouse theory is the glitches. To see how the model could produce glitches, we must consider the internal structure of a neutron star. The theory predicts that the interior consists of degenerate neutrons. Near the surface, where the pressure is less, protons, electrons, and atomic nuclei can exist, locked in a crystalline crust a few kilometers thick. As the spinning neutron star slows, centrifugal force decreases and gravity squeezes the neutron star tighter. Because the crystalline crust is rigid, stresses build until the crust breaks in a "starquake," the neutron star's version of an earthquake. Once the crust is broken, the neutron star can shrink slightly, and rotation speeds up as the neutron star conserves its angular momentum. This increase in the rate of rotation would increase the pulse rate.

The Crab nebula pulsar adds more evidence supporting the lighthouse theory. First, it is the fastest known pulsar, blinking 30 times a second. This high pulse rate is evidently a product of its youth—it formed only 900 years ago and hasn't had time to slow very much. Second, it pulses not only at radio wavelengths, but also at visible light, x-ray, and gamma-ray wavelengths (Figure 7-21). Circumstantial evidence suggests that this high energy radiation is also connected with the pulsar's youth. The only other pulsar known to emit light pulses is the Vela pulsar, a fast pulsar associated with a supernova remnant in the constellation Vela. From the age of the remnant, we can guess that the Vela pulsar is also young.

*Figure 7-21. High speed photographs show the Crab nebula pulsar blinking at visible wavelengths as it rotates. During each rotation lasting one-thirtieth of a second, the pulsar blinks twice as its two beams sweep over the earth. (Kitt Peak National Observatory.)*

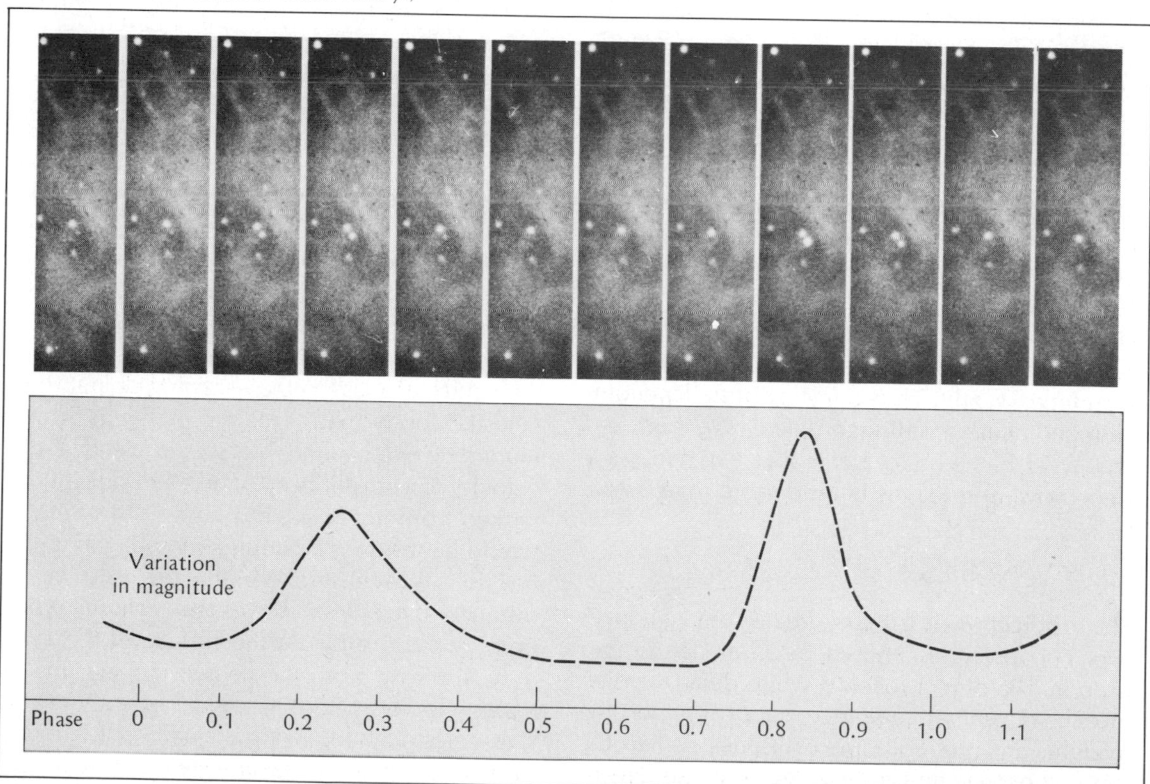

There is some evidence that a pulsar may emit more than just electromagnetic radiation. The earth is constantly bombarded by protons, electrons, and the nuclei of atoms traveling at nearly the speed of light. These **cosmic rays** are absorbed by earth's atmosphere, so they must be studied by equipment on high mountain tops or aboard high altitude balloons or rockets. Since cosmic rays are charged particles, they are deflected by the magnetic fields in the galaxy and come at the earth from all directions. Thus we cannot be sure where they originate. Some come from solar flares, but many may be produced in supernova explosions or the rapid expansion of the resulting supernova remnants. Some theorists suggest that highly magnetic neutron stars spinning on their axes could emit charged particles that we would detect on earth as cosmic rays. Thus the study of cosmic rays may someday reveal details of the structure of pulsars.

Although many mysteries remain, one thing seems plain. Stars dying in supernova explosions may leave behind spinning neutron stars, some of which we see as pulsars. However, the pulsars are only a partial answer to our speculation about the fate of a collapsing star that exceeds the Chandrasekhar limit. Theoretical calculations are difficult because we don't understand the nature of pure neutron material, but the models do agree that there is a maximum mass for neutron stars somewhere between 1.6 to 3 $M_\odot$. The degenerate neutrons cannot support more weight. Thus our story of stellar evolution and death is still incomplete. We must find out what happens to a star that explodes as a supernova and leaves behind more than 3 $M_\odot$.

## BLACK HOLES

If an object with a mass greater than about 3 $M_\odot$ collapses, no known force in nature can stop it. The object reaches white dwarf density, but the degenerate electrons cannot support the weight, and the collapse continues. When the object reaches neutron star density, the degenerate neutrons cannot support the weight, and the collapse goes on. The object quickly becomes smaller than an electron. No one knows of anything that can stop the object from reaching zero radius and infinite density.

An object of zero radius is called a **singularity,** the mathematician's term for a point. Astronomers, accustomed to working with objects of finite density, find singularities distressing. Nature, too, seems embarrassed by this final state of collapse, and hides it from us inside a region of space called a **black hole.**

Although black holes are difficult to discuss without general relativity and sophisticated mathematics, we can use common sense and some simple physics to see why they form. Finding the velocity we need to escape from the gravity around a celestial body will help explain how black holes were first predicted theoretically and how they might be detected.

*Escape Velocity.* Suppose we threw a baseball straight up. How fast must we throw it if it is not to come down? Of course gravity would pull back on the ball, slowing it, but if the ball were traveling fast enough to start with, it would never come to a stop and fall back. Such a ball would escape from the earth. The **escape velocity** is the initial velocity an object needs to escape from a celestial body.

Whether we are discussing a baseball leaving the earth or a photon leaving a collapsing star, the escape velocity depends on two things, the mass of the celestial body and the distance from the center of mass to the escaping object. If the celestial body had a large mass, its gravity would be strong and we would need a high velocity to escape, but if we began our journey farther from the center of mass, the velocity would be less. For example, if we began a journey in a space ship orbiting the earth at the distance of the moon, the escape velocity would be only 3200 mph. At the surface of the earth, only 4000 mi from the center of mass, the escape velocity is 25,000 mph. Evidently a massive celestial body with a small radius would have a tremendous escape velocity at its surface.

In 1798, Pierre Laplace, a French astronomer and mathematician, followed this common sense argument and concluded that there could be stars so massive that the escape velocity at their surface was greater than the speed of light. Thus no light should be able to escape from them, and they should look perfectly black. First called "frozen stars," such objects are now called black holes.

*The Structure of Black Holes.* Current theory predicts that if the core of a star collapses and is more massive than about 3 $M_\odot$, it will continue to collapse to a singularity and will form a black hole. Some scientists believe a singularity is impossible and that when we better understand the laws of physics we will discover that the collapse halts before the diameter reaches zero.

Whether a singularity forms or not, if the object becomes small enough, the escape velocity nearby is so large that no light can escape. We can receive no information about the object nor the region of space near it, and we refer to the region as a black hole. The boundary of this region is called the **event horizon** (Figure 7-22), because any event that takes place inside the surface is invisible to an outside observer.

The radius of the event horizon is called the **Schwarzschild radius $R_s$**, after the astronomer Karl Schwarzschild who predicted black holes from the theory of general relativity in 1916. The Schwarzschild radius depends only on the mass of the object. A 5 $M_\odot$ black hole would have a Schwarzschild radius of about 15 km (9 mi). Nothing within this radius would be visible to an outside observer.

Any object could be a black hole if it were smaller than its Schwarzschild radius. For example, if we could compress the earth to a radius of about 1 cm, its gravity would be so strong it would become a black hole. Fortunately, the earth will not collapse spontaneously into a black hole because its mass is less than the critical mass of about 3 $M_\odot$. Only objects more massive than this limit can form black holes under the sole influence of their own grav-

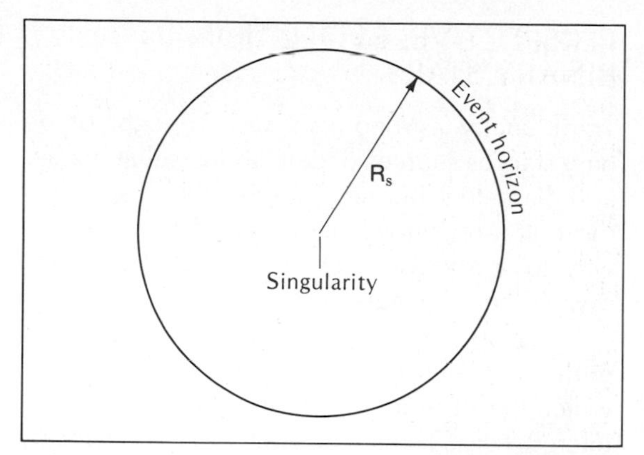


*Figure 7-22. A black hole forms when an object collapses to a small size (perhaps to a singularity) and the escape velocity in its neighborhood is so great light cannot escape. The boundary of this region is called the event horizon because any event that occurs inside is invisible to outside observers. The radius of the region **$R_s$** is the Schwarzschild radius.*

ity. In this chapter, we are interested in black holes that might originate from the deaths of massive stars. These would have masses larger than 3 $M_\odot$. In Chapter 11, we will deal with much less massive black holes that may have originated at the beginning of the universe.

Thus far our discussion of black holes has been entirely theoretical. We have said they must exist because we know of nothing that could prevent them from forming. Obviously, astronomers would like to confirm the existence of black holes by finding one, but how can we observe something that is totally dark? The answer may be to look for a black hole swallowing matter. As the matter falls in, it converts gravitational energy into thermal energy and grows very hot. Its temperature may exceed 1,000,000°K. Such hot material will emit a flood of x-rays, and if the x-rays are emitted before the matter crosses the event horizon, the x-rays can escape. Thus to find a black hole we must look for objects emitting x-rays.

A solitary black hole floating in space is unlikely to absorb much matter, but a black hole orbiting a star in a binary system might pull matter from the star. Thus our search for black holes leads to binary systems.

## PERSPECTIVE: EVOLUTION IN BINARY STARS

Traditionally astronomers have thought of binary stars as sources of data about stellar masses and diameters, but new observations, especially those at x-ray wavelengths, are giving us a new way to study compact objects—that is, white dwarfs, neutron stars, and black holes.

Over 60 percent of all stars are members of multiple systems, but only those binaries in which the stars are close together are of special interest here. The small separation is necessary to assure the transfer of mass from one star to the other. This mass transfer can produce violent explosions and generate x-ray radiation that may help us locate neutron stars and black holes.

*Mass Transfer.* Two processes transfer mass from one star to the other in a binary system. Hot, massive stars have strong stellar winds similar to, but stronger than, the solar wind. If the massive star has a close companion, some of the stellar wind can be swept up, generating high temperatures as it falls in (Figure 7-23a).

The second mass transfer process occurs when a star loses gravitational control over its surface layers. A star in a binary system controls a region of space around it called its **Roche lobe** (Figure 7-23b). Any gas that enters the Roche lobe falls into the star, and any gas that leaves the lobe escapes. The size of the lobes depends on the masses of the stars and their separation. If we outline the Roche lobes in a binary system, the drawing is likely to look like a lopsided dumbbell with one lobe larger than the other.

The theory of stellar evolution says that the more massive star will evolve fastest and will swell into a giant. If the stars are far apart, the Roche lobes are very large and nothing peculiar happens when one star becomes a giant. However, if the stars are close together, the lobes are small and the swelling giant soon fills its lobe. When its surface layers cross the lobe, the giant loses control over them and they escape into space. Some of that matter flows through the point of contact between the lobes and is captured by the companion star. Thus matter can be transfered from an evolving star to its companion (Figure 7-23b).

Mass transfer can drastically affect the evolution of the stars. A 15 $M_\odot$ star could swell into a giant and transfer 10 $M_\odot$ onto a 1 $m_\odot$ companion. Eventually the massive star would die, leaving behind a compact object. The companion, grown to 11 $M_\odot$, would evolve rapidly and become a giant, transferring matter back to the compact object. If that object were a white dwarf, this might cause an explosion. If the object were a neutron star or a black hole,

*Figure 7-23. Mass transfer in a close binary may occur when (a) a stellar wind adds mass to a compact object, or (b) when a star swells to fill its Roche lobe and looses mass to its companion.*

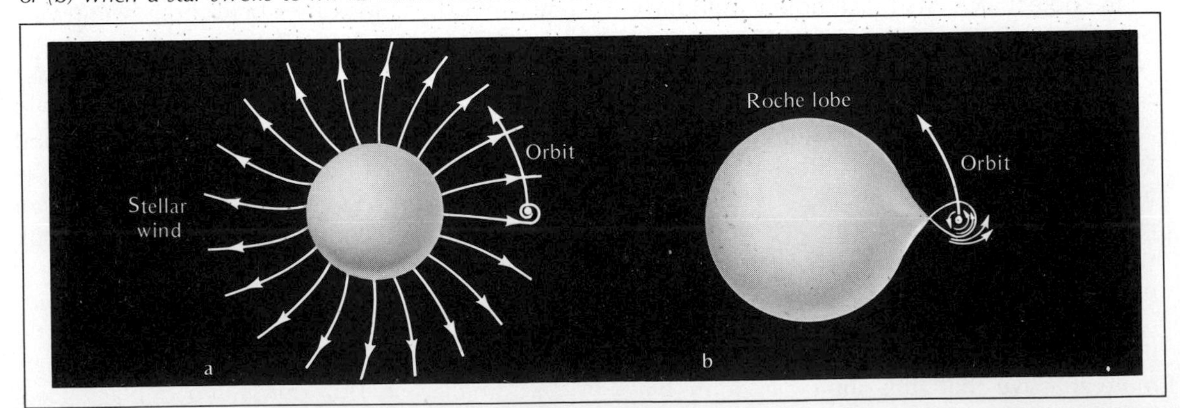

mass transfer could produce x-rays. All of these objects are related, but before we can discuss them, we must know how the transferred matter falls into the compact object.

*Accretion Disks.* Matter transferred to a compact object cannot fall directly into the object because of the conservation of angular momentum. Instead, it falls into a whirlpool around the object. For a common example of this effect, consider a bathtub filled with water. Gentle currents in the water give it some angular momentum, but its slow circulation is hardly apparent until we pull the stopper. Then, as the water rushes toward the drain, conservation of angular momentum forces it to form a whirlpool. This same effect forces matter falling into a compact object to form a whirling disk of gas called an **accretion disk** (Figure 7-24 and Color Plate 13).

The impact of matter falling on the disk can cause a hot spot on the rim, which can reach very high temperatures and may radiate more energy than the compact object itself.

Friction within the accretion disk has two important effects. First, it heats the matter. In some cases the disk may reach 1,000,000°K and radiate x-rays. Besides heating the gas, the friction also slows its motion and allows it to fall toward the inner edge of the disk where it can eventually fall into the compact object. Thus the disk acts as a brake, ridding the infalling gas of its angular momentum and allowing it to fall into the compact object. The disk around a neutron star or black hole can become very hot, but the disk around a white dwarf does not get as hot. Nevertheless, a white dwarf accreting matter can erupt in a violent explosion.

*Novae.* **Nova** is Latin for "new," and in astronomy it refers to the appearance of what seems to be a new star. A nova can appear in the sky and brighten in a few days, and then fade back to obscurity during the next few months. However, a nova is not a new star, but the eruption of an old star, a white dwarf (Figure 7-25).

These explosions are believed to be caused by the transfer of matter from a normal star, through an accretion disk, onto the surface of a white dwarf. Because the matter is drawn from the surface of a normal star, it is rich in unburnt fuel, and when it accumulates on the surface of the white dwarf, it forms a layer of unprocessed fuel. As the layer deepens, it becomes denser and hotter until the fuel ignites in a sudden explosion.

The nova explosion blows off the surface of the white dwarf in an expanding shell of gas that becomes very luminous. As the shell grows larger and less dense, it becomes transparent and the nova fades.

The white dwarf and its companion are hardly disturbed by the nova explosion. Mass transfer quickly resumes, and a new layer of fuel

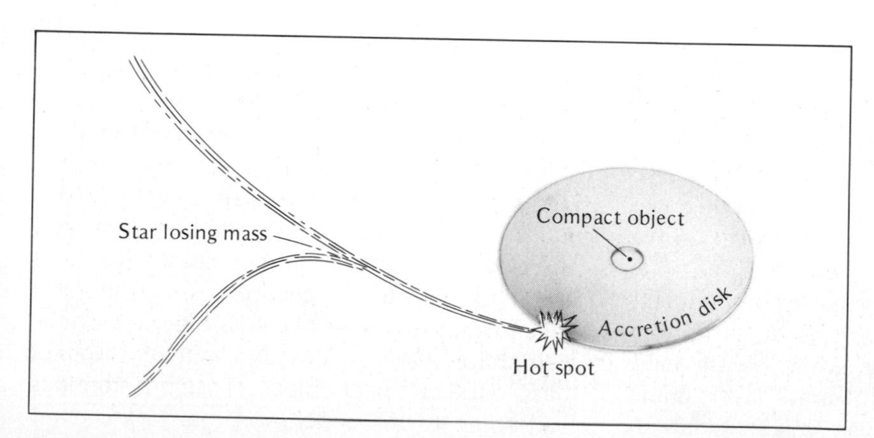


*Figure 7-24. Matter falling into a compact object forms a whirling accretion disk. Where matter impacts the disk, it may be heated to high temperatures.*

*Figure 7-25. Nova Cygni 1975 photographed near maximum when it was 2nd magnitude (top) and later when it had declined to about 11th (bottom). (Lick Observatory photograph.)*

begins to accumulate. How fast the fuel builds up depends on the rate of mass transfer. According to this theory, some novae might need hundreds of years to accumulate another explosive layer. Others might need only a few weeks. **Recurrent novae** are stars that erupt every few dozen years, and **dwarf novae** undergo small novalike explosions every few days or weeks. Though there are important differences between these types of stars, they appear to be related to the same process—accretion into a disk. For example, observations of dwarf novae such as Z Camelopardalis show irregular flickering believed to be due to the impact of matter on the hot spot at the edge of its accretion disk.

Though many novae appear to be members of binary systems, some may be single stars. A star shrinking toward the white dwarf stage might heat up so rapidly its surface layers could erupt. Another way novae might occur is through single white dwarfs attracting matter from the interstellar medium and building explosive surface layers. Thus we cannot be sure that every nova is associated with a binary system.

*Black Holes.* If the compact object in a binary system is a black hole, any material that falls inside the event horizon vanishes without a trace, but such a system would still be detectable because of the accretion disk. The gravity is so powerful that the inner parts of the disk would reach temperatures of millions of degrees and give off powerful x-ray radiation.

Cygnus X-1, the first x-ray object discovered in Cygnus, is apparently such a binary. Looking in the direction of the x-ray object, telescopes reveal a hot supergiant about 3000 pc from earth. The x-rays evidently originate from the accretion disk around an invisible companion orbiting the supergiant. Doppler shifts in the spectrum of the supergiant show that it is a member of a spectroscopic binary in which the other star is not detectable at the wavelengths of visible light. Analyzing the radial velocity variations of the supergiant and assuming it has the mass of about 30 $M_{\odot}$ normal for a star of its type, astronomers conclude that the mass of the invisible companion is about 8 $M_{\odot}$. This is well above the maximum mass for a neutron star, implying that Cygnus X-1 is a black hole.

Another system, V861 Scorpii, is an x-ray binary with a compact object of more than 5 $M_{\odot}$, implying that it too is a black hole. One or two other x-ray binaries are black hole suspects, but no evidence conclusively eliminates the possibility that they contain neutron stars instead.

Unfortunately, measuring the mass of the compact objects in these systems is very difficult. The spectra show the radial velocity variations of the visible star, but not of the nonluminous objects. To get masses, astronomers have to guess the mass of the visible star with the assumption that it is normal in spite of its loss of mass to the compact object. These uncertainties lead some to the conclusion that Cygnus X-1

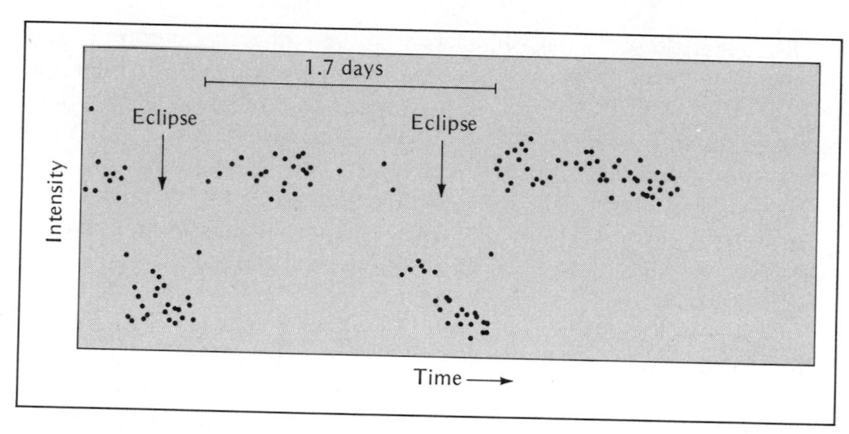


Figure 7-26. The x-rays from Hercules X-1 disappear as the x-ray pulsar is eclipsed behind its stellar companion.

and V861 Scorpii might contain neutron stars instead of black holes. Their arguments, however, require special circumstances that seem unlikely.

We have seen how close binary systems containing white dwarfs and black holes might behave. When such a system contains a neutron star, it can produce x-ray pulses.

*Pulsed X-ray Binaries.* Hercules X-1 is one of a few highly complex x-ray binaries whose x-radiation is pulsed like the radio signals from a pulsar. The pulses from Hercules X-1 have a period of 1.24 seconds, and vanish entirely for 5.8 hours every 1.7 days (Figure 7-26).

The origin of the pulses is still uncertain, but the most popular theory supposes that matter transferred from a companion star to a neutron star becomes entangled in its magnetic field and flows along the field to the magnetic poles, where its impact generates beams of x-rays. The spinning neutron star sweeps the x-ray beams around the sky, producing the observed pulses.

Hercules X-1 contains a 2 $M_{\odot}$ star whose surface temperature is about 7000°K. The star is variable because the x-rays from the neutron star heat one side of the star to about 20,000°K. As the system revolves, we see first the hot side and then the cool side of the star (Figure 7-27).

Doppler shifts in the star's spectrum show that the orbital period of the system is 1.7 days, which explains the disappearance of the x-ray pulses for 5.8 hours every 1.7 days. The system is an eclipsing binary, and the neutron star disappears behind the companion every 1.7 days. Analysis of the orbit tells us that the mass of the compact object is about 1.3 to 1 $M_{\odot}$, well within the range for neutron stars.

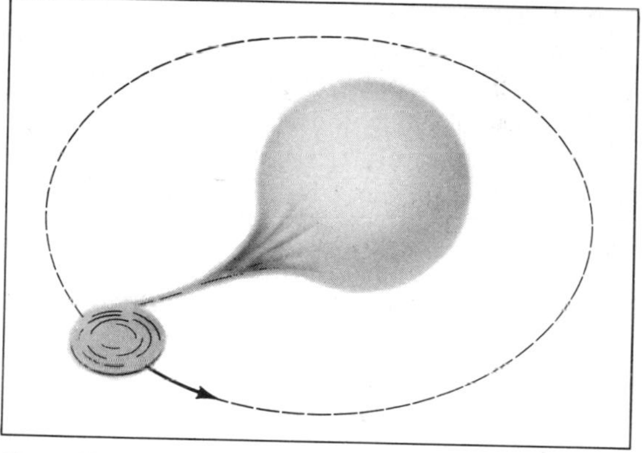

Figure 7-27. In Hercules X-1, x-rays from the x-ray pulsar heat the near side of the star to 20,000°K. The rotating system shows us the hot side and the cool side of the star alternately, varying the visual brightness.

*SS 433.* What could be more exotic than binary stars trading mass, white dwarfs exploding, black holes swallowing material ripped from their stellar companions, or neutron stars emitting powerful beams of x-rays? Our imaginations are pale compared to the diversity of nature, and the amazing complexity of x-ray binaries may yet hold more surprises.

One x-ray object recently discovered is so grotesque it has attracted the attention of the press and was even mentioned on TV's "Saturday Night Live." The news media reported that astronomers believed the object was simultaneously receding from and approaching the earth at one-fourth the speed of light. It is not unusual for the press to misquote technical material, but in this case the report is essentially correct.

Astronomers found that the optical spectrum of the object, known as SS 433, contains three sets of emission lines. One set is blue-shifted, one set is red-shifted, and one set is unaffected. If these are interpreted as Doppler shifts, they indicate that one part of the object is approaching earth at about one-fourth the speed of light while another part is receding at a similar velocity. The stationary lines indicate that a third part of the object is not moving at a high velocity.

By taking spectra for a number of months, astronomers found that the red- and blue-shifted lines were moving back and forth across each other with a period of 160 days, much like the lines of a spectroscopic binary (Figure 5-16). Careful measurements of the so-called stationary lines showed that they too moved, but with a small velocity and with a period of about 13 days. This lower velocity variation suggests that the object may be part of a binary system.

Two other facts hint that SS 433 contains a compact object. Astronomers have found that the object is a source of radio, infrared, and x-ray radiation. Also, the object lies near the center of a supernova remnant that is about 30,000 years old. It is possible that SS 433 contains a white dwarf, neutron star, or black hole left over from the supernova explosion.

At this point it is impossible to say what SS 433 is. Theories have ranged from a black hole of about $10^5$ $M_\odot$ to a close binary containing two neutron stars. A recent theory suggests that the object is a neutron star of a few solar masses orbiting a normal star every 13 days and pulling material away from the star into an accretion disk. Somehow the disk ejects some of this gas in two high-velocity beams pointing in opposite directions. If this spinning accretion disk precessed with a period of 160 days, the beams of gas would sweep around the sky like water jetted from a whirling lawn sprinkler. Light emitted from the high-velocity beams of gas could account for the red- and blue-shifted emission lines.

Whatever the explanation of SS 433's peculiarities, it is clear that the study of x-ray binaries has not yet become routine. We may find more objects like SS 433, perhaps some even more fantastic. Observations by new, orbiting telescopes at both optical and x-ray wavelengths are certain to reveal more about the eccentric behavior of x-ray binaries, compact objects, and the deaths of stars.

## SUMMARY

When a star's central hydrogen burning reactions cease, its core contracts and heats up, igniting a hydrogen-burning shell and swelling the star into a cool giant. The contraction of the star's core ignites helium first in the core and later in a shell. If the star is massive enough, it can eventually burn carbon and other elements.

As the giant star burns these fuels, it moves through the red giant region of the H-R diagram. Each time it moves through the instability strip, it becomes unstable and pulsates as a variable star. Examples are stars such as the RR Lyrae and Cepheid variables, both important indicators of distance.

If a star's mass lies between about 0.4 and 3 $M_\odot$, its helium core becomes degenerate before the helium ignites. In degenerate gas, pressure does not depend on temperature so there is no pressure-temperature thermostat to control the reactions. As a result, the core explodes in a helium flash. All of the energy produced is absorbed by the star. A similar thing happens when carbon ignites in stars between about 3 and 9 $M_\odot$, except that it is much more powerful. This carbon detonation is one possible cause of supernovae.

How a star evolves depends on its mass. Stars less massive than about 0.4 $M_\odot$ are completely mixed and have very little hydrogen left when they die. They cannot ignite a hydrogen shell or a helium core, so they become white dwarfs. Medium-mass stars between about 0.4 and 3 $M_\odot$ become giants and burn helium, but cannot burn carbon. They produce planetary nebulae and become white dwarfs. Stars more massive than approximately 3 $M_\odot$ burn helium, carbon, and heavy elements up to iron. When their iron core collapses, they explode in a supernova, leaving behind a neutron star or a black hole.

Thus the three end-states of stellar evolution are compact objects—white dwarfs, neutron stars, and black holes. A white dwarf is a degenerate star about the size of the earth, containing no nuclear reactions. It eventually cools into a black dwarf.

The Chandrasekhar limit says no white dwarf can have a mass greater than 1.4 $M_\odot$. If the object is more massive than this, it will collapse into a neutron star—a star of degenerate neutrons only about 10 km in radius. A spinning neutron star can produce beams of energy that sweep around the sky, producing pulses when they pass over the earth. The Crab nebula, a supernova remnant, contains such a pulsar.

If the collapsing core of a star is more massive than about 3 $M_\odot$, no known force can prevent it from contracting to very small size. It may even become a singularity, an object of zero radius. Near such an object, gravity is so strong not even light can escape and we term the region a black hole. The surface of this region, called the event horizon, marks the boundary of the black hole. The Schwarzschild radius is the radius of this event horizon, amounting to only a few kilometers for black holes of stellar mass.

In a binary system, mass can be transferred from a normal star to an orbiting compact object. The mass forms an accretion disk around the compact object where friction slows and heats the gas as it falls in. If the object is a white dwarf, the accumulation of unburnt fuel on its surface can lead to nova explosions. A black hole in the midst of an accretion disk has such an intense gravitational field that the infalling matter can become hot enough to emit x-rays. Such objects may have been located. Cygnus X-1 and V861 Scorpii are believed to be black holes in binary systems. A neutron star at the center of an accretion disk can absorb enough matter from the disk to produce beams of x-rays that sweep around the sky and produce x-ray pulses as they pass over us. Hercules X-1 is such an object.

## NEW TERMS

helium flash
degenerate
instability strip
RR Lyrae variable stars
Cepheid variable stars
period-luminosity relation
planetary nebula
black dwarf
Chandrasekhar limit
carbon detonation
supernova remnant
turn-off point
open star cluster
neutron star
pulsar
glitch
lighthouse theory
cosmic rays
singularity
black hole
escape velocity
event horizon
Schwarzschild radius ($\mathbf{R_s}$)
Roche lobe
accretion disk
nova
recurrent nova
dwarf nova

## QUESTIONS

1. How does the contraction of a star's helium core make the star swell into a giant?
2. Why does the expansion of the star's envelope make the star cooler and more luminous?
3. What causes the helium flash? How do we know that it occurs?
4. Describe the two properties of degenerate matter that are important in stars.
5. Why do slow changes in the periods of Cepheid variables show that they are evolving?
6. Why is a Cepheid variable's period related to its luminosity?
7. Why don't red dwarfs become giants?
8. Why can't a white dwarf contract as it cools? What is its fate?
9. Describe step by step how the sun will die.
10. Give four possible causes for the supernova explosion.
11. How can we estimate the age of a star cluster?
12. Describe the lighthouse theory of pulsars. What observational evidence supports the theory?
13. Why should neutron stars spin fast and have powerful magnetic fields?
14. Why do astronomers find physical singularities difficult to accept? Why are we unlikely to ever see a singularity?
15. What do novae, Cygnus X-1, and Hercules X-1 have in common? How do they differ?
16. Review the evidence that Cygnus X-1 might contain a black hole.

## PROBLEMS

1. Draw and label an H-R diagram showing the instability strip and the evolutionary paths of a massive star and a medium mass star. Use the diagram to explain the period-luminosity relation among Cepheid variables.
2. Draw a series of diagrams like Figure 7-1 to show how the sun will change as it burns a hydrogen shell, helium core, and then a helium shell.
3. Suppose a planetary nebula is 2 pc in diameter. If the Doppler shifts in its spectrum show it is expanding at 30 km/sec, how old is it? (Hints: 1 pc equals $3 \times 10^{13}$ km, and 1 year equals $3.15 \times 10^{7}$ seconds.)
4. If the Crab nebula is now 1.35 pc in radius and is expanding at 1400 km/sec, about when did it begin? (Hints: 1 pc equals $3 \times 10^{13}$ km and 1 year equals $3.15 \times 10^{7}$ seconds.)
5. Sketch H-R diagrams of a young star cluster, a medium-age cluster, and an old cluster.
6. If, in a star cluster, the most massive star still on the main sequence has a mass of 5 $M_\odot$, how old is the cluster? (See Box 6-2.)
7. Draw a diagram showing a binary star consisting of a giant star and a compact object. Sketch the Roche lobes and the accretion disk.
8. If the inner accretion disk around a black hole has a temperature of 1,000,000°K, at what wavelength will it radiate the most energy? What part of the spectrum is this in? (Hint: See Box 3-1.)

## RECOMMENDED READING


Block, D. L. "Black Holes and their Astrophysical Implications." *Sky and Telescope* 50 (July/Aug. 1975), pp. 20 and 87.

Bortle, J. E. "Does Anyone Understand SS 433?" *Sky and Telescope* 58 (Dec. 1979), p. 510.

Bradt, H. "The Crab Nebula: A Unique Astrophysical Laboratory." *Technology Review* 78 (Dec. 1975), p. 35.

Charles, P. A., and Culhane, J. L. "X-Rays from Supernova Remnants." *Scientific American* 233 (Dec. 1975), p. 38.

Chevalier, R. A. "Supernova Remnants." *American Scientist* 66 (Nov./Dec. 1978), p. 712.

Green, L. C. "Star Quakes: Have They Been Observed?" *Sky and Telescope* 41 (Feb. 1971), p. 76.

Green, L. C. "Some New Developments in X-Ray Astronomy." *Sky and Telescope* 53 (May 1977), p. 340.

Gursky, H., and Van den Heuvel, E.P.J. "X-Ray-Emitting Double Stars." *Scientific American* 232 (March 1975), p. 24.

Hopkins, J. "Supernovae!" *Astronomy* 5 (April 1977), p. 6.

Kershner, R. P. "Supernovae in Other Galaxies." *Scientific American* 235 (Dec. 1976), p. 88.

McClintok, J. E. "The Marriage of X-Ray and Optical Astronomy." *Technology Review* 78 (Dec. 1975), p. 27.

Maran, S. P. "A Star in Trouble." *Natural History* 89 (Jan. 1980), p. 86.

Peltier, L. C. "Hunting Variable Stars." *Astronomy* 3 (Feb. 1975), p. 51.

Penrose, R. "Black Holes." *Scientific American* 226 (May 1972), p. 38.

Percy, J. R. "Pulsating Stars." *Scientific American* 232 (June 1975), p. 66.

Schramn, D. N., and Arnett, W. D. "Supernovae." *Mercury* 4 (May/June 1975), p. 16.

Seeds, M. A. "Stellar Evolution." *Astronomy* 7 (Feb. 1979), p. 6. Reprinted in *Astronomy: Selected Readings,* ed. M. A. Seeds. Menlo Park, Calif.: Benjamin/Cummings, 1980.

Stephenson, F. R., and Clark, D. H. "Historical Supernovas." *Scientific American* 234 (June 1976), p. 100.

Thorne, K. S. "The Search for Black Holes." *Scientific American* 231 (Dec. 1974), p. 32.

Verschuur, G. L. "X-Ray Astronomy." *Astronomy* 3 (April 1975), p. 34.

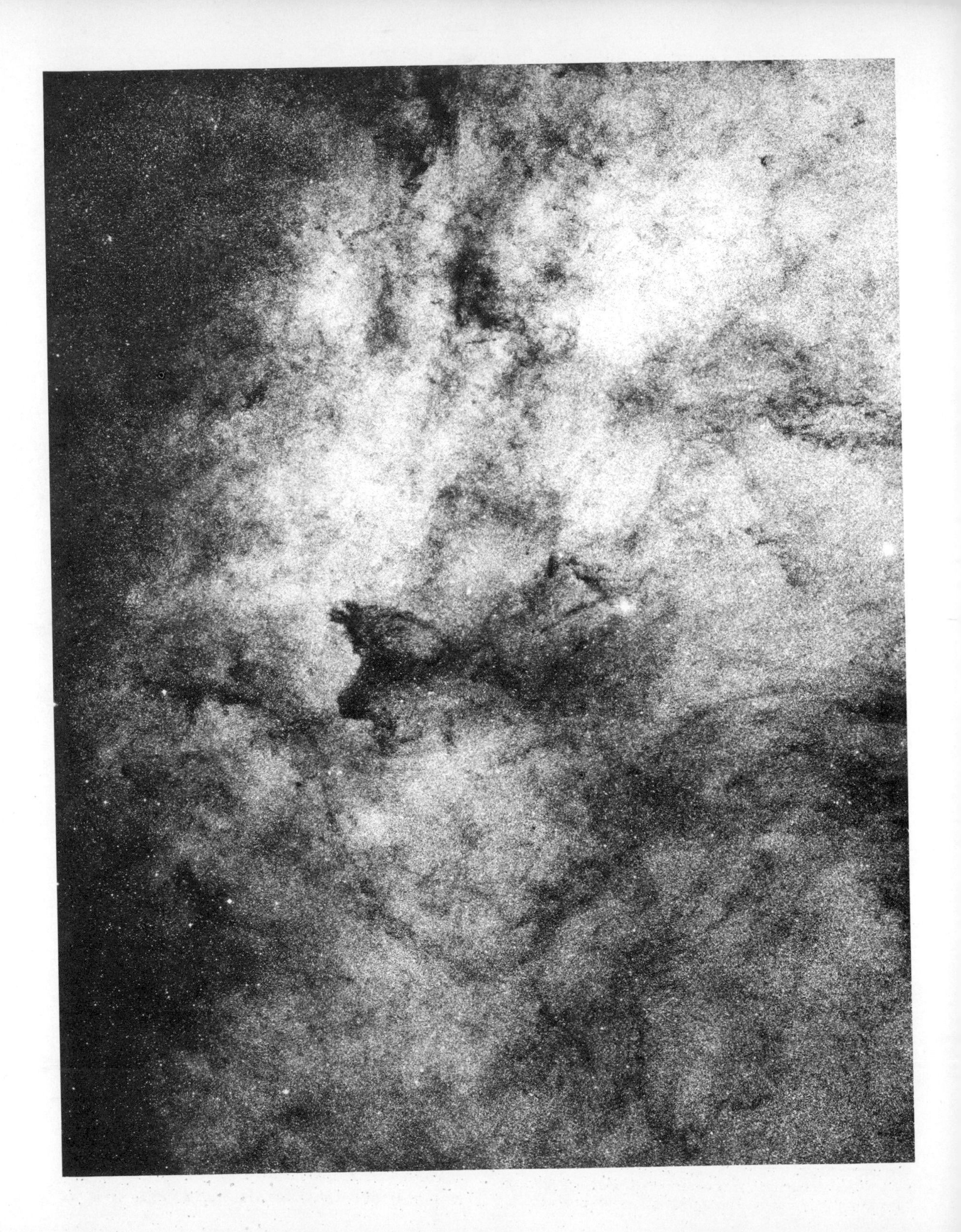

*Uncountable stars and thick clouds of dust obscure our view of the center of our galaxy. (Hale Observatories.)*

# Chapter 8 THE MILKY WAY

The ancient Greeks named the faint band of light that stretches around the sky *galaxies kuklos,* the "milky circle." The Romans changed the name a bit to *via lactea,* the "milky road" or "milky way." Today we recognize the Milky Way as the glow of 100 billion stars whirling in a great wheellike system that includes our sun. Millions of other such systems dot the sky, and, drawing on the Greek word for milk, we call them galaxies.

Almost every celestial object visible to our naked eyes is a member of our Milky Way galaxy. Two exceptions are the **Magellanic clouds,** small irregular galaxies located in the southern sky (Color Plate 17). Another exception is the Andromeda galaxy, just visible to our unaided eyes as a faint patch of light in the constellation of Andromeda.* Our galaxy probably looks much like the Andromeda galaxy (Figure 8-1 and Color Plate 16).

Unfortunately we cannot see much of our own galaxy from our position inside it. Clouds of dust block our view, like dark thunderclouds against a sunset (Figure 8-2). Photographs taken from earth can explore only about 10 percent of our galaxy, but that is enough to tell us what it is like and to hint at its history.

The key to understanding our galaxy is the separation of its stars into two regions. Most stars, including the sun and all the gas and dust, lie in a thin, rotating disk. Some stars and star clusters, however, are scattered through a spherical volume that completely encloses the galactic disk. The motions, compositions, and ages of the stars in these two distributions can help us visualize the formation of our galaxy.

As we explore our galaxy, we will discover two important themes. First, galaxies, like stars, are dominated by their own gravity. Each of the approximately $2 \times 10^{11}$ stars in our galaxy must move in its own orbit to balance the galaxy's gravity, and that motion influences the appearance and evolution of the galaxy as a whole. Second, the formation and evolution of stars determines the composition of the galaxy. There is no hope that we could understand our galaxy if we did not first grasp the stellar life cycle.

We begin our study of the Milky Way galaxy by taking a quick inventory to see where stars are located and how they move. Then we can investigate the way the birth and death of stars slowly alters the chemical composition of the galaxy.

## AN INVENTORY OF THE MILKY WAY GALAXY

Any galaxy as large as the Milky Way is a highly complex system of stars, gas, and dust moving in orbits and evolving from one form to another. To organize our data, we do two things. First we

* Consult the star charts at the end of this book to locate the Milky Way and the Andromeda galaxy in your night sky.

*Figure 8-1. The Great Galaxy in Andromeda is very similar to our own Milky Way galaxy. (Lick Observatory photograph.)*

*Figure 8-2. Vast dust clouds (black on this photograph) block our view of the Milky Way. This photograph of the region of Sagittarius looks toward the center of our galaxy (marked by the small box). (Hale Observatories.)*

divide the galaxy into two regions and describe the kinds of objects we find within these regions. Then we study the motions of the stars, gas, and dust. That will complete our inventory and prepare us to investigate the evolution of the galaxy.

*Components of the Galaxy.* The **disk component** of our galaxy consists of all matter confined to the plane of rotation—that is, everything in the disk itself. This includes stars, open star clusters, and nearly all of the galaxy's gas and dust.

We cannot quote a single number for the thickness of the disk for two reasons. First, the disk does not have sharp boundaries. The stars become gradually less crowded as we move away from the plane of the galaxy. Second, the thickness of the disk depends on the kind of object we study. The O stars and dense dust clouds lie within 300 ly of the galactic plane, but stars like the sun are much less confined: The disk defined by such stars is roughly 3000 ly thick (Figure 8-3).

The diameter of the disk and the position of the sun are also uncertain. Not only does the disk lack a sharp outer edge, but also it is heavily obscured by dust clouds. Estimates place the diameter at 100,000 ly. Such distances are often expressed in **kiloparsecs** (kpc). Since 1 kpc equals 1000 pc (3260 ly), the diameter of our galaxy is about 30 kpc.

Because the dust blocks our view of the galactic center, we can't measure our position directly, but radio telescopes and studies of star clusters place the sun about 10 kpc from the core. Thus we are about two-thirds of the way from the center to the edge.

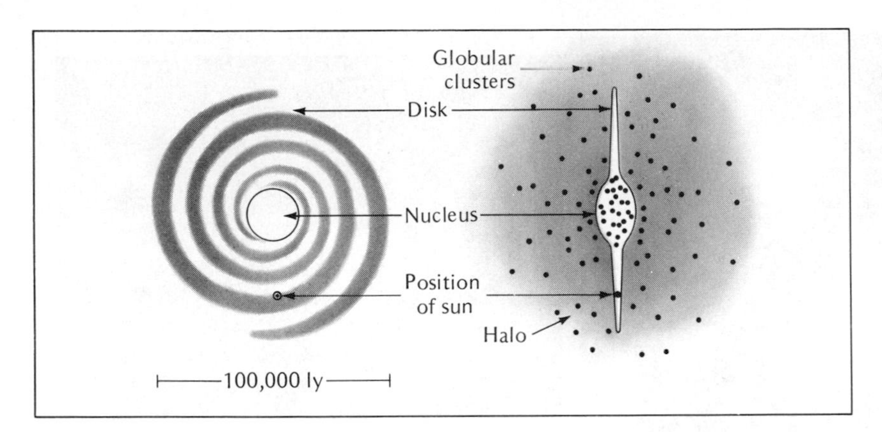


*Figure 8-3. The Milky Way galaxy seen face-on and edge-on illustrates the shape and location of the disk, halo, and nucleus. Note the position of the sun.*

The most striking features of the disk component are the **spiral arms,** long spiral patterns of bright stars, star clusters, gas, and dust. Such spiral arms are easily visible in photographs of other galaxies, and we will see on pages 190–195 that our own galaxy contains a similar spiral pattern.

The disk component contains two types of star clusters—associations and open clusters. **Associations** are groups of 10 to 100 stars so widely scattered in space their mutual gravity cannot hold the association together. We find the stars moving together through space (Figure 8-4) because they formed from a single gas cloud and have not yet wandered apart. However, associations are shortlived and the stars eventually go their separate ways. Two types of young associations—O and B associations and T Tauri associations (see Chapters 6 and 7)—are located along the spiral arms.

**Open clusters** (Figure 8-5) contain 100 to 1000 stars in a region 10 to 100 ly in diameter. Because they have more stars and occupy less space than associations, open clusters are more firmly bound by their gravity. Such clusters do lose stars occasionally as close encounters between cluster members eject stars from the group. However, it takes tens of millions of years for this process to affect the cluster. Thus open clusters are much more stable and longer lived than associations.

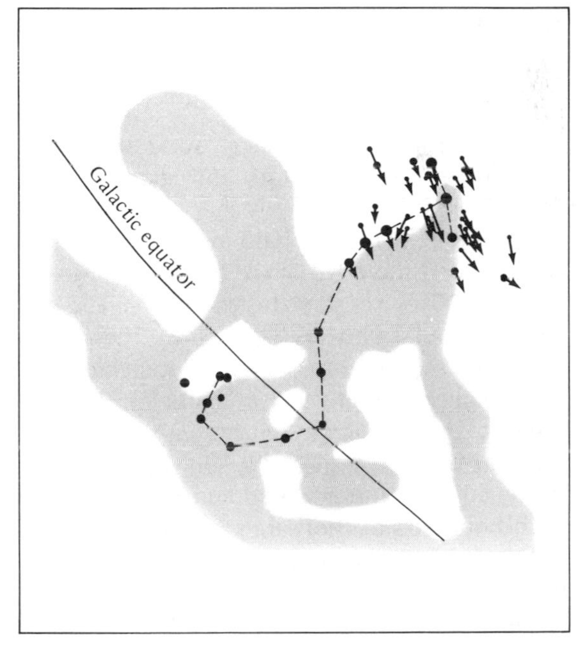


*Figure 8-4. Many of the stars in the constellation Scorpio are members of an O and B association. They have formed recently from a single gas cloud, and are moving together southwest along the Milky Way (shaded).*

The second component of our galaxy is the **spherical component,** which includes all matter in the galaxy scattered in a spherical distribution around the center. This includes the halo and the nuclear bulge.

The **halo** contains a thin scattering of stars, star clusters, and almost no gas and dust. It con-

*Figure 8-5. The Pleiades, an open cluster, contains about 500 stars. (Lick Observatory photograph.)*

tains no young bright stars, but rather cool, lower-main-sequence stars and giants. These stars are scattered so thinly that it is difficult to gauge the extent of the halo, but it must extend well beyond the edge of the disk.

The 100 or so star clusters in the halo are called **globular clusters** (Figure 8-6). These clusters contain 50,000 to 1,000,000 stars crowded into a sphere about 75 ly in diameter. Because stars in a globular cluster are 10 times closer together than those near the sun, people living inside such a cluster would find their sky filled with hundreds of thousands of stars. For comparison, only a few thousand stars are visible to the naked eye in our night sky.

Because globular clusters contain so many stars in such small regions, the clusters are very stable and have survived for billions of years. From the turn-off points in their H-R diagrams (Figure 8-6b), we can estimate their ages as 10 to 14 billion years, making them the oldest clusters in our galaxy. Box 8-1 describes the role they played in the discovery of the true size of our galaxy.

The stars of the nuclear bulge are similar to the halo stars, although evidence suggests the center of the bulge contains some young hot stars. The dust clouds in the disk and the tremendous crowding of stars in the central regions

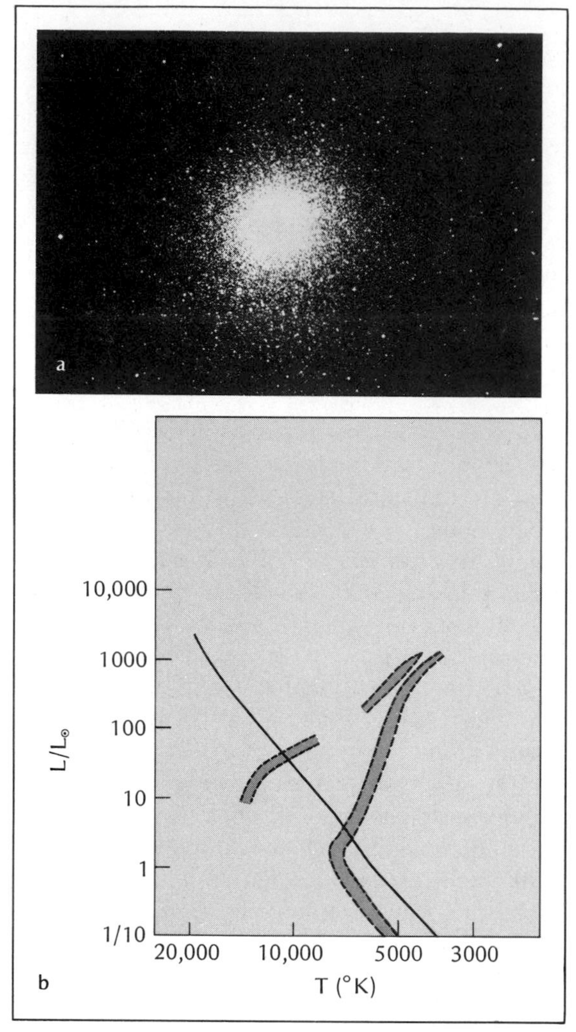


*Figure 8-6. (a) The globular cluster in Hercules, M 13, contains well over 100,000 stars in a region only about 25 pc in diameter. (b) The H-R diagrams of globular clusters show by their turn-off points that they are very old. M 13 is over 10 billion years old. (Hale Observatories photograph.)*

hopelessly obscures the galactic center at visual wavelengths. Pages 195–198 will review the radio and infrared observations of the galaxy's core.

So far our inventory has been static, merely listing the contents and sizes of the disk and spherical components. Now we must add motion and study galactic rotation.

*The Rotation of the Galaxy.* Clearly the galaxy must rotate. Each of the billions of stars exerts gravitational attraction on all the others, and the resulting mutual gravity pulls all stars toward the center. To keep from falling inward, each star must move in its own orbit around the center of the galaxy.

We can detect the motions of the stars in two ways. Doppler shifts in their spectra show how fast a star is moving along the line of sight. Motion across the line of sight is detectable if we compare photographs taken many years apart (Figure 8-7). The rate at which a star moves across the sky, termed its **proper motion**, is quite small, usually less than 0.1 seconds of arc per year. Nevertheless it can often be measured and, combined with the radial velocity and distance, reveals the shape of the star's orbit.

Disk stars move in nearly circular orbits that lie in the plane of the galaxy (Figure 8-8a). The sun, for example, is a disk star and moves about 250 km/sec in the direction of Cygnus, carrying the earth and other planets along with it. Since its orbit is a circle with a radius of 10 kpc, it takes about $250 \times 10^6$ years to make one circuit around the galaxy.

*Figure 8-7. Photographs taken 10 years apart reveal the proper motion of one of the sun's nearest neighbors, Barnard's star, which is only 1.8 pc away. (Lick Observatory photograph.)*

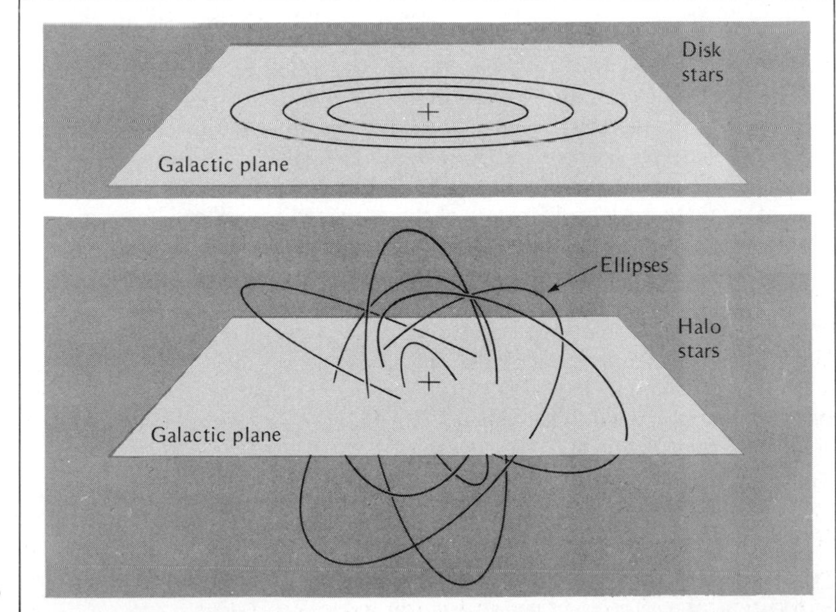


*Figure 8-8. (a) Stars in the galactic disk have nearly circular orbits that lie in the plane of the galaxy. (b) Stars in the halo have randomly oriented, elliptical orbits.*

The motion of stars near the sun shows that the disk does not rotate as a solid body. Stars orbiting just inside the sun's orbit must move faster because they are closer to the center, where gravity is stronger. Stars just outside the solar orbit move more slowly because they are farther from the center. Thus the galactic disk near the sun does not rotate like a solid disk. Instead, the outer stars move more slowly than the inner stars in what is called **differential rotation** (Figure 8-9).

Contrast this with the rigid rotation of a record on a turntable. Three spots lined up on the record would stay together as the record turned. Since the outermost spot had the farthest to go, it would move fastest, and the innermost spot would move slowest. In the galaxy, three stars lined up as in Figure 8-9 do not stay together. Because the galaxy rotates differentially, the innermost star pulls ahead and the outermost star falls behind.

The central parts of the disk do not rotate differentially. In the outer disk we find that orbital velocity increases as we travel inward, but as

## BOX 8-1 THE SIZE OF OUR GALAXY

Until the early part of this century astronomers had a poor understanding of our galaxy. They could see the Milky Way as a cloudy band of light across the sky, and they could count stars in various directions to gauge its extent, but, because most of the galaxy was hidden behind dust clouds, they failed to realize how limited their observations were. They thought the galaxy was at most about 10 kpc in diameter (Figure 8-10a) with the sun located at the center. In the years following World War I, a young astronomer named Harlow Shapley discovered that the galaxy was much bigger than anyone suspected.

Shapley began by studying the RR Lyrae variable stars (Box 7-2). These stars all have about the same absolute magnitude, +0.5, and are common in most globular clusters. In the nearer globular clusters, Shapley could identify the RR Lyrae stars by their variation (Figure 8-11). Measuring their apparent brightness and comparing that with their true luminosity gave him the distance to the cluster (see Box 5-2). Thus he used RR Lyrae stars to derive distances to the nearest globular clusters.

The variable stars in the more distant clusters were too faint to detect, but Shapley estimated the distances to these clusters from their angular diameters. For the clusters whose distance he knew, he could calculate diameters (in parsecs) with their observed angular diameters and the small angle formula (see Box 1-3). He found that the nearby clusters were about 75 ly in diameter, which he assumed was the average diameter of all globular clusters. He then used the angular diameters of the more distant clusters and the small angle formula to find their distances.

When he plotted the direction and distance to the globular clusters, he found they were distributed in a swarm centered not on the sun, but on a point thousands of parsecs away in the direction of Sagittarius (Figure 8-10b). He reasoned that since the motions of the clusters were dominated by the gravity of the entire galaxy, the center of the swarm should coincide with the center of the galaxy. Measuring the distance to this center revealed that the galaxy was much bigger than anyone had supposed.

Shapley's result established that our galaxy is much larger than the region we can see unobscured by dust clouds. More important, Shapley helped change the way we think of our place in nature. Previous to Shapley, our solar system appeared to be the center of the galaxy, perhaps the center of the universe. Shapley's result placed us in the suburbs of a very large galaxy. As he said, "[It] is a rather nice idea because it means that man is not such a big chicken. He is incidental—my favorite term is 'peripheral.'"*

* Harlow Shapley, *Through Rugged Ways to the Stars.* New York: Scribner's, 1969, p. 60.

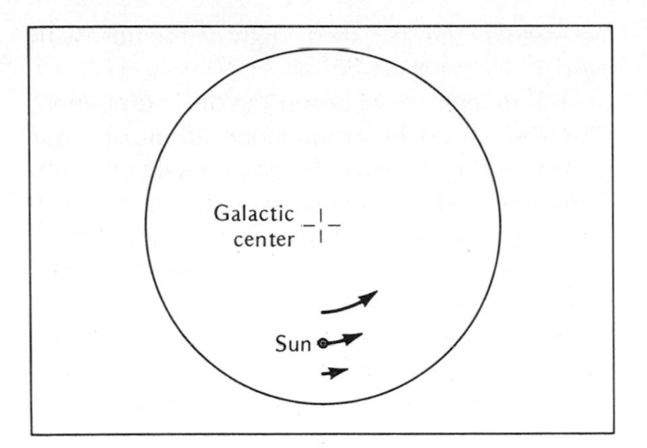


*Figure 8-9.* *The differential rotation of the galaxy's disk means that stars nearer the center orbit at higher velocities than stars farther from the center. Thus the star just inside the sun's orbit gains on the sun, and the star just outside falls behind.*

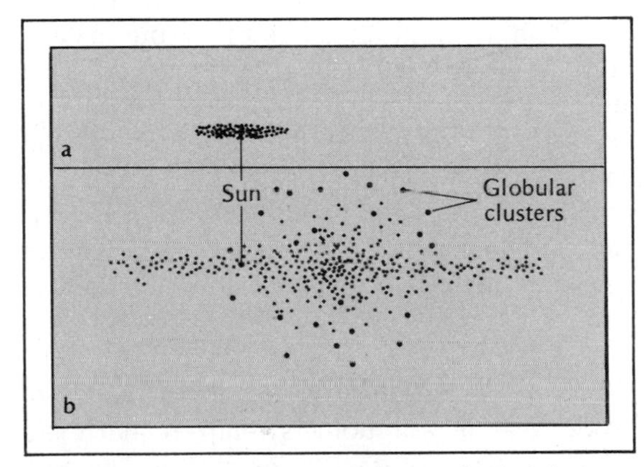


*Figure 8-10.* (a) *Before Shapley studied globular clusters, astronomers thought the galaxy was only about 10 kpc in diameter.* (b) *Shapley's work determining the distribution of the globular clusters showed that the galaxy was much larger and that the sun was not located at the center.*

*Figure 8-11.* *Numbered stars in this photograph of globular cluster M22 are variable. Studies of their apparent magnitude and period of pulsation can reveal the distance to the cluster. (Courtesy Amelia Wehlau and Helen Sawyer Hogg. Photograph taken with the 24-in telescope of the University of Toronto on Las Campanas, Chile.)*

we approach the center, orbital velocity begins to decrease. Stars near the galactic center enclose only a fraction of the galaxy's mass inside their orbits, and thus feel a weak gravitational force pulling them inward. They need not orbit rapidly to balance this weaker force. A graph of orbital velocity versus radius, called a **rotation curve** (Figure 8-12), shows how the galactic disk is divided into two parts: an outer region that rotates differentially, and an inner region in which orbital velocity decreases as we approach the center.

Motion in the halo is quite different. Each halo star and globular cluster follows its own randomly tipped elliptical orbit (Figure 8-8b). These orbits carry the stars and clusters far out into the spherical halo, where they move slowly, but when they fall back into the inner part of the galaxy, their velocities increase. Thus motions in the halo do not resemble a general rotation, but are more like the random motions of swarming bees.

As halo stars pass through the disk at steep angles, they move across the orbits of disk stars. Because they do not move in the same direction as the sun, they seem to have unusually high velocities. Thus we may find halo stars in the disk, but they are only passing through and can

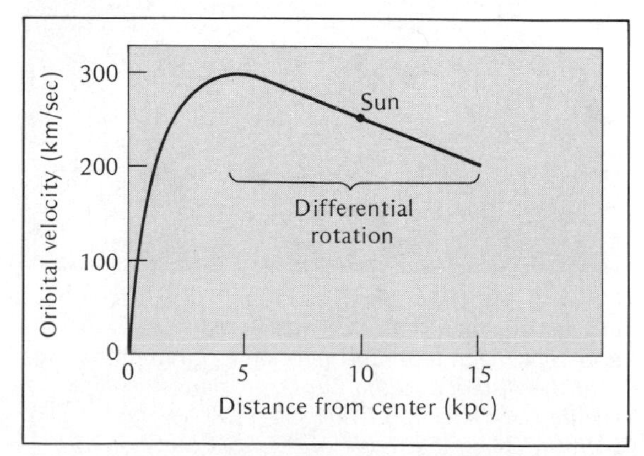


*Figure 8-12. The rotation curve of the galaxy shows that the outer region rotates differentially, while the inner region does not.*

be recognized by their high velocities with respect to the sun.

The differences between the disk component and the spherical component illuminate our galaxy's past. We have already seen that the two components differ in number of stars, amount of gas and dust, and orbital shape and orientation. In the next section we will discover that they also differ in chemical composition. That clue, combined with our knowledge of stellar evolution, will suggest a model for the formation of the galaxy.

## THE ORIGIN OF THE MILKY WAY

Just as paleontologists reconstruct the history of life on earth from the fossil record, astronomers can reconstruct the past of our galaxy from the fossil it left behind as it evolved. That fossil is the spherical component of the galaxy. The stars we see now in the nuclear bulge and halo formed when the galaxy was young. By studying those stars, their chemical composition, and their distribution in the galaxy, we can devise a theory to explain how the galaxy formed.

*Stellar Populations.* Near the end of World War II, astronomers realized that there were two types of stars in the galaxy. The type they were accustomed to studying were those located in the disk, like those near the sun. These they called **population I** stars. The second type, called **population II** stars, are usually found in the halo, in globular clusters, or in the central bulge. In other words, the two stellar populations are associated with the two components of the galaxy.

The stars of the two populations are very similar. They burn nuclear fuels and evolve in nearly identical ways. They differ only in the abundance of atoms heavier than helium, atoms that astronomers refer to collectively as **metals.** (Note that this is not the way the word *metal* is commonly used by nonastronomers.) Population I stars are metal-rich, containing 2 to 3 percent metals, while population II stars are metal-poor, containing only about 0.1 percent metals. The metal content of the star defines its population.

Population I stars belong to the disk component of the galaxy and are sometimes called disk population stars. They have circular orbits in the plane of the galaxy and are relatively young stars that formed within the last few billion years. The sun is a population I star, as are the type I Cepheids discussed in the previous chapter.

Population II stars belong to the spherical component of the galaxy and are sometimes called the halo population stars. These stars have randomly tipped, elliptical orbits, and are old stars. The metal-poor globular clusters are part of the halo population, as are the RR Lyrae and type II Cepheids.

Since the discovery of stellar populations, astronomers have realized that there is a gradation between populations (Table 8-1). Extreme population I stars, like the stars in Orion, are found only in the spiral arms. Slightly less metal-rich population I stars, called intermediate population I stars, are located throughout the disk. The sun is such a star. Stars even less metal-rich, such as stars in the nuclear bulge, belong to the intermediate population II. The most metal-poor stars are those in the halo and in globular clusters. These are extreme population II stars.

Why do the disk and halo stars have different metal abundances? The answer to this question is the key to the history of our galaxy. We must begin by discussing the cycle of element building in the Milky Way galaxy.

*The Element-Building Cycle.* We saw in the previous chapter how elements heavier than helium are built up by the nuclear reactions inside evolving stars. Theory suggests that atoms less massive than iron are built before the supernova explosion, and a small number of atoms heavier than iron could be made by nuclear reactions during the high-density, high-temperature phase of the supernova explosion. A chart of the abundance of the elements in the universe (Figure 8-13) illustrates the low abundance of atoms heavier than iron.

When the galaxy first formed, there should have been no metals because stars had not yet manufactured any. The gas from which the galaxy condensed must have been almost pure hydrogen (80 percent) and helium (20 percent). (Where the hydrogen and helium came from is a mystery we will save till Chapter 11.)

The first stars to form from this gas were metal-poor, and now, 10 to 14 billion years later, their spectra still show few metal lines. Of course, they may have manufactured some atoms heavier than helium, but since the stars' interiors are not mixed, those heavy atoms stay

**Table 8-1 Stellar Populations**

| | Population I | | Population II | |
|---|---|---|---|---|
| | Extreme | Intermediate | Intermediate | Extreme |
| Location | Spiral arms | Disk | Nuclear bulge | Halo |
| Metals | 3% | 1.6% | 0.8% | Less than 0.8% |
| Shape of orbit | Circular | Slightly elliptical | Moderately elliptical | Highly elliptical |
| Ave. age (yr) | 100 million and younger | 0.2–10 billion | 2–10 billion | 10–14 billion |

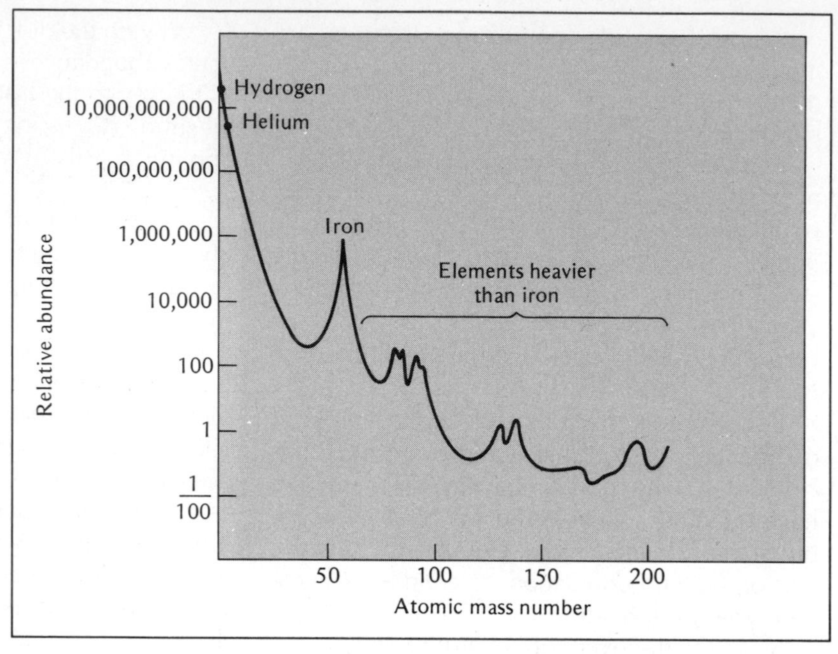


*Figure 8-13. The abundance of the elements. Because they are made only during supernova explosions, the elements heavier than iron are rare.*

trapped at the centers of the stars where they were produced and do not affect the spectrum (Figure 8-14). The population II stars that we see today are the survivors of the first generation of stars to form in the galaxy.

Most of the first stars evolved and died, and the most massive became supernovae and enriched the interstellar gas with metals. Successive generations of stars formed from gas clouds that were more enriched, and each generation added to the enrichment through the supernovae of massive stars. By the time the sun formed, 5 billion years ago, the element-building process had added about 1.6 percent metals. Since then the metal abundance has increased further, and stars forming now, in the Orion nebula for example, incorporate 2 to 3 percent metals and represent the metal-rich population I stars. Thus metal abundance varies between populations because of the production of heavy atoms in successive generations of stars.

Globular-cluster stars do contain a few metals, and that fact raises one of the interesting mysteries of modern astronomy: Where did the metals come from? We will see in a later chapter that the universe should have made very few metals when it began, so the gas from which our galaxy formed should have been nearly metal-free. Some theories suggest that there was an age of star formation before our galaxy took shape, and that these stars manufactured some metals that were later incorporated in the extreme population II stars. Other theories suggest that our galaxy began with a burst of star formation that created massive, highly luminous stars that lived short lives and enriched the forming galaxy with metals. In any case, the presence of metals in globular-cluster stars may someday tell us about conditions before the formation of our galaxy.

*Figure 8-14. Spectra of population II and population I stars of similar spectral type show hydrogen lines of equal strength, but the lines of heavier atoms are conspicuously weak in the population II star's spectrum. (Lick Observatory photograph.)*

The population II stars are in the halo and the population I stars are in the disk because the population II stars formed long ago when the galaxy was young and drastically different from its present form. By studying the distribution of the stellar populations we can visualize a process by which our galaxy could have formed.

*The Formation of the Galaxy.* A number of theories have been put forward to try to account for the origin of the Milky Way galaxy, and it is not possible at present to choose the correct version. However, we can summarize these theories by discussing a traditional hypothesis popular for some decades and a newer idea only a few years old.

The traditional theory supposes that the Milky Way galaxy began 10 to 14 billion years ago as a swirl of hydrogen and helium gas that began to contract under the influence of its own gravity. As it contracted, density grew until the gas began to fragment into individual clouds. Because the original gas was turbulent, the clouds formed with random velocities. When the gas density grew high enough, some of the gas began to form clusters of metal-poor stars. Because the galaxy was approximately spherical at this time, these first star clusters formed in a spherical distribution that we see today as the halo.

These star clusters may not have looked like globular clusters. Many probably contained too few stars and were too scattered through space to hold themselves together. They gradually dissociated, freeing stars that wandered through the spherical cloud. Clusters that had more stars packed in a smaller volume survived, although they, too, lost members occasionally. Today, over 10 billion years since they formed, the surviving clusters are the highly stable globular clusters.

Although the galaxy began as a spherical distribution of gas, it immediately began to collapse into a rotating disk. Randomly moving eddies in the cloud collided with each other and the turbulent motions canceled out, leaving the galaxy with a uniform rotation. A low-density cloud of gas rotating uniformly around an axis cannot maintain a spherical shape. A star can remain spherical because it has high internal pressure to balance the weight of the gas, but in a cloud where the density is low, this pressure is not effective. Like a blob of pizza dough spun into the air, the cloud must flatten into a rotating disk (Figure 8-15).

According to the traditional theory, this collapse into a disk took billions of years and did not alter the orbits of the stars that had formed when the galaxy was spherical. The halo population stars were left behind as a fossil of the early galaxy. Subsequent generations of stars formed in flatter distributions. The intermediate population I stars, for instance, are scattered hundreds of light-years above and below the plane of the disk. The gas distribution in the galaxy has now become so flat that the newest stars, the extreme population I stars, are confined to a disk only 300 ly thick.

In addition to a change in distribution, the shapes of stellar orbits also changed as the galaxy flattened. When the galaxy was young, the turbulent gas moved at random and the stars

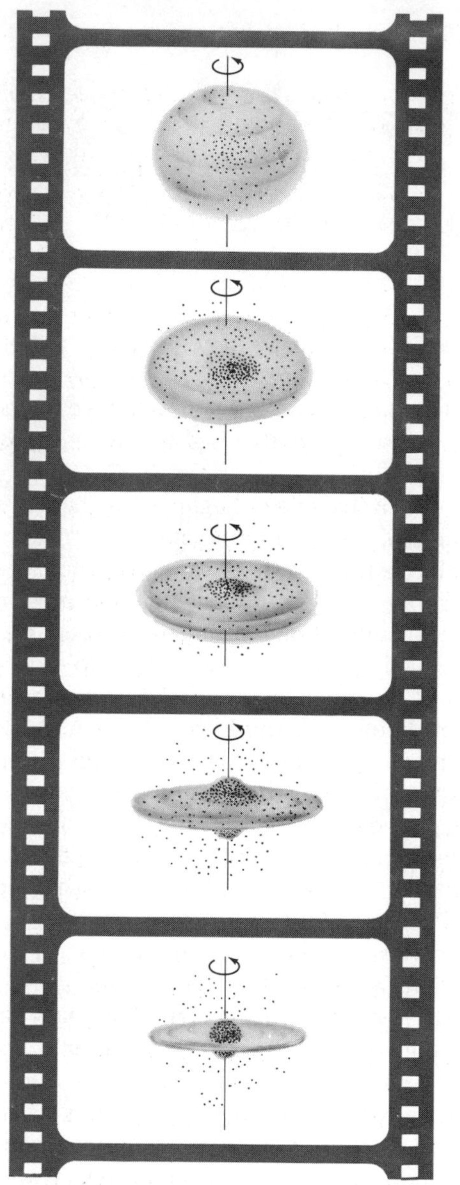

*Figure 8-15. According to the traditional theory, the galaxy began as a spherical cloud of gas (shaded) in which stars and star clusters (dots) formed. As the rotating gas cloud collapsed into a disk, the halo stars were left behind as a fossil of the early galaxy.*

that formed then took up orbits with random orientations and random shapes. As the galaxy collapsed into a disk, the random gas motion canceled out, and the gas took up more circular orbits, as did the forming stars. This explains why the oldest stars in the galaxy have the most elliptical orbits.

Simultaneously with these changes in distribution and orbital shape, the metal abundance in the stars grew with every generation, producing the stellar populations summarized in Table 8-1. The populations and their properties provide a permanent record of our galaxy's past.

This traditional view of the formation of the Milky Way sounds good, but new theories differ from it in some important respects. We will summarize these as a single hypothesis.

Some studies suggest that the hydrogen and helium gas produced large star clusters some time before the formation of the Milky Way galaxy. These star clusters may have contained $10^8$ stars, some of which were massive and exploded as supernovae, thus slightly enriching the remaining gas with metals. Collisions between these star clusters began to build a larger cluster, which attracted more clusters and eventually grew into a galaxy-size swarm of stars of approximately spherical shape. As this cluster grew massive, the remaining gas in the neighborhood fell into the galaxy and gradually collapsed to form a disk. Like the traditional theory, the new theory explains the distribution of populations by the gradual change in the shape of the galaxy. The important difference is in the process by which the galaxy first took form.

Even if we knew how the galaxy formed and evolved, there would still be mysteries hidden in the Milky Way's star clouds. In the following sections, we will consider two of these problems—spiral arms and the galactic nucleus.

## SPIRAL ARMS

The most striking feature of galaxies like the Milky Way is the system of spiral arms that wind outward through the disk. These arms contain swarms of hot blue stars, clouds of dust and gas, and young star clusters. The spiral pattern presents astronomers with two problems. First, how can we study the spiral arms of our galaxy when our view is obscured by dense clouds of dust? Second, what are the spiral arms? In this section we will see that mapping the location of the arms is easier than explaining their origin.

*Tracing the Spiral Arms.* Studies of other galaxies show us that spiral arms contain hot, blue stars. Thus one way to study the spiral arms of our own galaxy is to locate these stars. Fortunately, this is not difficult since O and B stars are often in associations, and, being very bright, they are easy to detect across great distances. Unfortunately, at these great distances their parallax is too small to measure, so their distances must be found by other means, usually by spectroscopic parallax (see Box 5-3).

O and B associations near the sun are not located randomly (Figure 8-16). They form three bands, indicating that there are three segments of spiral arms near the sun. If we could penetrate the dust clouds, we could locate other O and B associations and trace the spiral arms farther, but, like a traveler in a fog, we can see only the region near us.

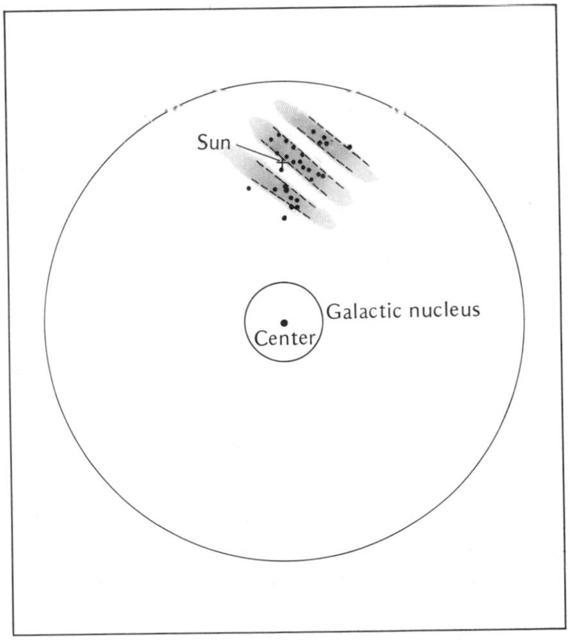

Objects used to map spiral arms are called **spiral tracers.** O and B associations are good spiral tracers because they are bright and easy to see at great distances. Other tracers include young open clusters, clouds of hydrogen ionized by hot stars (so-called **emission nebulae**), and certain kinds of variable stars.

Notice that all spiral tracers are young objects. O stars, for example, live only a few million years. If their orbital velocity is about 250 km/sec, they cannot have moved more than about 500 pc since they formed. This is less than the width of a spiral arm. Since they don't live long enough to move away from the spiral arms, they must have formed there.

The youth of spiral tracers gives us an important clue to the nature of the arms. Somehow the arms are associated with star formation. Before we can follow this clue, however, we must extend our map of spiral arms to show the entire galaxy.

Sun

Galactic nucleus

Center

*Figure 8-16. The O and B associations near the sun lie along three bands (shaded). These are segments of spiral arms.*

*Radio Maps of Spiral Arms.* The dust clouds that block our view at visual wavelengths are transparent at radio wavelengths because radio waves are much longer than the diameter of the dust particles. When we point a radio telescope at a section of the Milky Way, we receive 21-cm radio signals coming from cool hydrogen in a number of spiral arms at various distances across the galaxy. Fortunately, the signals can be unscrambled by measuring the Doppler shifts of the 21-cm radiation. The result is a map of the distribution of cool hydrogen gas clouds throughout the disk of our galaxy (Figure 8-17). Figure 8-18 shows how radio and optical data can be combined.

Radio maps reveal a number of things. First, the spiral pattern we see near the sun continues throughout the disk. Second, the spiral arms are rather irregular and interrupted by bends, spurs, and gaps. The stars we see in Orion, for example, appear to be a detached segment of a spiral arm. There are significant sources of error in the radio mapping method, but many of the ir-

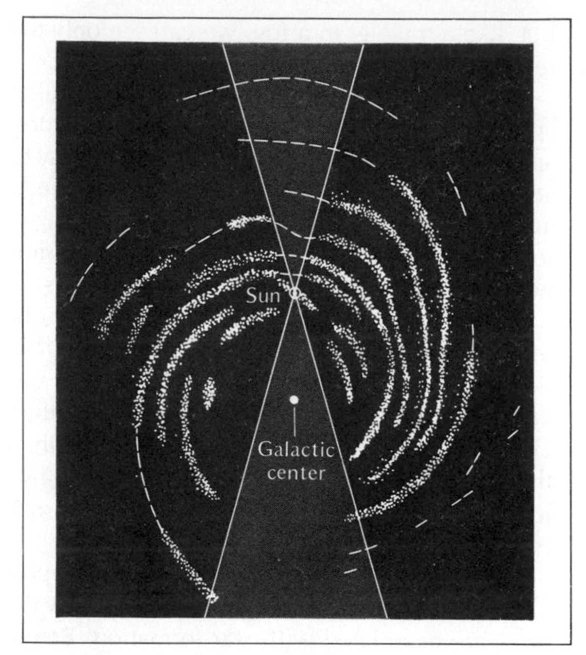


*Figure 8-17. A map of our galaxy at 21 cm shows traces of a spiral pattern in the distribution of cool hydrogen. (Courtesy Gerrit Verschuur.)*

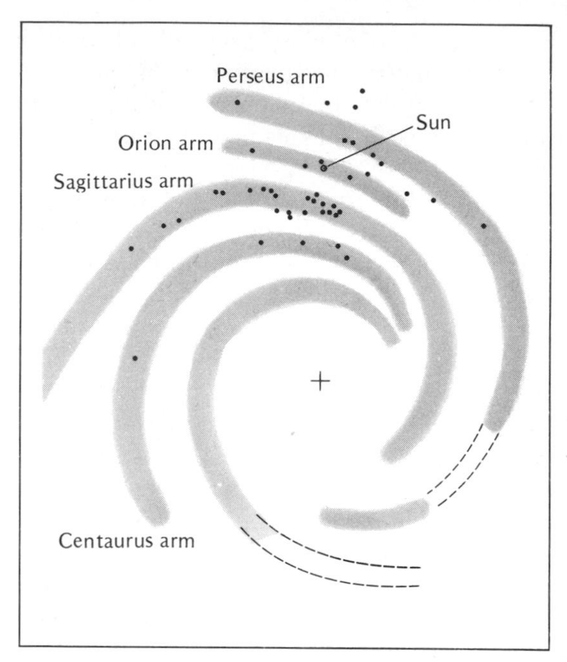


*Figure 8-18. Combined optical and radio data reveal the overall spiral pattern of the Milky Way. Compare with Figure 8-24.*

regularities along the arms seem real and photographs of nearby spiral galaxies show similar features. Recently a team of astronomers combined all available data on our galaxy's spiral pattern and compared their model with other galaxies. The result of the study was a sketch of what the Milky Way galaxy probably looks like to a distant observer (Figure 8-19).

The most important feature revealed in the radio maps is easy to overlook—spiral arms are regions of high gas density. Spiral tracers told us that the arms were regions of young objects, and we therefore suspected that the arms were regions of star formation. The radio maps confirm our suspicion by telling us that the material needed to make stars is abundant in spiral arms.

*The Density Wave Theory.* Having mapped the spiral pattern, we can ask, "What is a spiral arm?" We can be certain they are not physically connected structures like bands of

*Figure 8-19. An observer a few million light-years away might have this view of our galaxy. The cross marks the position of the sun. (Adapted from a study by G. De Vaucouleurs and W. D. Pence,* The Astronomical Journal, *October 1978, Volume 83, No. 10, p. 1163.)*

magnetic field holding the gas in place. If they were, they would be destroyed by the strong differential rotation. Because the inner parts of the disk spin so much faster than the outer parts, a spiral arm would get stretched out and obliterated in a few tens of millions of years. Yet spiral arms are very common in disk-shaped galaxies and must be reasonably permanent features.

To explain how the arms could persist in the differentially rotating disk, some astronomers suggest that they are a disturbance in the moving material of the disk. A boulder in a rushing mountain stream can create an eddy that seems permanent even though the water in the eddy constantly changes. The spiral arms could be regions of compression through which the gas of the disk moves like water through an eddy. Because the regions of compression that we see as spiral arms would have higher density and would move around the galaxy just as a sound wave moves through air, this idea is called the **density wave theory.**

Computer models indicate that the density wave would look very much like the spiral patterns we see in spiral galaxies. The density wave should take the form of a two-armed spiral winding outward from the nuclear bulge to the edge of the disk. In addition, the theory predicts that the arms should be stable and not wind up. The most important prediction, however, is that star formation should occur along the arms.

To see how star formation could occur in the density wave, we must study the motion of gas clouds as they move into the arm. Although the arms move around the galaxy in the same direction that the galaxy rotates, they move rather slowly and the orbiting stars and clouds of gas overtake the arms from behind (Figure 8-20).

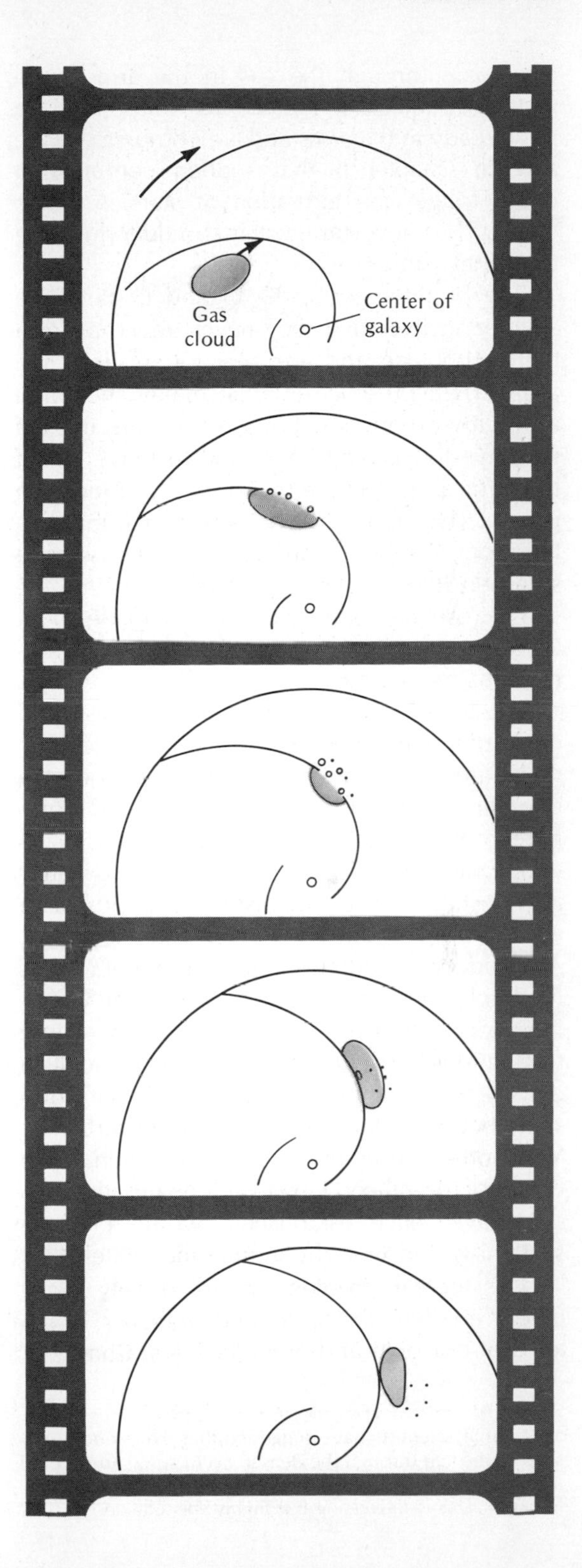


*Figure 8-20. According to the density wave theory, gas clouds overtake the spiral arm from behind and smash into the density wave. The compression triggers the formation of stars. The massive stars (open circles) are so shortlived that they die before they can leave the spiral arm. The less massive stars (dots) emerge from the front of the arm with the remains of the gas cloud.*

Stars pass through the gas in the arms unaffected, but incoming gas clouds smash into the gas already in the arms and are compressed. We saw in Chapter 6 that sudden compression could trigger the formation of stars in a gas cloud. Thus new star clusters should form along the spiral arms.

The brightest stars, the O and B stars, live such short lives that they never travel far from their birthplace and are only found along the arms. Their presence is what makes the spiral arms glow so brightly. Lower-mass stars, like the sun, live longer and have time to move out of the arms and continue their journey around the galaxy. The sun may have formed in a star cluster about 5 billion years ago when a gas cloud smashed into a spiral arm. Since that time the sun has escaped from its cluster and made about 20 trips around the galaxy, passing through spiral arms many times.*

The density wave theory is very successful in explaining the properties of spiral galaxies, but there are two problems. First, how does the complicated spiral disturbance originate? Computer models are not very helpful with this problem. One possibility is that the galaxy is naturally unstable to certain disturbances, just as a guitar string is unstable to certain vibrations. Any sudden disturbance—the rumble of a passing truck, for example—can set the string vibrating at its natural frequencies. Similarly, minor fluctuations in the galaxy's disk might generate a density wave. Another suggestion is that collisions between galaxies excite the disturbance. Yet another theory involves an explosion at the center of the galaxy. However it begins, the density wave, once established, would be self-sustaining and would continue indefinitely.

The second problem in the density wave theory involves the spurs and branches in the arms of our own and other galaxies. Computer models of density waves produce regular, two-armed spiral patterns. We must conclude that to produce spurs and branches, the density of the disk material has to be highly irregular and that the arms sometimes break into segments and branch when they encounter such irregularities. This is not a very satisfactory explanation, however, when we consider the uniform motion of the disk material. Perhaps the explanation lies in a new theory of spiral structure that involves the compressional triggering of star formation.

*Self-Sustaining Star Formation.* The formation of massive stars along a spiral arm could lead to the formation of more stars in a self-sustaining process that could maintain the arm as a semipermanent feature of the galaxy. Recent computer studies of this idea have been only partially successful in producing spiral patterns, but the theory is promising and may be an important addition to the density wave theory.

**Self-sustaining star formation** may occur if the birth of a massive star compresses neighboring clouds of gas (Figure 8-21). This compression could be caused by the outward rush of heated gas as the massive star turns on, or by the expansion of a supernova remnant produced by the death of a massive star. Since massive stars

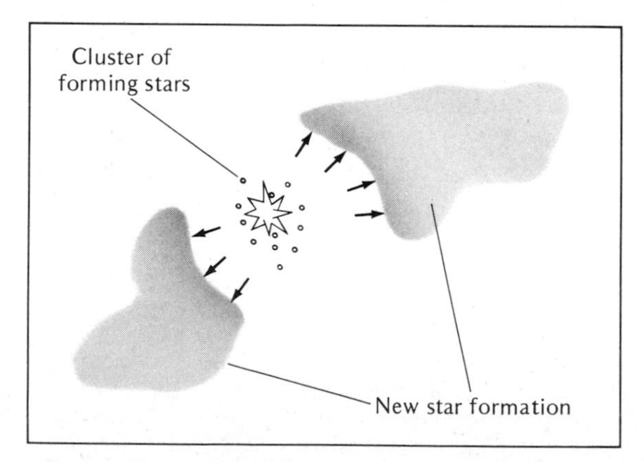


*Figure 8-21. Self-sustaining star formation may occur when a cluster of forming stars develops a massive member. The energetic turn-on or sudden supernova can compress nearby gas clouds and trigger new star formation.*

* Some scientists have suggested that the periodic passage of the sun through the denser gas in spiral arms could affect the amount of sunlight reaching earth and thus cause ice ages. This is interesting but highly speculative.

live such short lives, their birth and the production of a supernova are separated by a mere instant in the life of a galaxy. Whichever process compresses the neighboring clouds, star formation could begin in those clouds if the compression was sufficient. If that star formation created additional massive stars, the process might be self-sustaining.

If self-sustaining star formation works, we should expect star formation to take place in patches throughout the disk of the galaxy. Differential rotation would tend to pull those patches into segments of spiral arms because the inner edges of the patches would orbit about the center of the galaxy faster than the outer edges (Figure 8-22).

Computer models of self-sustaining star formation do produce patterns that have spiral features (Figure 8-23), but they do not have the symmetrical, two-armed patterns so common in spiral galaxies. While self-sustaining star formation may not work alone to produce spiral arms, it may modify arms generated by the density wave. Segments of the density wave in regions favorable for self-sustaining star formation may grow rapidly and become spurs and branches. The multiarmed pattern we see in the radio maps of our own galaxy might be due to this combination of processes.

Spiral arms are graceful streamers of light in spiral galaxies like the Milky Way, and the processes that create them are fascinating in their complexity. Another region of our galaxy, the nucleus, is equally fascinating and complex, but it substitutes raw power for grace.

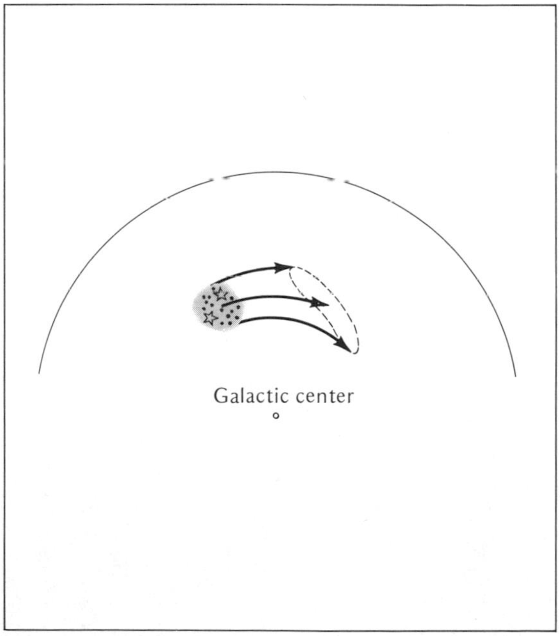

*Figure 8-22. A cloud of gas in which self-sustaining star formation is active can be stretched into a spiral shape by differential rotation. The inner edge of the cloud orbits faster than the outer edge.*

*Figure 8-23. A computer model of spiral arms produced by self-sustaining star formation is superimposed on the galaxy M 81. The computer model has been tipped to match the inclination of the galaxy. (Reprinted courtesy of Humberto Gerola and* The Astrophysical Journal, *published by the University of Chicago Press; © 1978 The American Astronomical Society.)*

## THE NUCLEUS

The most mysterious region of our galaxy is its very center, the nucleus. Hidden behind thick clouds of gas and dust, it is totally invisible at optical wavelengths. But radio, infrared, x-ray, and gamma-ray signals penetrate the clouds and give us a picture of stars crowded together, a disk of gas spinning at the center, and clouds of gas rushing outward as if from an explosion.

*Observations.* If we examine the Milky Way on a dark night, we might notice a slight thickening in the direction of the constellation Sagittarius, but nothing specifically identifies that as the direction toward the heart of the galaxy (Figure 8-24). Even Shapley's study of globular clusters (Box 8-1) identified the center of the galaxy only approximately. When radio astronomers turned their telescopes toward Sagittarius, they found a collection of radio sources, the most powerful of which, **Sagittarius A,** lies at the expected location of the galactic core (Figure 8-25).

Before we explore the galactic center in detail, let us survey the kinds of information astronomers can get at different wavelengths. Signals at radio, infrared, x-ray, and gamma-ray wavelengths originate in different ways, and consequently tell us about conditions in different areas of the core.

The galactic center emits thermal radiation, which comes from hot gas, most of which lies in clouds of ionized hydrogen around hot stars. Since hot stars live short lives and must have formed recently, the thermal radiation is evidence that some star formation is still going on in the nucleus of the galaxy.

In addition to 21-cm radiation from clouds of cool hydrogen, the galactic center is also a source of **synchrotron radiation.** This radiation is produced when high-energy electrons move through a magnetic field. Thus the signals from the nucleus evidence a magnetic field and high-energy electrons. No one knows what excites the electrons to such high speeds. They might be produced by supernova explosions or in a single object such as a massive black hole.

Infrared radiation with a wavelength of about $2 \times 10^{-6}$ m (20,000 Å) comes mainly from cool stars. Since most of the stars in the nucleus are cool population II stars, a map of this radiation reveals the distribution of stars. The results of such studies show that the stars near the center are astonishingly crowded.

Infrared radiation at wavelengths longer than about $4 \times 10^{-6}$ meters (40,000 Å) comes mainly from dust heated by stars. The strength of these signals indicates that the dust is dense in some regions and heated by closely spaced stars.

Orbiting space telescopes have detected

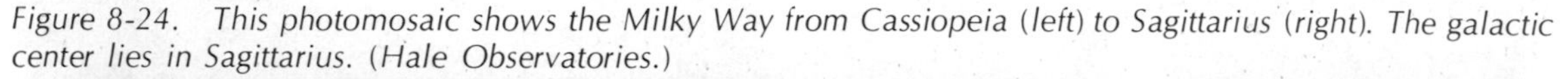

*Figure 8-24. This photomosaic shows the Milky Way from Cassiopeia (left) to Sagittarius (right). The galactic center lies in Sagittarius. (Hale Observatories.)*

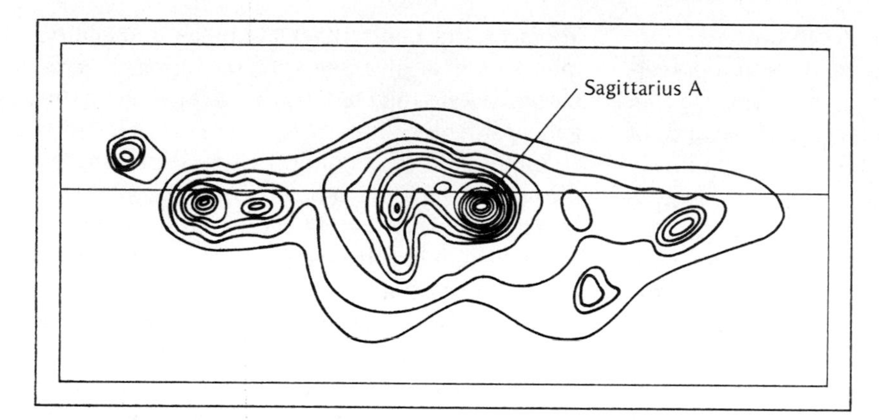


*Figure 8-25. A radio map of the Sagittarius region of the Milky Way reveals an intense radio source, Sagittarius A, at the expected location of the galactic core (see box in Figure 8-2). (Adapted from Downes, D., Maxwell, A., and Meeks, M. L., 1965,* Nature, *Volume 208, p. 1189.)*

x-rays and gamma rays coming from the galactic center. The x-rays probably originate in large numbers of x-ray binaries among the millions of stars near the center. The gamma rays may be produced by interactions between high-energy particles.

Though the galactic nucleus is undetectable at visual wavelengths, it is observable at these other wavelengths. To see what this radiation tells us about the conditions in the nucleus, let's take an imaginary journey from the sun to the center of the galaxy.

*Conditions in the Center.* If we could suspend the laws of physics and journey from the earth to the center of our galaxy at the speed of light, we would find drastic changes in the average separation between stars. Near the sun we would find about 0.06 stars/pc$^3$, making the stars about 1.6 pc apart. Observers on earth would see our spacecraft pass near a star every five years or so. After 25,000 years we would enter the fringe of the nucleus and find the stars crowded closer together. As we neared the core, the star density would rapidly grow to about $4 \times 10^6$ stars/pc$^3$. Stars would be less than 800 AU apart, and observers on earth, if they could see us at all, would note that we passed a star every four days.

Radio observations at 21 cm predict that we would find a disk of gas about 750 pc in radius and about 100 pc thick rotating around the galactic center (Figure 8-26). Within this disk, about 250 pc from the center is an irregular ring of dense clouds of gas and dust. Deep inside these clouds, protected from ultraviolet radiation, molecules can survive and emit the radio signals whose Doppler shifts tell us the ring is

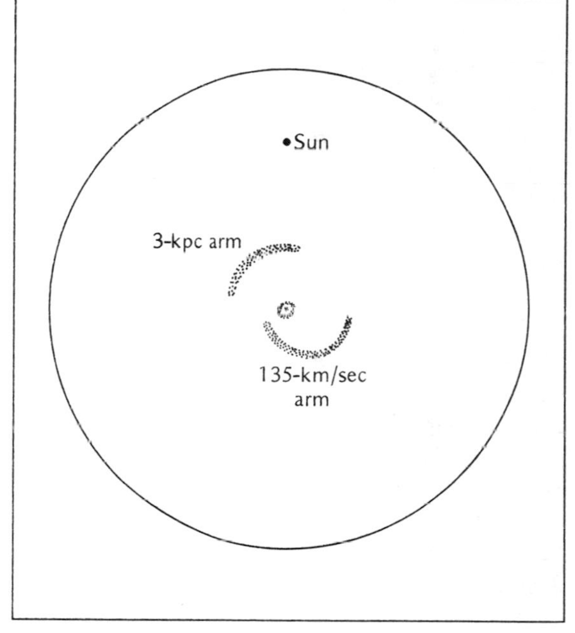


*Figure 8-26. Near the center of our galaxy we find a rotating disk of gas containing a ring of dense clouds. Between the center and the sun lies the 3-kpc arm and beyond the center the 135-km/sec arm. All of these features are expanding outward.*

expanding outward at about 140 km/sec.

Very near the center we would find a powerful source of infrared radiation about 1 pc in diameter. It probably consists of a swarm of small dust clouds heated by the tremendously crowded stars.

At the very heart of the galaxy lies a mysterious source of radio signals only a few parsecs in diameter—Sagittarius A. The central core of the source is less than 10 AU in radius and contains turbulent, ionized gas that produces broad spectral lines at infrared and radio wavelengths. If gravity holds this turbulent gas at the center, then the mass of the region must be about $5 \times 10^6 \; M_\odot$. It is exciting to think that this may be a supermassive black hole generating energy by swallowing infalling matter, but since no one knows how the matter is distributed through the region, we cannot be sure what lies at the precise center.

*Explosions.* The amount of energy radiated and the outward expansion of gas clouds suggest a core explosion, so we must look at our galaxy again, at the disk as well as the nucleus, to search for additional evidence of past eruptions. If such evidence exists, we must then ask how a galactic nucleus could explode.

Radio observations at 21 cm reveal a cloud containing $10^7 \; M_\odot$ of neutral hydrogen moving outward from the nucleus toward us at 53 km/sec (Figure 8-26). This cloud is called the **3-kpc arm** because it lies 3 kpc from the galactic center. Whether it is a segment of a spiral arm, a segment of a ring, or simply a large cloud of gas is unknown. However, it could be the remains of gas ejected millions of years ago in an explosion at the center of the galaxy.

Similar radio observations reveal a cloud of neutral hydrogen about 2.5 pc from the center on the far side of the nucleus. Its Doppler shift indicates that it is receding at 135 km/sec. Since its distance is uncertain, it is not called the 2.5-kpc arm, but rather the **135-km/sec arm.** It too may be an arm, a segment of a ring, a large cloud, or gas ejected by an explosion in the galaxy's center.

Other galaxies contain evidence of core explosions. The Andromeda galaxy contains a small, intensely bright core only 5 pc in diameter—about the same size as Sagittarius A. In the next chapter we will discuss Seyfert galaxies, which contain small, bright nuclei that seem to be exploding. We will also examine galaxies that now radiate strong radio signals apparently from the remains of past explosions in their centers.

No one knows what could have caused an explosion in the nucleus of our galaxy, but there are a number of interesting theories. For example, the stars in the center are so crowded that collisions should not be unusual, and large superstars of 500 to 1000 $M_\odot$ might accumulate by the coalescence of stars. These should be highly unstable and might explode violently, or collapse into black holes and release energy in the process. Just the collision of two stars might produce large amounts of energy. Another suggestion supposes that large numbers of neutron stars and black holes exist in the nucleus, left over from the collapse of massive stars. Matter falling into these compact objects could radiate tremendous amounts of energy. Yet another theory suggests that the center of our galaxy contains a single supermassive black hole of more than 1 million $M_\odot$. Gas, dust, and the debris of star collisions falling into such an object could release enormous amounts of energy. Although this theory is still speculative, the next chapter will review evidence that a super black hole may exist at the center of a nearby galaxy much larger than our own.

Clearly we cannot learn much about the Milky Way galaxy if we ignore other galaxies around us. We must compare our own galaxy with others to see how it fits into the pattern.

## SUMMARY

The Milky Way galaxy contains two components, the disk component and the spherical component. The disk is about 100,000 ly in diameter and the sun is about two-thirds of the way from the center to the edge. Disk stars are metal-rich population I stars moving in circular orbits that lie in the plane of the disk. Because stars farther from the center move slower in their orbits, the disk rotates differentially.

The spherical component consists of a nuclear bulge at the center and a halo of thinly scattered stars and globular clusters that completely envelops the disk. Halo stars are metal-poor population II stars moving in random, elliptical orbits.

The distribution of populations through the galaxy suggests a way the galaxy could have formed from a spherical cloud of gas that gradually flattened into a disk. The younger the stars, the more metal rich they are, and the more circular and flat their orbits are.

The very youngest objects lie along spiral arms within the disk. They live such short lives they don't have time to move from their place of birth in the spiral arms. Maps of these spiral tracers and cool hydrogen clouds reveal the spiral pattern of our galaxy.

The spiral density wave theory suggests that the spiral arms are regions of compression that move through the disk. When an orbiting gas cloud smashes into the compression wave, the gas cloud forms stars. Another process, self-sustaining star formation, may act to modify the arms as the birth of massive stars triggers the formation of more stars by compressing neighboring clouds.

The nucleus of the galaxy is invisible at optical wavelengths, but radio, infrared, x-ray, and gamma ray radiation can penetrate the dust clouds. These wavelengths reveal crowded central stars and heated clouds of dust. Some of the central features are expanding outward. The 3-kpc arm and the 135-km/sec arm are both moving outward from the center, as is a ring of dense molecular clouds. These expanding features suggest the nucleus may have exploded millions of years ago.

The very center of the Milky Way is marked by a radio source, Sagittarius A, which is also a source of infrared radiation. The core of the source is only 10 AU in diameter and contains about $5 \times 10^6$ $M_\odot$. The nature of the object is unknown.

## NEW TERMS

Magellanic clouds
disk component
kiloparsec
spiral arm
association
open cluster
spherical component
halo
globular cluster
proper motion
differential rotation
rotation curve
population I
population II
metals
spiral tracers
emission nebula
density wave theory
self-sustaining star formation
Sagittarius A
synchrotron radiation
3-kpc arm
135-km/sec arm

## QUESTIONS

1. Why is it difficult to specify the dimensions of the disk and halo?
2. Why didn't astronomers before Shapley realize how large the galaxy was?
3. Explain why some star clusters lose stars more slowly than others.
4. Contrast the motion of the disk stars and halo stars. Why do their orbits differ?
5. Why are metals less abundant in older stars than in younger?
6. Why are all spiral tracers young?
7. Why couldn't spiral arms be physically connected structures? What would happen to them?
8. Why does self-sustaining star formation produce clouds of stars that look like segments of spiral arms?
9. Describe the kinds of observations we can make to study the galactic nucleus.
10. What evidence do we have that the nucleus has exploded in the past?

## PROBLEMS

1. Make a scale sketch of our galaxy in cross-section. Include the disk, sun, nucleus, halo, and some globular clusters. Try to draw the globular clusters to scale size.
2. Because of dust clouds, we can see only about 5 kpc into the disk of the galaxy. What percentage of the galactic disk can we see? (Hint: Consider the area of the entire disk and the area we can see.)
3. If the fastest passenger aircraft can fly 1000 mph (1600 km/hr), how long would it take to reach the sun? The galactic center? (Hint: 1 pc = $3 \times 10^{13}$ km.)
4. If the RR Lyrae stars in a globular cluster have apparent magnitudes of 14, how far away is the cluster? (Hint: See Box 5-2.)
5. If a globular cluster is 10 minutes of arc in diameter and 8.5 kpc away, what is its diameter? (Hint: Use the small angle formula, Box 1-3.)
6. If we assume that a globular cluster 4 minutes of arc in diameter is actually 25 pc in diameter, how far away is it? (Hint: Use the small angle formula, Box 1-3.)
7. Make a series of sketches to show how differential rotation would alter the positions of three stars initially lying in a straight line pointing toward the galactic center.
8. Draw a series of sketches to show how the galaxy began from a spherical cloud and collapsed into a disk. Explain why it could not remain spherical.
9. If the sun is 5 billion years old, how many times has it orbited the galaxy?
10. Draw a series of sketches to show how a cloud of gas moves through a spiral arm and forms stars.
11. Infrared radiation from the center of our galaxy with a wavelength of about $2 \times 10^{-6}$ meters (20,000 Å) comes mainly from cool stars. Use this wavelength as $\lambda_{max}$ and find the temperature of the stars.

## RECOMMENDED READING

Bok, B. J. "The Spiral Structure of Our Galaxy." *Sky and Telescope* 39 (Jan. 1970), p. 21.

Bok, B. J. "Updating Galactic Spiral Structure." *American Scientist* 60 (Nov./Dec. 1972), p. 709.

Bok, B. J., and Bok, P. *The Milky Way*. 4th ed. Cambridge, Mass.: Harvard University Press, 1974.

Chaisson, E. J. "Gaseous Nebulas." *Scientific American* 239 (Dec. 1978), p. 164.

Geballe, T. R. "The Central Parsec of the Galaxy." *Scientific American* 241 (July 1979), p. 60.

Gordon, M. A., and Burton, W. B. "Carbon Monoxide in the Galaxy." *Scientific American* 240 (May 1979), p. 54.

Herbst, W., and Assousa, G. E. "Supernovas and Star Formation." *Scientific American* 241 (Aug. 1979), p. 138.

Holzinger, J. R., and Seeds, M. A. *Laboratory Exercises in Astronomy*. Ex. 34–37. New York: Macmillan, 1976.

Jaki, S. L. *The Milky Way, an Elusive Road for Science*. New York: Science History Publications, 1972.

Larson, R. B. "The Origin of Galaxies." *American Scientist* 65 (March/April 1977), p. 188.

Lesh, J. R. "Swarms of Stars: Cosmic Calibrators." *Astronomy* 6 (March 1978), p. 6. Reprinted in *Astronomy: Selected Readings*, ed. M. A. Seeds. Menlo Park, Calif.: Benjamin/Cummings, 1980.

Maran, S. P. "Strung Out Stars." *Natural History* 87 (Feb. 1978), p. 30.

Maran, S. P. "Deep in the Heart of the Milky Way." *Natural History* 87 (Oct. 1978), p. 142.

Pasachoff, J. M., and Goebel, R. W. "Laboratory Exercises in Astronomy—Cepheid Variables and the Cosmic Distance Scale." *Sky and Telescope* 57 (March 1979), p. 241.

Sanders, R. H., and Wrixon, G. T. "The Center of the Galaxy." *Scientific American* 230 (April 1974), p. 66.

Verschuur, G. L. *The Invisible Universe*. New York: Springer-Verlag, 1974.

Verschuur, G. L. "Inside Gould's Belt." *Astronomy* 2 (Oct. 1974), p. 16.

Whitney, C. A. *The Discovery of Our Galaxy*. New York: Alfred A. Knopf, 1971.

*The Sombrero galaxy in Virgo is about 40,000,000 ly away. The dark lane is dust in the disk of the galaxy, and globular clusters are faintly visible surrounding the nucleus. (Kitt Peak National Observatory.)*

# Chapter 9 GALAXIES

Previous chapters took us to the stars, star clusters, spiral arms and finally to the core of the Milky Way galaxy. Yet we have hardly begun our exploration of the universe. If our journey were compared to a trip around the world, we would not yet have traveled 1000 feet. In this chapter, we leave behind the familiar Milky Way and penetrate deep into space, out among the galaxies.

However, we do not leave behind the tools and methods of previous chapters. The tools we developed to determine the properties of stars (Chapter 5) are useful for determining the properties of galaxies. The most important properties are distance, size, luminosity, mass, and motion. As in the case of stars, distance is the key. Once we know the distance to a galaxy, its size and luminosity are easy to find. Mass is more difficult, but it too can be measured with methods like those we used to find the mass of binary stars and the Milky Way.

The motions of the galaxies give us additional clues to their distance. The lines in the spectra of galaxies have Doppler shifts proportional to distance. The more distant a galaxy is, the farther toward the red its spectral lines are shifted. Chapter 11 will interpret these red shifts as evidence that the universe is expanding, but in this chapter we will use the red shift phenomenon as a way of estimating the distance to a galaxy.

Galaxies are not randomly scattered through space, but are grouped into clusters ranging from isolated pairs up to clusters of thousands. Our own galaxy is a member of a small cluster of approximately two dozen galaxies.

Although millions of galaxies are visible on long exposure photographs of the sky, there are only a few types of galaxies. The shapes of these galaxy types, the kinds of stars they contain, and their distribution in clusters of galaxies are hints to their origin and evolution.

## MEASURING THE PROPERTIES OF GALAXIES

Unfortunately, the diameter, luminosity, mass, and motion of a galaxy are not obvious in photographs, but, to understand galaxies, astronomers must measure these properties. Just as in our study of stellar characteristics (Chapter 5), the first step is to find the distance. Once we know a galaxy's distance, its size and luminosity are relatively easy to find. Later in this section, we will see that finding mass and motion is more difficult.

*Distance, Diameter, and Luminosity.* The distances to galaxies are so large it is not convenient to measure them in light-years, parsecs, or even kiloparsecs. Instead we will use the unit **megaparsec (Mpc),** or 1,000,000 pc.

One Mpc equals 3,260,000 ly or approximately $2 \times 10^{19}$ mi.

To find the distance to a galaxy we must search among its stars for a familiar object whose luminosity or diameter is known. Such objects are called **distance indicators.**

Because their period is related to their luminosity, Cepheid variable stars are fairly reliable distance indicators. Box 7-2 explained how a giant star evolving through the instability strip in the H-R diagram could pulsate with a period between one and 60 days. Since the period of pulsation and the star's average luminosity depend on its mass, a period-luminosity relation exists (Figure 9-1). If we know the star's period, we can use the period-luminosity diagram to learn its absolute magnitude. By comparing its apparent magnitude with its absolute magnitude we can find its distance (Box 9-1).

Cepheids can reveal the distance to any galaxy in which individual Cepheids are visible (Figure 9-2). Unfortunately only about 30 galaxies are close enough to have visible Cepheids. Galaxies beyond about 6 Mpc are too distant to have detectable Cepheids, and we must find their distance from other indicators.

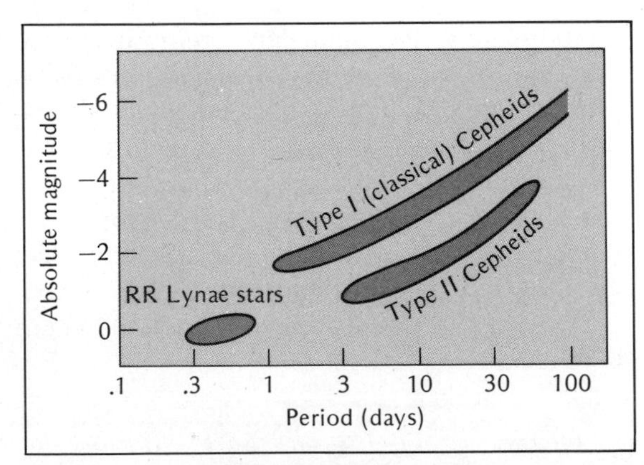


*Figure 9-1. The period-luminosity diagram relates the period of pulsation to luminosity. The classical Cepheids are population I stars; the type II Cepheids and RR Lyrae variables are population II stars.*

*Figure 9-2. The Andromeda galaxy is relatively close and can be resolved into individual stars. In these two views of a section of the galaxy, variable stars are visible. A determination of the stars' periods and average apparent magnitudes can establish the distance to the galaxy. (Hale Observatories.)*

BOX 9-1 USING CEPHEID DISTANCE INDICATORS

Cepheid variables are useful distance indicators because the period of their variation indicates their luminosity. Comparing the variable's luminosity with its apparent brightness tells us its distance.

For example, suppose we located a Cepheid variable in the disk of a galaxy. From measurements of photographs, we find its period is 60 days and its average apparent magnitude is 19. Since it is in the disk, we can assume it is a population I star, and therefore a type I (classical) Cepheid. The period-luminosity diagram (Figure 9-1) shows that 60-day classical Cepheids have absolute magnitudes of about −6. Subtracting absolute from apparent magnitude gives us a distance modulus of 25.

$$m - M_v = 19 - (-6) = 25$$

Using the table of distances in Box 5-2, we find that this distance modulus corresponds to a distance of 1 Mpc.

Studies of the few dozen galaxies whose distances are known from Cepheids reveal large numbers of bright giants and supergiants, large globular clusters, and occasional novae at maximum brightness. These objects have absolute magnitudes of about −9, so they too can be used as distance indicators. If we found bright giants and supergiants in a galaxy, we could measure their apparent magnitude and find the distance to the galaxy. The same is true for novae and large globular clusters. These distance indicators are good out to about 25 Mpc, beyond which objects of this brightness are invisible (Figure 9-3).

Another distance indicator is the cloud of ionized hydrogen called an **H II region** that forms around a very hot star. The Great Nebula in Orion is an H II region. Studies of the nearest galaxies, whose distances are known from other distance indicators, show that these H II regions have predictable diameters. If we detect H II regions in a distant galaxy, we can measure their angular diameter and assume they have the same diameter as the H II regions in other galaxies. Then we can find the distance from the small angle formula (Box 1-3). The H II regions are important distance indicators because they are easy to detect and because they give us distances beyond the limit of 25 Mpc. Unfortunately, the diameter of an H II region depends on the kind of star at its center and the density of its gas. Thus different kinds of galaxies may have different size H II regions. This limits the dependability of the method.

To measure distances to the farthest galaxies we must use the galaxies themselves as distance indicators. Studies of nearby galaxies indicate that an average galaxy like the Milky Way has a luminosity of about $16 \times 10^9$ solar luminosities, corresponding to an absolute magnitude of −20.5. If we see a similar galaxy and measure its apparent magnitude, we can find its distance (Box 5-2). Other types of galaxies have different luminosities, so the astronomer must know what kind of galaxy he is observing. Unfortunately there is considerable error in this method because of differences in luminosity among galaxies of the same type. Averaging the brightest galaxies in a cluster reduces some of the error and determines distances out to about 400 Mpc.

In spite of the uncertainties in distance measurements, it is clear that galaxies are far apart and scattered through the universe to tremendous distances. The nearest large galaxy, the Andromeda galaxy, is 0.66 Mpc (2,150,000 ly) distant. The most distant objects identifiable on photographs as galaxies are about 2200 Mpc

*Figure 9-3. Variable stars are good distance indicators to 6 Mpc, the brightest objects in galaxies to 25 Mpc, and the brightest galaxies in clusters to 400 Mpc. Beyond that the Hubble Law, discussed later in the text, is the only source of distances.*

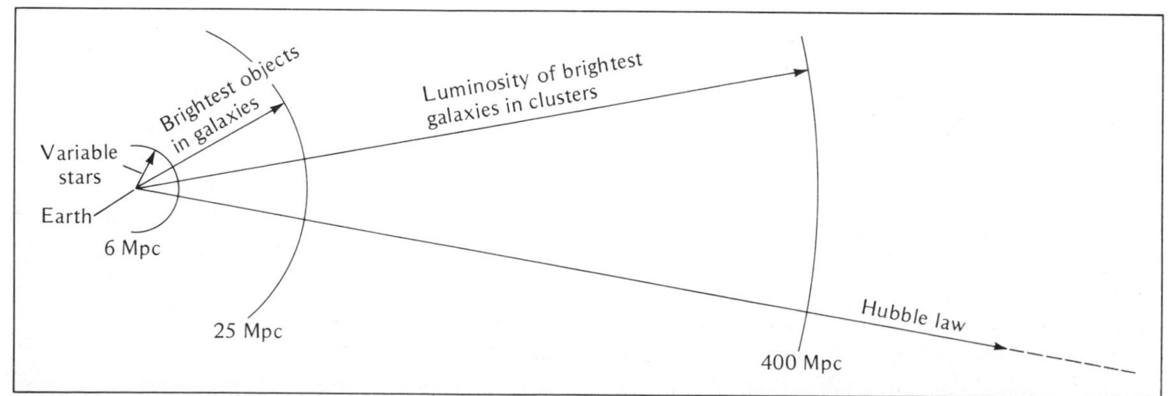

(7,000,000,000 ly) away from us (Figure 9-4). Radio telescopes can detect faint sources of radio signals that are evidently distant galaxies much more luminous at radio wavelengths than at visible light wavelengths. Thus the radio telescopes can detect them even though they are beyond the limit of the largest optical telescopes.

The tremendous distance between the galaxies produces an affect akin to time travel. When we look at a galaxy millions of light-years away, we do not see it as it is now, but as it was millions of years ago when its light began the journey toward earth. Thus when we look at a distant galaxy we look into the past by an amount called the **look-back time,** a time in years equal to the distance to the galaxy in light-years.

The look-back time for nearby galaxies is not significant since galaxies change very slowly. The Andromeda galaxy, for example, has a look-back time of about 2 million years, a mere eye blink in the life of a galaxy. But when we look at more distant galaxies, the look-back time becomes an appreciable part of the age of the universe. We will see evidence in Chapter 11 that the universe began 10 to 20 billion years ago. Thus when we look at the most distant visible galaxies, we are looking back 7 billion years to a time when the universe may have been significantly different. This effect will be important in this and the next chapter.

Clearly the distance to a galaxy is the key to finding its diameter and luminosity. If we can measure its distance by some reliable indicator, we can use its apparent brightness to find its luminosity (Box 5-2). This will work so long as we did not estimate the galaxy's luminosity in the first place to find its distance. In addition, once we know the distance, we can measure a galaxy's angular diameter, and then use the small angle formula (Box 1-3) to find its diameter in parsecs.

Finding a galaxy's diameter and luminosity is easy when the distance is known, but finding a galaxy's mass is a challenging puzzle.

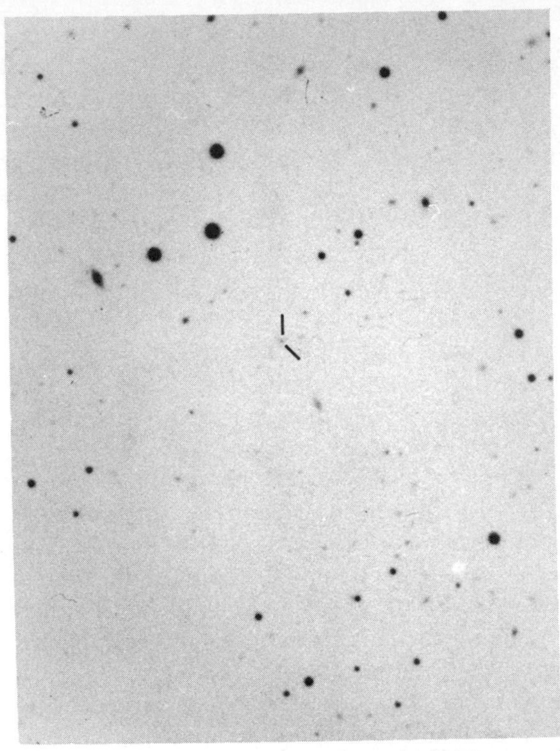

*Figure 9-4. A photographic plate of 3C275, one of the most distant galaxies ever photographed, shows little detail in the galaxy's image (marked by bars). Its red shift is 0.480, suggesting it is roughly 7,000,000,000 ly (2200 Mpc) distant. (Cerro Tololo Inter-American Observatory/Hyron Spinrad.)*

*Mass.* Although the mass of a galaxy is difficult to determine, it is an important quantity. It tells us how much matter the galaxy contains, which gives us clues to the galaxy's origin and evolution. In this section, we will examine four ways to find the masses of galaxies.

One way is called the **rotation curve method.** We begin by photographing the galaxy's spectrum at different points along its diameter and plotting the Doppler shift velocities in a rotation curve like that in Figure 9-5. This tells us how fast the galaxy is rotating. The sizes of the orbits the stars follow around the galaxy's center are related to the size of the galaxy, easily found from its angular diameter

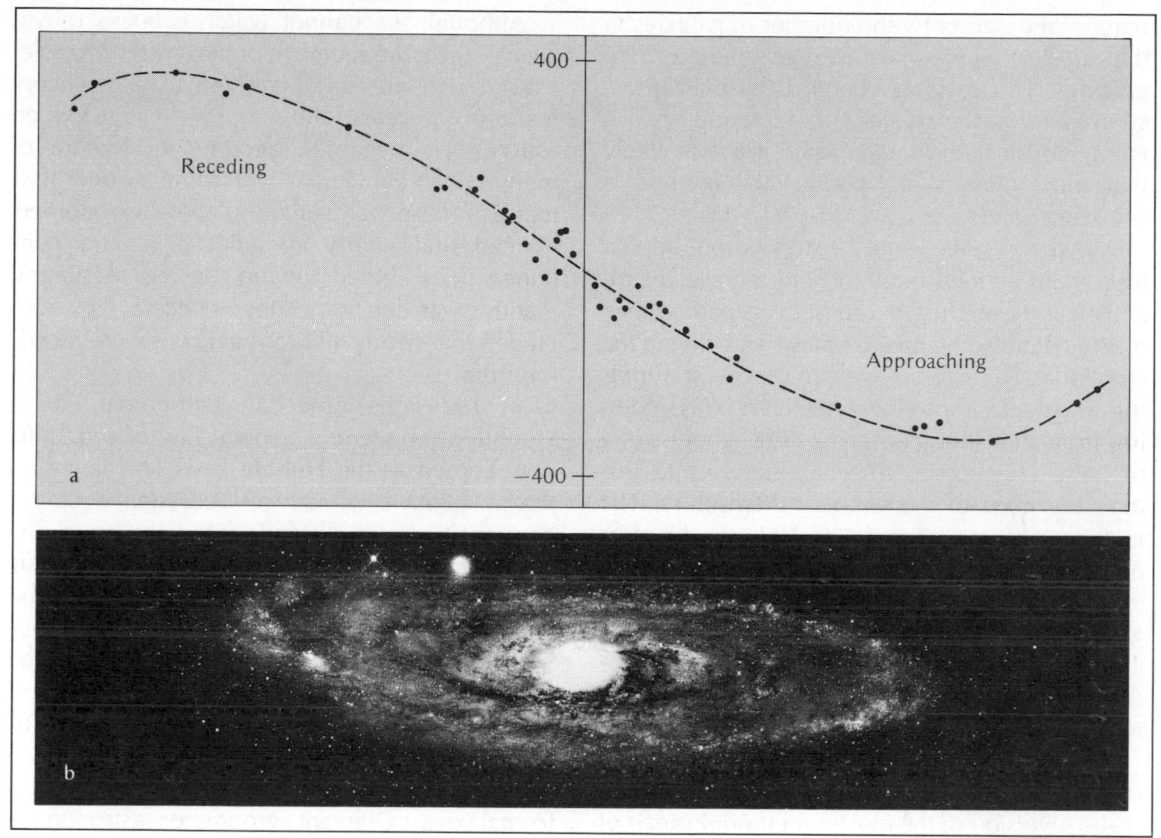


*Figure 9-5. The data points show Doppler velocities at various points along the Andromeda galaxy. The dashed line is the rotation curve. (Lick Observatory photograph.)*

and its distance. We then ask how massive the galaxy must be to hold stars in orbits of that size at that velocity.

The rotation curve method is very accurate, but it only works for the nearer galaxies, whose rotation curves can be observed. Because the more distant galaxies appear too small to analyze in this detailed way, we must use other methods to find their masses.

It is not unusual to find pairs of galaxies orbiting each other. In such cases, we analyze them as if they were a pair of binary stars. The only difficulty with this **double galaxy method** is that the galaxies take hundreds of millions of years to orbit once, so over periods of thousands of years we would see no motion. Because we can't observe the shapes of the orbits, we can't determine the orbital inclination, and we can't find their true masses. The solution to this problem is to average the results for many pairs of galaxies. Different orbital inclinations should average out and give good average masses.

The **cluster method** of finding galactic mass depends on the motions of galaxies within a cluster. If we measure the radial velocities of many galaxies in a cluster, we find that some velocities are larger than others because of the orbital motions of the individual galaxies in the cluster. Given the range of velocities and the size of the cluster, we can ask how massive a cluster of this size must be to hold itself together with this range of velocities. Dividing the total mass of the cluster by the number of galaxies in the cluster yields the average mass of the galaxies. This method contains the built-in assumption that the cluster is not expanding. If it is, our result is too large. Since it seems likely that most clusters are bound, the method is probably valid.

However, some large clusters do not appear to contain enough mass to hold themselves together. If these clusters are not dissipating, they must contain some **missing mass**—missing in the sense that it's there but we can't locate it. Orbiting space telescopes have detected x-rays coming from thin, hot gas spread throughout some of these clusters, but the gas amounts to less than the mass of the galaxies themselves. The missing mass must equal 5 to 10 times the mass of the galaxies. No one has yet detected the missing mass, but some astronomers suggest that galaxies are surrounded by extensive halos of gas and low luminosity stars that would make the total mass of a galaxy much larger than what we would estimate from photographs.

The fourth way of measuring a galaxy's mass is called the **velocity dispersion method.** It is really a version of the cluster method. Instead of observing the motions of galaxies in a cluster, we observe the motions of matter within a galaxy. In the spectra of some galaxies, broad lines indicate that stars and gas are moving at high velocities. If we assume the galaxy is bound by its own gravity, we can ask how massive it must be to hold this moving matter within the galaxy. This method, like the one before, assumes that the system is not coming apart.

The measured masses of galaxies cover a wide range. The smallest contain about $10^{-6}$ as much mass as the Milky Way galaxy. The largest contain as much as 50 times more.

*Motion.* Since many galaxies look like great whirlpools of stars and Doppler shifts prove that they are rotating, we might hope to see some motion. Unfortunately galaxies are too large and too distant and move too slowly to show any visible change over hundreds or thousands of years.

Although we cannot watch galaxies move, we can measure radial velocities by the Doppler effect. Such measurements led to a discovery made in the years following World War I when astronomers began to accumulate spectra of many faint nebulae. As it became evident that many of these were galaxies, a few astronomers noticed that nearly all galactic spectra contained lines shifted toward the red. Assuming that this was due to the Doppler effect, they concluded that nearly all of the galaxies were receding from us.

In 1929, the American astronomer Edwin Hubble announced a general law of red shifts now known as the **Hubble Law.** This law says that a galaxy's velocity of recession equals a constant times its distance. Thus the more distant a galaxy is, the faster it recedes from us (Figure 9-6). The constant **H**, now known as the **Hubble constant,** is very difficult to determine.

One important study of the recession of the galaxies suggests that **H** equals about 50 km/sec/Mpc. However, other studies yield values as high as 100 km/sec/Mpc. The uncertainty arises from the difficulty of determining the distances to galaxies. Different groups of astronomers have found distances in different ways and have arrived at different values of **H.** In Chapter 11 we will see that this has an important effect on our understanding of the history of the universe, but here it is sufficient to recognize that

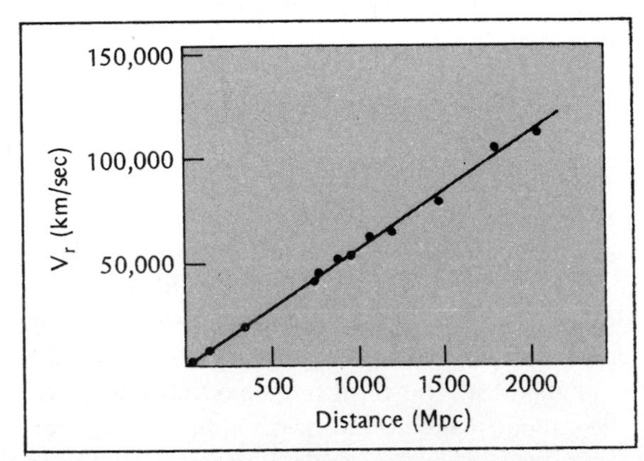


*Figure 9-6. The more distant a galaxy is, the larger its velocity of recession. The line represents the Hubble Law.*

the Hubble constant is poorly known and to adopt a provisional value of about 50 km/sec/Mpc.

The Hubble Law is important because it is commonly interpreted to show that the universe is expanding. In Chapter 11 we will discuss the implications of this expansion, but here we will use the Hubble Law as a practical way to estimate the distance to a galaxy. As Box 9-2 explains, a galaxy's velocity of recession divided by the Hubble constant equals its distance. This makes it relatively easy to find galactic distances, since large telescopes can photograph the spectrum of a distant galaxy and reveal its red shift even though distance indicators such as variable stars are totally invisible.

However, we cannot abandon distance indicators and use the Hubble Law exclusively. In Chapter 10 we will discuss peculiar galaxies that, according to some astronomers, may not obey the Hubble Law. In addition, Chapter 11 will describe how distant galaxies may recede faster than the Hubble Law predicts. To detect such departures from the law, we must measure distances with distance indicators and use the red shifts only for estimates.

It is clearly impossible to find the distance, diameter, luminosity, and mass of each of the millions of galaxies in the sky. Instead, we must classify the galaxies into categories, and then study the properties of the galaxies in each category. Understanding the differences between these types of galaxies will lead us to theories of their origin and evolution.

---

BOX 9-2 THE HUBBLE LAW

The Hubble Law relates a galaxy's radial velocity $\mathbf{V_r}$ in kilometers per second to its distance **D** in megaparsecs. We can visualize this relation as a graph in which we plot radial velocity and distance for a number of galaxies (Figure 9-6). Points that represent galaxies fall along a straight line, showing that the more distant a galaxy, the faster it recedes from us. A galaxy's radial velocity in kilometers per second equals the Hubble constant times its distance in megaparsecs.

$$V_r = H\,D$$

The best measurements of distance and velocity suggest that **H** is approximately equal to 50 km/sec/Mpc. This tells us that for every million parsecs that separate two galaxies, they recede from each other at 50 km/sec. Thus two galaxies 10 Mpc apart move away from each other at 500 km/sec.

If we can measure the radial velocity of a galaxy, we can estimate its distance from our galaxy by dividing by the Hubble constant. For example, the Virgo cluster of galaxies has a radial velocity of 1180 km/sec. To find its distance, we divide by H, giving 24 Mpc.

---

## GALACTIC MORPHOLOGY

Morphology, meaning form or structure, is often used by biologists to refer to the variation in structure among related organisms. Since we must compare different types of galaxies that appear to be related but differ in form, we refer to our study as galactic morphology. Just as the morphology of organisms may tell a biologist how a species evolved, galactic morphology may give us clues to how galaxies evolve.

*Galaxy Classification.* We will begin our study of galaxy shapes by classifying them into three broad classes: elliptical, spiral, and irregular. We can then subdivide these classes to account for small variations in form among similar galaxies. To organize these classes in an easily remembered system, astronomers usually arrange them in a **tuning fork diagram** as first devised by Edwin Hubble (Figure 9-7).

About 70 percent of all galaxies are **elliptical** (Figure 9-8). They are round or elliptical in shape, have almost no visible gas or dust, lack hot bright stars, and have no spiral pattern. The stars in elliptical galaxies are more crowded toward the center, and the outer parts of some larger ellipticals are peppered by hundreds of globular clusters (Figure 9-8a).

The largest ellipticals are about five times

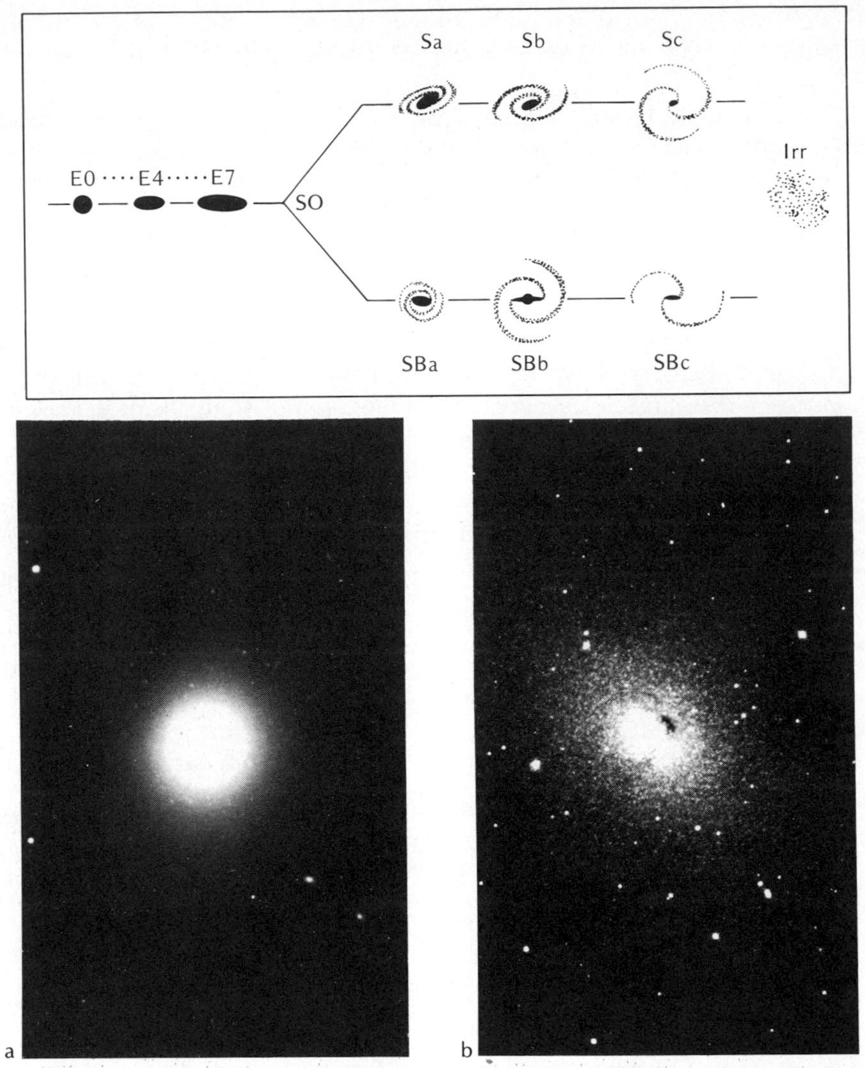


*Figure 9-7. The tuning fork diagram organizes the galaxy types according to star formation history. It is not an evolutionary diagram.*

*Figure 9-8. (a) M 87 is a giant elliptical galaxy surrounded by a swarm of over 500 globular clusters. (b) NGC 185 is a dwarf elliptical peculiar for the dust cloud seen against its bright center. It is a satellite of the Andromeda galaxy and is resolved into individual stars. (Lick Observatory photographs.)*

larger than our galaxy, and the smallest are 100 times smaller. This wide range is also reflected in their mass, with the most massive being 50 times the mass of our galaxy and the least massive $10^6$ times less (Table 9-1). The largest of these galaxies are called giant ellipticals and the smallest, dwarf ellipticals (Figure 9-8).

Elliptical galaxies are identified by the letter E followed by a number ranging from 0 to 7 indicating the apparent shape of the galaxy. Galaxies that look circular are classified E0; the more elliptical the galaxy's outline, the larger the number. No elliptical galaxy is known that is more elliptical than E7.

**Table 9-1 Normal Galaxies Compared to the Milky Way***

| | Elliptical | Spiral | Irregular |
|---|---|---|---|
| Mass | 0.000001 to 50 | 0.005 to 2 | 0.0005 to 0.15 |
| Diameter | 0.01 to 5 | 0.2 to 1.5 | 0.05 to 0.25 |
| Luminosity | 0.00005 to 5 | 0.005 to 10 | 0.00005 to 0.1 |

* In units of the mass, diameter, and luminosity of the Milky Way galaxy.

The spectra of elliptical galaxies indicate that they are limited to a single generation of stars. A few studies have detected small dust and gas clouds where a few new stars are forming but for the most part all of their stars are old. Lack of extensive gas and dust means that few new stars can form, so the only stars we see are old, lower-mass stars like those in our galaxy's spherical component. The massive stars would have died long ago. Indeed, we can think of an elliptical galaxy, with its flattened spherical shape and swarm of globular clusters, as a galaxy that lacks a disk component.

The dwarf ellipticals are small, containing only a few million stars (about 0.1 percent the number in our galaxy), and extending only a few hundred parsecs in diameter. Since these galaxies are small and contain few stars, they are not very luminous and are thus hard to find. Of the approximately two dozen galaxies near the Milky Way galaxy, nine are dwarf ellipticals. If dwarf ellipticals are that common throughout the universe, they are the most common form of galaxy.

*Figure 9-9. This edge-on view of a spiral galaxy dramatically points up the dust in the galaxy. (Lick Observatory photograph.)*

Though **spiral galaxies** make up only 15 percent of all galaxies, they are the most striking. Their distinguishing characteristic is an obvious disk component that contains gas and dust, and hot, bright stars. The dust is especially obvious in those galaxies we see edge on (Figure 9-9).

The largest spirals are about 1.5 times larger than our galaxy, and the smallest are about 5 times smaller. The most massive are slightly more massive than the Milky Way, and the least massive are less than 1 percent of that mass. Our Milky Way galaxy is larger and more massive than average.

Among spiral galaxies we identify three distinct types: S0 galaxies, normal spirals, and barred spirals. Unlike our galaxy, the S0 galaxies show no obvious spiral arms, have very little gas and dust, and contain very few hot, bright stars (Figure 9-10). However, they have an obvious disk component with a large nucleus at

*Figure 9-10. NGC 5866 is an S0 galaxy with a disk of stars but little gas and dust. (Hale Observatories.)*

the center. They appear to be intermediate between elliptical and spiral galaxies.

Normal spiral galaxies can be further subclassified into three groups according to the size of their nucleus, and the degree to which their arms are wound up (Figure 9-11). Spirals that have little gas and dust, larger nuclei, and tightly wound arms are classified Sa. Sc galaxies have large clouds of gas and dust, small nuclei, and very loosely wound arms. The Sb galaxies are intermediate between Sa and Sc. Since there is more gas and dust in the Sb and Sc galaxies, we find more young, hot, bright stars along their arms. The Andromeda galaxy (Figure 8-1) and our own Milky Way galaxy are Sb galaxies.

A minority of spirals have an elongated nucleus with spiral arms springing from the ends of the bar. These **barred spiral galaxies** are classified SBa, SBb, or SBc according to the same criteria listed for normal spirals (Figure 9-12). The elongated shape of the nucleus is not well understood, but some astronomers working with computer models have succeeded in imitating the rotating bar structure (Figure 9-13). It appears to occur when an instability develops in the stellar distribution within the rotating galaxy. The gravitational field of the bar alters the orbits of the inner stars and generates a stable, elon-

*Figure 9-11. Normal spirals are classified according to the size of the nucleus, and the tightness of the arms. (Hale Observatories.)*

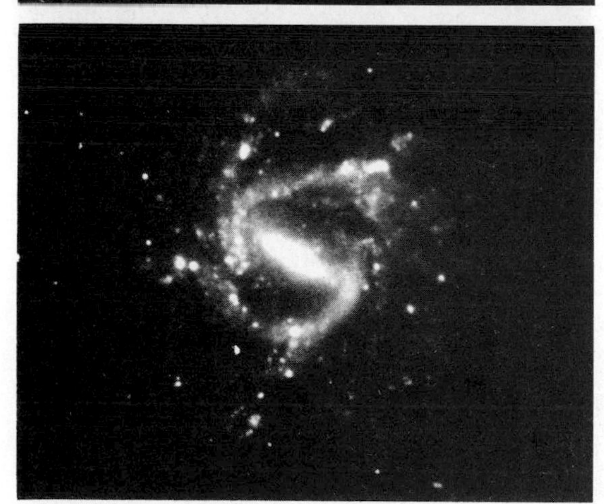

*Figure 9-12. Barred spiral galaxies have an elongated nucleus from which the arms spring. (Hale Observatories.)*

gated nucleus. Thus, except for the peculiar rotation in the nucleus, barred spirals are similar to normal spiral galaxies.

About 15 percent of all galaxies are **irregular** (Figure 9-14 and Color Plate 17). They have a chaotic appearance with large clouds of gas and dust mixed with both young and old stars. They have no obvious spiral arms or nuclei. Since they are small, ranging from 5 percent to 25 percent the diameter of our galaxy, they are difficult to detect. The Magellanic clouds are the best-studied examples of irregular galaxies.

From an analysis of the galaxies near us in space, we can estimate that most galaxies are elliptical and that spiral and irregular galaxies represent a minority. However, elliptical galaxies lack bright stars and many are dwarf ellipticals, so they are not easily visible. Spiral galaxies, on the other hand, contain many bright stars and are much easier to see. Thus most of the galaxies we can conveniently observe are spiral. In one list of the brightest galaxies in the sky, 68 percent were spiral and only 18 percent were elliptical. Compare these figures with the estimated true abundances in Table 9-2.

**Table 9-2 Frequency of Galaxy Types**

| | |
|---|---|
| Elliptical | 70% |
| Spiral | 15% |
| Irregular | 15% |

Galaxies do not occur in isolation. Some are colliding and interacting with each other as described in Box 9-3. Most galaxies, including our own, are part of clusters of galaxies. To understand how galaxies form and evolve we must study them in these communities.

*Clusters of Galaxies.* Nature seems to mass-produce galaxies in clusters rather than turning them out one at a time (Figure 9-15). Even the rare single galaxy may have formed in

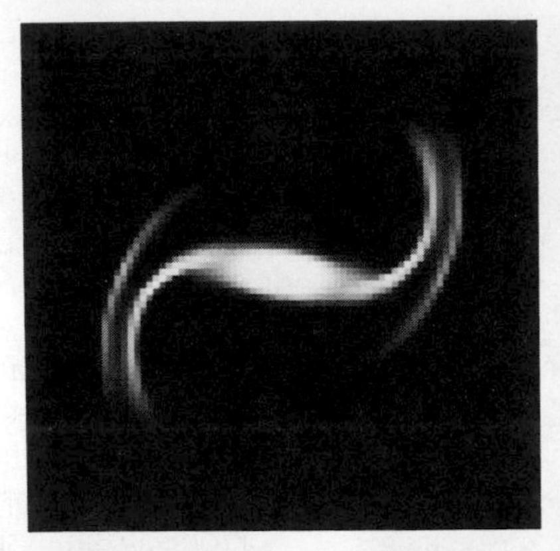
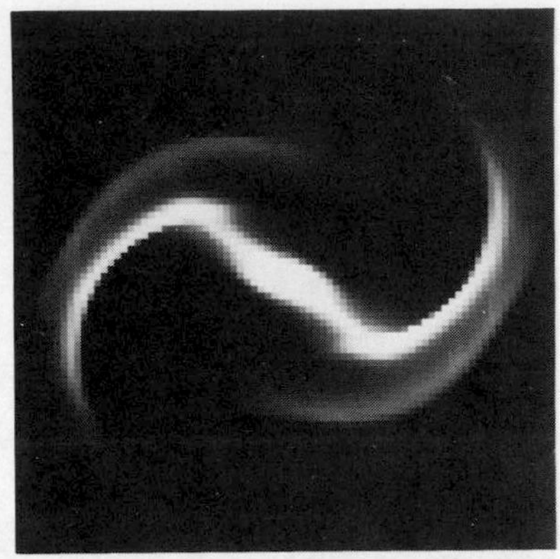

*Figure 9-13. Two computer models of a rotating galaxy successfully imitate the elongated nucleus so characteristic of barred spiral galaxies. (Reprinted courtesy of James M. Huntley and* The Astrophysical Journal, *published by the University of Chicago Press; © 1978 The American Astronomical Society.)*

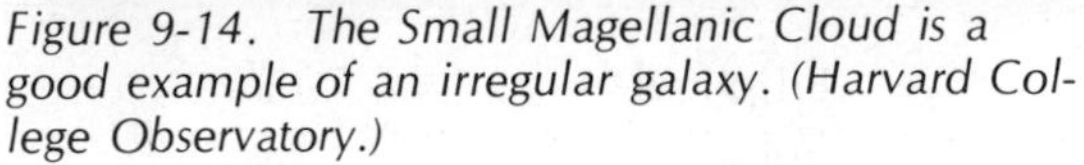

*Figure 9-14. The Small Magellanic Cloud is a good example of an irregular galaxy. (Harvard College Observatory.)*

*Figure 9-15. The Hercules cluster contains about 300 galaxies and is approximately 175 Mpc distant. (Hale Observatories.)*

a cluster and later escaped. Our Milky Way is a member of a cluster containing over 20 galaxies, and over 2700 other clusters have been catalogued within 4,000,000,000 ly.

We can sort clusters of galaxies into two groups: rich clusters and poor clusters. **Rich clusters** contain over a thousand galaxies, mostly elliptical. These are scattered through a spherical volume about 3 Mpc ($10^7$ ly) in diameter. Such a cluster is nearly always condensed; that is, the galaxies are concentrated toward the cluster center. Continuing this trend toward central condensation, rich clusters often contain one or more giant elliptical galaxies at their centers.

One example of a rich cluster is the Virgo cluster, a group of over 2500 galaxies located about 20 Mpc ($65 \times 10^6$ ly) from us. At this distance no Cepheid variable stars are visible, so the distance has been found from the brightest stars, globular clusters, and H II regions. As expected, the Virgo cluster is approximately spherical and centrally condensed. The giant elliptical galaxy M 87 lies at the center (Figure 9-8a).

Some astronomers suggest that the giant el-

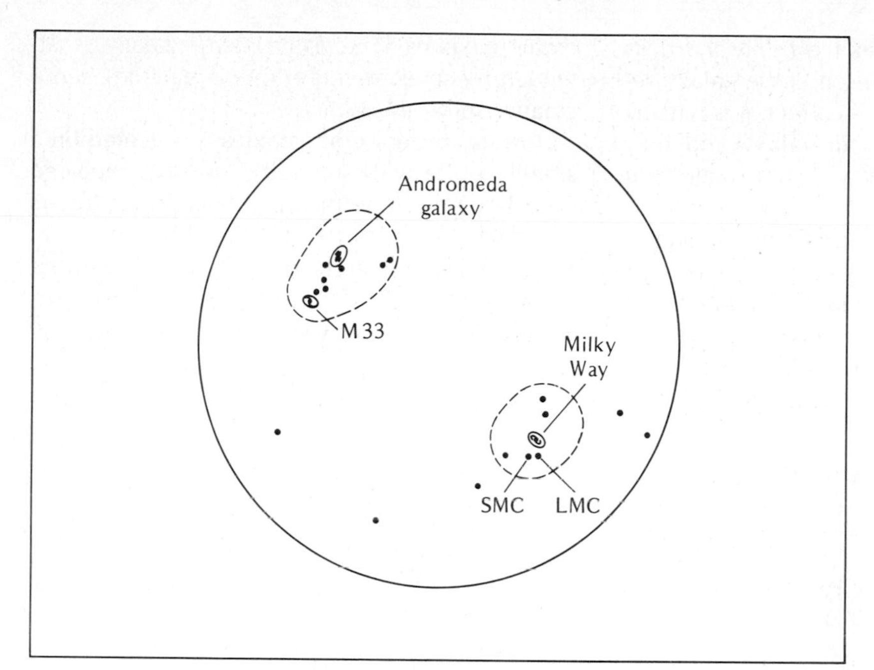


*Figure 9-16. A map of the Local Group of galaxies showing relative distances. Note the two subclusters. The outer circle is 4,000,000 ly in diameter.*

liptical galaxies found in rich clusters are the product of the coalescence of galaxies. Undoubtedly, galaxies must collide often in the crowded interior of a rich cluster, and that may build giant elliptical galaxies. In fact, some suggest that giant galaxies may eventually absorb all lesser companions, forming tremendously massive super galaxies. Since we would have to wait at least a billion years to see the actual merging of galaxies, we can only speculate about such galactic cannibalism, but it is clear that collisions are important in rich clusters of galaxies.

Collisions may be less important in the second type of cluster, the **poor clusters.** These contain fewer than 1000 galaxies and are irregularly shaped. Instead of being condensed toward the center, poor clusters tend to have subcondensations, small groupings within the cluster. Since these clusters are less condensed, collisions may be less frequent.

Our own Local Group, which contains the Milky Way, is a good example of a poor cluster (Figure 9-16). It contains slightly more than 20 members scattered irregularly through a volume about 1 Mpc in diameter. Of the brighter galaxies, 14 are elliptical, 3 are spiral, and 4 are irregular.

The total number of galaxies in the Local Group is uncertain because a few galaxies lie in the plane of the Milky Way and are hidden by the obscuring dust clouds. For instance, the Italian astronomer Paolo Maffei, studying infrared photographs, discovered two galaxies hidden behind the dust clouds of the Milky Way in the constellation of Cassiopea. Maffei I is a giant elliptical that would be visible to the naked eye if there were no dust in the way, and Maffei II is a spiral galaxy that probably contains many bright stars. Neither object is detectable at visual wavelengths because of the dust clouds in the Milky Way. Because the Maffei galaxies are so heavily obscured it is difficult to determine their distances, but the best estimates place them just beyond the edge of the Local Group.

Radio observations suggest that another galaxy is hidden behind the Milky Way in the constellation of Gemini. If the data are correct, the galaxy is a spiral about 0.001 the mass of our galaxy, and it is only about 20 kpc (65,000 ly) from the sun. It has been dubbed Snickers because, though it resembles the Milky Way, its mass is peanuts in comparison. If the galaxy really exists and its distance is correct, it is passing just beyond the edge of our galaxy and may be distorting the Milky Way's spiral pattern in that direction.

The clumpiness of the Local Group illustrates the subclustering in poor galaxy clusters. The two largest galaxies, the Milky Way and the Andromeda galaxy, are the centers of two subclusters. The Milky Way galaxy is accompanied by the Magellanic clouds and about three other dwarf galaxies. The Andromeda galaxy is attended by seven dwarf elliptical galaxies and a smaller spiral, M 33.

Just as clustering of galaxies is repeated on a smaller scale with subclustering, it is repeated on a larger scale with superclustering. Clusters of galaxies seem to be associated in clusters of clusters called superclusters. The Local Group is a part of the Local Supercluster, an approximately disk-shaped swarm of galaxy clusters 50 to 75 Mpc in diameter.

## BOX 9-3 COLLIDING GALAXIES

Galaxies should collide fairly often. The average separation between galaxies is only about 20 times their diameter. Like two elephants blundering about at random under a circus tent, galaxies should bump into each other once in a while. Stars, on the other hand, almost never collide with each other. In the region of the galaxy near the sun, the average separation between stars is about $10^7$ times their diameter. Thus collision between two stars is about as likely as collision between two gnats flitting about at random in a football stadium.

Large telescopes reveal hundreds of galaxies that appear to be colliding with other galaxies. One of the most famous pairs of colliding galaxies, NGC 4676A and NGC 4676B, is called The Mice because of the taillike deformities (Figure 9-17). In addition to tails, some interacting galaxies seem to be connected by a bridge of matter.

When two galaxies collide, they actually pass through each other, and there is no direct contact between the stars of one galaxy and the stars of the other. The gas in the galaxies does interact but it is mainly the galaxies' gravitational fields that twist and deform the galactic shapes, producing peculiar tails and bridges. Since this interaction may last hundreds of millions of years, it is impossible for us to watch peculiar features develop in real galaxies. But computer models can simulate such a collision and display it on a TV screen. In a large modern computer, such a simulation takes only five minutes or so. Although highly simplified, these models produce tails and bridges very similar to those seen in peculiar galaxies.

By repeated adjustments of the masses of the galaxies, their velocities, the angle of their encounter, and other parameters, the computer can produce a convincing imitation of The Mice. The tails visible in the photograph are apparently due to tidal interactions as the galaxies swing past each other along narrow orbits.

With different initial parameters, collision forms a bridge that seems to connect the galaxies. A striking example of this is M 51, the Whirlpool galaxy (Figure 9-18). Although the Whirlpool is often used as an example of a normal spiral, it is peculiar in that its long arm appears to connect with a smaller galaxy. The computer model of this collision suggests that the two galaxies have collided recently. We now see the small galaxy beyond its larger neighbor, and the spiral arm that seems to connect them only passes in front of the smaller galaxy.

Since collisions between galaxies are fairly common, we should expect to see an occasional head-on collision. Computer models suggest that such an interaction could produce a **ring galaxy** such as those shown in Figure 9-19. The low expected frequency of head-on collisions agrees with the rarity of ring galaxies.

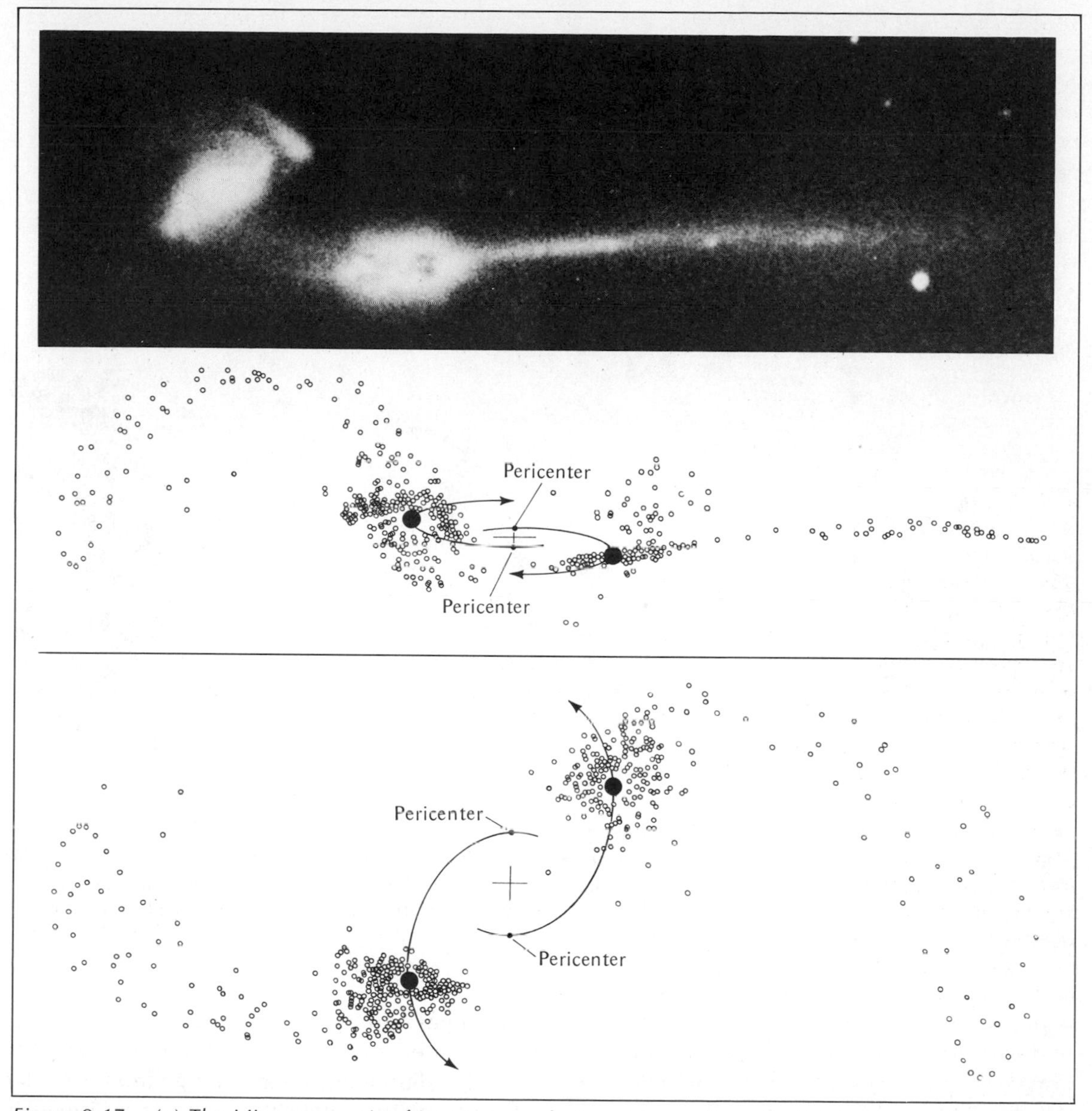


*Figure 9-17.* *(a) The Mice are a pair of interacting galaxies with peculiar tails. (b) A computer simulation of a close encounter between normal galaxies produces similar tails. (From "Violent Tides Between Galaxies" by A. Toomre and J. Toomre. Copyright © 1973 by Scientific American, Inc. All rights reserved. Photograph courtesy of Halton Arp/Hale Observatories.)*

*The Origin and Evolution of Galaxies.* Clues to the history of galaxies lie in two different places. First, galactic morphology gives us clues to the influence of different initial conditions such as mass, rotation, and so forth. The classification of galaxies is the first step in the

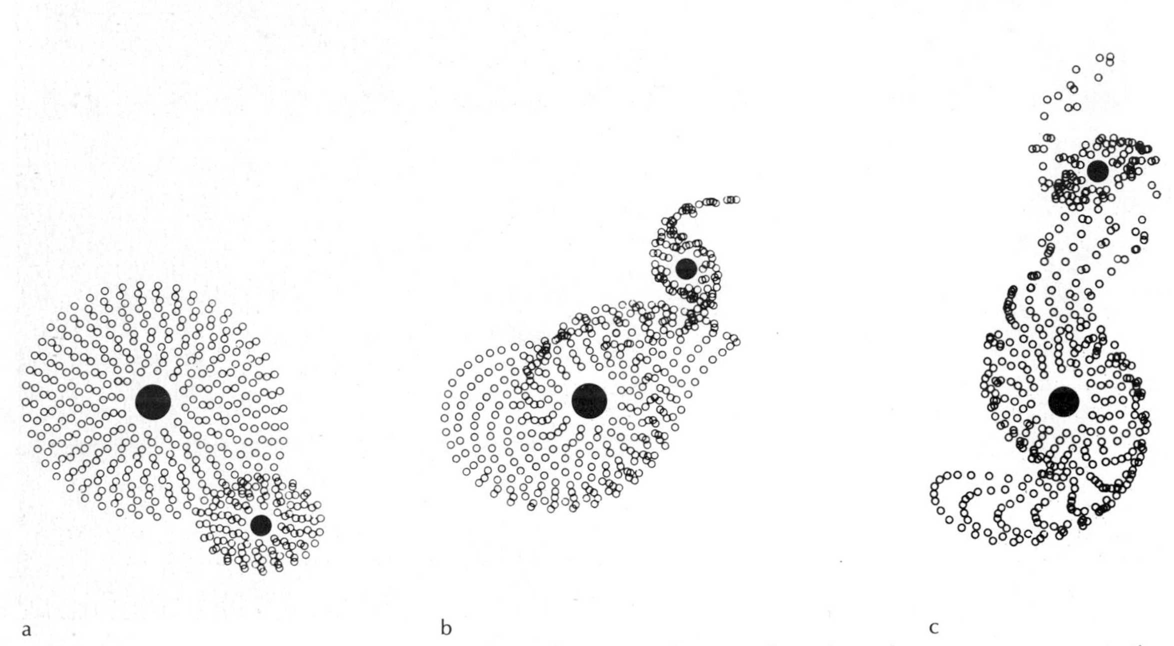


*Figure 9-18. A computer simulation of a collision between galaxies* (*a, b, c, d*) *produces a pattern very similar to M 51, the Whirlpool galaxy, and its small companion* (*e*). (*From "Violent Tides Between Galaxies" by A. Toomre and J. Toomre. Copyright © 1973 by Scientific American, Inc. All rights reserved. Lick Observatory photograph.*)

exploration of their origin and evolution. Second, different kinds of clusters of galaxies hint at how external processes such as collisions affect the shape of a galaxy.

Before we try to build a theory of galactic evolution, we must pause long enough to eliminate a tempting but discredited idea. When the tuning fork diagram first appeared, astronomers theorized that it was an evolutionary diagram showing how elliptical galaxies evolved into spiral and then into irregular galaxies. Unfortunately, this simple theory doesn't work. Studies of elliptical galaxies show that they contain little or no gas and are not making many new stars. Thus an elliptical can never become a spiral.

In addition, evolution can't go from irregular to spiral to elliptical. Only four irregular galaxies, those in the Local Group, are close enough to study in detail, but all four contain both young and old stars. Thus they cannot be young in comparison with spiral and elliptical galaxies.

The tuning fork diagram is not an evolutionary diagram. Galaxies do not evolve from elliptical to spiral or vice versa, any more than cats evolve into dogs. The tuning fork is a star

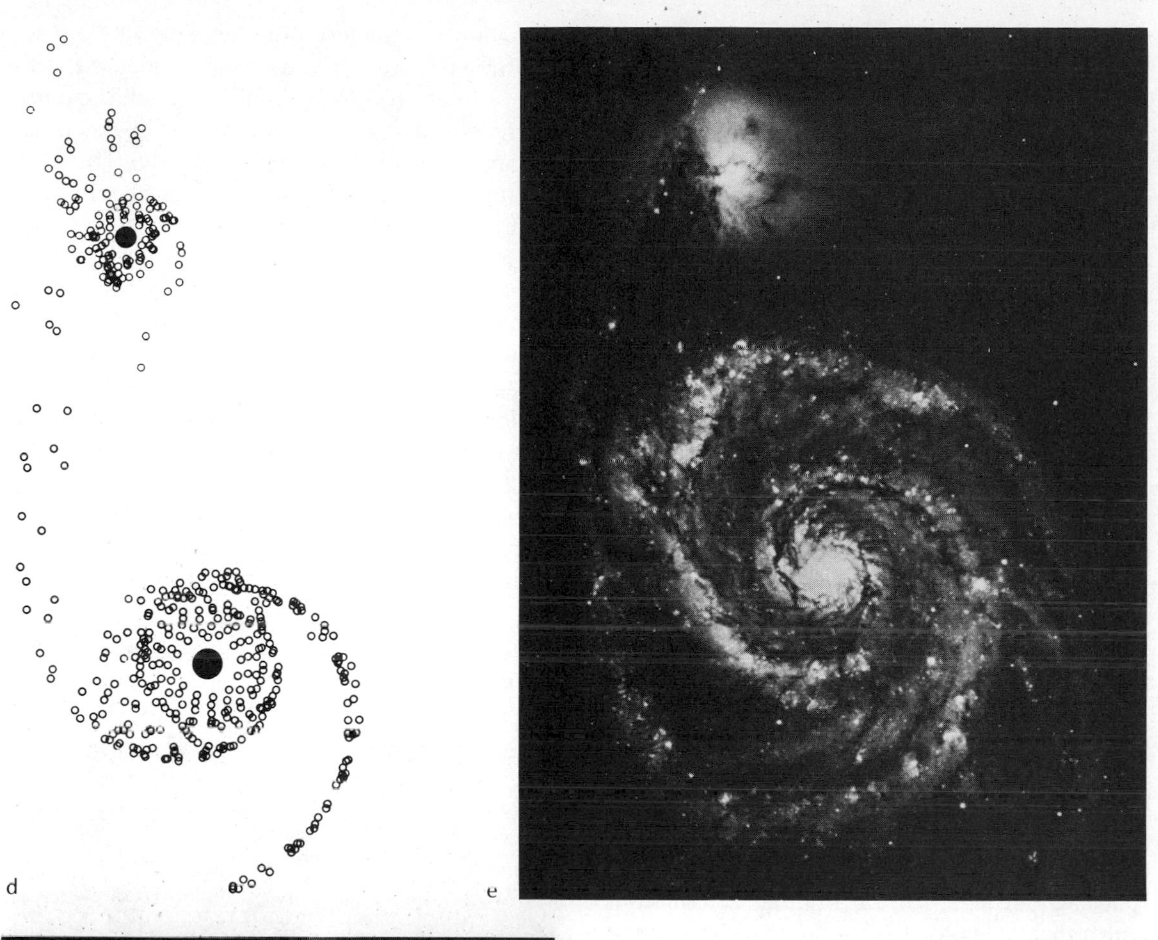


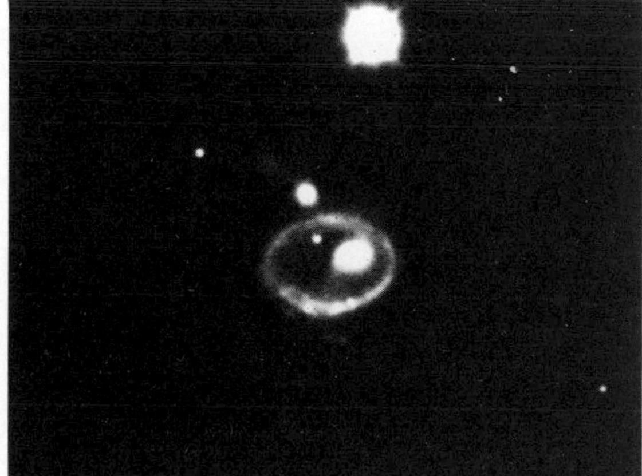

*Figure 9-19. Ring galaxies may be the products of head-on collisions between galaxies. (Kitt Peak National Observatory and Cerro Tololo Inter-American Observatory.)*

formation diagram showing that different types of galaxies have had different histories of star making.

Current theory suggests that elliptical galaxies have been wastrels, squandering their gas and dust in a sudden burst of star formation when they were young. For a few billion years they must have been brilliant with young stars, but, having used up their supply of gas, they are now in decline. They cannot make many new stars and those that remain are cool, lower main-sequence stars and red giants.

Although most astronomers agree that the elliptical galaxies used up their star-making supplies while they were still young, the cause is unclear. Perhaps elliptical galaxies formed from gas clouds that had little rotation or turbulence. These stagnant clouds would contract rapidly and reach star-forming density quickly. The resulting rapid star formation could leave the galaxy without gas and dust.

The predominance of elliptical galaxies in rich clusters leads to the speculation that collisions are important. Rich clusters seem to have formed from denser clouds of gas that might favor the formation of elliptical galaxies. In addition, a galaxy forming in such a cluster should suffer a higher than usual number of collisions with other galaxies. Collisions could compress the gas in the galaxies and trigger star formation, driving the galaxies to consume their supplies rapidly. Also, collisions may strip gas and dust away from smaller galaxies, leaving them with nothing from which to make new stars.

This discussion seems to imply that ellipticals are dead, inactive galaxies in which nothing much happens. Observations of some ellipticals, however, reveal emission lines of ionized gas in their cores, indicating the presence of a few hot stars. Other ellipticals are sources of radio signals, indicating they are undergoing powerful eruptions. We will discuss these active elliptical galaxies in the next chapter.

Unlike ellipticals, spiral galaxies have conserved their gas and are still in the process of forming large numbers of stars. Because the star formation proceeded more leisurely, the galaxy had time to flatten into a disk. (See Figure 8-15.) Star formation began rapidly at the center, where density was highest, while in the outer regions stars formed less frequently. Thus the spiral galaxies we see now still have supplies of gas from which to make new stars (Figure 9-20).

How do irregular galaxies form? One suggestion is that irregulars form from turbulent clouds. Turbulence may prevent rapid contraction and allow star formation to occur more slowly. At the same time, turbulence may prevent the formation of a massive nucleus with its attendant differential rotation and spiral arms.

Our discussion concludes that the overall shape of a galaxy depends on the rate of star formation and the influence of interactions between galaxies in clusters. However, this is only one interpretation. The origin and evolution of galaxies are not nearly as well understood as the formation and evolution of stars. The subject is a new and exciting field of astronomical research.

*Figure 9-20. The clouds of gas and dust are striking in Sc galaxies such as NGC 598. (Hale Observatories.)*

Though the past history of galaxies is hazy, their future evolution seems clear. The future of a galaxy is dominated by its star formation. When gas and dust are used up, star formation must stop. Hot, bright stars are massive and die quickly. Since they mark the spiral arms, the spiral pattern must vanish. "Spirals" will then have a smooth, featureless appearance like the S0 galaxies, though their disks will be thin. Lower main-sequence stars can live for hundreds of billions of years, but, if the universe lasts that long, they too must die, and leave behind a dark, cold universe filled with galaxies of dead stars.

## SUMMARY

To measure the properties of galaxies we must first find their distances. For the nearer galaxies, we can judge distances using distance indicators, objects whose luminosity or diameter is known. The most accurate distance indicators are Cepheid variable stars. Other distance indicators are bright giants and supergiants, globular clusters, and novae. Another type of distance indicator, the H II regions, have known diameters. To use them we compare their angular diameter with their known linear diameter. In addition, we can estimate the distance to the farthest galaxy clusters using the average luminosity of the brightest galaxies.

The Hubble Law shows that the radial velocity of a galaxy is proportional to its distance. Thus we can use the Hubble Law to estimate distances. The galaxy's radial velocity divided by the Hubble constant is its distance in megaparsecs.

The masses of galaxies can be measured in four ways—the rotation curve method, the double galaxy method, the cluster method, or the velocity dispersion method. The first method is the most accurate, but it is only applicable to nearby galaxies.

We can divide galaxies into three classes—elliptical, spiral, and irregular—with subclasses giving the galaxy's shape or the amount of gas and dust. These galaxy types probably formed in different ways. The ellipticals used up all of their gas and dust in a sudden burst of star formation when they were young. The spirals formed more slowly and conserved their gas and dust and flattened into disks. The irregulars may have formed from turbulent gas clouds.

## NEW TERMS

megaparsec (Mpc)
distance indicator
H II region
look-back time
rotation curve method
double galaxy method
cluster method
missing mass
velocity dispersion method
Hubble Law
Hubble constant (**H**)
tuning fork diagram
elliptical galaxy
spiral galaxy
barred spiral galaxy
irregular galaxy
ring galaxy
rich galaxy cluster
poor galaxy cluster

## QUESTIONS

1. Why wouldn't white dwarfs make good distance indicators?
2. Why isn't the look-back time important among nearby galaxies?
3. Explain how the rotation curve method of finding a galaxy's mass is similar to the method used to find the masses of binary stars.
4. Explain how the Hubble Law permits us to estimate the distances to galaxies.
5. Draw and label a tuning fork diagram. Why can't evolution go from elliptical to spiral? From spiral to elliptical?
6. What is the difference between an Sa and and Sb galaxy? Between an S0 and an Sa galaxy? Between an Sb and an SBb galaxy? Between an E7 and an S0 galaxy?
7. Why might we describe elliptical galaxies as "wastrels"?

## PROBLEMS

1. If a galaxy contains a 30-day type I (classical) Cepheid with an apparent magnitude of 20, what is the star's absolute magnitude and how far away is the galaxy?
2. Sketch a galaxy and its rotation curve; indicate which parts are approaching and which are receding.
3. If a galaxy has a radial velocity of 4000 km/sec, how far away is it?
4. Sketch the Milky Way galaxy and the Andromeda galaxy to scale in Figure 9-16.

## RECOMMENDED READING


Bok, B. J. "The Star Clouds of Magellan." *Natural History* 88 (June/July 1979), p. 86.

Groth, E. J., Peebles, P. J. E., Seldner, H., and Soneira, R. M. "The Clustering of Galaxies." *Scientific American* 237 (Nov. 1977), p. 76.

Hausman, M. A. "Galactic Cannibalism." *Mercury* 8 (Nov./Dec. 1979), p. 119.

Holzinger, J. R., and Seeds, M. A. *Laboratory Exercises in Astronomy*. Ex. 38. New York: Macmillan, 1976.

Larson, R. B. "The Origin of Galaxies." *American Scientist* 65 (March 1977), p. 188.

Larson, R. B. "The Formation of Galaxies." *Mercury* 8 (May/June 1979), p. 53.

Maran, S. P. "Ring Galaxies." *Natural History* 86 (Nov. 1977), p. 106.

Strom, S. E., and Strom, K. M. "The Evolution of Disk Galaxies." *Scientific American* 240 (April 1979), p. 72.

Toomre, A., and Toomre, J. "Violent Tides Between Galaxies." *Scientific American* 233 (December 1973), p. 38.

Van den Bergh, S. "Golden Anniversary of Hubble's Classification System." *Sky and Telescope* 52 (Dec. 1976), p. 410.

*The pair of galaxies NGC 4038-9 appear to be interacting producing peculiar shapes and powerful radio emission. (Hale Observatories.)*

# Chapter 10 PECULIAR GALAXIES

Of the millions of galaxies visible through earth-based telescopes, a surprising number do not fit the standard categories of elliptical, spiral, and irregular. Some of these peculiar galaxies appear to result from interactions between galaxies as described in Box 9-3. Other peculiar galaxies have brilliant nuclei filled with high-temperature gas churning at high velocities, and a few show direct evidence of the ejection of matter in long jets and filaments racing out of the nucleus. This suggests that events of titanic violence are occurring in the centers of some galaxies.

Many peculiar galaxies are sources of radio emission. Even a "normal" galaxy like our own Milky Way emits a little radio energy, but some peculiar galaxies are tremendously powerful sources of radio waves. A few emit over 10 million times more radio energy than a normal galaxy. How this energy is produced is unknown, but the process may be related to violent events within the nuclei.

The suggestion that the center of a galaxy can explode seems fantastic, but it may help explain the mysterious objects called QSOs. These objects have large red shifts that are usually interpreted as evidence that they are at tremendous distances. If that is true, they must be 10 to 1000 times more luminous than a normal galaxy. How an object could produce such tremendous energy is one of the deepest mysteries of astronomy.

## RADIO GALAXIES

A **radio galaxy** is a galaxy that is a strong source of radio signals. Normal galaxies radiate some radio energy because of interstellar gas, supernova remnants, pulsars, planetary nebulae, and so on, but only about 1 percent of a normal galaxy's luminosity falls within the radio spectrum. A radio galaxy can emit $10^7$ times more radio energy than a normal galaxy.

We can divide radio galaxies into two types—double-lobed sources, and active-core sources. Although they appear to be different, these two types may be related.

*Double-Lobed Radio Galaxies.* A **double-lobed radio galaxy** emits radio energy from two regions called lobes located on opposite sides of the galaxy. In some cases the galaxy itself is silent at radio wavelengths.

Radio lobes are generally much larger than the galaxy they accompany. Many are as large as 60 kpc (200,000 ly) in diameter, twice the size of the Milky Way galaxy. From tip to tip the radio lobes span hundreds of kiloparsecs. The largest known radio galaxy is 3C236 (the 236th object in the *Third Cambridge Catalogue of Radio Sources*). It may be the largest object in the universe, with radio lobes that reach across 5.8 Mpc ($19 \times 10^6$ ly) (Figure 10-1).

Radio lobes have two properties that hint at their origin: They radiate synchrotron radiation

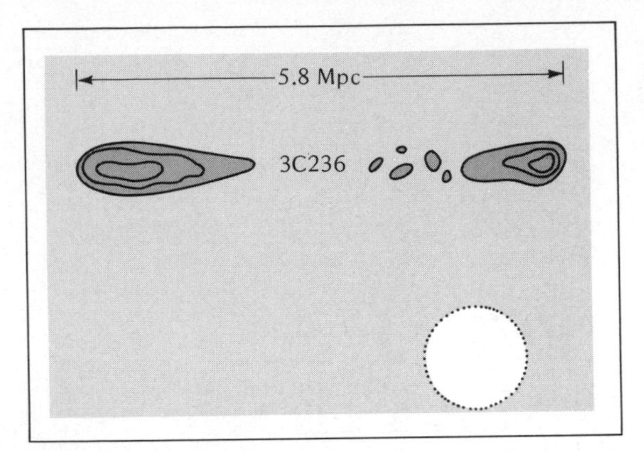


Figure 10-1. The radio lobes of 3C236, the largest radio galaxy known, span 5.8 Mpc. The circle illustrates the approximate size of the Local Group.

and are often most intense on the side away from the central galaxy. Synchrotron radiation is produced when high-speed electrons move through a magnetic field. Though the field is at least 1000 times weaker than the earth's, it fills a tremendous volume and represents vast stored energy. In addition, the high-speed electrons also contain energy. The total energy in a radio lobe equals about $10^{60}$ ergs, approximately the energy equivalent of $10^6$ $M_\odot$. Clearly the process that creates radio lobes is powerful.

The second hint to the nature of the radio lobes is that many radio lobes emit more intensely from the side away from the galaxy (Figure 10-2). This suggests that the radio lobes are expanding away from the galaxy, and that their

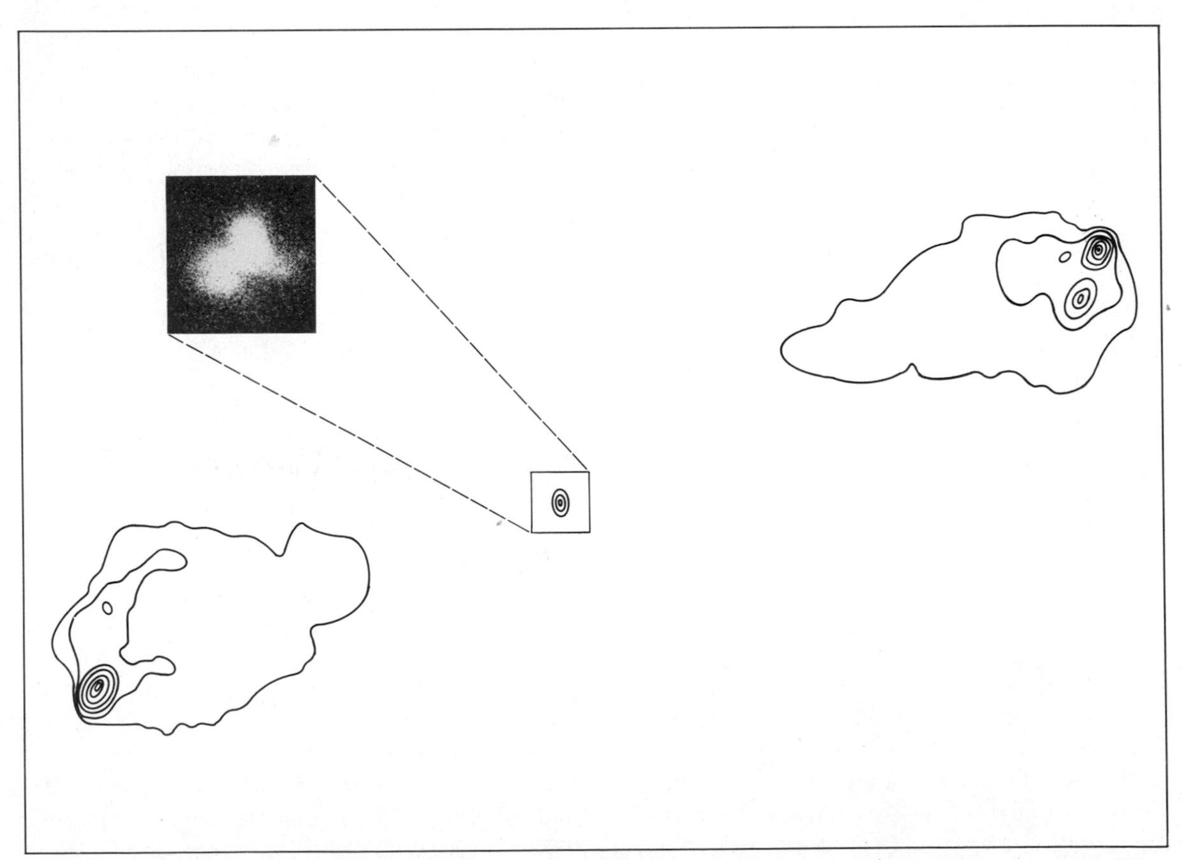

Figure 10-2. Cygnus A is a powerful double-lobed radio galaxy. The central galaxy (inset) is deformed. Note that the edges of the lobes farthest from the galaxy are brightest. (Adapted from a map by Hargrave and Ryle, Hale Observatories photograph.)

leading edges are colliding with intergalactic gas. Such a collision could compress and concentrate the magnetic field and the hot gas, producing stronger radio emission from the region.

These two hints suggest that radio lobes originate when the cores of galaxies explode, ejecting gas clouds in opposite directions. If the gas is ionized, it would drag along the galaxy's magnetic field, and, when the lobes expanded beyond the bounds of the galaxy, they would emit synchrotron radiation.

The central galaxy in a double-lobed source is usually a giant elliptical, and it is often located in a cluster. These galaxies are often deformed by strange tails, jets, and dust clouds, and their spectra show emission lines of highly excited low-pressure gas. All of these properties suggest explosive violence.

The prime example of a double-lobed radio galaxy is called Cygnus A, the second brightest radio source in the sky (Figure 10-2). It radiates about $10^7$ more radio energy than the Milky Way galaxy, all from two lobes containing tangled magnetic fields and bright leading edges. The central galaxy is oddly distorted into a double appearance. Unfortunately, Cygnus A is about 200 Mpc ($6.5 \times 10^8$ ly) away and even the best photographs do not show any detail.

Another fascinating radio galaxy is Centaurus A (Figure 10-3), a double-lobed source so close it covers 10° in the sky—as large as the bowl of the Big Dipper. Located between the radio lobes is NGC 5128, a giant elliptical galaxy encircled by a mysterious ring of dust in which new stars appear to be forming. Doppler shifts in the galaxy's spectrum show that the galaxy is rotat-

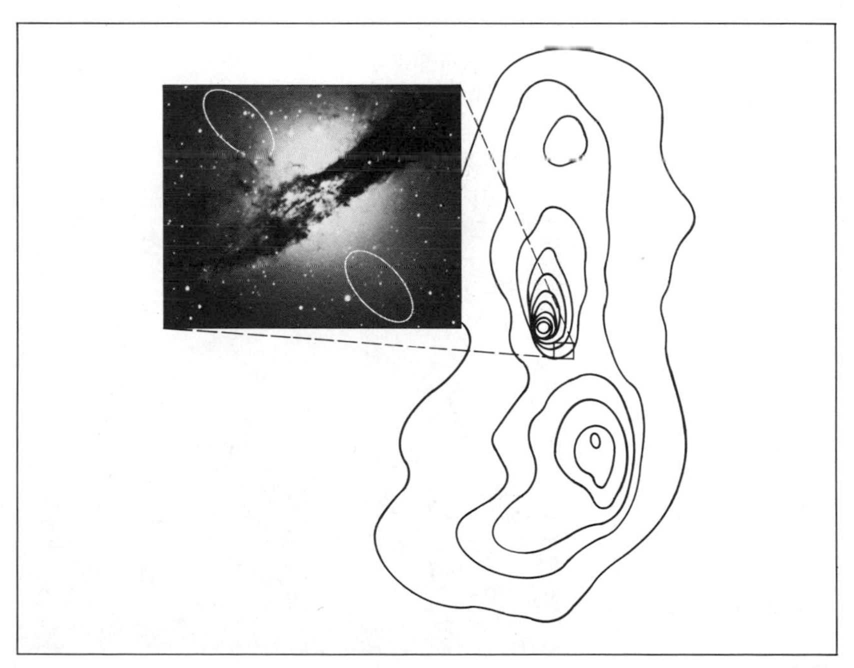

*Figure 10-3. Centaurus A and the peculiar galaxy NGC 5128. Note the dust ring and the two smaller radio lobes in the photograph. (Adapted from a map by Cooper, Price, and Cole; Hale Observatories photograph.)*

ing about an axis perpendicular to the dust lane, and long-exposure photographs reveal a faint jet extending 40 kpc ($1.3 \times 10^5$ ly) along the axis of rotation toward the northern lobe—as if matter were still being ejected from the galaxy.

High-resolution radio maps of this galaxy reveal a second pair of radio lobes, small and intense, located within the bounds of the galaxy itself. Perhaps the galaxy has exploded twice, once long ago, producing the large pair of lobes, and once more recently to produce the smaller pair that have not yet expanded beyond the edge of the galaxy. Other radio galaxies show multiple sets of lobes, suggesting that however violent the eruption is, it does not destroy the galaxy. It may survive to erupt again and again.

This theory of the explosive origin of radio lobes raises many problems. For example, how can the lobes keep emitting such floods of energy without cooling off and fading away? The high-speed electrons must slow as they emit synchrotron radiation, so the lobes must have a continuing source of energy. Some have suggested that the lobes are powered by energy from a continuous eruption in the center of the galaxy, while others propose that the lobes contain neutron stars and black holes that supply fresh high-speed electrons.

Another problem is how the lobes stay together. In some cases the radio lobes are small compared to the distance they have traveled from their parent galaxy. The magnetic field is too weak to hold the gas together, so why don't the lobes dissipate? Some research suggests the lobes are compressed and contained by their collision with the intergalactic gas.

Still another mystery is how a galaxy ejects lobes. Whatever the energy source, it somehow chooses two opposite directions in space and hurls the hot, massive lobes outward. This suggests that the eruption is channeled along some axis of symmetry within the galaxy, perhaps along a magnetic axis or, as in the case of NGC 5128, along the axis of rotation.

The deepest mystery is, of course, what makes a galaxy explode. Although there are no answers yet, the key may lie in the second type of radio galaxy, the active-core sources.

*Active-Core Radio Galaxies.* In an **active-core radio galaxy,** radio energy comes not from radio lobes but from the core of the galaxy, as if it were in the process of exploding. Since active cores are common, they indicate that galaxy explosions are not unusual events.

**Seyfert galaxies** show symptoms of core explosions. They are normal spiral galaxies with unusually bright, tiny cores that fluctuate in brightness (Figure 10-4). Spectra of these cores reveal emission lines from hot gas clouds moving at high velocities. Most Seyfert galaxies are powerful sources of infrared radiation and some are radio sources; at least one is a source of x-rays. It certainly seems that Seyfert galaxies are undergoing a central eruption.

Since 2 percent of all spiral galaxies are Seyferts, we can conclude one of two things: Either explosions occur in 2 percent of all spiral galaxies, or all spiral galaxies explode about 2 percent of the time. The peculiar heart of our own galaxy suggests that all spirals explode now and then. We may be living in a galaxy that was

*Figure 10-4. Active-core source Perseus A is a Seyfert galaxy with an intensely bright nucleus. (Kitt Peak National Observatory.)*

a Seyfert only a few hundred million years ago.

Occasionally a galaxy with an active core shows other evidence of an explosion. M 87, for example, is a giant elliptical located in the nearby Virgo cluster and has a peculiar bright core visible at optical and radio wavelengths (Figure 9-8). A weaker radio halo extends beyond the visible edge of the galaxy. The impressive feature of M 87, however, only appears on short-exposure photographs, which reveal a tremendous jet of matter 1500 pc long squirting out of the core. The jet emits synchrotron radiation, proving that it contains a magnetic field and high-speed electrons (Figure 10-5).

The suggestion that M 87 is exploding leads us back to our fundamental problem. What is the energy source? New observations of M 87 may hold the answer. Astronomers studying the bright core found that the stars were tremendously crowded (Figure 10-6) and that the spectral lines were very broad. The broad lines are evidence that the stars in the core are moving at high velocities. If the rapidly moving stars are held in the small core by gravity, the core must be very massive—$5 \times 10^9$ $M_\odot$ according to the measurements. This mass could be a dense swarm of stars, but the astronomers point out that their results are consistent with a $5 \times 10^9$

*Figure 10-5. The core of M 87 (inset) contains a central bright spot and a jet of matter ejected from the center. (Kitt Peak National Observatory.)*

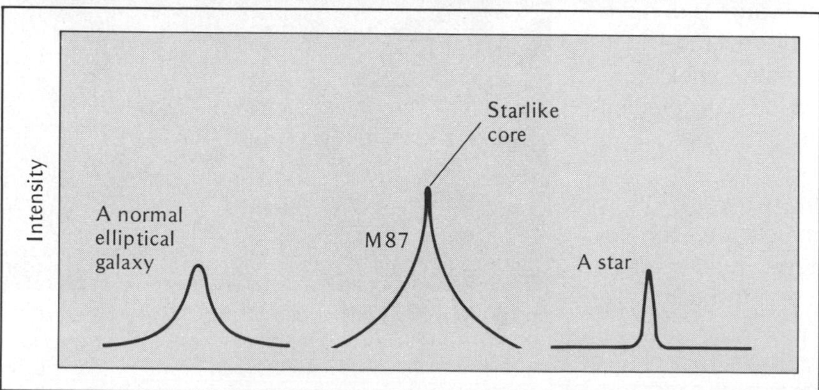


*Figure 10-6. The intensity across M 87 reveals a brilliant starlike core, quite unlike the intensity across a normal elliptical galaxy.*

$M_{\odot}$ black hole at the center of the galaxy. If this is true, then matter falling into the black hole might grow hot enough to provide the energy for the radio galaxy.

Another team of astronomers studying M 87 has detected radio pulses coming from the nucleus about once a second. The pulses are only 0.001 second long, but in that short time the nucleus must emit about 10,000 times the energy that the sun radiates in an entire second. The pulses' discoverers suggest that a 1,000,000,000 $M_{\odot}$ black hole lying in the heart of the galaxy might produce such pulses.

It is much too early to conclude that exploding galaxies contain massive black holes at their centers, but this suggestion illustrates the dilemma posed by the radio galaxies. How can a galaxy's core generate tremendous energy in a small central region?

Radio galaxies may even be related to other kinds of objects. Our own Milky Way galaxy shows evidence of a past core explosion, and other nearby galaxies such as the Andromeda galaxy have similar small cores. Violence in the cores of galaxies may be very common, though it may only rarely approach the power of a radio galaxy. Radio galaxies may even be related to the mysterious QSOs we will now examine.

## QUASI STELLAR OBJECTS

**Quasi stellar objects** (also called **QSO**s, or quasars) are small, powerful objects that seem to lie far away. Many QSOs are sources of radio signals. When they were first discovered in the early 1960s, their properties seemed unexplainable, but more recent observations are piecing together a picture of these objects.

*Discovery.* The discovery of QSOs shocked astronomers. The objects were unlike anything that had been observed before. They seemed to lie at great distances and have large luminosities.

Even the largest optical telescopes cannot see very far into space. The nearest galaxies are faint but they can be photographed easily. Farther out in space, the galaxies and galaxy clusters become very hard to detect. Galaxies 4,000,000,000 ly away are mere wisps on photographs, and galaxies beyond 8,000,000,000 ly are nearly invisible (Figure 9-4). Radio telescopes can locate radio sources more distant than this, but the galaxies associated with such sources are not detectable at visual wavelengths because of the tremendous distances.

In the early 1960s radio interferometers (Chapter 3) showed that a number of radio sources were much smaller than normal radio galaxies. Photographs of the location of these radio sources did not reveal a central galaxy, not even a faint wisp, but rather a single starlike point of light (Figure 10-7). The first of these objects so identified was 3C48, and later the source 3C273 was added. Though these objects

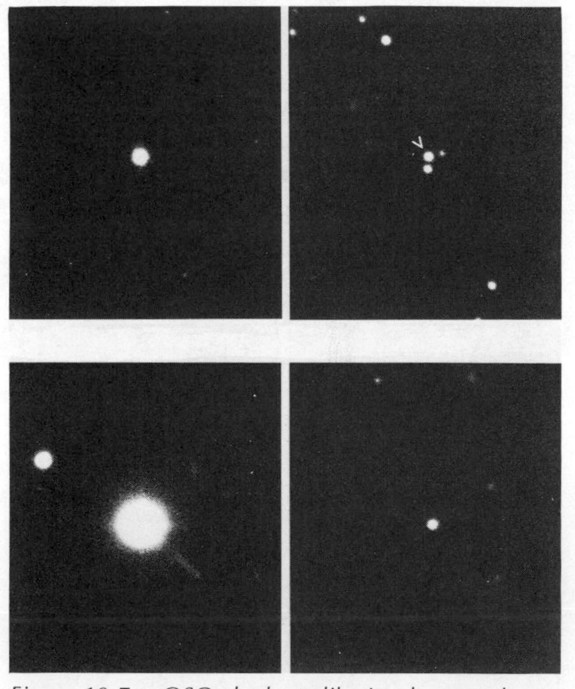

*Figure 10-7. QSOs look starlike in photographs and are obviously not galaxies. Note the jet extending away from 3C273 (see Figure 10-10). (Hale Observatories.)*

emitted radio signals like those from radio galaxies, they were obviously not galaxies. Even the most distant photographable galaxies look fuzzy but these seemed like stars, so they were called quasi stellar objects.

At first the spectra of these objects only added to the mystery. A jumble of unidentifiable emission lines was superimposed on a continuous spectrum. But in 1963 Maarten Schmidt at Hale Observatories tried red shifting the hydrogen Balmer lines to see if they could be made to agree with the lines in 3C273's spectrum. At a red shift of 15.8 percent, three lines clicked into place (Figure 10-8). Other QSO spectra quickly yielded to this approach, revealing even larger red shifts.

The QSOs have large red shifts. The red shift **Z** is the change in wavelength **Δλ** divided by the unshifted wavelength $\lambda_0$.

$$\text{red shift} = Z = \frac{\Delta\lambda}{\lambda_0}$$

If these red shifts arise from Doppler shifts because of velocities of recession, then according to the Hubble Law larger red shifts imply larger distances. Thus the red shift of a QSO may indicate its distance (see Box 9-2).

The red shift of 3C273 is 0.158 and 3C48's is 0.37. These are large red shifts but not as large as the largest observed galaxy red shift, about 1.0. However, QSOs were soon found with red shifts much larger than that of any known galaxy. These large red shifts imply that QSOs are at great distances. A galaxy that far away would not be detectable, yet the QSOs are easily photographed. Their large distances and moderate apparent brightness imply that they are 10 to 1000 times the luminosity of a large galaxy.

The terrific luminosity of QSOs is difficult to explain by itself, but another observation makes the problem even harder to crack. Soon after QSOs were discovered, astronomers detected fluctuations in brightness over time intervals as short as a few days. To further explore this variation, the astronomers went back into photographic plate collections housed in various observatories and measured the brightness of the QSOs on old plates. The results showed erratic, rapid fluctuations in brightness. For example, 3C273 changed its brightness by 2500 percent in one month in 1936.

These rapid fluctuations turned QSOs into an astronomer's nightmare. Recall from our discussion of pulsars (Chapter 7) that an object cannot change its brightness appreciably in a time less than the interval light needs to cross its diameter. The rapid fluctuations occurring in QSOs over periods of days or weeks show that QSOs are small, not more than a few light-days or light-weeks in diameter. How can QSOs generate 10 to 1000 times more energy than a galaxy, and do it in tiny regions only a few

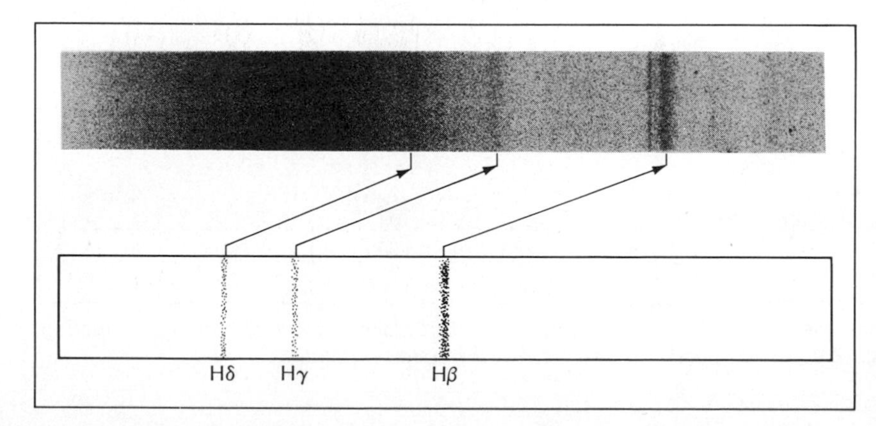


*Figure 10-8. The spectrum of 3C273 contains three hydrogen Balmer lines red shifted by 15.8 percent. The drawing shows the unshifted positions of the lines. (Courtesy Maarten Schmidt/Hale Observatories.)*

light-weeks in diameter? No one has yet answered that question.

Since their discovery, QSOs have yielded some secrets that may point to a solution. Searches have found hundreds of QSOs, most of which have red shifts larger than 1. As explained in Box 10-1, the large red shifts imply large velocities of recession. The QSO with the largest known red shift is OQ 172. Its red shift is 3.53, indicating that it is receding at 91 percent the speed of light.

If these high velocities are due to the expansion of the universe, then dividing by the Hubble constant translates the QSO velocities into dis-

## BOX 10-1 THE RELATIVISTIC RED SHIFT

The classical red shift equation discussed in Box 4-3 is only an approximation. It works quite well so long as the radial velocity of the star or galaxy is much smaller than the speed of light, and, since the velocities of planets, binary stars, and clusters of stars in our galaxy are hardly ever more than a few hundred kilometers per second, the classical red shift equation is accurate enough for most astronomical problems.

However, when the velocity of an object is an appreciable fraction of the speed of light, we must use the correct equation. Because this equation is derived by the theory of relativity, it is called the **relativistic red shift** equation. It relates the radial velocity $\mathbf{V_r}$ to the speed of light **c** and the red shift **Z.**

$$\frac{V_r}{c} = \frac{(Z + 1)^2 - 1}{(Z + 1)^2 + 1} \qquad \text{where } Z = \frac{\Delta\lambda}{\lambda_0}$$

For example, suppose a QSO has a red shift of 2. What is its velocity? First, Z + 1 equals 3, and thus $(Z + 1)^2$ equals 9. Then the velocity in terms of the speed of light is

$$\frac{V_r}{c} = \frac{9 - 1}{9 + 1} = \frac{8}{10} = 0.8$$

Thus the QSO has a radial velocity of 80 percent of the speed of light.

Figure 10-9 illustrates how $\mathbf{V_r/C}$ depends on **Z.** For small red shifts, the velocity is the same in both the classical and relativistic case, but as **Z** gets larger, the difference between the classical approximation and the true velocity increases. No matter how large **Z** becomes, the velocity can never quite equal the speed of light.

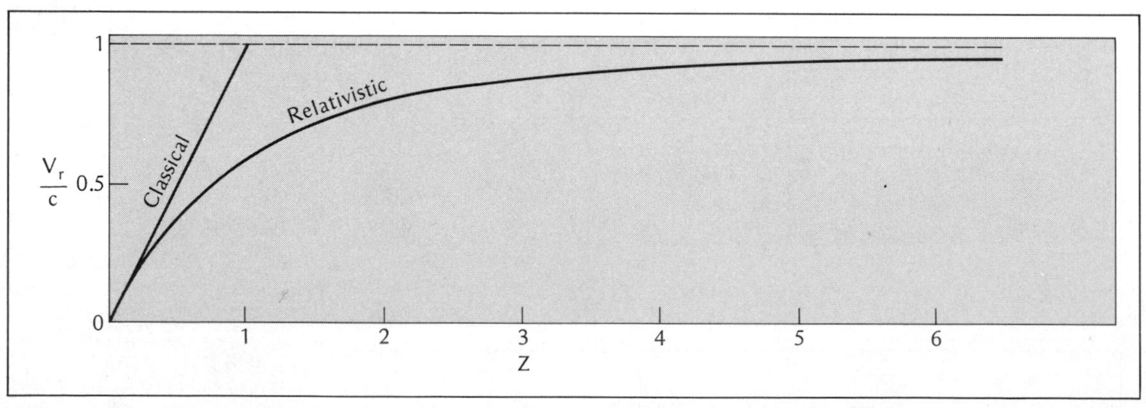


*Figure 10-9. At high velocities the relativistic red shift must be used in place of the classical approximation. Note that no matter how large **Z** gets, the speed can never equal the speed of light.*

tances. The results are staggering. Most QSOs are much farther away than the farthest observable galaxy, and the most distant, the QSO called OQ 172, is about 15,500,000,000 ly away. Thus when we look at distant QSOs we are looking back in time at the universe soon after it formed. Though most astronomers accept this explanation of the red shifts, we will consider some alternatives in the next section.

Although the QSOs are distant, they do reveal some structure. For example, 3C273 shows a jet of material (Figure 10-10) much like the jet in M 87. Many QSOs show a slight fuzziness and some have surrounding wisps. In addition, radio maps of some QSOs reveal double radio lobes. These structural properties lead some to think of QSOs as the brilliant cores of distant, powerful radio galaxies.

The most important clue to QSOs may lie in the absorption lines found in their spectra. QSOs have emission lines with a very high red shift, but some also have sets of absorption lines with slightly smaller red shifts. The QSO called OF 097 has eight sets of absorption lines, each red-shifted by a different amount. They suggest that the QSO has repeatedly ejected clouds of matter, and the clouds ejected in our direction, with radial velocities slightly smaller than the QSO's, form absorption lines at smaller red shifts. This ejection of matter recalls the explosive ejections of radio galaxies.

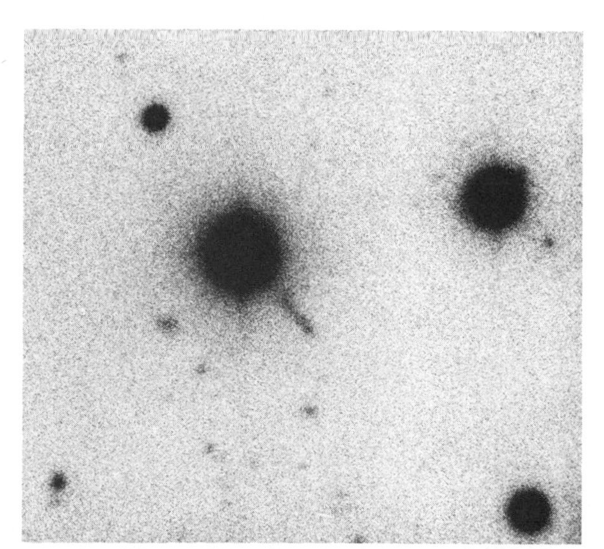

*Figure 10-10. A negative print of the QSO 3C273 shows the jet of material. (Courtesy Maarten Schmidt/Hale Observatories.)*

Another possible source of absorption lines is intervening galaxies. QSOs are so distant their light could pass through a number of galaxies on its way to us, yet the galaxies would be invisible because of their distance. Nevertheless, these galaxies could form sets of absorption lines with red shifts that correspond to the distances to the galaxies. The only objection to this hypothesis is that there don't seem to be enough galaxies to explain all of the observed line sets. Although a few absorption line sets may be due to galaxies, many may arise from ejected gas clouds.

If this is true, it leads us back to the basic energy puzzle. Some absorption line sets could only be produced by gas clouds ejected from the QSO at 60 percent the speed of light. The amount of energy required for such an expulsion is titanic.

*The Local Hypothesis.* Faced by the difficulty of finding what could generate sufficient energy to power a QSO, some astronomers turned the problem around. They theorized that QSOs were not at the great distances indicated by their red shifts, but were local objects, scattered among the nearer galaxies. If they were that close, they would not be superluminous, and it would not be so difficult to explain their energy production.

This solution doesn't really solve the problem, however. It only trades the energy puzzle for a red shift puzzle. If the QSOs are not at great distances, then what produces their fantastic red shifts? Astronomers know of only two processes that can produce red shifts, the Doppler effect and the gravitational red shift.

If the red shifts in QSO spectra are due to the Doppler shift, then the QSOs must be traveling at high velocities. Some astronomers have suggested that the QSOs are objects ejected from galactic nuclei at very high velocities, but this theory has a serious shortcoming—the red shifts show that the QSOs are receding from us. If galaxies eject QSOs, we should expect approximately half to be coming toward us. Yet no QSO has ever been found with a blue shift, even though blue-shifted QSOs should be brighter and easier to locate. If QSOs are local, then the Doppler effect can't explain their red shifts.

The second possibility, the gravitational red shift, requires that QSOs have powerful gravitational fields. Light leaving such a field loses energy and shifts toward the red end of the spectrum. This is a well-established phenomenon observed in the spectra of white dwarfs. However, all of the emission lines in a QSO's spectrum have the same red shift. If the red shift were gravitational, then the strength of the gravitational field would have to be the same throughout the QSO, which is impossible, or the QSO would have to be much smaller than its rapid fluctuations predict. Thus it doesn't seem that gravity could produce QSO red shifts.

If the local hypothesis is correct, then we must assume that some unknown process is producing the large red shifts in QSO spectra. Though some astronomers find that difficult to accept, it is not out of the question. In the Perspective at the end of this chapter, we will consider evidence that the traditional interpretation of the red shift may be incorrect. For now, however, we must proceed with the discussion of QSOs as superluminous objects at very large distances.

*A Model QSO.* The hypothesis that QSO red shifts are due to the expansion of the universe has two advantages. It explains the large red shifts via the Doppler effect and it relates the QSOs to the phenomena observed in radio galaxies.

In recent years, astronomers have begun to consider a model of what a QSO must be like. Although this model seems to explain the spectral features, the source of energy is still a mystery. For our purposes we can divide the QSO spectrum into three components: the continuous spectrum, the emission lines, and the absorption line sets. Each of these arises in a different part of the QSO model.

At the center of our model an unknown source of energy accelerates electrons to high speeds. These electrons, spiraling through a weak magnetic field, emit synchrotron radiation, both light and radio. This accounts for the continuous spectrum. This region must be no larger than a few light-weeks in diameter in order to permit the rapid, irregular fluctuations observed.

The best evidence is that QSO emission lines do not fluctuate rapidly, so they need not arise in the small core. Rather, they could occur in hot gas surrounding the core in a region many light-years in diameter. This gas is kept hot by the synchrotron radiation streaming out of the core, and the width of the emission lines indicates that the gas is turbulent, apparently because of the intense radiation.

Surrounding the small heart of the QSO we find clouds of ejected gas, now cooled and forming absorption lines in the QSO spectrum (Figure 10-11). Many of these clouds may have been ejected in successive eruptions, and some may be traveling at high velocities. The total extent of this cloud system may exceed 1000 pc.

This QSO model does not explain the energy source, and many theories have attacked that mystery. Some have suggested chain reactions of supernovae or collisions between matter and antimatter. Whatever the final answer, there is a growing suspicion among astronomers that the answer will involve gravity. Gravitational theories include massive spinning neutron stars and super black holes with masses up to hundreds of millions of solar masses surrounded by accretion disks generating tremendous temperatures.

Connecting this model QSO to the evolution of galaxies is a task that the next decade may see to a successful conclusion. QSOs may represent an early stage in the history of matter, a stage in which large gas clouds collapse and form massive black holes, releasing tremendous amounts of energy as they accrete matter. The whirlpools

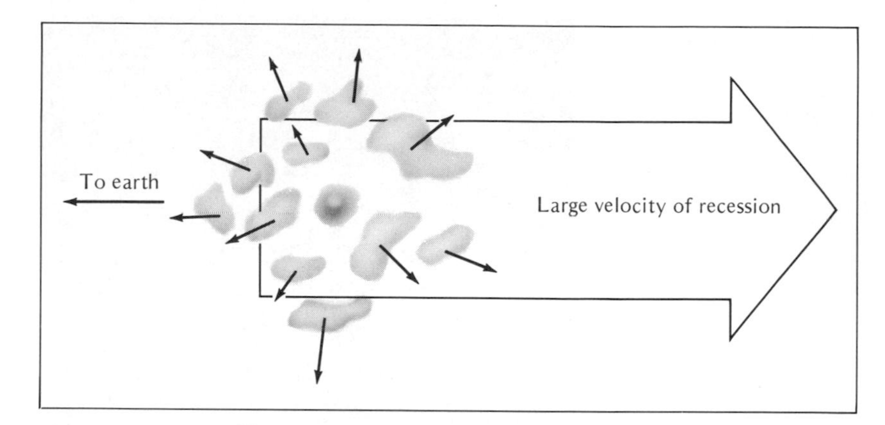


*Figure 10-11. A model QSO contains a tiny source of energy, surrounded by a cloud of hot gas. Ejected clouds of cooler gas rush outward and form absorption lines when earthbound light passes through them.*

that form around these gravitational powerhouses may eventually form radio galaxies. In fact, QSOs could be the ancestors of both radio and normal galaxies.

Our study of galaxies has led us far out in space and back in time through radio galaxies to QSOs. We are now at the threshold of the study of the universe itself. Before we begin that study in the next chapter, we must take time out to reexamine one of the basic laws of astronomy, the Hubble Law. A few astronomers are suggesting that some galaxies and QSOs don't obey the Hubble Law.

## PERSPECTIVE: THE RED SHIFT PROBLEM

The QSO mystery spurred astronomers to reexamine the origin of red shifts. The Hubble Law is based on the assumption that the red shifts of the galaxies arise from the Doppler shift. Even if that is generally true, the assumption may not be true for all objects—QSOs for instance. Or, an object's red shift may not arise entirely from its radial velocity. Part of the red shift may originate in some other process. As we have seen, gravitational red shifts could not explain the QSOs, so astronomers began to consider other possibilities, usually called **non-Doppler red shifts.**

The key observation in the search for non-Doppler red shifts would be a galaxy or QSO whose distance was known and whose red shift differed significantly from that predicted by the Hubble Law. Although no conclusive proof is available yet, a few objects do seem to have the wrong red shifts.

*QSOs Associated with Galaxies.* Because QSOs do not contain any visible distance indicators, we cannot find their distances without the Hubble Law. But if a QSO could be found that was connected to a normal galaxy, then the two objects would have to be at the same distance. If the galaxy had a small red shift and the QSO had a large red shift, it would prove that some non-Doppler red shift was at work.

Astronomer Halton Arp of Hale Observatories located one of the best candidates for a non-Doppler red shift, an object called Markarian 205. It is a QSO located very near the galaxy NGC 4319. In Figure 10-12 the QSO appears to be linked to the galaxy by a bridge of matter, which would imply they are at the same distance. However, the red shifts of the two objects are very different. The galaxy's red shift is 0.006 and the QSO's is 0.07. The Hubble Law says the QSO is 12 times farther away than the galaxy and just happens to be lined up so it appears to be connected. Such a chance alignment seems very remarkable.

*Red Shifts in Galaxy Clusters.* Another exception to the Hubble Law might be found in

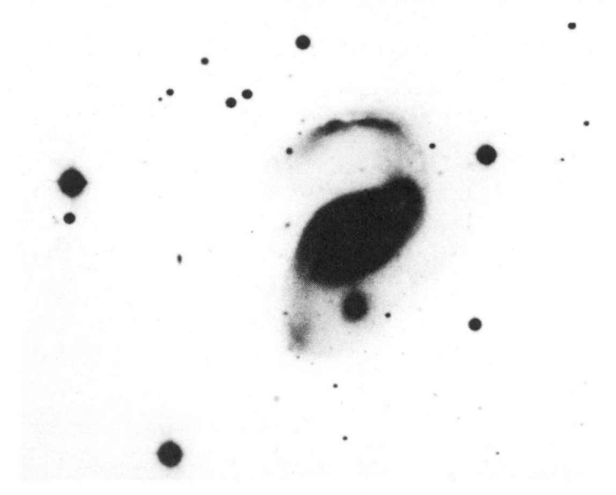

*Figure 10-12. The QSO Markarian 205 appears to be connected to the galaxy NGC 4139. (Courtesy Halton Arp/Hale Observatories).*

*Figure 10-13. Stephan's Quintet contains four galaxies with red shifts of 0.021 and one galaxy (at lower left) whose red shift is 0.0027. (Kitt Peak National Observatory.)*

clusters of galaxies. If observations revealed that some galaxies in a cluster had red shifts significantly different from the rest, the red shift would no longer dependably indicate distance.

One of the most striking candidates for a non-Doppler red shift lies in a small cluster of five galaxies called Stephan's Quintet (Figure 10-13). Four of these galaxies have red shifts of 0.021, but one has a red shift of 0.0027. If these red shifts are interpreted by the Hubble Law, the odd galaxy must be only one-eighth as far away as the rest. However, optical filaments seem to envelop all five galaxies, and the H II regions have about the same angular diameter in the galaxies. This seems to place the five galaxies at the same distance.

Another startling group, VV 172, is a chain of galaxies (Figure 10-14). Such chains are not unusual and may represent galaxies that formed long ago from a single eddy of gas. The red shift of four of these galaxies is 0.053, but one of them has a red shift of 0.12. If these red shifts are due to the expansion of the universe, then the odd galaxy is 2.3 times farther away than the rest and just happens to line up with a gap in the chain.

All of these examples depend on the probability that a nearby and a distant object lie so nearly in the same direction that they appear to

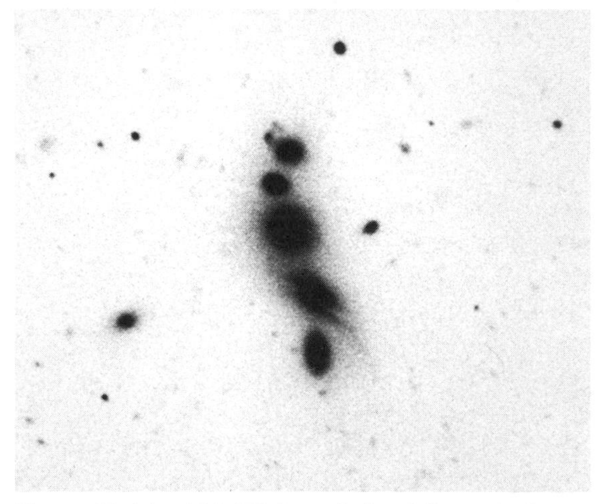

*Figure 10-14. VV 172 is a chain of five galaxies. The second from the top has a red shift 2.3 times larger than the other four. (Courtesy Halton Arp/Hale Observatories.)*

be connected. A few astronomers argue that the probability of such alignments is so small that the galaxies and QSOs must actually be associated with each other. If they are, they are at the same distance and the peculiar red shifts must arise from some unknown process. To evaluate these groups of objects we must consider the probability that such alignments might occur by chance.

*The Statistics of Alignment.* Those who point to two apparently connected objects with different red shifts all argue that the probability of a chance alignment is tiny, but in each case they apply statistics to a single event. This misuse of statistics can lead to startling misunderstandings. For example, what is the chance that you will read this sentence at this particular second of the year? Since there are 86,400 seconds in a day and about 365.24 days in a year, the probability is about 31,557,000 to 1 against. Yet the event did occur. Clearly we cannot apply statistical arguments safely to single events. Instead of asking what the probability is that a particular pair of objects are aligned by chance, we must ask what the chances are of any such alignments anywhere in the sky.

We can make a rough estimate by noting that there are over 3 million galaxies bright enough to photograph. That amounts to 80 galaxies per square degree of sky. If we assume the galaxies are randomly scattered around the sky, a statistical analysis suggests that there should be about 40 chance alignments of distant galaxies with nearer galaxies over the whole sky. Therefore, that we see a few such alignments is not significant. The critical question is whether we see more than could occur by chance. Some astronomers claim the answer is no.

The alignment of QSOs with nearer galaxies should also be possible. Recent surveys suggest there may be thousands of QSOs scattered over the sky, and we should expect to see some of these lined up with nearby galaxies. Could Markarian 205 be such an alignment? The bridge that connects the QSO and galaxy might be an optical effect caused by the overlap of the images on a photographic plate. This effect has been reproduced in a darkroom, and photographs taken with different telescopes don't show any bridge. None of this proves that Markarian 205 is a chance alignment, but it does show that the object is not conclusive evidence of non-Doppler red shifts.

Even the galaxies in Stephan's Quintet may be a chance grouping. A recent study shows that the galaxy with the low red shift is probably a member of a nearby loose cluster, and the other four galaxies are a subgroup in a larger and more distant cluster. Because galaxy clusters are large, they often overlap each other, and in such cases we should expect to see some peculiar alignments.

Perhaps the strongest evidence for non-Doppler red shifts is galaxies such as VV 172. It seems difficult to believe this is a chance alignment, but, since galaxy chains are fairly common, we should expect a few such alignments. In any case, we must not apply statistical arguments to single events. Only a survey of hundreds or thousands of such groupings will tell us whether we see more peculiar red shifts than we should.

*The Dependability of the Red Shift.* Most astronomers trust the red shift as a distance indicator for two reasons. First, it covers a tremendous range of distance, from nearby galaxies to the most distant visible. Hubble's original study only probed 10,000,000 ly. Since then modern astronomers have confirmed the Hubble Law out to 6,200,000,000 ly. A recent study of QSOs suggests that if we could find dependable distance indicators for these objects, they too would obey the Hubble Law.

The second reason for confidence in the Hubble Law is that it is a general law that fits our current conception of how the universe works. A physical law, principle, or theory does not exist in a vacuum; it must fit with all other knowledge, just as a piece in a jigsaw puzzle must match all adjacent pieces. At present the Hubble Law fits. It applies to all types of galaxies, and it meshes with current theories of the origin and evolution of the universe (Chapter 11). That does not prove the Hubble Law is correct, but it means that astronomers will hesitate to abandon it until someone finds a conclusive example of a non-Doppler red shift.

## SUMMARY

Some galaxies have peculiar properties. A few are probably the result of collisions. Others may be the result of explosions in the cores of the galaxies. Active-core galaxies emit powerful radio signals from their cores and appear to be in the process of exploding. Double-lobed radio galaxies emit radio signals from the lobes on either side of the galaxy. Some of these galaxies show jets and filaments of matter rushing out of the core.

The source of energy for exploding galaxies is unknown, but it may involve gravity. Observations suggest that M 87, an active core radio galaxy, may contain a super massive black hole. Matter falling into such an object could release tremendous energy and might force a core explosion.

The QSOs may be related objects. Their spectra show emission lines with large red shifts. Their large red shifts imply they are very distant, and that they are visible at all implies they are superluminous. Because they fluctuate rapidly, they must be no larger than a few light-weeks in diameter.

Since they lie at great distances, we see them as they were long ago. The look-back time to the most distant QSO is about 15½ billion years. Thus they may represent an early stage in the formation of galaxies. The first collapse of large gas clouds may form massive black holes and release tremendous energy in a small region. Hot gas around the region could produce the observed emission lines, and ejected clouds of cooler gas could form the absorption lines.

Because of the QSO controversy, astronomers are reexamining the red-shift–distance relation. Searches have found a number of objects with red shifts that do not agree with their presumed distances. None of these, however, is conclusive. They could be chance alignments of distant objects with large red shifts and nearby objects with small red shifts. Until some unambiguous example is found, astronomers will continue to depend on the Hubble Law.

## NEW TERMS

radio galaxy
double-lobed radio galaxy
active-core radio galaxy
Seyfert galaxy
quasi stellar object (QSO)
relativistic red shift
non-Doppler red shift

## QUESTIONS

1. What evidence do we have that galaxies explode more than once?
2. What evidence do we have that M 87 might contain a massive black hole?
3. Why do we conclude that QSOs must be small?
4. How does our model QSO account for the three components of a QSO spectrum?
5. What is the local hypothesis of QSOs?
6. What is VV 172? Do you think it represents a real non-Doppler red shift? Defend your position.

## PROBLEMS

1. The total energy stored in a radio lobe is about $10^{60}$ ergs. How many solar masses would have to be converted to energy to produce this energy? (Hints: Use $E = mc^2$. One solar mass equals $2 \times 10^{33}$ gm.)
2. Use the small angle formula (Box 1-3) to find the linear diameter of a radio source with an angular diameter of 0.0015 seconds of arc and a distance of 3.25 Mpc.
3. What is the radial velocity of 3C48 if its red shift is 0.37? (Hint: See Box 10-1.)
4. The hydrogen Balmer line H$\beta$ has a wavelength of 4861 Å. It is shifted to 5639 Å in the spectrum of 3C273. What is the red shift? (Hint: What is $\Delta\lambda$?)
5. Plot the red shifts and velocities of the QSOs 3C48, 3C273, and OQ 172 on Figure 10-9.
6. The Hubble Law has been confirmed out to $6.2 \times 10^9$ ly. At what velocity are galaxies at this distance receding?

## RECOMMENDED READING

Goldsmith, D. "Exploding Galaxies." *Mercury* 6 (Jan./Feb. 1977), p. 2.

Hamilton, D., Keel, W., and Nixon, J. F. "Variable Galactic Nuclei." *Sky and Telescope* 55 (May 1978), p. 372.

Hopkins, J. "Quasars: Oddities of Space." *Astronomy* 4 (May 1976), p. 6.

Jastrow, R. "The Quasar Controversy." *Natural History* 83 (May 1974), p. 74.

Kraus, J. *Big Ear.* Powell, Ohio: Cygnus-Quasar Books, 1976.

Maran, S. P. "The Cygnus A Conundrum." *Natural History* 86 (May 1977), p. 84.

Morrison, P. "Resolving the Mystery of the Quasars." *Physics Today* 26 (May 1973), p. 23.

Schmidt, M., and Bello, F. "The Evolution of Quasars." *Scientific American* 224 (May 1971), p. 55.

Schnopper, H. W., and Delvaille, J. P. "The X-Ray Sky." *Scientific American* 233 (Aug. 1975), p. 26.

Strom, R. G., Miley, G. K., and Oort, J. "Giant Radio Galaxies." *Scientific American* 233 (Aug. 1975), p. 26.

Thomsen, D. E. "Convocation to Contemplate Quasars." *Science News* 116 (Sept. 29, 1979), p. 217.

*The Coma Cluster of galaxies lies in the constellation Coma Berenices. It is about 370,000,000 ly away and contains about 800 galaxies. The light that formed this photographic image left the Coma Cluster about the time life emerged from earth's oceans. (Kitt Peak National Observatory.)*

# Chapter 11 COSMOLOGY

How did the universe begin? How big is it? Does it have an edge? Will the universe ever end, and if so, how? These are some of the questions we will consider as we discuss **cosmology,** the study of the nature, origin, and evolution of the universe.

We will begin our study by reviewing some of our basic assumptions about the universe—for example, the assumption that the laws of physics are the same everywhere in the universe. Assumptions are important in all branches of science, but they are especially important in cosmology. Cosmologists have few observations to analyze, so they must devise theoretical models from the laws of nature and basic assumptions. To understand these models and judge their validity, we must understand the assumptions that lie behind them, and then compare them with the few observations available.

One of the basic observations of cosmology is that the night sky is dark. We are so accustomed to a dark night sky it does not immediately strike us as significant, but we will see that this simple fact can tell us a great deal about the universe. Another important observation is the red shift apparent in the spectra of galaxies. If we assume that these red shifts are due to the Doppler effect, then the galaxies must be receding from us, and the universe must be expanding.

A large portion of this chapter will concern the **big bang,** the theory that the universe began in a violent explosion that expelled gas clouds. According to this theory, the gas clouds formed clusters of galaxies that we see today still rushing away from each other. We will find that one form of the big bang theory suggests that the universe will eventually slow to a stop and fall back, smashing everything together, and, perhaps, generating another big bang.

Another portion of this chapter will deal with a theory that proposes that the universe is not changing, that it has existed forever much as it is now, and that it will never end. Though the theory is interesting, a number of observations make it unlikely.

## METHODS OF COSMOLOGY

We have built into our personalities such deeply felt expectations about the universe that it is easy for us to go astray. To avoid being misled, we must proceed with extreme care and analyze our assumptions and methods for errors and misunderstandings.

Ideally we would like to study cosmology with no assumptions at all. We would prefer to begin with those facts of which we are certain and deduce the nature, origin, and fate of the universe by logical, foolproof arguments. Unfortunately, this is impossible because there are too few facts of which we can be certain. To make progress in cosmology, we must assume some things just to cut the problem down to manageable size. We can analyze these assumptions with extreme care, but we cannot yet prove they are correct.

*Basic Assumptions.* Although we could make many assumptions, three are basic—homogeneity, isotropy, and universality.

**Homogeneity** is the assumption that matter is uniformly spread through space. Obviously this is not true on the small scale, because we can see matter concentrated in planets, stars, and galaxies. Homogeneity refers to the large-scale distribution. If the universe is homogeneous, we should be able to ignore individual galaxies and think of matter as an evenly spread gas in which each particle is a galaxy. The best observations suggest that the universe is homogeneous.

**Isotropy** is the assumption that the universe looks the same in every direction, that it is isotropic. On the small scale this is not true, but if we ignore local variations like galaxies and clusters of galaxies, then the universe should look the same in any direction. For example, we should see roughly the same number of galaxies in every direction. In the next section we will see evidence that the universe is isotropic.

The most easily overlooked assumption is **universality,** which holds that the physical laws we know on earth apply everywhere in the universe. Though this may seem obvious at first, some astronomers challenge universality by pointing out that when we look out in space we look back in time. If the laws of physics change with time, we may see peculiar effects when we look at distant galaxies. For now we will assume that the physical laws observed on earth apply everywhere in the universe.

The assumptions of homogeneity and isotropy lead to an assumption so fundamental it is called the **cosmological principle.** According to this principle, any observer in any galaxy sees the same general features of the universe. For example, all observers should see the same kinds of galaxies. As in previous assumptions, we ignore local variations and consider only the overall appearance of the universe, so the fact that some observers live in galaxies in clusters and some live in isolated galaxies is only a minor irregularity.

Evolutionary changes are not included in the cosmological principle. If the universe is expanding and the galaxies are evolving, then observers living at different times may see galaxies at different stages. The cosmological principle says that once observers correct for evolutionary changes, they should see the same general features.

The cosmological principle is actually a statement that we are not in a special place, that our location in the universe is average. This is a fundamental assumption to a scientist, but it is not necessarily valid for a philosopher or a theologian. In fact, the basic lesson of nearly all religions is that humanity is special and favored.

Once we establish our basic assumptions, we face a choice between two methods of attack. We can observe the universe and try to deduce its properties from its appearance, or we can build a theoretical universe based on our assumptions and the laws of nature and then compare our theory with reality. Both methods are valid.

*The Red Shift Assumption.* The basic observation of cosmology is that galaxies have red shifts in their spectra, and that the red shifts of the more distant galaxies are larger than those of nearer galaxies (Figure 11-1). Relating these red shifts to Doppler shifts due to the recession of the galaxies is an assumption with an overwhelming impact on modern cosmology.

If the red shifts are due to the Doppler effect, then the galaxies are receding from us as described by the Hubble Law, and the universe is expanding uniformly. To see how the Hubble Law implies uniform expansion, imagine a rubber band with dots painted on it at 1-cm intervals (Figure 11-2). As we stretch the rubber band, each dot moves away from its neighbors. Two adjacent dots originally 1 cm apart move

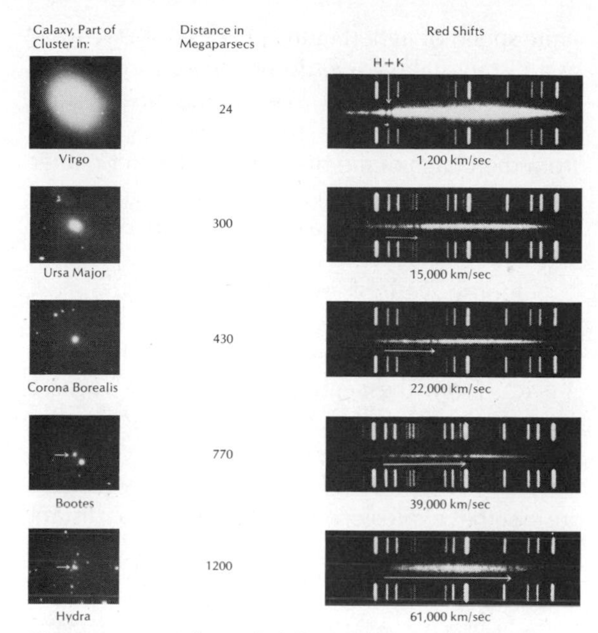


*Figure 11-1.* *The red shifts in these galaxy spectra appear as displacements of the H and K lines of calcium toward the red (arrows). The red shifts are assumed to be Doppler shifts due to the expansion of the universe. (Hale Observatories.)*

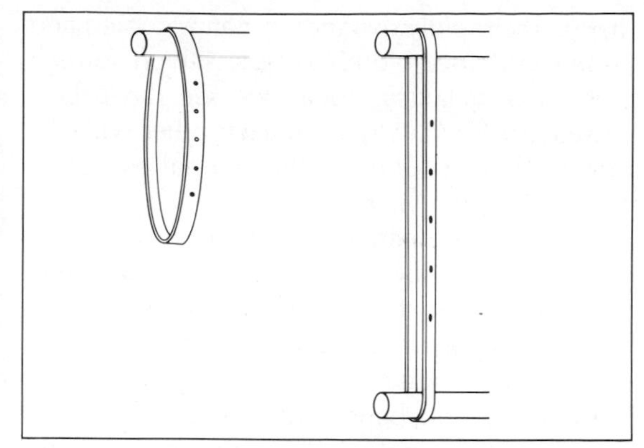

*Figure 11-2.* *The uniform expansion of the universe can be represented with dots on a rubberband. Initially 1 cm apart, the dots recede from each other when the rubber band is stretched. The velocity of separation depends on the distance between the dots.*

only slightly farther apart so their velocity of separation is small, but two dots that were 4 cm apart separate much more, and their velocity of separation is four times larger. Thus the uniform stretching of the rubber band causes the dots to move away from each other at velocities proportional to their distances. Compare the motion of the dots with the motion of galaxies. According to the Hubble Law, the larger the distance between two galaxies, the faster they recede from each other. This is exactly the result we expect from uniform expansion.

The rubber band also illustrates that there is no identifiable center to the expansion. A colony of bacteria living on any of the dots will see the same thing—dots receding at velocities proportional to distance. We see galaxies receding from us, but we cannot identify any galaxy or point in space as the center of the expansion of the universe. Any observer in any galaxy should see the same expansion that we see.

One reason for accepting the assumption that the red shifts are Doppler shifts is that the results of that assumption agree with our other basic assumptions. There can be no center to the expansion, just as there can be no special places in the universe. It all fits together.

If our assumptions are correct, we live in an expanding universe. The simple fact that the night sky is dark can tell us that the universe is not only expanding, but that it has a finite age (Box 11-1).

## THE BIG BANG THEORY

The big bang theory supposes that the universe began with a violent explosion from which all matter and energy originated. How the universe behaves after the explosion depends on the amount of matter. It may expand forever, or it may slow to a stop and fall back.

*The Beginning.* The big bang theory suggests that the universe began with an explosion, a big bang, as matter and energy burst out of a highly dense state and began the expansion we see continuing today. According to the theory, the rapidly expanding matter cooled and fragmented into clouds of gas, which formed clusters of galaxies. Today we see the galaxy clusters rushing away from each other with the motion they acquired in the original explosion 16 to 20 billion years ago.

Though our imaginations try to visualize the explosion as a localized event, we must keep firmly in mind that the big bang did not occur at a single locality, but filled all of space. We are now riding a fragment of that eruption, so we are inside the big bang, and the explosion is still going on all around us. We cannot point in any particular direction and say, "The big bang occurred over there." The big bang occurred everywhere (Figure 11-4).

Although the explosion took place long ago, it is possible for us to detect it because of the finite speed of light (Figure 11-4). When we look at a distant galaxy, we do not see it as it is now but as it was long ago. We see the most distant galaxies as they were soon after they formed from the chaos of the big bang. It doesn't matter what direction we look because we are inside the explosion and galaxies surround us on all sides.

Suppose we look even farther, past the most distant objects, backward in time to the fiery clouds of matter in the exploding fireball (Figure 11-4c). From these great distances we receive light that was emitted by the hot gas soon after the explosion began. Again, it does not matter what direction we look in space because we are inside the explosion surrounded by a distant wall of fire.

The radiation that comes to us from this dis-

---

## BOX 11-1 WHY DOES IT GET DARK AT NIGHT?

We have all noticed that the night sky is dark. However, apparently reasonable assumptions about the universe lead to the conclusion that the entire sky should glow as brightly as the surface of a star. This conflict between observation and theory is called **Olbers' paradox,** after Heinrich Olbers, a Viennese physician and astronomer who first publicized the paradox in 1826.

Suppose we assume that the universe is static, infinite, and eternal. Suppose we also assume that it is uniformly filled with stars. (The aggregation of stars into galaxies makes no difference to our argument.) If we look in any direction, our line of sight must eventually reach the surface of a star (Figure 11-3). Consequently, every point on the surface of the sky should be as bright as the surface of a star, and it should not get dark at night.

Imagine the entire sky glowing with the brightness of the surface of the sun. The glare would be overpowering. In fact, the radiation should rapidly heat the earth and all other celestial objects to the average temperature of the surface of the stars, 1000°K at least. Thus we can pose Olbers' paradox in another way: "Why is the universe so cold?"

We might try to resolve the paradox by supposing that interstellar matter absorbs the radiation from the distant stars. However, if the universe has existed forever, the interstellar medium should have heated up to the average temperature of a stellar surface, and the gas and dust clouds should be glowing just as brightly as the stars. Interstellar matter is helpful only if we suppose the universe is not infinitely old.

Olbers' paradox tells us that at least one of our assumptions is wrong. In fact, we have made two errors. The universe is not static, nor is it infinitely old.

Because the universe is expanding, distant stars are receding from us at high velocities, and their movement away produces large red shifts and reduces the energy of the arriving photons. Thus we can't see the light from the more distant stars in the universe because their radiation is red-shifted to very long wavelengths. If we can't see the light, it can't make the sky glow. In addition, the long-wavelength photons carry so little energy they do not

tance has a tremendous red shift. The farthest visible galaxies have red shifts of about 1.0, and the farthest QSOs about 3.5, but the radiation from the big bang fireball has a red shift of about 1000. Thus, the light emitted by the fireball arrives at the earth as infrared radiation and short radio waves. We will see in a later section how astronomers detected this radiation from the big bang.

*The Origin of Matter.* Tracing the explosion, expansion, and cooling of the big bang fireball is difficult because it begins from a fantastically hot, dense material. Theory predicts that the density was at least as great as that of a neutron star, and its temperature was somewhere between $10^{10}$ and $10^{12}$°K. At these densities and temperatures normal atoms and atomic particles cannot exist, so the physics of the material is difficult to understand.

Once the expansion began, temperature and density dropped rapidly (Figure 11-5). Only 25 seconds after the beginning, the temperature had fallen to about $4 \times 10^{9}$°K and the density was about 1000 gm/cm$^3$. Almost all of this density was due to radiation from the tremendous heat. Only about 0.01 gm/cm$^3$ was true matter in the form of protons, neutrons, electrons, positrons, and neutrinos. No atoms could form under these conditions.

The young universe was dominated by radiation. The matter was so dense it was opaque, and radiation could not move freely. As the universe expanded, the temperature fell and the radiation became weaker, but it remained the

contribute much heat to the universe, and thus the universe is not very hot.

Recent theoretical work suggests that although expansion is part of the solution, the dominant effect is the finite age of the universe. When we look out to large distances, we look back in time and see the universe at an earlier epoch. If we look far enough, we look back to an age before stars began to shine. We can't receive any light from greater look-back times because stars and galaxies did not yet exist. The universe may be infinite in extent, but its finite age means that light from beyond 16 to 20 billion light-years has not yet reached us. Consequently, the night sky is dark and the universe is cold.

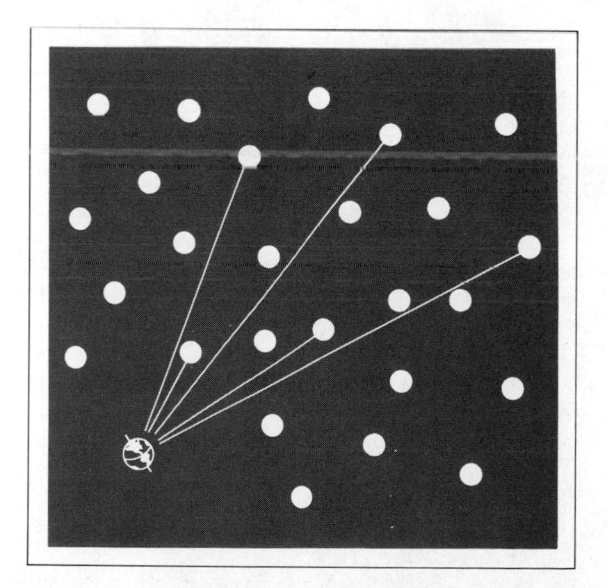

*Figure 11-3. If the universe is infinitely large and infinitely old and uniformly filled with stars, any line from the earth should eventually reach the surface of a star. This predicts that it should not get dark at night, a puzzle commonly referred to as Olbers' paradox.*

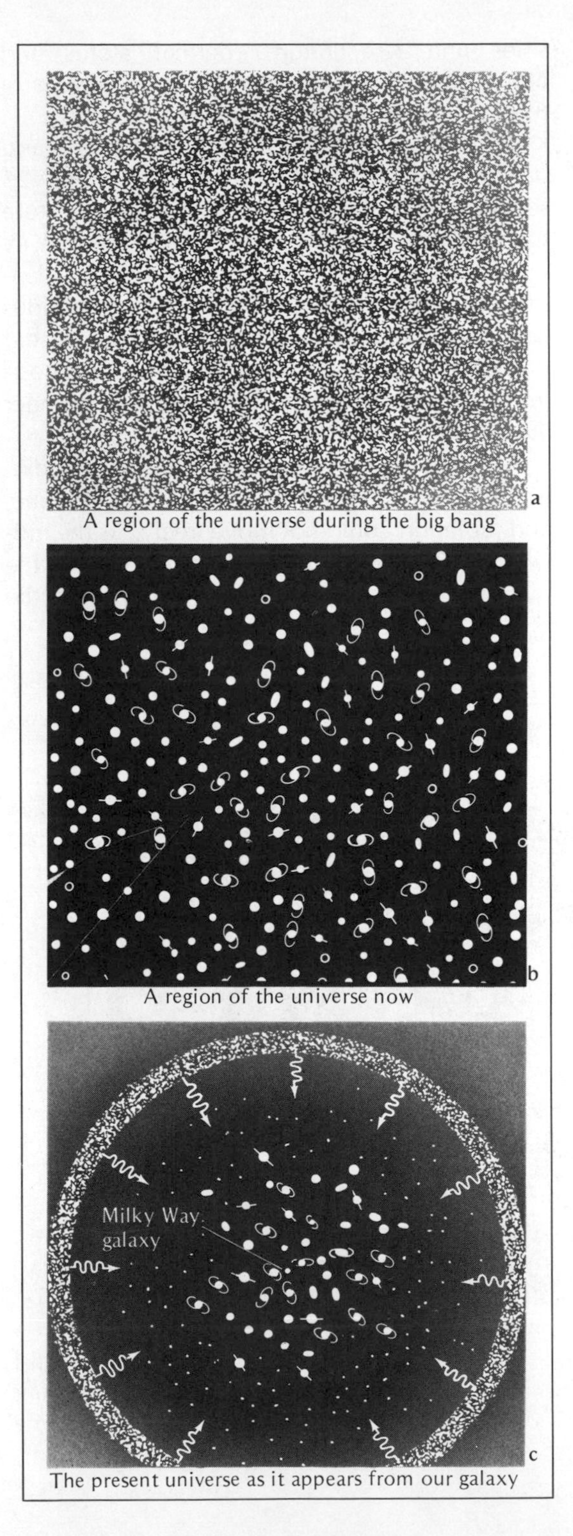


*Figure 11-4. Three views of a small region of the universe centered on our galaxy. (a) During the big bang explosion the region is filled with hot gas and radiation. (b) Later the gas forms galaxies, but we can't see the universe this way because the look-back time distorts what we see. (c) Near us we see galaxies, but farther away we see young galaxies (dots) and at great distance we see radiation (arrows) coming from the hot clouds of the big bang explosion.*

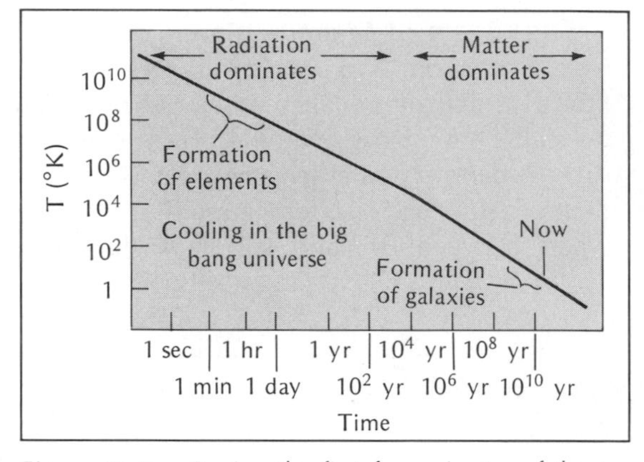


*Figure 11-5. During the first few minutes of the big bang, some matter became deuterium and helium and a few heavier atoms. Later, when radiation was no longer dominant, galaxies formed, and nuclear reactions inside stars made the rest of the chemical elements.*

dominant feature of the universe for the first million years or so (Figure 11-5).

About two minutes after the beginning, the expansion had cooled the universe to about $10^{9}$°K and radiation became weak enough to permit the formation of atomic nuclei. Neutrons and protons joined to form deuterium nuclei, and most of the deuterium captured more neutrons to become helium in a process similar to the proton-proton chain in hydrogen-burning stars. A few nuclei heavier than helium formed, but they were rare because no stable nuclei exist with masses of five or eight hydrogen masses. Such nuclei break up almost as fast as they form, so while the big bang made plenty of helium, it could not span the gaps at five and eight mass units to make large numbers of heavier nuclei. (We saw in Chapters 7 and 8 that heavy atoms are made in stars and supernovae.)

A few hours after the big bang, the temperature had fallen to about $10^7$°K and the building of atomic nuclei ground to a halt. About 20 percent of the mass had become helium, and only a few heavier elements had formed. A small amount of deuterium was left over, and the rest was hydrogen. These are the abundances the galaxies had when they first formed.

The young universe was composed of ionized material because the high temperature prevented the nuclei from capturing electrons. Not until the expanding universe reached an age of $10^4$ to $10^6$ years did the temperature fall low enough for nuclei to capture electrons and form neutral atoms. This neutralization occurred at a temperature of about 4000°K and marks the end of the reign of radiation. Once the matter became neutral, it became relatively transparent and the radiation could move independently, releasing the matter to continue its expansion under the impetus it acquired during the dominance of radiation.

Cosmology cannot deal precisely with the formation of galaxies and galaxy clusters because we don't know why the expanding matter coagulated into blobs. One suggestion is that the original expansion was not perfectly uniform. Regions of higher density became gravitationally unstable and collapsed to form clusters of galaxies. In any case, galaxies did not begin to form until the universe was roughly a billion years old.

*The Oscillating Universe.* The big bang theory, although it explains the origin of the elements and the recession of the galaxies, leaves us with an obvious problem: What came before the big bang? Most cosmologists refuse to speculate on the origin of the cosmic explosion and deal only with its evolution. It is probably meaningless to ask what happened before the beginning.

Another difficulty concerns the eventual fate of the universe. Will it go on expanding forever, with stars burning out, galaxies exhausting their star-forming gas and becoming cold, dark systems, expanding forever through an endless, dark universe? Certainly there is no scientific reason why the universe could not end this way, however unattractive it may sound.

These questions lead some astronomers to consider a modified version of the big bang theory. This **oscillating universe theory** supposes that the universe begins with a big bang, expands outward, making atoms and galaxies, and then, drawn by gravity, falls back, smashing everything together to a high density state from which a new universe may emerge in a new big bang (Figure 11-7). If this is the case, the big bang we see is not the true beginning of the universe, but only a stage of high density through which the universe passes periodically as it expands and contracts.

The reason the expanding universe might slow and eventually fall back is the gravitational attraction of every particle of matter for every other particle. Of course, gravity is most effective at slowing the expansion when the universe is young and the fragments of the big bang are close together. As the matter expands, gravity's braking effect grows weaker, but gravity is a long range force and even across billions of light-years it can slow the outward rush.

Whether gravity can bring the expansion to a halt depends on the amount of gravity in the universe, which in turn depends on the density of matter. Thus to decide whether our universe will expand forever or will eventually fall back, we must try to measure the average density of the universe. If it exceeds the critical density of $4 \times 10^{-30}$ gm/cm$^3$, then gravity will be strong enough to slow the universe to a stop and make it fall back. In that case the universe is said to be **closed.** If the average density is less than $4 \times 10^{-30}$ gm/cm$^3$, the universe will expand forever and it is said to be **open.** (See Box 11-2 for the origin of the terms *closed* and *open*.)

To measure the amount of matter in the universe, we can count the galaxies we see in a given volume of space, multiply by the average mass of a galaxy, and divide by the volume. In this method it is important to add in the mass equivalent to the energy present in the universe, since mass and energy are different forms of the same thing. Because of the uncertainties in the average mass of galaxies and other factors, this method is not too precise. Nevertheless, it yields a density of about $5 \times 10^{-32}$ gm/cm$^3$, about 1 percent of the mass needed to make the universe fall back.

Perhaps, though, the universe contains large amounts of undetected mass. In Chapter 9 we saw that some large clusters of galaxies do not appear to contain enough mass to hold themselves together (Figure 11-8). If we assume that this missing mass is present as hot gas in the cluster or as massive halos around the galaxies, then the average density of the universe may be as high as 20 percent of the critical density. About a decade ago, the first orbiting x-ray telescopes detected a uniform background of x-rays

## BOX 11-2 THE CURVATURE OF SPACE

Most modern cosmological theories are based on Einstein's general theory of relativity. To get a valid picture of modern cosmology we must examine general relativity and its main feature, curved space-time. The concept of space-time may seem esoteric, but it is a common part of our lives. If you make an appointment to meet someone at the library at 6:00 PM, you are specifying a point in space-time—a location and a time. Thus we can specify the path of an object through space-time by describing its successive locations in space at successive moments of time.

According to the general theory of relativity, the presence of a mass curves space-time in the region near the mass. The word *curvature* can mean distortion. In curved space-time the relation between space and time is distorted so that what appears to be a straight line may in fact be curved as seen by an observer far from the region of curved space-time.

This curvature of space-time is important in cosmology because it can affect the behavior of the expanding universe through gravity. When a particle finds itself in a curved region of space-time, it experiences a force that we call gravity. Thus we feel gravity on the earth's surface because the earth's mass curves space-time in its neighborhood. All of the effects we attribute to gravitational forces are actually the effects of curved space-time acting on the motions of particles. In fact, the combined effects of all of the masses in the universe can curve space-time on a large scale and affect the overall expansion of the universe.

To discuss space curvature, we can use a two-dimensional analogy of our three-dimensional universe. Suppose a civilization of bacteria was confined to a two-dimensional surface. This surface could be flat (zero curvature), spherical (positive curvature), or saddle-shaped (negative curvature) (Figure 11-6). Since the bacteria could not leave the surface, they might be unaware of the true curvature of their universe. One way they could detect the curvature of their universe would be to draw circles and measure their areas. On the flat surface a circle would always have an area of exactly $\pi r^2$, but on the positively curved surface its area would be less than $\pi r^2$, and on the negatively curved surface, more. Drawing small circles would not suffice because the area of small circles would not differ noticeably from $\pi r^2$, but if the bacteria could draw big enough circles, they could actually measure the curvature of their two-dimensional universe.

We are three-dimensional creatures, but our universe might still be curved, and we could measure that curvature by measuring the volume of spheres. If space-time is flat, then no matter how big a sphere we measure, its volume will always be $^4/_3\pi r^3$. But in positively curved space-time, spheres would have less volume than this; in negatively curved space-time, they would have more. The spheres must be many megaparsecs in radius for the results to be significant.

We could measure the volume of such spheres by counting the number of galaxies within a certain distance **r** from the earth. If

that were thought to be coming from a hot, thin gas filling intergalactic space. If this were true, the gas might exert sufficient gravitational force to slow the expansion and make the universe fall back. However, recent observations with the HEAO satellites suggest that much of this x-ray background comes from discrete objects, perhaps distant galaxies or QSOs. Thus the existence of the intergalactic gas is in question, and we cannot tell whether the universe is open or not.

Another possibility involves the elusive neutrinos. Recently physicists have speculated that neutrinos, rather than being massless, may have a very small mass. If this speculation proves to be true, then the universe is probably closed because there are about 100,000,000 neutrinos in the universe for every atom. Even if the neutrinos have a tiny mass, they are so numerous their total mass could close the universe.

However, the presence of deuterium in the earth's oceans and in interstellar space suggests

galaxies are homogeneously scattered through space, then the number within distance **r** should be proportional to the volume of the sphere. If, as we count to greater and greater distances, we find the number of galaxies increasing proportional to $\frac{4}{3}\pi r^3$, space-time is flat. However, if we find an excess of distant galaxies, space-time is negatively curved, and if we find a deficiency, space-time is positively curved (Figure 11-6).

Recall that this space curvature is caused by the presence of mass. Each particle of mass causes some curvature, and the sum of all space curvature in the universe determines how the universe will move. Thus the ultimate fate of the universe depends on its density. If the average density is greater than $4 \times 10^{-30}$ gm/cm³, space-time is positively curved, meaning that there is enough gravity to slow the expansion and make the universe fall back. Because it is finite and can only expand so far, such a universe is termed closed. If the average density is less than the critical value, space-time is negatively curved, gravity is weak, and the universe will expand forever without limit. Such a universe is termed open. If the mass is equal to the critical density, space-time is flat, and the universe will just slow to a stop after an infinite time (dotted line in Figure 11-7).

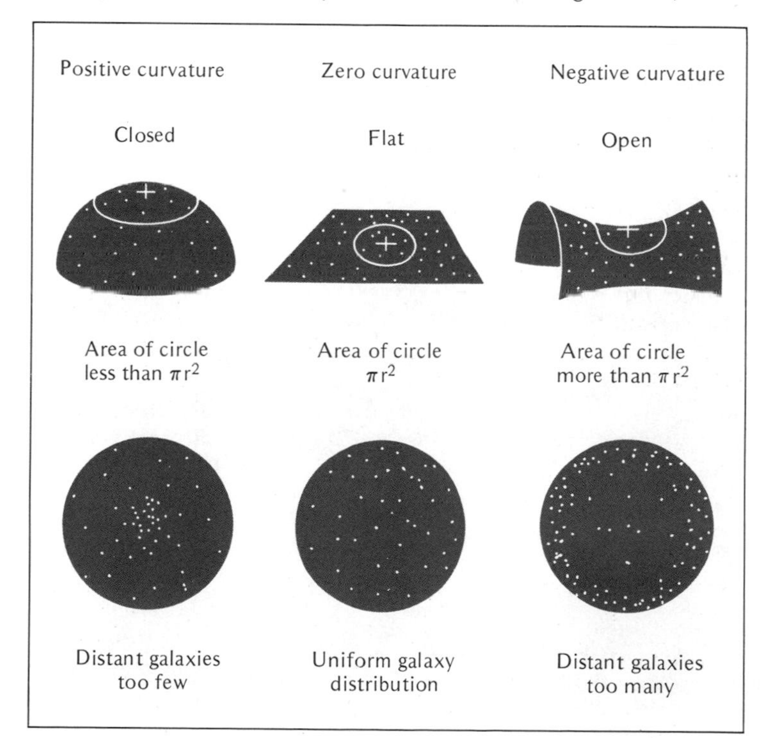


*Figure 11-6. In two-dimensional space, curvature distorts the area of a circle; in three-dimensional space, it distorts the volume of a sphere. We can measure distortion and detect curvature by counting galaxies.*

*Figure 11-7. The expansion of the universe as a function of time.* **R** *is some measure of the size of the universe. Open universe models expand without end (shaded region), but closed models pass through repeated big bangs (curved solid line). Dotted line marks the dividing line between open and closed models. Note that the estimated age of the universe depends on the rate at which the expansion is slowing down.*

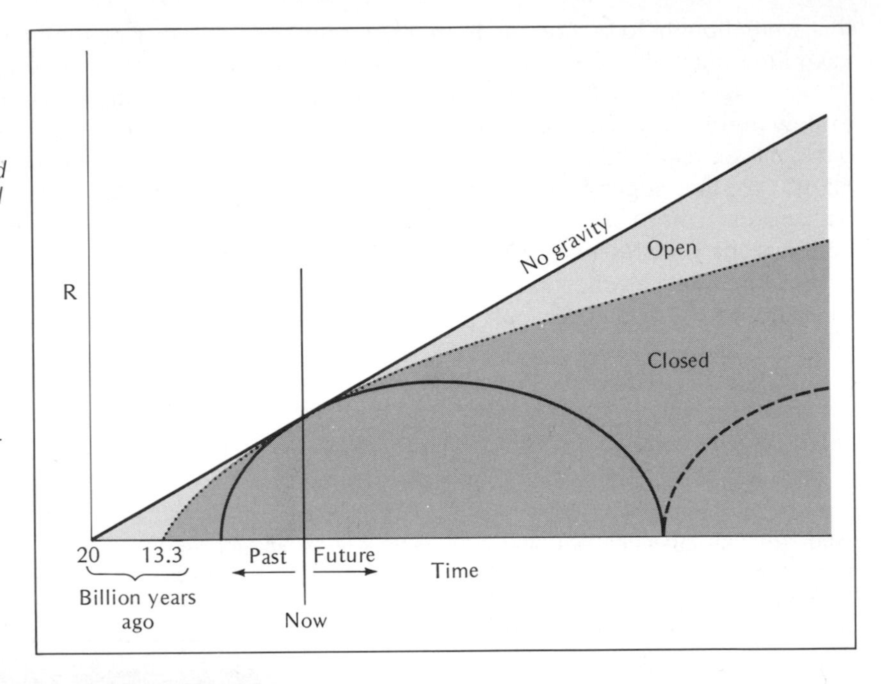

*Figure 11-8. Some clusters of galaxies do not appear to contain enough matter to hold themselves together. If they contain hidden mass, then its gravitation can affect the expansion of the universe. (Hale Observatories.)*

that the universe is not closed. Deuterium is an isotope of hydrogen in which the nucleus contains a proton and a neutron. One out of every few million hydrogen atoms is actually deuterium. This element is easily converted into helium so little or no deuterium can be produced in stars. In fact, stars destroy what deuterium they have to start with. Thus the deuterium in the earth's oceans must have been made in the first few minutes of the big bang explosion. The amount that was produced depends critically on the nature of the explosion, and places a lower limit on the rate of expansion of the universe. The abundance of deuterium in interstellar space suggests that the present density of the universe can't be more than about 10 percent of the critical density.

The observations are not yet precise enough to tell us whether the universe is open or closed. If it is closed, then some time in the future, roughly 100 billion years from now, all matter and energy will come crashing back together in an event sometimes called the "big crunch." Certainly all traces of our universe would be destroyed as atoms are smashed to fragments and compressed to high densities. Of course, this also means that we could never locate traces of previous universes.

If the universe is closed, it will collapse and no known force could prevent it from falling into a black hole. In fact, if the universe is closed, then it *is* a black hole, and we are living during a temporary expansion that must be followed by a collapse. Whether that collapse could produce a new expansion and a new universe is a total mystery.

*The Age of the Universe.* If we assume the universe began in a big bang explosion, then the present rate of expansion gives us a hint to the age of the universe. We merely ask how long the galaxies, moving at their present velocities, would have taken to get to their present separations. If the Hubble constant is 50 km/sec/Mpc, then the age of the universe is about $20 \times 10^9$ years (Box 11-3).

This is how long the galaxies would have taken to reach their present separations if there were no gravity slowing the expansion (solid line in Figure 11-7). However, the mutual gravitation of all of the galaxies slows the expansion, so the galaxies must have had higher velocities in the past. If they were moving more rapidly when the universe was young, then they would not have taken as long to reach their present separation, and $20 \times 10^9$ years overestimates the age of the universe. To find the true age, we must know the extent to which gravity has slowed the expansion, and that depends on the average density of the universe. As we have seen in the previous section, the average density is highly uncertain.

If the average density of the universe is equal to the critical density, the density needed to close the universe, then the universe must be two-thirds as old as the estimate above—about $13.3 \times 10^9$ years. If the density is less than that, the universe is open, and its age is between $13.3 \times 10^9$ and $20 \times 10^9$ years. If the density is more than the critical value, then the universe is younger than $13.3 \times 10^9$ years. The best current data suggest that the universe is open and has an age of 16 to $20 \times 10^9$ years.

Recent studies by different groups of astronomers have yielded different values for the Hubble constant. One group arrived at a value of about 50 km/sec/Mpc, but two other groups suggest **H** is about 100 km/sec/Mpc. If this larger value is correct, then the universe cannot be older than about $10 \times 10^9$ years, and if the universe is closed, it cannot be older than two-thirds of this, or about $6.7 \times 10^9$ years. This finding creates a difficulty because astronomers believe our galaxy is at least $10 \times 10^9$ years old and may be even older. The new value of **H** does not seem to be consistent with what we know about stellar evolution and the age of our galaxy, but it is not yet possible to decide where the error lies.

In this book, we will use a value of about 50 km/sec/Mpc because it yields an age for the universe consistent with most astronomical knowledge. However, keep the uncertainties in mind. The next ten years will see some exciting debate centered on the value of **H** and the age of the universe.

## THE STEADY STATE THEORY

Of all the cosmological theories, one of the most popular was the steady state theory. Its contention that the universe is eternal and unchanging held a powerful appeal for those who found big bangs unpalatable. Unfortunately, observations did not support the theory and it was generally abandoned in the early 1970s. The idea is worthy of attention, though, because of the important role it played in modern astronomy and because its downfall illustrates the fate of an unsuccessful theory.

*The Perfect Cosmological Principle.* Both the big bang theory and the oscillating universe theory propose that the universe changes. According to them the universe was much different only 10 to 20 billion years ago.

Because they found a changing universe philosophically unattractive, the English astronomers Bondi, Gold and Hoyle reexamined the basic assumption of cosmology, and in 1948 they suggested that the cosmological principle was not complete. The principle stated that the universe looked the same from any location in space, but it neglected time. They modified the cosmological principle by giving time an equal footing with space. The result is called the **perfect cosmological principle**—the universe looks about the same from every location in space at any time. As in the case of the cosmological principle, the perfect cosmological principle deals only with the general properties of the universe. Thus individual stars and galaxies may change and die, but the general appearance of the universe does not change. It is in a steady state.

The distinguishing characteristic of the steady state theory is that the universe does not evolve, but it is almost impossible to deny the fundamental observation of cosmology—red shifts in the spectra of galaxies—and the common interpretation of that observation—that the universe is expanding. The combination of the perfect cosmological principle and the expansion of the universe leads to a startling conclusion.

*Continuous Creation.* If the universe is expanding, then galaxies are receding from each other and the average density of the universe should be decreasing. However, according to the perfect cosmological principle, the general properties of the universe, including its density, must remain constant. Thus the steady state theory asks us to believe that matter is created continuously out of nothing to keep the average density of the universe constant. According to the theory, the newly created matter, probably in the form of hydrogen atoms, ac-

---

BOX 11-3 THE HUBBLE CONSTANT AND THE AGE OF THE UNIVERSE

The Hubble constant **H** is a measure of the rate of expansion of the universe. It contains enough information to estimate the age of the universe.

To discover how long ago the universe began expanding, divide the distance **D** to a galaxy by the velocity $\mathbf{V_r}$ with which that galaxy recedes. The result is the time the galaxy took to travel the distance, assuming it has maintained constant velocity.

Since **D** is measured in megaparsecs and $\mathbf{V_r}$ in kilometers per second, we must convert **D** into kilometers by multiplying by $3.085 \times 10^{19}$, the number of kilometers in one megaparsec. Then dividing by $\mathbf{V_r}$ yields the age of the universe in seconds. To convert to years we must divide by $3.15 \times 10^7$, the number of seconds in a year. Thus the age of the universe in years is

$$T = \frac{D}{V_r} \times \frac{3.085 \times 10^{19}}{3.15 \times 10^7}$$

or approximately

$$T \approx \frac{D}{V_r} \times 10^{12} \text{ years}$$

However, we don't have to measure **D** and $\mathbf{V_r}$ if we know the Hubble constant. To see why, recall that **H** is a galaxy's radial velocity divided by its distance.

$$H = \frac{V_r}{D}$$

So we can simplify our formula for the age of the universe.

$$T = \frac{1}{H} \times 10^{12} \text{ years}$$

If **H** is 50 km/sec/Mpc, the universe is about $20 \times 10^9$ years old, assuming there is no gravity to slow the expansion. To find the true age of the universe, we must know the extent to which gravity has slowed the expansion, and that depends on the average density of the universe.

---

cumulates in vast clouds between the galaxies. As the galaxies recede from each other, grow old, and die, new galaxies form from these intergalactic clouds.

Although the continuous creation of matter violates the conservation of mass and energy, many theorists found it less objectionable than a big bang explosion for two reasons. First, the rate of creation is less than one hydrogen atom per cubic meter every 11 billion years. In a volume the size of an average classroom, we would have to wait 65 million years for the creation of a single hydrogen atom. Thus the rate of creation is so low that we could never detect any violation of the conservation laws that lie at the heart of modern physics.

In addition, rival theories also contain hidden problems with creation. For instance, big bang cosmologists usually refuse to speculate on the origin of the explosion. For some astronomers it was easier to conceive of creation spread over all time as a continuous process than as a single event that occurred at a specific instant.

However, the steady state theory is no longer accepted as a reasonable description of the universe. To see why it fell out of favor, we must compare the various models of the universe with observations and try to choose the theory that fits best.

## COSMOLOGICAL TESTS

A **cosmological test** is a measurement or observation whose result can help us choose between rival cosmological theories. Although there are many different theories, we will limit our discussion to the major three—the open big bang universe, the closed big bang universe, and the steady state universe.

Almost all cosmological tests are searches for evidence that the universe is evolving. The discovery of any evolution in the general properties of the universe would be a severe blow to the steady state theory, and, properly interpreted, such evidence might tell us whether we are in an open or a closed universe.

*The Evolution of Galaxies.* Although galaxies are visible at optical wevelengths out to distances of about 8,000,000,000 ly, astronomers must concentrate on galaxies within about 2,000,000,000 ly to see any detail (Figure 11-9). Thus present optical telescopes can look back no more than about 2 billion years, and what they show us implies that galaxies have not changed drastically in that short interval. To see the effects of evolution in the universe, we must look much farther.

If the QSO red shifts arise from the expansion of the universe, then the QSOs are much more distant than normal galaxies, and the farthest QSO is about 15,500,000,000 ly away. This implies that QSOs represent some early state in the history of the universe, perhaps the beginning of galaxy formation. Unfortunately, there are no accurate distance indicators for the QSOs, so their distance cannot be conclusively established and their nature remains uncertain. Consequently QSOs are not conclusive evidence that the universe has evolved.

Radio telescopes can probe much farther than optical telescopes, mainly because radio galaxies are much more luminous at radio wavelengths than at optical wavelengths. Consequently, radio astronomers can perform a cosmological test by counting the number of

*Figure 11-9. A cluster of galaxies slightly more than 4,000,000,000 ly away shows little detail. (Hale Observatories.)*

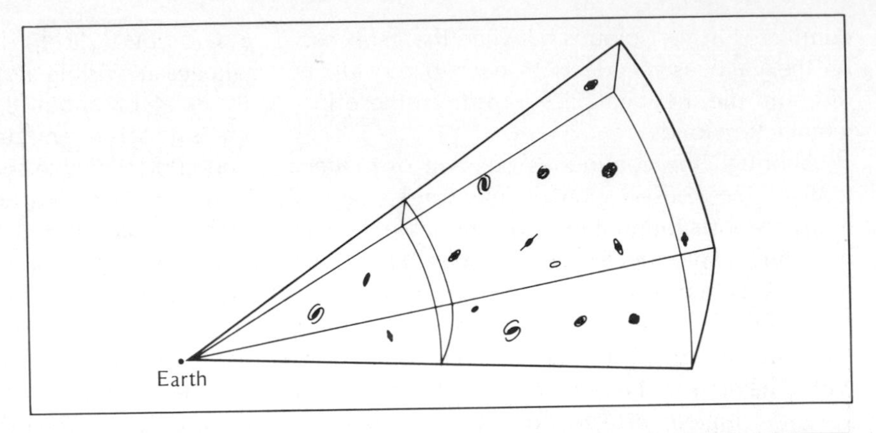


*Figure 11-10. The volume we sample increases as we look farther away. The number of galaxies we count in a flat universe should be proportional to the distance cubed.*

radio galaxies at various distances. If distant regions of space contain more radio galaxies than nearby regions, it would show that the universe has changed.

Unfortunately, this process requires that we know the distance of the radio galaxies, a quantity that radio astronomers can only estimate. However, if we assume that all radio galaxies radiate the same power, we can assume that the weaker signals come from those more distant.

If the universe is in a steady state, then the number of radio galaxies per unit volume must remain the same, and as we look out into space to large look-back times, we should see the same density of radio galaxies. However, the volume in which we count increases as the cube of the distance (Figure 11-10), so we should see more faint radio galaxies even if the density is constant. The critical question is, Do we see more than we should?

In 1968, the English radio astronomer Sir Martin Ryle published the results of extensive counts of radio sources, most of them presumably radio galaxies. He found that the number of sources increased with distance as expected, but that there were too many faint sources (Figure 11-11). Apparently there is an excess of distant radio galaxies. This implies that radio galaxies were more common in the distant past, or that they were more powerful and thus easier to detect. In either case, the universe was different in the past, a condition the steady state theory cannot explain.

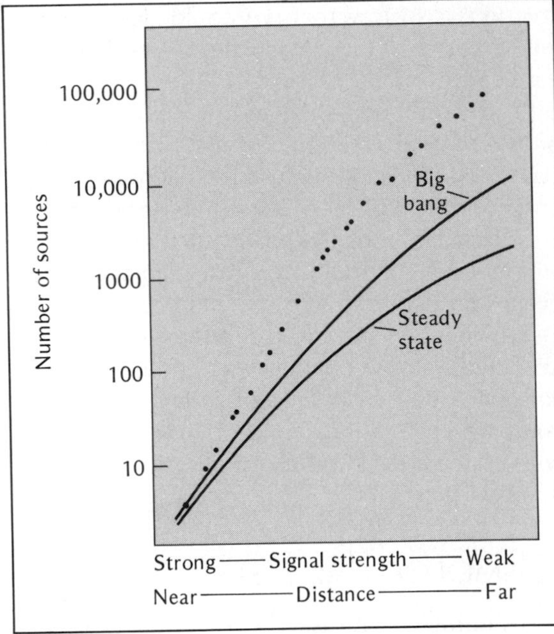


*Figure 11-11. The excess of faint radio sources shows that the universe has changed and is not in a steady state. Observations do not agree with any current model, suggesting that the evolution of radio galaxies is distorting the counts.*

The excess of faint radio sources does not agree with the predictions of either the open or the closed big bang universe. This does not mean that these two theories are wrong, only that we don't know enough about how radio galaxies have evolved.

The excess of faint radio sources was a serious blow to the steady state theory since it showed that the universe had changed. But the steady state theory had received another shock shortly before when scientists discovered radiation coming from the big bang fireball.

*The Primordial Background Radiation.* As we concluded previously, the big bang theory predicts the existence of radiation from the original explosion. If we look far enough into space, we should eventually see this radiation coming from the explosion at look-back times of about 16 to 20 billion years. This radiation should come at us from all sides, and owing to the large red shift, it should lie in the radio and infrared parts of the spectrum (Figure 11-4).

In 1948, physicist George Gamow completed a theoretical analysis of the conditions during the early stages of the big bang explosion. His calculations showed that the universe would have been terrifically hot and dominated by radiation. A year later, the physicists Alpher and Herman pointed out that the very large red shift of the big bang fireball would make it seem cool; they predicted a temperature of about 5°K. At that time there was no way to detect the radiation, but in 1965, a team led by Princeton physicist Robert Dicke refined the calculations, concluded that the radiation should be detectable, and began developing the sensitive microwave detector required.

At about the same time two physicists, Arno Penzias and Robert Wilson, were working on a microwave receiver at Bell Labs in New Jersey (Figure 11-12) only a few miles from Princeton. They found a peculiar source of noise in their equipment, and after months of testing they concluded the noise did not originate in their equipment but came isotropically from the entire sky. Once they heard of Dicke's work, Penzias and Wilson quickly identified the noise as radiation from the hot clouds of the big bang explosion, the so-called **primordial background radiation.** Penzias and Wilson won the 1978 Nobel Prize for Physics for their discovery.

*Figure 11-12. Arno Penzias (right) and Robert Wilson pause before the horn antenna with which they discovered the primordial background radiation. (Bell Laboratories.)*

Since its discovery, the primordial background radiation has been measured at many wavelengths (Figure 11-13). Ground-based radio telescopes have measured its strength in the radio and microwave parts of the spectrum, and rocket and balloon-borne instruments have measured it in the far infrared. These tests confirm that it is black body radiation with an apparent temperature of about 2.7°K.

Although the radiation is almost perfectly isotropic, recent observations show a small departure from complete isotropy. The primordial background radiation seems to have slightly shorter wavelengths—it seems hotter—in the direction of the constellation Leo. In the opposite direction, the radiation seems slightly cooler. This difference is evidently due to the motion of our galaxy through space. The Milky Way galaxy is moving about 540 km/sec in the direction of Leo, causing a slight blue shift in the background radiation that makes it appear slightly hotter. Conversely, radiation from the opposite side of the universe is slightly red-shifted and looks cooler.

The importance of the primordial background radiation lies in its interpretation as radiation from the big bang fireball. Many as-

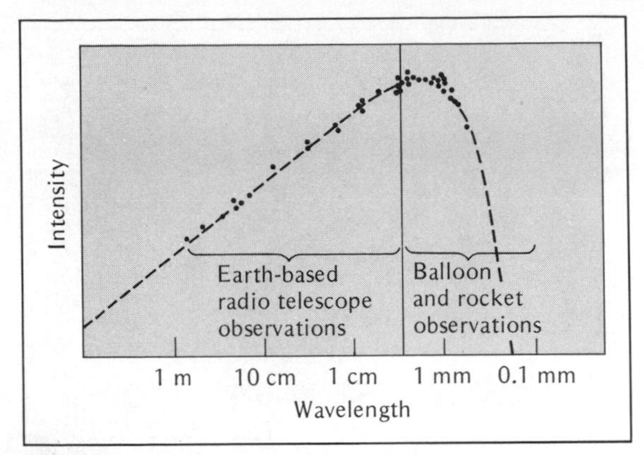


*Figure 11-13. The primordial background radiation is black body radiation with a temperature of about 2.7°K. It lies in the infrared and radio part of the spectrum.*

tronomers cite it as evidence that there was a violent explosion long ago. If this is true, the steady state theory must be abandoned. Unfortunately, we cannot say whether the big bang was a single event or merely one in a series.

The excess of faint sources and the primordial background radiation show that the universe has changed. Faced with this evidence, the steady state theory collapsed. Most cosmologists now assume that the universe had an explosive beginning and is evolving.

*The Evolution of the Hubble Constant.* All three of our cosmological theories predict that the universe is expanding, but they differ in how its expansion changes with time. The steady state theory says it must expand at a constant rate. However, the open big bang theory says gravity is slowing the expansion slightly, and the closed big bang theory says gravity is slowing the expansion rapidly. If we could measure a change in the rate of expansion, then we might be able to choose the correct theory.

The Hubble constant **H** describes the rate of expansion of the universe, and if the universe is decelerating, **H** must be decreasing. This decrease would be much too slow to detect in hundreds or even thousands of years, but it may be detectable if we extend the velocity-distance diagram to great distance.

Galaxies plotted according to their velocity and distance fall along a line indicating that the more distant galaxies are receding faster than those nearby. The line's slope (steepness) equals **H** (Figure 11-14). Thus the faster the universe expands, the steeper the line will be. For nearby galaxies the line has a slope of about 50 km/sec/Mpc. If we look at distant galaxies, we look back in time to an earlier age when **H** may have been larger. If that is the case, the slope of the line should get steeper as we look farther into space.

If the steady state theory is correct, the general properties of the universe, including its rate of expansion, should never change. Thus the steady state theory predicts that the slope of the

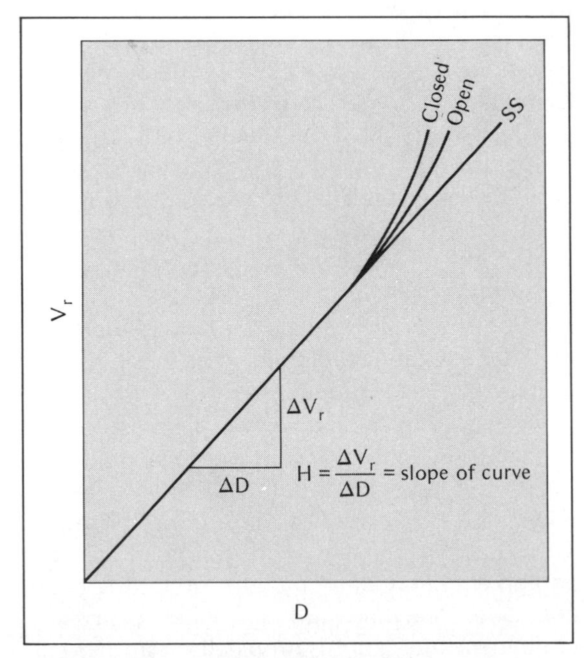


*Figure 11-14. The slope of the relation in the velocity-distance diagram is equal to the Hubble constant **H.** The steady state theory predicts a straight line, but the open and closed big bang universe theories predict higher velocities in the past.*

line should be constant, and the line should be straight. The other two theories predict that gravity is slowing the expansion, so if they are correct the line should curve upward, indicating faster expansion in the past. Because the open universe theory supposes that gravity has slowed the universe only slightly, it predicts a small increase in **H** as we look back in time. But the closed universe theory says gravity is strong and the universe has been slowed down drastically. Thus it predicts a much larger departure from a straight line (Figure 11-14). Clearly we can choose the correct theory if we can plot galaxies out to the point where the lines diverge.

Unfortunately, getting the distances to these galaxies is difficult because they are too far away to have visible distance indicators such as Cepheid variables, luminous supergiants, or globular clusters. Instead, the astronomer must judge the distances by calibrating the properties of the galaxies themselves. Nearby galaxies, whose distances can be found from the indicators mentioned above, provide the luminosities of different kinds of galaxies. By observing the apparent magnitude of similar, more distant galaxies, the astronomer can calculate distances.

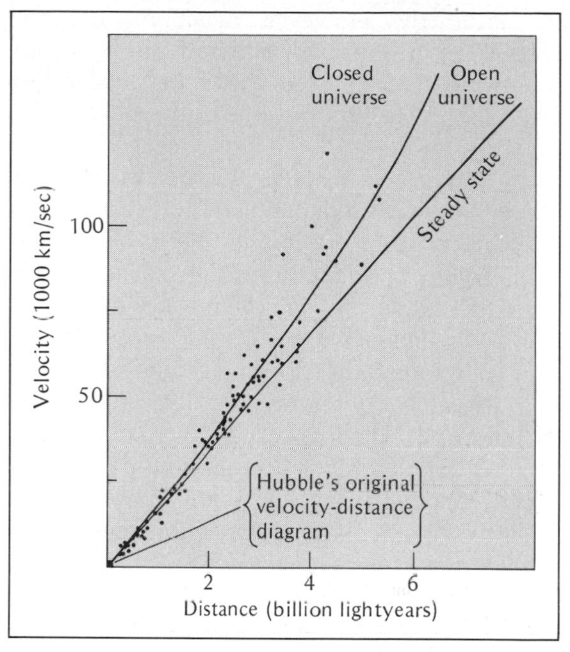


*Figure 11-15. Hubble's original velocity-distance diagram (lower left) has now been extended to great distance. The data suggest that the universe was expanding faster in the past. It is not possible to choose between open and closed models on the basis of these data. (Adapted from data by Kristian, Sandage, and Westphal.)*

These distances are uncertain because we don't know how galaxies evolve. We might suspect that long ago, when galaxies had more gas and dust, they could make more hot, bright stars and were more luminous. To judge the distance to a galaxy on the basis of its luminosity we must correct for this excess of bright stars and for other evolutionary effects. Unfortunately, the exact form of this correction is unknown, and thus the distances are uncertain.

The best results to date have been obtained by Allan Sandage and his coworkers. They have revised the distance calibration, starting with the luminosity of the variable stars used as distance indicators in the nearest galaxies and working their way outward to more distant galaxies. These results, plotted in Figure 11-15, show the galaxies falling above the steady state's straight line, indicating that the universe is slowing down. However, the uncertainties in the data do not permit a choice between the open big bang and the closed big bang theories. In the future, better observations made from space telescopes and a better understanding of the luminosities of young galaxies may reduce the uncertainties and extend the data to greater distances.

The evidence now favors an open universe. The steady state theory can't cope with the observations of the primordial background radiation, the excess of faint radio galaxies, or the increase in **H** at large look-back times. Nor does the universe appear to contain enough mass to make it oscillate.

*Primordial Black Holes.* Ordinarily we think of the big bang fireball as an event that took place so long ago that it can only be studied indirectly. However, radiation from the fireball has been found, and now physicists are suggesting that pieces of the fireball, locked in tiny black holes, may still exist throughout the universe.

In 1974, the theoretical physicist Stephen Hawking discovered that black holes are not totally black. His work implies that a black hole leaks energy and that the smaller a black hole's mass, the larger the leak. If he is correct, black holes with masses approximately equal to the mass of a mountain ($10^{15}$ gm) should be detectable from the gamma rays they leak into space.

Hawking's theory would have little importance if the only black holes were those formed by the deaths of massive stars. As we saw in Chapter 7, such black holes should have masses a few times larger than the sun's. The miniscule amount of energy expected to leak from these massive black holes would be totally undetectable. However, some cosmologists believe that low-mass black holes may have formed during the first instants of chaos following the big bang, and these low-mass black holes might be detectable.

According to the theory, the black hole's mass should decrease as it leaks energy. In a sense, it is evaporating as its mass leaks away as energy. However, the size of the leak grows larger as the mass of the black hole shrinks, so the black hole will explode in a burst of gamma rays as its mass grows very small (Figure 11-16). How long a black hole takes to evaporate depends on its mass. If the big bang did make black holes of various masses, those of about $10^{15}$ gm should be evaporating now and filling space with bursts of gamma rays.

Unfortunately, these gamma-ray bursts may be very difficult to detect. The known laws of physics do not describe the precise process by

*Figure 11-16. A mini black hole leaks energy faster as it grows smaller. As it reaches very small mass, it evaporates explosively emitting a burst of gamma rays.*

which such a mini black hole might evaporate. It may release its energy over a period of 0.1 second, or it may concentrate it in an outburst lasting only 0.0000001 second. This means we can't be sure how intense the bursts might be.

Another complication is earth's atmosphere which absorbs gamma rays. Thus observations must be made from earth orbit or by indirect means. The orbiting Solar Observatories and the second Small Astronomy Satellite (SAS-2) discovered a uniform background glow of gamma rays along the Milky Way. This gamma-ray background apparently comes from the interaction of cosmic rays with the interstellar matter. If our galaxy contains many primordial black holes, their gamma-ray bursts should add to the gamma-ray background in the form of black body radiation with a temperature of about $70 \times 10^{9}$°K. SAS-2 did not detect this radiation, but its detectors were not quite sensitive enough to rule out its existence (Figure 11-17).

An indirect way to look for the gamma-ray bursts is to use a large radio telescope to listen for pulses from the interstellar gas near the explosion. Such searches have not identified any such pulses, but again, the observations are not sensitive enough to rule out the existence of primordial black holes.

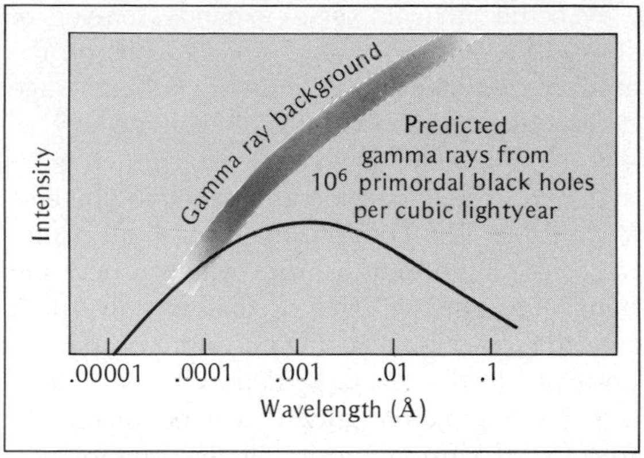

*Figure 11-17. The intensity of the gamma-ray background (shaded) places an upper limit on the number of primordial black holes. There can be no more than $10^6$ per cubic light year (solid curve). (Adapted from a figure by S. Hawking.)*

That all attempts to detect the explosion of primordial black holes have failed does not mean they do not exist. No method thus far is sufficiently sensitive to rule them out, and the physics of the explosion is so uncertain we can't be sure how detectable they actually are. As new detectors are put into orbit and more time is spent listening for radio pulses we may eventually find the primordial black holes and get a new insight into the origin of the universe.

## SUMMARY

The basic assumptions of cosmology are homogeneity, isotropy, and universality. Homogeneity says the matter is spread uniformly through the universe. Isotropy says the universe looks the same in any direction. Both deal only with general features. Universality assumes that the laws of physics known on earth apply everywhere. In addition, the cosmological principle asserts that the universe looks the same from any location.

It is a fact that the spectra of galaxies contain red shifts. It is an assumption that this red shift is related to the Doppler effect. If this is correct, then the universe is expanding.

The Hubble Law implies that the universe is expanding. Tracing this expansion backward in time we come to an initial high-density state commonly called the big bang, the explosion that started the expansion. From the Hubble constant we can conclude that the expansion began 10 to 20 billion years ago.

During the first few minutes of the big bang, about 20 percent of the matter became helium, and the rest remained hydrogen. Very few heavy atoms were made. As the matter expanded, instabilities caused the formation of clusters of galaxies, which are still receding from each other with the impetus they acquired during the explosion.

Whether the universe expands forever or slows to a stop and falls back depends on the amount of matter in the universe. If the average density is greater than the critical density of $4 \times 10^{-30}$ gm/cm$^3$, it will provide enough gravity to slow the expansion and force the universe to collapse. The collapse may smash all matter back to a high density from which a new big bang may emerge. This oscillating universe is termed closed because the curvature of space is positive. If the average density is less than $4 \times 10^{-30}$ gm/cm$^3$, gravity will be unable to stop the expansion and it will continue forever. Such a universe is termed open because the space curvature is negative.

The steady state theory is based on the perfect cosmological principle—the universe looks the same from any location at any time. It predicts that the universe is unchanging. Since the universe is expanding, the steady state theory requires that matter be continuously created to keep the average density constant.

We can choose between rival theories by performing cosmological tests. The observation that the night sky is dark is a significant test, since it shows that the universe cannot be infinite in age and extent. Another cosmological test is the counting of radio galaxies. The excess of faint radio sources shows that radio galaxies were more common or more luminous in the past. In either case, the universe has changed, contrary to the steady state theory.

In another cosmological test, scientists have detected the primordial background radiation. It is isotropic black body radiation with a temperature of about 2.7°K. It is commonly interpreted as evidence of a big bang.

Yet another test shows that the distant galaxies are receding from each other faster than the nearer galaxies. That is, the Hubble constant was larger in the past. This shows that the universe is slowing its expansion and is not in a steady state, but the data are not good enough to distinguish between a closed universe and an open universe.

Measurements of the average density of matter in the universe reveal that there is only about 1 percent of the matter needed to halt the expansion. Though there may be undetected mass in the universe, it may not be sufficient to halt the expansion. Thus present evidence implies we are in an open universe that will expand forever.

If there was a big bang explosion, it may have created low-mass black holes that can explode in gamma-ray bursts. If these primordial black holes are found, they will give us clues to conditions during the first seconds of the big bang explosion.

## NEW TERMS

cosmology
big bang theory
homogeneity
isotropy
universality
cosmological principle
Olbers' paradox
oscillating universe theory
open universe
closed universe
perfect cosmological principle
cosmological test
primordial background radiation
primordial black hole

## QUESTIONS

1. What are the three basic assumptions of cosmology? What do they mean?
2. What is the difference between the cosmological principle and the perfect cosmological principle?
3. Why can't we find a center to the expansion of the universe?
4. Explain how the big bang explosion could have made helium. Why couldn't it make much carbon?
5. What determines whether the universe is open or closed?
6. Why did the steady state theory require the continuous creation of matter?
7. Why was the detection of evolution in the universe a blow to the steady state theory? Give examples.
8. Why is the primordial background radiation so cool?
9. Why does it get dark at night?
10. How does our ignorance of the evolution of radio and normal galaxies affect cosmological tests?
11. How might we detect primordial black holes?

## PROBLEMS

1. Use the data in Figure 11-1 to plot a velocity-distance diagram, find **H**, and determine the approximate age of the universe.
2. If a galaxy 8 Mpc away from us recedes at 456 km/sec, how old is the universe?
3. If the value of the Hubble constant were found to be 60 km/sec/Mpc, how old would the universe be?
4. Use a diagram to explain how counting galaxies could determine the curvature of space.
5. Explain why we can see the big bang explosion in any direction we look. (Hint: Use Figure 11-4.)
6. Explain why a deceleration in the expansion of the universe means **H** must increase with distance.
7. Why can't we detect primordial black holes of $10^{14}$ gm? Of $10^{16}$ gm?

## RECOMMENDED READING


Burnstein, J. "Physics and the Cosmos." *Natural History* 86 (Oct. 1977), p. 106.

Goldsmith, D. "The Cosmic Bomb." *Astronomy* 2 (April 1975), p. 6.

Gott, J. R., Gunn, J. E., Schramm, D. N., and Tinsley, B. M. "Will the Universe Expand Forever?" *Scientific American* 234 (March 1976), p. 62.

Green, L. C. "Cosmology Today." *Sky and Telescope* 54 (Sept. 1977), p. 180.

Harrison, E. R. "Why the Sky is Dark at Night." *Physics Today* 27 (Feb. 1974), p. 30.

Hartline, B. K. "Double Hubble, Age in Trouble." *Science* 207 (Jan. 11, 1980), p. 167.

Hawking, S. W. "The Quantum Mechanics of Black Holes." *Scientific American* 236 (Jan. 1977), p. 34.

Irwin, J. B. "Radio Astronomy and Cosmology." *Sky and Telescope* 52 (Dec. 1976), p. 425.

Islam, J. N. "The Ultimate Fate of the Universe." *Sky and Telescope* 57 (Jan. 1979), p. 13.

Krogdahl, W. S. "The Creation of the Universe." *Sky and Telescope* 45 (March 1973), p. 140.

Lang, K. R., and Mumford, G. S. "A New Look at the Hubble Diagram." *Sky and Telescope* 51 (Feb. 1976), p. 83.

Narlikar, J. *The Structure of the Universe*. London: Oxford University Press, 1977.

Parker, B. "The First Second of Time." *Astronomy* 7 (Aug. 1979), p. 6.

Porter, N. A., and Weekes, T. C. "The Search for Exploding Black Holes." *Sky and Telescope* 55 (Feb. 1978), p. 113.

Overbye, D. "Out From Under the Cosmic Censor: Stephen Hawking's Black Holes." *Sky and Telescope* 54 (Aug. 1977), p. 84.

Tinsley, B. M. "The Cosmological Constant and Cosmological Change." *Physics Today* 30 (June 1977), p. 32.

Trimble, V. "Cosmology: Man's Place in the Universe." *American Scientist* 65 (Jan./Feb. 1977), p. 76.

Schramm, D. N. "The Age of the Elements." *Scientific American* 230 (Jan. 1974), p. 69.

Schwartzenburg, D. "Does Cosmology Have a Future?" *Astronomy* 7 (July 1979), p. 35.

Silk, J. *The Big Bang*. San Francisco: W. H. Freeman, 1980.

Webster, W. "The Cosmic Background Radiation." *Scientific American* 231 (Aug. 1974), p. 26.

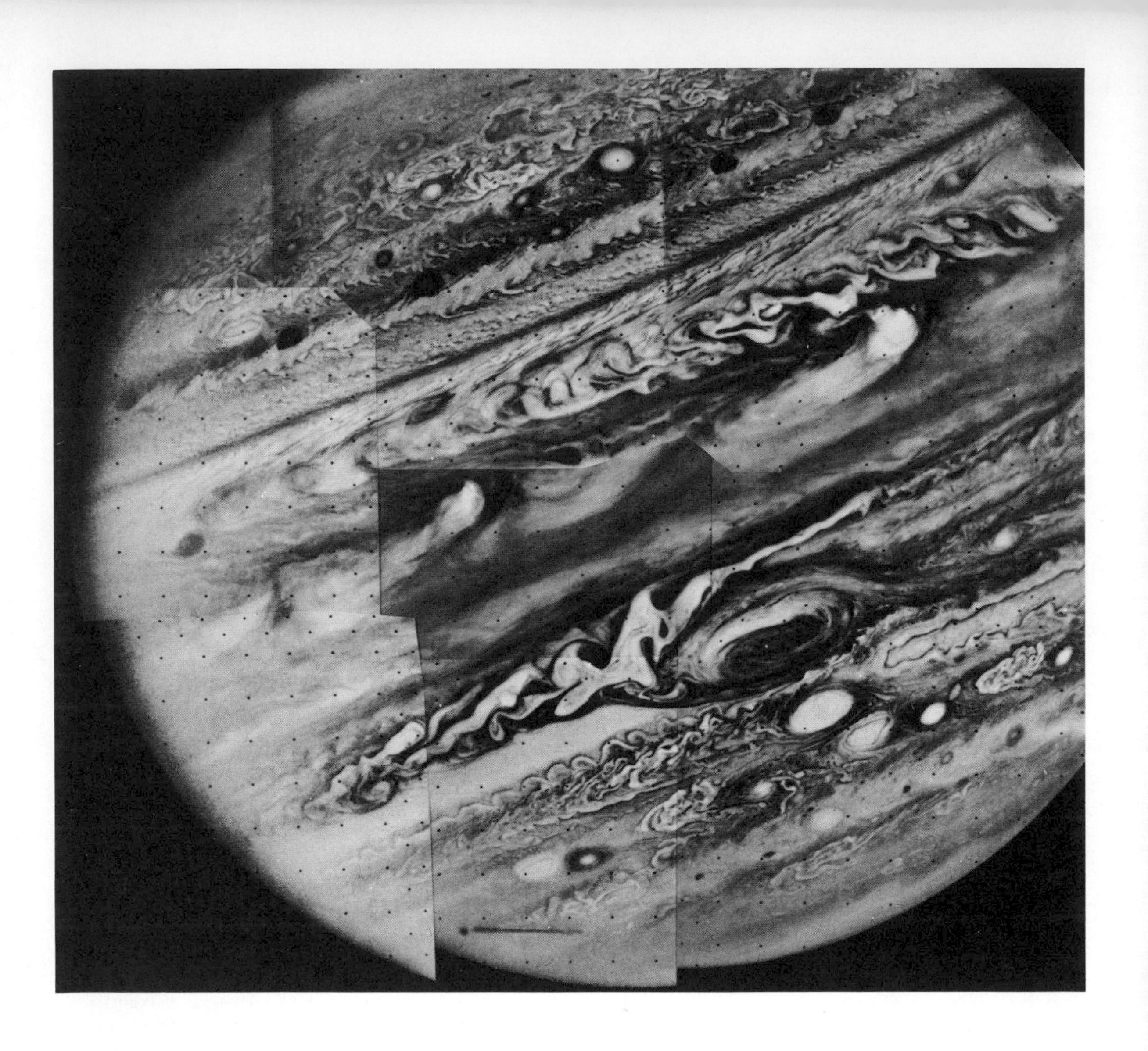

*Jupiter's turbulent cloud belts and its Great Red Spot photographed by Voyager 1. The planet is 11.2 times the diameter of earth. (NASA.)*

# Chapter 12 THE ORIGIN OF THE SOLAR SYSTEM

Our solar system, consisting of our sun, nine planets, and some scattered debris, occupies no more than $10^{-30}$ of the volume of the observable universe. If the fraction of this book devoted to the solar system were proportional to the volume of the universe the solar system takes up, this chapter would contain less than a trillionth of a trillionth of one letter of the alphabet. However, the significance of the solar system—for us at least—far exceeds its relative size.

An obvious reason for studying our solar system is that it is our immediate neighborhood in the universe, and as a matter of curiosity and self-preservation we should be familiar with our surroundings. By exploring the planets around the sun, we may pave the way for their future colonization and utilization. However, before humans can make practical use of the planets, we must survive the next century or so on earth. Understanding how planets form and evolve teaches us more about our own planet. The barren wastes of Mars and the sulfuric acid fogs of Venus may teach us how to preserve our more comfortable climate.

Astronomers need no justification for studying the solar system, for it is the only collection of nonluminous bodies in the universe that we can study in detail. If there are planets orbiting other stars, they are too far away and reflect too little light to be visible. Yet there may be more planets in the universe than stars, so, if only to complete our study of celestial objects, we must examine the solar system.

An intriguing reason for studying the planets of our solar system is the search for life beyond the earth. As if enchanted, some matter on the earth's surface lives and is aware of its existence. To search for other living beings we must examine the surfaces of planets—stars are too hot and space is too cold for the evolution of life forms. Life seems to require the moderate conditions found only on the surfaces of some planets.

Our search for life on other planets follows a sequence of ideas through the following chapters. This chapter describes how the present characteristics of the solar system originated in a cloud of gas—the so-called solar nebula. The next chapter explains how the planets grew from solid condensations in the solar nebula, and Chapters 14 and 15 describe the nature of the planets of our solar system. Finally, having considered the formation of planets and the events that shape their surfaces and atmospheres, we will discuss the origin of life on other worlds in Chapter 16.

## A SURVEY OF THE SOLAR SYSTEM

The solar system is almost entirely empty space (Figure 12-1). Imagine that we reduce the solar system until the earth is the size of a grain of table salt, about 0.3 mm (0.01 in) in diameter.

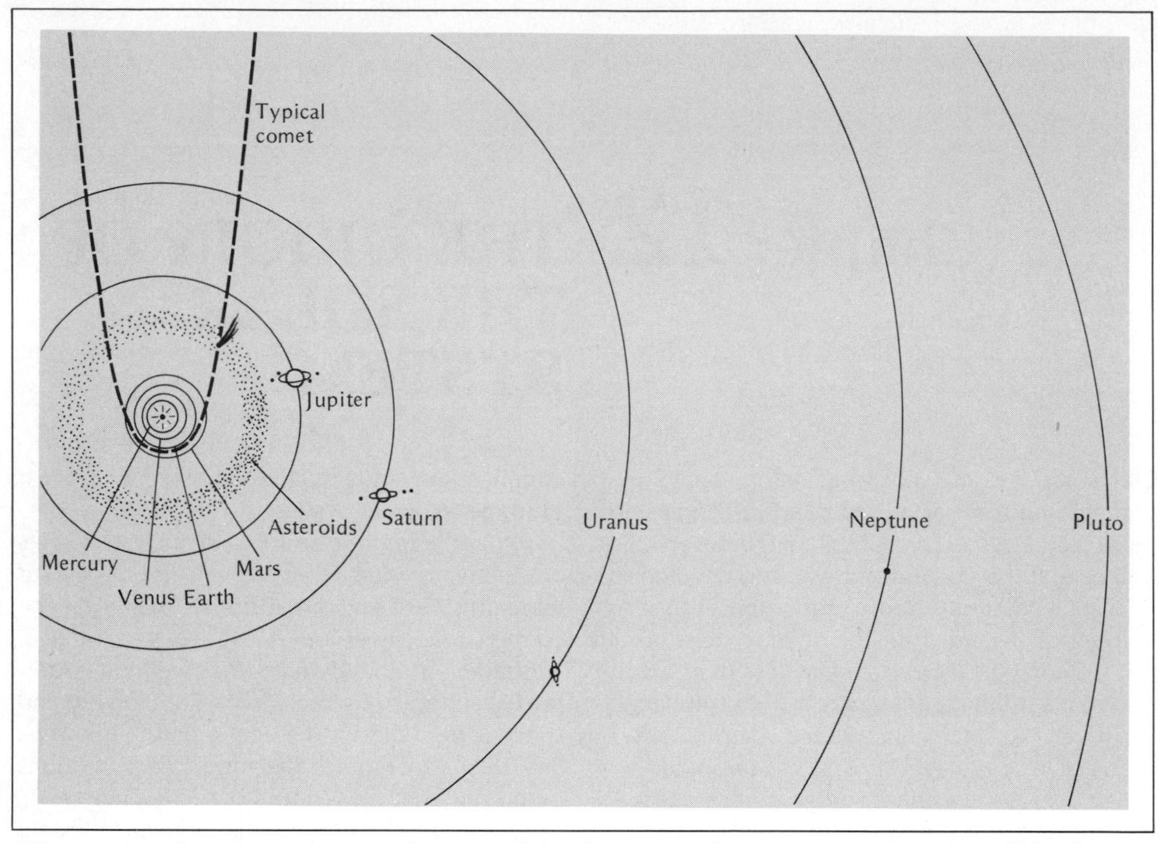


*Figure 12-1. The solar system. The diameters of the planets have been exaggerated by about 4000. Even so, Mercury, Venus, Earth, the moon, Mars, and Pluto are too small to be visible.*

The moon is a speck of pepper about 1 cm (0.4 in) away, and the sun is the size of a small plum 4 m (13 ft) from the earth. Mercury, Venus, and Mars are grains of salt. Jupiter is an apple seed 20 m (66 feet) from the sun, and Saturn is a smaller seed over 36 m (120 ft) away. Uranus and Neptune are slightly larger than average salt grains, and Pluto, the farthest planet, is a speck of pepper over 150 m (500 ft) from the central plum.

Though the planets seem isolated, they are, in fact, earth's nearest neighbors. Astronomers can get vast amounts of data about them, not only by telescope but also by space probe. We cannot discuss all such data here. Instead, we will survey the solar system, listing its most obvious characteristics in Table 12-2. Later, as we discuss the origin of the sun and planets, we can compare the various theories with the table. An adequate theory should account for these general properties of the solar system.

*Revolution and Rotation.* The planets revolve around the sun in orbits that lie close to a common plane. Mercury, the closest planet to the sun, follows an orbit tipped 7° to the earth's, and Pluto's orbit is tipped 17.2°. The rest of the planets' orbital planes are inclined by no more than 3.4°. Thus the solar system is basically disk-shaped.

The rotation of the sun and planets on their axes also seems related to this disk shape. The sun rotates with its equator inclined only 7.25° to the earth's orbit, and most of the other planets' equators are tipped less than 30°. Since Venus rotates in the opposite direction to the earth's rotation, we could say its equator is tipped 180° (we will discuss this problem later). Another planet with a peculiar inclination is Uranus. It is inclined 97.7° in its orbit, and thus its axis of rotation lies nearly in the plane of its orbit. The origin of the inclination of Uranus's axis of rotation is poorly understood, but it may be related to the way Uranus formed. With these exceptions, the planets of the solar system rotate with their equators near the plane of the earth's orbit.

The disk shape of the solar system is apparent in the preferred direction of motion—counterclockwise as seen from the north. All of the planets revolve counterclockwise around the sun, and, with the exception of Venus and Uranus, they rotate* counterclockwise on their axes.

The rotation of a few solar system bodies seems to have been altered by tidal interactions. The moon's rotation is a good example of how tidal forces can lock the rotation of one celestial body to another. The earth's gravitational field has pulled the moon into an egg shape, and as the moon orbits the earth, it rotates on its axis, keeping the large end of the egg pointed toward earth (Figure 12-2). It may have spun faster in the past, but that motion has been slowed to its present state by tidal forces (see Chapter 2).

A similar tidal process may have altered the rotation of Mercury and Venus. Mercury is in an elliptical orbit close to the sun, and tidal forces have slowed its rotation until it now rotates exactly 1.5 times during each orbit. Thus it turns one side toward the sun at one close approach, and the other side at the next close approach. Thus Mercury's rotation appears to have been altered by the sun's gravity.

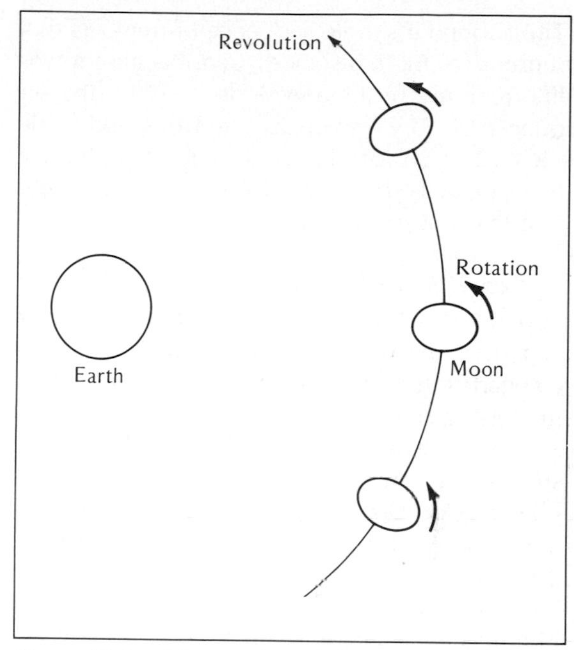


*Figure 12-2. The rotation of the moon is tidally locked to the earth. Thus it keeps the same side toward the earth as it moves along its orbit. The lunar shape is exaggerated for clarity.*

The rotation of Venus is a bit more mysterious. Radar observations show that it rotates backwards—clockwise as seen from the north—once every 243.01 days. This motion turns nearly the same face toward earth each time the two planets come near each other, suggesting that the rotation of Venus is tidally locked to earth. However, if Venus were perfectly locked to earth, its period would be 243.16 days. No one knows why Venus rotates backwards, nor why its period of rotation is so nearly in resonance with the earth. It may simply be a coincidence.

With a few exceptions, most of which are understood, revolution and rotation in the solar system follow a disk theme. But sharp eyes might detect another significant pattern in the planetary orbits shown in Figure 12-1: Each planet is a little less than twice as far from the sun as its inward neighbor. In 1766, Johann

* Astronomers distinguish between the words *revolve* and *rotate*. A planet revolves around the sun, but rotates on its axis.

Titius found a simple sequence of numbers that reproduces these distances, and, because it was first reported by Johann Bode in 1772, the sequence is now known as the **Titius-Bode rule** (Box 12-1). No one knows why it works, but we list it in Table 12-2 because it appears to show that the solar system did not form randomly.

*Two Kinds of Planets.* Perhaps the most important clue we have to the origin of the solar system is the division of the planets into two categories: terrestrial or earthlike, and jovian or Jupiter-like (Figure 12-3). The **terrestrial planets** are small, dense, rocky worlds with less atmosphere than the jovian planets. The terrestrial planets—Mercury, Venus, Earth, and Mars—lie in the inner solar system. The **jovian planets** are different in every respect. They are large, gaseous, low-density worlds. The jovian planets—Jupiter, Saturn, Uranus, and Neptune—lie in the outer solar system beyond the asteroids. Note that Pluto does not fit either category very well. It is small like the terrestrial planets but lies far from the sun and has a low density like the jovian planets.

All of the terrestrial planets are scarred by craters. Mercury's surface looks much like the moon's, with thousands of overlapping craters (Figure 12-4). Because Mercury is very hot and only 40 percent larger than the moon, it has held no atmosphere. Venus is nearly as large as the earth and has a thick atmosphere. In fact, its surface is perpetually hidden below a dense layer of clouds, so it is only by mapping its surface with radar that astronomers have found craters on Venus. The earth, too, has some craters, though erosion rapidly wears such features away. Mars is about half the earth's diameter and has a much thinner atmosphere, which permits us to see its surface easily. Space probe photos show that some of the markings visible from earth are, in fact, craters (Figure 12-5). In addition, planetary probes have found craters on both satellites of Mars and on the satellites of Jupiter. This suggests that craters are a characteristic of every object in the solar system with a surface capable of retaining such features.

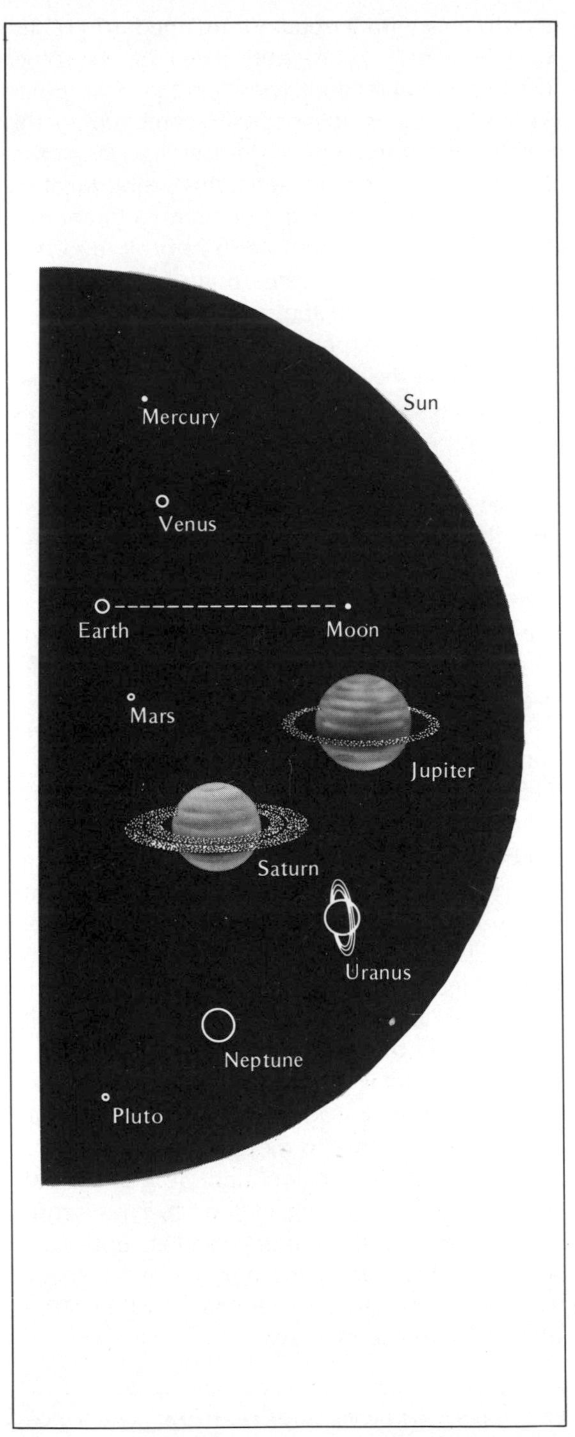


*Figure 12-3. The relative sizes of the planets compared to the solar disk.*

All of the jovian planets are similar in structure (Figure 12-6). They are massive and rich in hydrogen and helium. Jupiter is over 318 times as massive as the earth, and Saturn is about 95 earth masses. Uranus and Neptune are smaller, 15 and 17 earth masses respectively. Most of this material is hydrogen and helium.

Mathematical models of the interiors of the jovian planets predict that they are mostly hydrogen in two different states. The swirling cloud belts of Jupiter's atmosphere hide deeper layers where the pressure forces the hydrogen into a liquid state (Figure 12-7). Even deeper inside, the pressure is so high the hydrogen atoms can no longer hold their electrons. With the electrons free to move about, the material is an excellent conductor of electricity and is called liquid metallic hydrogen. The centers of the jovian planets are believed to be occupied by rocky cores about the size of the earth.

Although photographs of jovian planets reveal swirling cloud patterns and although the planets are sometimes called gas giants, their atmospheres are not very deep. They are in fact liquid giant planets, composed almost entirely of liquid hydrogen. If these gas-giant planets were magically shrunk to a few centimeters in diameter, the gaseous atmospheres would be no deeper than the fuzz on a badly worn tennis ball.

Because the jovian planets are rich in hydrogen and helium, their average density is low—less than 1.75 gm/cm$^3$. Saturn's very low density, 0.7 gm/cm$^3$, is less than the density of water, so the planet would float if we could find a bathtub big enough. In contrast, the terrestrial planets have densities ranging from about 3 to 5 gm/cm$^3$. Thus the planets are divided into two strikingly different groups: the inner, high-density, terrestrial planets, and the outer, low-density, jovian planets.

Another characteristic of the jovian planets is

BOX 12-1 THE TITIUS-BODE RULE

The Titius-Bode rule specifies a simple series of steps that produce a list of numbers matching the sizes of the planetary orbits. To construct this list, we write down the number 0 and below that 3. Continuing downward, we make each number twice the preceeding number—0, 3, 6, 12, 24, 48, and so forth. After adding 4 to each number, we divide by 10. The result represents the sizes of the planetary orbits in astronomical units (see Table 12-1.)

The rule describes the inner planets well, and even includes 2.8 AU for the asteroid belt between Mars and Jupiter. But there is some disagreement in the outer solar system. If we leave out Neptune, Pluto fits well, but the omission of Neptune is not justified by any physical evidence.

Some astronomers contend that the agreement is mere chance. There is no physical reason for any of the steps in the Titius-Bode rule. In fact, no matter what the sizes of the orbits, a mathematician could find a sequence of steps that would produce matching numbers. However, the simplicity of the Titius-Bode rule leads many astronomers to view it as significant. It may be telling us that there was nothing random about the formation of the planets.

**Table 12-1 Planetary Distance from the Sun**

| Planet | Titius-Bode Prediction | Observed |
|---|---|---|
| Mercury | (0 + 4)/10 = 0.4 AU | 0.387 AU |
| Venus | (3 + 4)/10 = 0.7 | 0.723 |
| Earth | (6 + 4)/10 = 1.0 | 1 |
| Mars | (12 + 4)/10 = 1.6 | 1.524 |
| Asteroid belt | (24 + 4)/10 = 2.8 | 2.77 average |
| Jupiter | (48 + 4)/10 = 5.2 | 5.203 |
| Saturn | (96 + 4)/10 = 10.0 | 9.539 |
| Uranus | (192 + 4)/10 = 19.6 | 19.18 |
| Neptune | | 30.06 |
| Pluto | (384 + 4)/10 = 38.8 | 39.44 |

a

b

*Figure 12-4.* *Both Mercury (a) and the moon (b) are heavily cratered. ((a) NASA/JPL and (b) Lick Observatory photograph.)*

*Figure 12-5.* *(right) Spacecraft visiting Mars have shown that it is geologically complex. This Viking Orbiter 2 photograph taken in August 1976 shows Ascreaus Mons, an extinct volcano, with water ice clouds nearby. The vast rift canyon called Valles Marineris is at center right, and the south pole is at the bottom. (NASA/JPL.)*

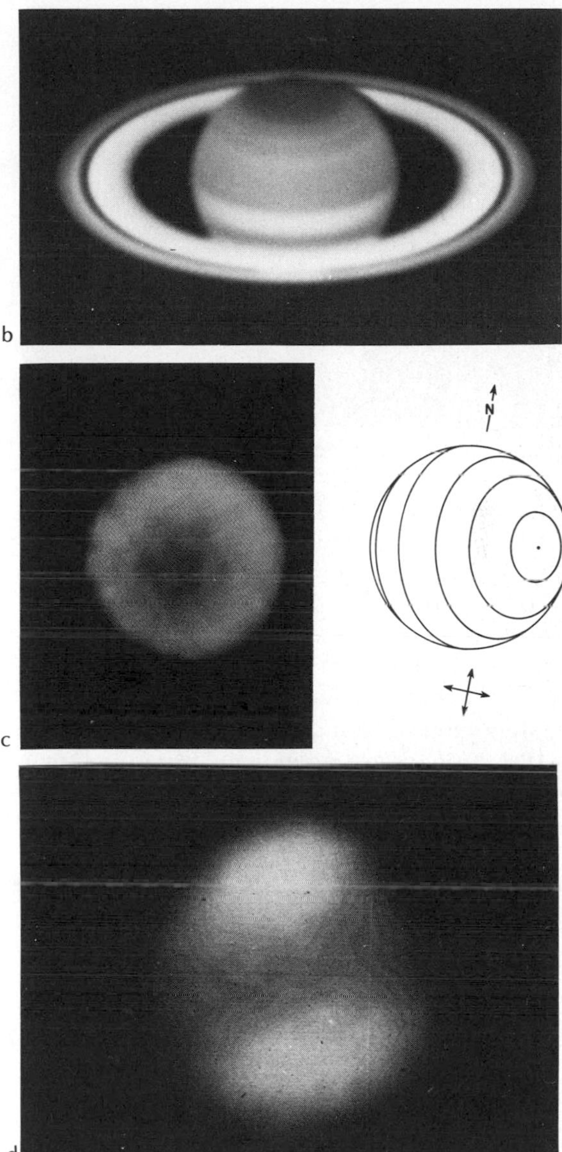


*Figure 12-6. (a) Jupiter, its cloud belts, and Great Red Spot photographed by Voyager 1. (b) Saturn, its belts and rings photographed by the 100-inch telescope. (c) Uranus photographed by an electronic imaging system in the light of the 8900 Å absorption of methane. The brightness of the image near the rim is due to a layer of methane ice-crystal haze high in the planet's atmosphere. The drawing shows the orientation of the planet's axis. (d) Neptune imaged in the same way as Uranus. The bright areas are reflective clouds of methane ice crystals. ((a) NASA/JPL; (b) Hale Observatories; (c) Lunar and Planetary Laboratories; (d) courtesy H. J. Reitsema, B. A. Smith, and S. M. Larson, University of Arizona.)*

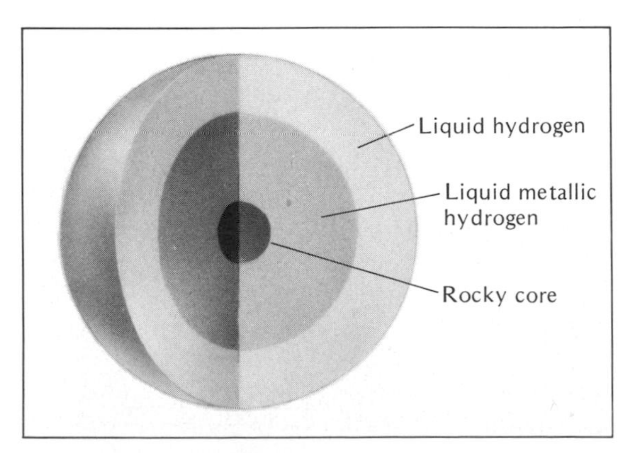


*Figure 12-7. (left) The hypothetical model for Jupiter's interior contains liquid hydrogen, liquid metallic hydrogen, and a silicate core.*

*Figure 12-8. Jupiter and its four Galilean satellites. Its ten smaller satellites are visible only in larger telescopes. (Lick Observatory photograph.)*

their large satellite systems. Jupiter has 15 known moons, of which 4, the **Galilean satellites**—named after Galileo who discovered them—are visible through a small telescope or even binoculars (Figure 12-8). Saturn has 11 known moons, Uranus 5, and Neptune 2. Five of these moons are larger than the earth's moon, and Saturn's Titan is believed to have an atmosphere.

In addition to satellite systems, at least three of the jovian planets have ring systems. Saturn's rings are a striking sight, even through a small telescope (Figure 12-9). These rings are apparently composed of swarms of snowball-sized chunks of water frost with imbedded impurities. This material orbits the planet in the plane of its equator and reflects about 80 percent of the sunlight that hits it, making the rings brilliant. Though the rings look substantial through a telescope, they are no more than 10 km (6 mi) thick. This is shown dramatically every 15 years when the rings are seen edge-on to earth and disappear completely.

The three bright rings commonly visible to earth-based telescopes are labeled A, B, and C, but careful measurements suggest a D ring lying between the inner edge of the visible rings and the top of the planet's atmosphere. Earth-based observations have also hinted at a very faint E ring extending as far as 21 times farther from the planet than the outer edge of the visible rings. In 1979, Pioneer 11 passed through the E ring twice with no serious damage, so it may consist of very thinly scattered material. The spacecraft found no evidence of the D ring, but it did discover another ring, dubbed the F ring, lying just beyond the edge of the A ring. In addition, the spacecraft found that the gaps between the rings are not totally clear of material, leading some astronomers to think of the rings as a single plane of orbiting material with some regions more densely populated than others.

*Figure 12-9. Two of Saturn's rings (A and B) are bright enough to be visible in a small telescope, as is the gap between them named after the astronomer Cassini. The Pioneer 11 spacecraft could not detect ring D, but it did discover ring F. (Hale Observatories photograph.)*

In contrast, the rings of Uranus are nearly invisible from earth. Some astronomers believe

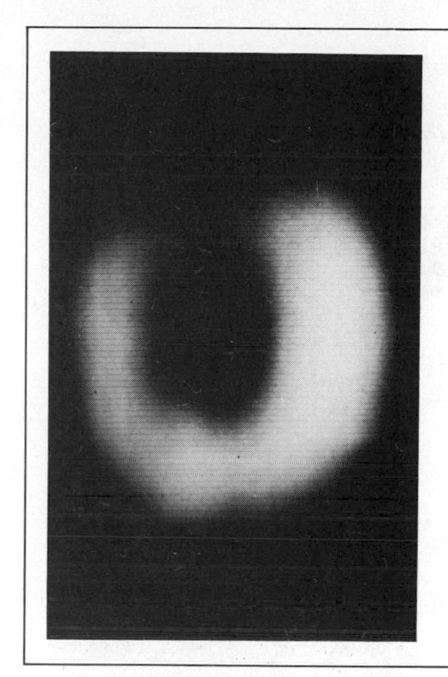

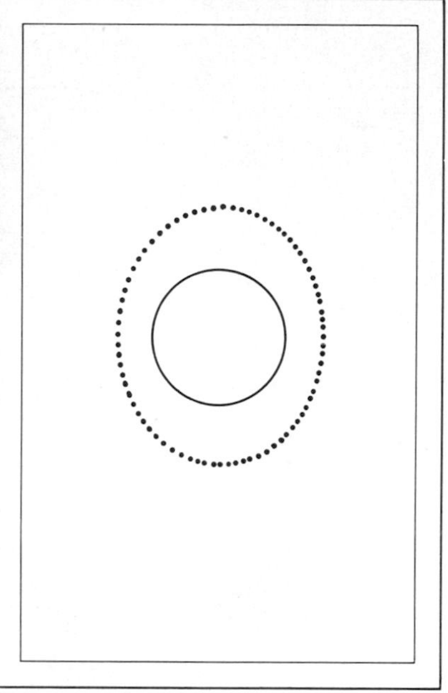

*Figure 12-10. Two infrared images of Uranus were combined to cancel out the planet (solid circle) and reveal light believed to be reflected by the planet's rings. (Keith Matthews and Gerry Neugebauer.)*

they have detected the rings by specially processing infrared photographs taken with large telescopes (Figure 12-10), but the detected glow may be light scattered from material distributed around the planet and not light from the rings themselves. The nine narrow rings now known were discovered when the orbital motion of Uranus carried it in front of a star and the unexpected dimming of the starlight revealed the rings. The material in these rings does not reflect more than about 5 percent of the light that hits it. It cannot be ice, but must be something about the color of coal.

When the Voyager spacecraft passed near Jupiter in 1979, scientists found that it, too, was surrounded by a ring (Figure 12-11). Again, this material is not ice, but darker, rocky dust orbiting in the plane of the planet's equator.

The presence of ring systems in the solar system raises an obvious question. Does Neptune have rings? No one knows, since the planet is so distant that rings like Jupiter's or Uranus's could probably not be detected from earth. However, since three of the four jovian planets are known to have rings, Neptune too may be surrounded by a disk of orbiting rubble.

Ring systems are unique to the jovian planets. The terrestrial planets have no rings, suggesting that the ring-forming processes occur only in the outer solar system. Thus planetary rings may be a significant property of the solar system and deserve a place in Table 12-2.

*Space Debris.* The sun, the planets, and their satellites are not the only bodies of the solar system. Interplanetary space is littered with three kinds of space debris: asteroids, comets, and meteoroids. Although this material represents a tiny fraction of the mass of the system, it is a rich source of information about the origin of the planets.

The **asteroids*** are small, rocky worlds, most of which orbit the sun between the orbits of Mars and Jupiter. A few asteroids follow orbits that bring them into the inner solar system, and

* Also called minor planets.

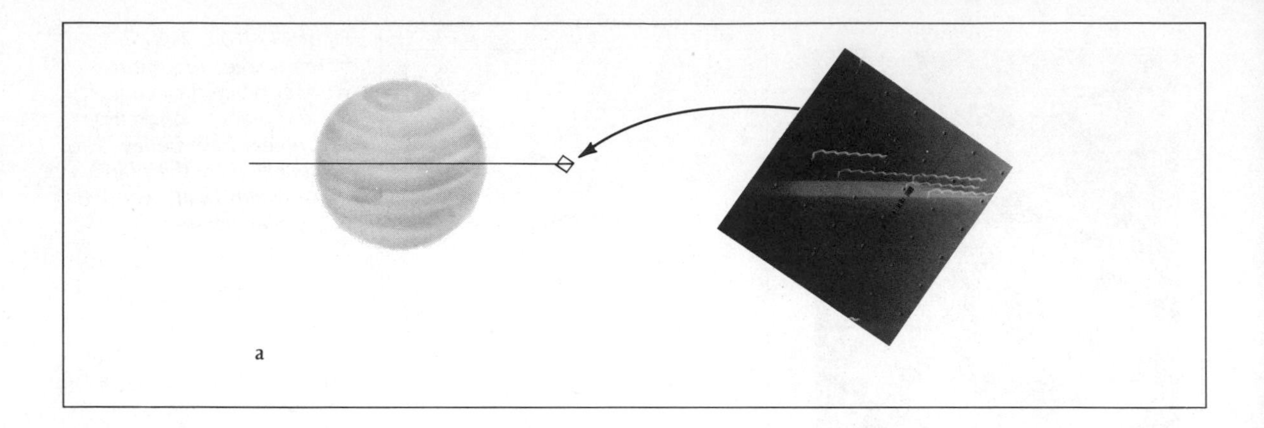


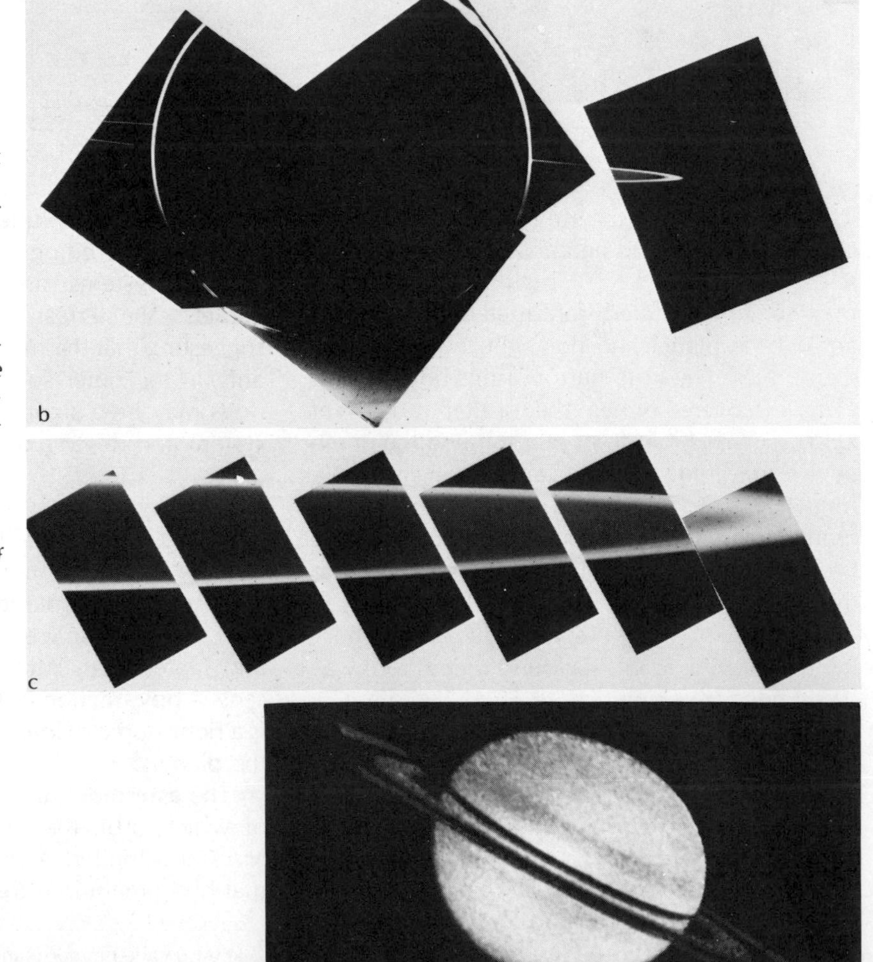


*Figure 12-11. (a) The photograph taken by Voyager 1 on which Jupiter's ring was discovered. (b) A Voyager 2 photomosaic taken from within the planet's shadow. The planet is outlined by sunlight scattered from haze in the upper atmosphere. The ring ends abruptly where it enters the planet's shadow. (c) High resolution Voyager 2 photographs show a hint of structure in Jupiter's ring. (d) The rings of Saturn look unfamiliar in this Pioneer 11 photograph because it was taken from above the ring plane while the inclination of the planet presents the underside to sunlight. This view of the "dark side" of the rings is only rarely visible from earth. Note the shadow of the rings on the planet. (NASA.)*

several occasionally pass within a few tens of millions of miles of earth. Some are located in Jupiter's orbit, and some have been found as far away as the orbit of Saturn.

About 200 of these objects are over 100 km (60 mi) in diameter, and more than 2000 are over 10 km (6 mi). There are probably 500,000 over 1 km (0.6 mi) and billions that are smaller. Because even the largest are only a few hundred kilometers in diameter, telescopes reveal no surface features. However, slow variations in the amount of reflected sunlight suggest they are irregular in shape and rotating as they orbit the sun. Only the largest are roughly spherical.

Earth-bound astronomers may have gotten a close look at a pair of asteroids when the Viking and Mariner space probes to Mars photographed the two small Martian satellites, Deimos and Phobos (Figure 12-12). These objects are only 12 and 28 km (7.5 and 18 mi) in diameter, and the photos show that they are irregular in shape and heavily cratered. Whether these are asteroids that Mars has captured is still open to debate, but many asteroids of the belt must look like Deimos and Phobos (see Box 13-2).

One theory of the origin of the asteroids proposes that they are the remains of a planet that broke up. The majority lie around 2.77 AU from the sun, just where the Titius-Bode rule predicts a planet, but two factors make this theory unlikely. First, it takes a tremendous amount of energy to break up a planet. Disrupting a planet the size of the earth would consume all the energy generated by the sun in three weeks. In addition, the total mass of the asteroids is only about one-fiftieth the mass of the moon, hardly enough to be the remains of a planet.

*Figure 12-12. The asteroids must look much like Phobos, one of the tiny satellites of Mars. (NASA/JPL.)*

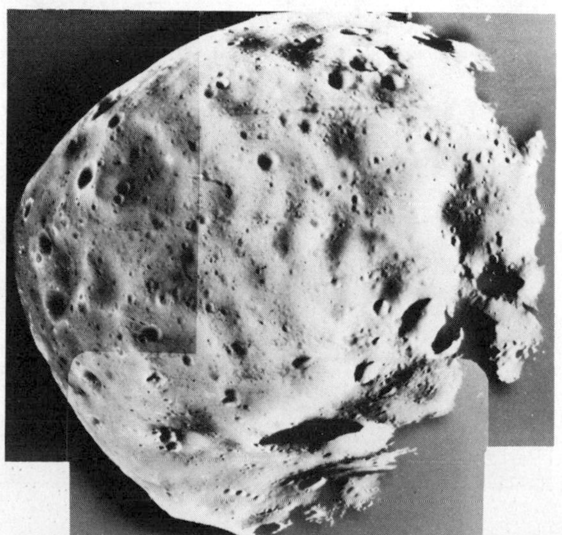

If the asteroids are not the remains of a planet, then they must be the debris left over from the origin of our solar system. Probably they are the remains of a planet that tried to form 2.8 AU from the sun but never succeeded. Clearly this is a clue to how our planetary system began, and we must add the asteroids to Table 12-2.

In contrast to asteroids, bright **comets** are impressively beautiful objects (Figure 12-13). A comet may take months to move through the solar system, during which time it appears as a glowing head, called the **coma,** accompanied by a long tail of gas and dust. Unfortunately, most comets are quite faint and are difficult to locate even at their brightest.

The most popular comet model is called the "dirty snowball" theory. In this model, a comet is a ball of frozen water, carbon dioxide, and other gases, with imbedded bits of dust and rock. This snowball may be only a few kilometers in diameter. When the snowball enters the inner solar system, the sun's radiation begins to vaporize the ices, producing a coma of gas and dust. The released material is pushed away by the pressure of sunlight and the solar wind into a long tail that always points away from the sun like a weather vane (Figure 12-13b).

The significance of comets lies in their composition and orbits. If comets formed at the same time as the planets, their icy composition hints at the composition of the early solar system. In addition, comets follow randomly tipped orbits that do not lie in the disk of the solar system. Whatever theory we propose for the origin of the solar system, it must account for these two properties of comets.

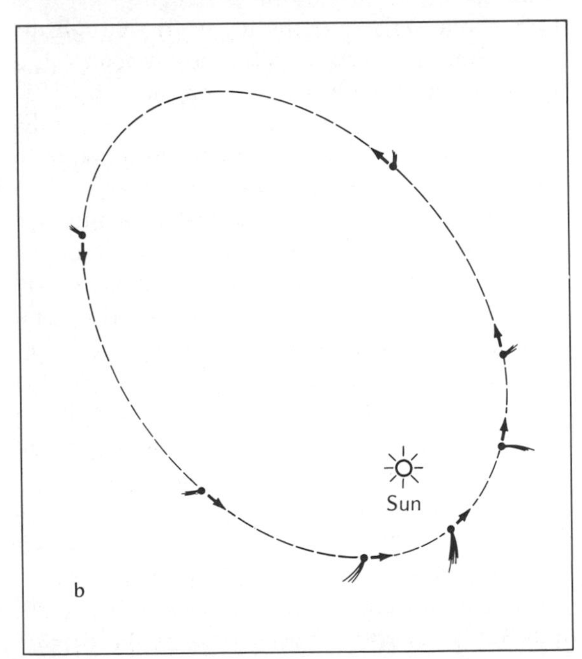


Figure 12-13. *(a) A comet may remain visible in the evening or morning sky for weeks as it moves through the inner solar system. Comet West was in the sky during March 1976. (b) A comet in a long elliptical orbit becomes visible when the sun's heat vaporizes its ices and pushes the gas and dust away in a tail.*

Unlike the stately comets, **meteors** flash across the sky in momentary streaks of light (Figure 12-14). They are commonly called "shooting stars." Of course, they are not stars, but small bits of rock and metal falling into the earth's atmosphere and bursting into incandescent vapor about 80 km (50 mi) above the ground because of friction with the air. The earth is constantly bombarded by such objects, gaining about $10^7$ gm (10 tons) of mass each day from infalling meteors. If you watch the sky, you can see from 3 to 15 meteors an hour on any dark night.

Technically the word *meteor* refers to the streak of light in the sky. In space, before its fiery plunge, the object is called a **meteoroid,** and, if any part of it survives its fiery passage to the earth's surface, it is called a **meteorite.** Most meteoroids are specks of dust, grains of sand, or tiny pebbles. Almost all the meteors we see in

Figure 12-14. *A meteor is the streak of glowing gasses produced by a bit of material falling into the earth's atmosphere. Friction with the air vaporizes the material about 80 km (50 mi) above the earth's surface. (Lick Observatory photograph.)*

the sky are produced by meteoroids that weigh less than 1 gm. Only rarely is one massive enough and strong enough to survive its plunge and reach earth's surface.

If the fiery track of a meteor is photographed from two or more locations, its direction and velocity can be found. Then we can backtrack and find the orbit the meteoroid had before it fell into the earth's atmosphere. Several of these tracked meteors seem to have come from the asteroid belt, but others seem to be debris from comets—the dirt in the dirty snowball.

Meteorites are either stony—looking much like earth rocks—or metallic. Both types can be traced to the asteroid belt, so the asteroids must consist of both stony and metallic material. Clearly our theory of the origin of the solar system must account for both cometary and asteroidal meteors and explain their compositions. Thus we include meteors in Table 12-2.

**Table 12-2 Characteristic Properties of the Solar System**

1. Disk Shape of the Solar System
   a. Orbits in nearly the same plane
   b. Common direction of rotation and revolution
2. Titius-Bode Rule
3. Two Planetary Types
   a. Terrestrial—inner planets
   b. Jovian—outer planets
4. Planetary Ring Systems
   a. Jupiter
   b. Saturn
   c. Uranus
5. Space Debris—Asteroids, Comets, and Meteors
   a. Composition
   b. Orbits
6. Common Ages of about 4.6 Billion Years
   a. Earth
   b. Moon
   c. Meteorites
   d. Sun

Our list of characteristic properties of the solar system is nearly complete. We have noted the disk shape of the system, the division of the planets into two types, and the abundance of space debris. The only item that remains is the age of the solar system's objects. We will add that in the next section, and then devise a theory to explain how our solar system began.

## THE SOLAR NEBULA

Astronomers generally agree that the sun and the planets formed at about the same time from a cloud, or nebula, of gas and dust. This solar-nebula theory raises three questions: When did the nebula give birth to the solar system? Of what was the nebula composed? And finally, how did solid matter evolve from a cloud of gas? This section will explore these questions.

*Ages.* If the solar nebula theory is correct, the planets should be about the same age as the sun. The most accurate way to find the age of a celestial body is to bring a sample into a laboratory and determine its age by radioactive dating. Unfortunately, the only celestial bodies that can so far be dated in this way are the earth, the moon, and meteorites.

The oldest earth rocks so far discovered and dated are about 3.9 billion years old. That does not mean that the earth formed 3.9 billion years ago. The surface of the earth is active, and the crust is continually destroyed and reformed from material welling up from beneath the crust (see Chapter 14). Thus the age of these oldest rocks tells us only that the earth is *at least* 3.9 billion years old.

One of the most exciting goals of the Apollo lunar landings was bringing lunar rocks back to earth's laboratories, where they could be dated. Since the moon's surface is not being recycled like earth's, some parts of it might have survived unaltered since early in the history of the solar system. Dating the rocks showed the oldest to be 4.48 billion years old. Thus the solar system must be *at least* 4.48 billion years old.

Another important source for determining the age of the solar system is meteorites. Radioactive dating of meteorites yields ages of about 4.6 billion years. Their age must correspond closely to the true age of the solar system.

One last celestial body deserves mention: the sun. Astronomers estimate the age of the sun as about 5 billion years, but this is not a radioactive date since they cannot obtain a sample of solar material. Instead, they estimate the sun's age from the radioactive ages of the earth, the moon, and meteorites. Computer models of the sun give only approximate ages, but they generally agree with the age of the earth.

Apparently all of the bodies of the solar system formed at about the same time some 4.6 billion years ago. This is the last item we add to the list of significant properties in Table 12-2. It also answers our first question—how long ago did the solar system form?

*The Nature of the Solar Nebula.* Although the solar nebula spawned the planets long ago, it left behind traces that allow us to reconstruct its chemical and physical composition. The present chemical composition of the sun and planets not only tells us the composition of the nebula, but also hints that it had a violent origin. Also, the present motions in the solar system outline the original shape and motion of the nebula.

According to the theory of stellar evolution (Chapter 6), stars originate from interstellar gas clouds rich in hydrogen and helium. The sun is mostly hydrogen and helium with small traces of carbon, nitrogen, oxygen, and other heavy elements, and the jovian planets have similar compositions. The terrestrial planets are too small to have retained their gaseous hydrogen and helium, but if their present compositions were augmented with the hydrogen and helium they have lost, they, too, would have compositions like the sun's. This abundance of light elements in the present solar system suggests that the solar nebula was a fragment of an interstellar gas cloud.

Chemical analysis of some meteorites has uncovered traces of short-lived radioactive elements that must have been formed no more than a few million years before the planets. These atoms may have been made in a supernova explosion that must have erupted within 60 ly (18 pc) of the gas cloud that was eventually to form the sun and planets. As we saw in Chapter 6, the shell of gas ejected by a supernova can compress the surrounding gas clouds and trigger star formation. Thus our solar system may owe its existence to a supernova explosion that occurred 4.6 billion years ago.

The study of stars now forming can tell us more about the physical composition of the solar nebula. T Tauri stars, for example, are young stars surrounded by the clouds of gas from which they have recently formed. The infrared radiation coming from these objects tells us that the gas contains large quantities of dust and suggests that the solar nebula contained both gas and dust. In addition, the T Tauri stars are blowing away their surrounding nebulae at speeds up to 200 km/second. Thus the young sun, while passing through this stage in its evolution, may have driven away large amounts of gas and dust and destroyed its nebula just as the T Tauri stars are now destroying theirs.

The present motions of the planets were inherited from the original motion of the solar nebula. The tendency of solar system bodies to revolve and rotate counterclockwise shows that the solar nebula was rotating. In Chapter 8 we saw that a disk-shaped galaxy forms from the contraction of a rotating cloud of gas. Though the solar nebula was much smaller, the physics is the same and the contracting cloud must have flattened into a swirling turbulent disk of gas and dust with the sun forming at its center. As the planets formed, they took up orbits within the nebular disk. Thus the present disk shape of the solar system is a result of the rotation of the solar nebula (Figure 12-15).

The solar nebula theory accounts for the present chemical composition of the sun and planets and their motions, but a problem re-

*Figure 12-15. Because the solar nebula was rotating (a), it contracted into a disk (b), and the planets formed with orbits lying in nearly the same plane (c).*

mains. How did nature convert the gas of the solar nebula into solid material from which to make planets?

*Condensation of Solids.* The key to understanding the process that converted the nebular gas into solid matter is the variation in density among solar system objects. We have already noted in Table 12-2 that the four inner planets are high-density terrestrial bodies, while the outermost planets are low-density giant planets. This division is due to the different ways gases condensed into solids in the inner and outer regions of the solar nebula.

Even among the four terrestrial planets, we find subtle differences in density. Merely listing the observed densities of the terrestrial planets is not helpful because the earth and Venus, being more massive, have stronger gravity and have squeezed their interiors to higher densities. We must therefore look at the **uncompressed densities**—the densities the planets would have if their gravity did not compress them. These densities (Table 12-3) show that the closer a planet is to the sun, the higher its uncompressed density.

**Table 12-3 Observed and Uncompressed Densities**

| Planet | Observed Density ($gm/cm^3$) | Uncompressed Density ($gm/cm^3$) |
|---|---|---|
| Mercury | 5.44 | 5.4 |
| Venus | 5.24 | 4.2 |
| Earth | 5.52 | 4.2 |
| Mars | 3.93 | 3.3 |
| (The moon) | 3.36 | 3.35 |

This density variation probably originated when the solar system first formed solid grains. The kind of matter that condensed would depend on the temperature of the gas, and, in the inner regions, the temperature may have been 1500°K or so. The only materials that could form grains at this temperature are compounds with high melting points such as metal oxides and pure metals, which are very dense. Farther out in the nebula it was cooler, and silicates (rocky material) could condense. These are less dense than metal oxides and metals. In the outer regions it was cold, and ices of water, methane, and ammonia could condense. These are low-density materials.

The sequence in which the different materials condense from the gas as we move away from the sun is called the **condensation sequence** (Table 12-4). It suggests that the planets, forming at different distances from the sun, accumulated from different kinds of materials. Thus the inner planets formed from high-density metal oxides and metals, and the outer planets formed from low-density ices.

The condensation sequence may account for the variation in planetary density with distance from the sun, but it does not explain how small grains of solid matter could come together to build a planet. In the next chapter, we will see how planets could have grown in the solar nebula.

**Table 12-4 The Condensation Sequence**

| Temperature (°K) | Condensate | Planet (Estimated Temperature of Formation), (°K) |
|---|---|---|
| 1500 | Metal oxides | Mercury (1400) |
| 1300 | Metallic iron and nickel | |
| 1200 | Silicates | |
| 1000 | Feldspars | Venus (900) |
| 680 | Troilite (FeS) | Earth (600) |
| | | Mars (450) |
| 175 | $H_2O$ ice | Jovian (175) |
| 150 | Ammonia-water ice | |
| 120 | Methane-water ice | |
| 65 | Argon and neon ice | Pluto (65) |

## SUMMARY

We can reconstruct the process by which the solar system formed by studying the characteristic properties of the system. These properties suggest that the sun and planets formed at about the same time from the same cloud of gas and dust—the solar nebula.

One of the most striking of the solar system's characteristic properties is its disk shape. The orbits of the planets lie in nearly the same plane, and they all revolve around the sun in the same direction, counterclockwise as seen from the north. This is also true of many of the satellites of the planets. With only a few exceptions the planets rotate counterclockwise around axes roughly perpendicular to the plane of the solar system. This disk shape and the motion of the planets appear to have originated in the solar nebula.

Another striking characteristic feature of the solar system is the division of the planets into two families. The terrestrial planets are small and dense, and they lie in the inner part of the system. The jovian planets are large, low-density worlds that lie in the outer part of the system. In general, the closer a planet lies to the sun, the higher its uncompressed density.

The solar system is now filled with smaller bodies such as asteroids, comets, and meteors.

The asteroids are small, rocky worlds, most of which orbit the sun between Jupiter and Mars. They appear to be material left over from the formation of the solar system.

According to the "dirty snowball" theory, comets are small bodies composed of frozen water, ammonia, methane, and other compounds, with imbedded bits of dust and rock. When a comet in a long elliptical orbit comes near the sun, the ices vaporize and the pressure of sunlight and the solar wind push the vapors and dust outward into a tail. Thus a comet's tail always points away from the sun.

The meteors that flash across the night sky are small bits of rock and metal that fall into earth's atmosphere and are incinerated by friction with the air. These bodies originate in the asteroid belt and in comets.

Another important characteristic of the solar system bodies is their similar ages. Radioactive dating tells us that the earth, moon, and meteorites are no older than about 4.6 billion years. Thus it seems our solar system took shape about 4.6 billion years ago.

The solar nebula theory proposes that the solar system began as a contracting cloud of gas and dust that flattened into a rotating disk. The center of this cloud became the sun, and the planets eventually formed in the disk of the nebula. Because the nebula was disk-shaped, the planetary orbits lie in nearly the same plane and most rotation in the solar system has the same direction—the direction in which the nebula was originally rotating.

According to the condensation sequence, the inner part of the nebula was so hot only high-density minerals could form solid grains. The outer regions, being cooler, condensed to form icy material of lower density. The planets grew from these solid materials with the denser planets forming in the inner part of the nebula and the lower-density jovian planets forming farther from the sun.

## NEW TERMS

Titius-Bode rule
terrestrial planets
jovian planets
Galilean satellites
asteroids
comet
coma
meteor, meteoroid, meteorite
uncompressed density
condensation sequence

## QUESTIONS

1. Why do the planets follow orbits that lie in nearly the same plane?
2. How does the rotation of Mercury differ from that of most of the planets?
3. Describe the differences among the rings of Jupiter, Saturn, and Uranus.
4. Why does it seem unlikely that the asteroids are the remains of a planet that broke up?
5. What is the "dirty snowball" theory?
6. From what two sources do meteors come? How do we know?
7. How can we estimate the age of the solar system?
8. How is the origin of our solar system related to the formation and evolution of stars?

## PROBLEMS

1. If a planet existed beyond Pluto, where would it lie according to the Titius-Bode rule? (Searches for such a planet have found nothing.)
2. In Table 12-3, which object's observed density differs least from its uncompressed density? Why?
3. What composition might we expect for a planet that formed in a region of the nebula where the temperature was about 100°K?

## RECOMMENDED READING


Cameron, A. G. W. "The Origin and Evolution of the Solar System." *Scientific American* 233 (Sept. 1975), p. 32.

Chapman, C. R. "The Nature of Asteroids." *Scientific American* 232 (Jan. 1975), p. 24.

Falk, S. W., and Schramm, D. N. "Did the Solar System Start with a Bang?" *Sky and Telescope* 58 (July 1979), p. 18.

Grossman, L. "The Most Primitive Objects in the Solar System: Carbonaceous Chondrites." *Scientific American* 232 (Feb. 1975), p. 30.

Hartmann, W. K. *Moons and Planets.* Belmont, Calif.: Wadsworth, 1972.

———. "In the Beginning." *Astronomy* 4 (June 1976), p. 6.

Jaki, S. L. "The Titius-Bode Law: A Strange Bicentenary." *Sky and Telescope* 43 (May 1972), p. 280.

Lewis, J. S. "The Chemistry of the Solar System." *Scientific American* 230 (March 1974), p. 50.

Maran, S. P. "Is It an Asteroid, a Comet, or a Moon?" *Natural History* 88 (Jan. 1979), p. 108.

Morrison, D. "Asteroids." *Astronomy* 4 (June 1976), p. 6.

Oberg, J. E. "Tunguska: Collision with a Comet." *Astronomy* 5 (Dec. 1977), p. 18.

Schramm, D. N., and Clayton, R. N. "Did a Supernova Trigger the Formation of the Solar System?" *Scientific American* 239 (Oct. 1978), p. 124.

Van Flandern, T. C. "Rings of Uranus: Invisible and Impossible?" *Science* 204 (8 June 1979), p. 1076.

Veverka, J. "Phobos and Deimos." *Scientific American* 236 (Feb. 1977), p. 30.

Wetherill, G. W. "The Allende Meteorite." *Natural History* 87 (Nov. 1978), p. 102.

Wood, J. A. *The Solar System.* Englewood Cliffs, N. J.: Prentice-Hall, 1979.

*The airless lunar surface has remained unchanged for billions of years. The lunar samples collected by the Apollo astronauts hold clues to the growth of planetary bodies around the forming sun 4.6 billion years ago. (NASA.)*

# Chapter 13 THE FORMATION OF PLANETS

If we could have explored the solar nebula 4.6 billion years ago when the planets were taking shape, we would have found it a very peculiar place. Within the disk-shaped cloud of gas lay small grains of solid material condensing from the gas. In the inner part of the nebula, where the temperature was high, these grains were dense metals and metal oxides. Farther from the center, near the present orbit of the earth, the grains were largely silicates. In the outer solar system we might have found ourselves caught in a blizzard of whirling ice crystals as the water, ammonia, methane, and other compounds froze out of the nebula. These tiny grains were the seeds from which the planets grew.

How these solid grains came together to form planets is a mystery that planetary scientists are only now unraveling. Though the details are obscure, it is clear that the grains collided and stuck together, producing larger bodies that in turn collided and built even bigger bodies. As the objects grew toward planetary dimensions they changed, melting to form dense cores and baking gases from their interiors to form atmospheres. Slowly, the objects acquired the characteristics we associate with planets.

Had we visited the solar nebula during the formation of the planets we might have seen little happening. The processes that built the planets probably worked slowly; the planets grew gradually during millions, perhaps hundreds of millions of years. Even if we could have spent an entire lifetime drifting through the solar nebula watching the planets form, we might have seen little change.

In a sense we can travel back in time and explore the solar nebula, even though the nebula vanished when the sun became a luminous star and blew away the remaining gas. Like a prehistoric beast, the solar nebula left behind fossils—meteorites, asteroids, and comets. The nature of these fossils can tell us of the processes that built the planets.

## PLANET BUILDING

In the development of a planet, three groups of processes operate. First, grains of solid matter grow larger, eventually reaching diameters ranging from a few centimeters to kilometers. These objects, called **planetesimals,** are believed to be the bodies that the second group of processes collect into planets. Finally, a third set of processes clears the solar nebula away. The study of planet building is the study of these three groups of processes.

*The Formation of Planetesimals.* According to the solar nebula theory, planetary development in the solar nebula began with the growth of dust grains. These specks of matter, whatever their composition, grew from microscopic size by two processes: condensation and accretion.

A particle grows by **condensation** when it adds matter one atom at a time from a surrounding gas. Thus snowflakes grow by condensation in the earth's atmosphere. In the nebula, dust grains were continuously bombarded by atoms of gas, and some of these stuck to the grains. A microscopic grain capturing a gas atom increases its mass by a much larger fraction than a gigantic boulder capturing a single atom. Thus condensation can increase the mass of a small grain rapidly, but as the grain grows larger, condensation becomes less effective.

The second process is **accretion,** the sticking together of solid particles. In building a snowman, we roll a ball of snow across the snowy ground so that it grows by accretion. In the solar nebula, the dust grains were, on the average, no more than a few centimeters apart, so they collided with each other frequently. Their mutual gravitation was too small to hold them to each other, but other effects may have helped. Static electricity generated by their passage through the gas could have held them together, as could compounds of carbon that might have formed a sticky surface on the grains. Ice grains might have stuck together better than some other types. Of course, some collisions might break up clumps of grains; on the whole, however, accretion must have increased grain size.

There is no clear distinction between a very large grain and a very small planetesimal, but we can consider an object a planetesimal when its diameter becomes a centimeter or so. Objects this size and larger were subject to new processes that tended to concentrate them. One important effect may have been the collapse of the growing planetesimals into the plane of the solar nebula. Dust grains could not fall into the plane because the turbulent motions of the gas kept them stirred up, but the larger objects had more mass and the gas motions could not have prevented them from settling into the plane of the spinning nebula. This would have concentrated the solid particles into a thin plane about 0.01 AU thick and would have made further planetary growth more rapid.

Though this collapse of the planetesimals into the plane is analogous to the flattening of a forming galaxy, an entirely new process may have become important once the plane of planetesimals formed. Computer models show that the rotating disk of particles should have been gravitationally unstable and would have broken up into small clouds (Figure 13-1). This would further concentrate the planetesimals and help them coalesce into objects up to 100 km (60 mi) in diameter. Thus the theory predicts that the nebula became filled with trillions of planetesimals ranging in size from pebbles to tiny planets. As the largest began to exceed 100 km in diameter, new processes began to alter them, and a new stage in planet building began, the growth of protoplanets.

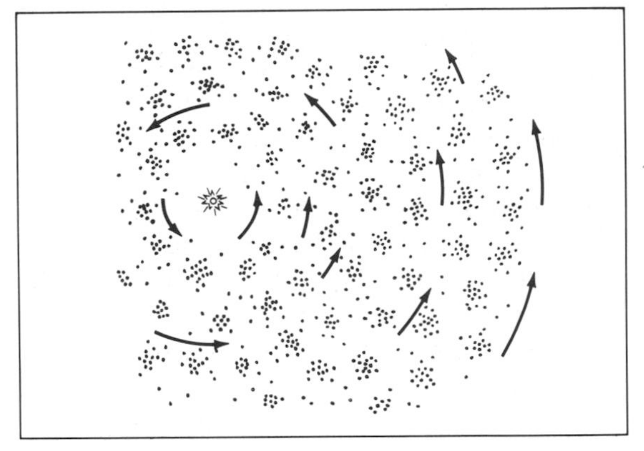

*Figure 13-1. Gravitational instabilities in the rotating disk of planetesimals may have forced them to collect in clumps, accelerating their growth.*

*The Growth of Protoplanets.* The coalescense of planetesimals eventually led to the formation of **protoplanets,** massive objects destined to become planets. As these larger bodies grew, new processes began making them grow faster and altered their physical structure.

If planetesimals collided with each other at orbital velocities, it is unlikely that they would have stuck together. The average orbital velocity in the solar system is about 30 km/second (67,000 mi/hour). Head-on collisions at this velocity would have pulverized the material. However, the planetesimals moved in the nebular plane and in the same direction, and thus avoided head-on collisions. Instead, they merely rubbed shoulders at low relative velocities. Such collisions would be more likely to fuse them together than to shatter them.

In addition, some adhesive effects probably helped. Sticky coatings and electrostatic charges on the surfaces of the smaller planetesimals probably aided formation of larger bodies. Collisions would have fragmented some of the surface rock, but, if the planetesimals were large enough, their gravity would have held onto some fragments, forming a layer of soil composed entirely of crushed rock. Such a layer on the larger planetesimals may have been effective in trapping smaller bodies.

The largest planetesimals would grow the fastest because they had the strongest gravitational field. Not only could they hold on to a cushioning layer to trap fragments, but their stronger gravity could attract additional material. These planetesimals probably grew quickly to protoplanetary dimensions, sweeping up more and more material. When massive enough, they trapped some of the original nebular gas to form primitive atmospheres. At some point, they crossed the boundary between planetesimals and protoplanets.

As protoplanets formed, new processes melted and transformed them into true planets. Melting occurred because of the terrific amount of energy given up by infalling material. Pulled by the protoplanet's gravity, the infalling material struck with high velocity, and most of this energy was transformed into heat in the surface layers. This heat of formation was sufficient to melt the surface and perhaps the entire planet. The decay of radioactive elements and the presence of magnetic fields may also have contributed to the heating.

Once the protoplanet began to melt, two effects changed it into a planet. The first was **differentiation,** the separation of material according to density. Iron ore differentiates when it melts in an open hearth furnace—the dense iron sinks to the bottom and the lower density silicates float to the top as slag. This process became important once the protoplanet melted. Denser material would sink to the center and form a dense metallic core, and lighter material would float to the surface and form a crust. This probably accounts for the earth's dense nickel-iron core and lighter silicate crust.

The second effect is **outgassing,** the release of gases from a planet's interior. When the protoplanets melted, they exhaled gases that merged with the nebular gases in their atmospheres. Most of the earth's present amosphere was produced by outgassing.

In contrast, the formation of the jovian planets did not involve differentiation and outgassing, though heat of formation did play a role. Jupiter and Saturn apparently formed in a region of the nebula where planetary growth was especially rapid. In the cooler outer regions the gas could form three times as much ice as it could silicates. Some theoretical models indicate that Jupiter formed from this material in less than a thousand years, though most theories predict millions of years for the formation of the jovian planets.

The heat of formation of these massive planets was tremendous. Jupiter must have grown so hot that it glowed with a luminosity of about 1 percent that of the present sun. However, since it never got hot enough to generate nuclear energy, as a star would, it cooled. Jupiter is still hot inside. In fact, both Jupiter and Saturn radiate more heat than they absorb from the sun, so they are evidently still cooling.

When the jovian planets grew sufficiently large, they attracted vast amounts of nebular gas and thus grew rich in light gases such as hydrogen and helium. The terrestrial planets could not do this because they never reached sufficient mass and because the gas in the inner nebula was hotter and more difficult to trap.

The origin of the jovian planets hides many

mysteries, but one is especially intriguing—the origin of the rings orbiting Jupiter, Saturn, and Uranus. These are probably not due to material left over from the formation of the planets. The pressure of sunlight and the solar wind would alter the orbits of small particles and force them to spiral into the planet, thus destroying any ring. In addition, ice exposed to the solar wind would slowly waste away as high-energy particles in the solar wind chipped off atoms and molecules. Thus it seems highly unlikely that Saturn's rings, composed of small ice fragments, could have survived for 4.6 billion years.

The only alternative is to suppose that the rings formed more recently. Any material scattered near a jovian planet might accumulate in the plane of the planet's equator and thus form a ring. In addition, any satellite whose orbit brought it too close to its planet would be torn apart by tidal forces, and its debris would be added to the rings. The limiting distance between planet and satellite is called the **Roche limit** and is equal to 2.44 planetary radii if the planet and satellite have similar densities. Any satellite whose orbit carries it inside the Roche limit will be unable to hold itself together with its own gravity. If the satellite were icy, it would produce icy rings. All of Saturn's bright rings lie within the planet's Roche limit.

Triton, Neptune's largest satellite, is gradually moving closer to Neptune and will probably be ripped apart within a few billion years. Since Triton is larger than Mercury, its destruction could produce a large amount of debris, which if it is icy, could give Neptune highly reflective rings like Saturn's.

---

## BOX 13-1 THE ORIGIN OF THE MOON

Theories of the moon's origin take one of three forms: fission, condensation, or capture. Fission theories suppose that the earth and moon were once a rapidly spinning object that was tidally deformed by the sun (Figure 13-2a). If the tides were large enough, fragments might break from the object and form the moon; what remained became the earth. If this separation occurred after the original object differentiated, then the earth might have retained most of the heavy iron core, and the moon would have formed from the low-density crust. Thus the theory explains the moon's low density.

Fission theories have two faults. First, tides due to the sun would not have been strong enough to pull the moon away. Second, the lunar surface material is richer in certain isotopes than the earth, implying that they were never part of the same body. For these reasons, fission theories are unacceptable.

Condensation theories suppose that the moon and earth condensed from the same cloud of material (Figure 13-2b). If this were true, they should now have similar densities and compositions. But the moon is significantly less dense than the earth, and the chemical abundances of its surface materials are different. Thus the earth and moon did not condense from the same cloud of material.

Capture theories suppose that the moon formed elsewhere in the solar system and was later captured by the earth (Figure 13-2c). One suggestion is that the moon formed just inside the orbit of Mercury. There the heat would have prevented the condensation of solid metallic grains, and only high-melting-point metal oxides could have solidified. Thus the moon would contain little iron and nickel and would have a low density. Calculations show that once formed, an encounter with Mercury could have kicked the moon outward as far as the earth.

Though the capture theory sounds attractive, it requires some coincidences. The required interaction with Mercury is improbable, and the subsequent encounter with earth is also unlikely. This doesn't prove it couldn't have happened, but scientists are always suspicious of theories that require a chain of coincidences.

We might get a better theory if we combined capture with condensation. The forming solar system was filled with planetesimal

We can now guess why the terrestrial planets do not have rings. They lie too close to the sun, and the sunlight and solar wind quickly blow away small orbiting particles.

The growth of planets in the solar nebula explains many of the properties listed in Table 12-2, but a few problems remain. One of these is the origin of the moon. Though there are many hypotheses, none lacks flaws. At present there is no adequate theory to explain how the earth got such a large satellite (Box 13-1). This has led some frustrated astronomers to joke that the moon does not exist!

*Clearing the Nebula.* The planets apparently were born as the sun formed from the solar nebula. As soon as the sun became a luminous star, it began to clear the nebula, blowing gas away and removing solid particles that had not become part of planets. This clearing of the nebula brought planet building to a halt.

Four effects helped to clear the nebula. The most important was **radiation pressure.** When the sun became a luminous object, light streaming from its surface pushed against the particles of the solar nebula. Large bits of matter like planetesimals and planets were not affected, but low-mass specks of dust and individual gas atoms were pushed outward and eventually driven from the system. This is not a sudden process, and it may not have occurred at the same time everywhere in the nebula. Before sunlight could begin clearing the outer nebula, it first had to push its way through the inner nebula.

The second effect that helped clear the

---

objects, and if some of these were captured into orbits around the earth, they might have combined to build the moon. If we suppose that these objects originally condensed in other parts of the solar nebula, then their composition and the subsequent composition of the moon would be unlike the earth's. This idea has the attraction of requiring no special coincidences.

*Figure 13-2. Three theories of the moon's origin. (a) Fission theories suppose that the earth and moon were once one body and broke apart. (b) Condensation theories suppose that the moon formed from material near the earth. (c) Capture theories suggest that the moon formed elsewhere and was captured by earth.*

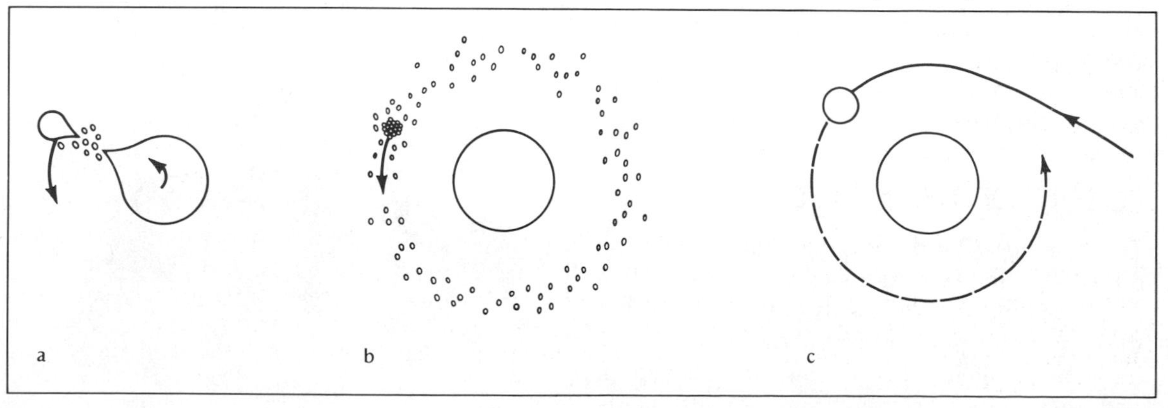

nebula was the solar wind, the flow of ionized hydrogen and other atoms away from the sun's upper atmosphere. This flow is a steady breeze that rushes past the earth at about 600 km/second (370 mi/second). When the sun was young it may have had an even stronger solar wind, and irregular fluctuations in its luminosity, like those observed in T Tauri stars, may have produced surges in the wind that helped push dust and gas out of the nebula.

The third effect for clearing the nebula is the sweeping up of space debris by the planets. Some astronomers believe that this was very efficient, while others suspect that it was less important than the first two processes. Certainly it helped. The extensive cratering of the terrestrial planets and the satellites of Earth, Mars, and Jupiter is evidence that the last of the planetesimals were gobbled up by the planets about 4 billion years ago. Since then the solar system has been relatively clear of solid debris.

The fourth effect is the ejection of material from the solar system by close encounters with planets. If a small object such as a planetesimal passes close to a planet, it can gain energy from the planet's gravitational field and be thrown out of the solar system. Ejection is most probable for encounters with massive planets, so the jovian planets were probably very efficient at ejecting the icy planetesimals that formed in their region of the nebula.

These four effects acted together to clear away the solar nebula. This may have taken considerable time, but eventually the solar system was relatively clear, the planets could no longer gain mass, and planet building ended.

## FOSSILS OF THE NEBULA

The solar nebula did not vanish entirely. Like the dinosaur, it left behind traces of its passing. In a sense, the planets are fossils of the nebula, but they are not especially revealing because they have been altered by melting, differentiation, and outgassing. Asteroids and comets, being smaller, have for the most part not been seriously altered, and traces of the solar nebula are clearly evident in their composition. Thus they are undisturbed fossils of the solar system's past.

However, it is not now possible to visit the asteroid belt or fly off across space to sample a comet. The only available samples of asteroids and comets are those fragments that fall into earth's atmosphere. Consequently, our study of asteroids and comets begins with a study of meteorites.

*Meteorites.* Meteorites can be divided into two broad categories. Iron meteorites are solid chunks of iron and nickel. Stony meteorites are silicate masses that resemble earth rocks. Both types contain clues to the history of the solar system.

When iron meteorites are sliced open, polished, and etched with nitric acid, they reveal regular bands called **Widmanstätten patterns** (Figure 13-3). The patterns arise from crystals of nickel-iron alloys that have grown very large, indicating that the meteorite cooled no faster than a few degrees per million years. Explaining how iron meteorites could have cooled so slowly will be a major step in analyzing their history.

In contrast to iron meteorites, most stony meteorites appear never to have been heated to melting. In fact, we can classify stony meteorites

*Figure 13-3. The Widmanstätten pattern in an iron meteorite. (Griffith Observatory.)*

into three types according to the degree to which they have been heated. **Chondrites** are stony meteorites that contain **chondrules,** rounded bits of glassy rock not much larger than a pea (Figure 13-4). Chondrules seem to have solidifed very quickly from molten drops of rock. Some astronomers suggest that they were the first droplets of matter to condense out of the solar nebula, while others suspect they are drops of molten rock splashed into space by collisions between planetesimals. In any case, the chondrules appear to be very old, and since melting would have destroyed them, their presence in chondrites indicates that these meteorites never melted.

Nevertheless, chondrites have been heated slightly. The rock in which the chondrules are imbedded is also old, but it contains no volatiles. The condensing solar nebula should have incorporated carbon compounds and water into the forming solids, but no such material is present in the chondrites. They have been heated enough to drive off these volatile compounds but not enough to melt the chondrules.

The **carbonaceous chondrites** contain both chondrules and volatile compounds. Had they been heated, even slightly, the volatiles would have evaporated. Thus the carbonaceous chondrites are the least altered remains of the solar nebula.

*Figure 13-4. A stony meteorite sliced open and polished to show the small, spherical inclusions called chondrules. (Smithsonian Institution.)*

Stony meteorites of the third type are called **achondrites** because they contain no chondrules. They also lack volatiles and appear to have been subjected to intense heat that melted chondrules and drove off volatiles, leaving behind rock with compositions similar to earth's lavas.

The characteristics of iron and stony meteorites hint at a process that could have formed them in large planetesimals (Figure 13-5). If the interior of the planetesimal grew hot enough to melt, differentiation would form an iron-rich core and a silicate mantle. The crust of such a body might never have melted and so might remain unaltered.

If an object were a hundred kilometers in diameter, then the outer layers of rock would insulate the iron core. It would lose heat slowly, and the iron would cool so gradually that large crystals could grow. A sample of such material sliced, polished, and etched would show Widmanstätten patterns. Thus the material in iron meteorites could have formed in the cores of such bodies.

The outer mantle of some objects could have formed achondrites, chondrites, and perhaps even carbonaceous chondrites. The mantle might easily be heated enough to melt the chondrules and form the rock types found in achondrites. In the outer mantle and crust of some planetesimals, chondrules might not have melted and fragments from those regions would resemble chondrites, or, if the volatiles were present and not driven off by heat, carbonaceous chondrites.

Another way the carbonaceous chondrites might have formed places them farther from the sun. Planetesimals in the outer asteroid belt would have been cooler and could have retained volatiles more easily. They could not have been large, or heating and differentiation would have altered them. But the smaller bodies that formed in the outer asteroid belt might have resembled carbonaceous chondrites.

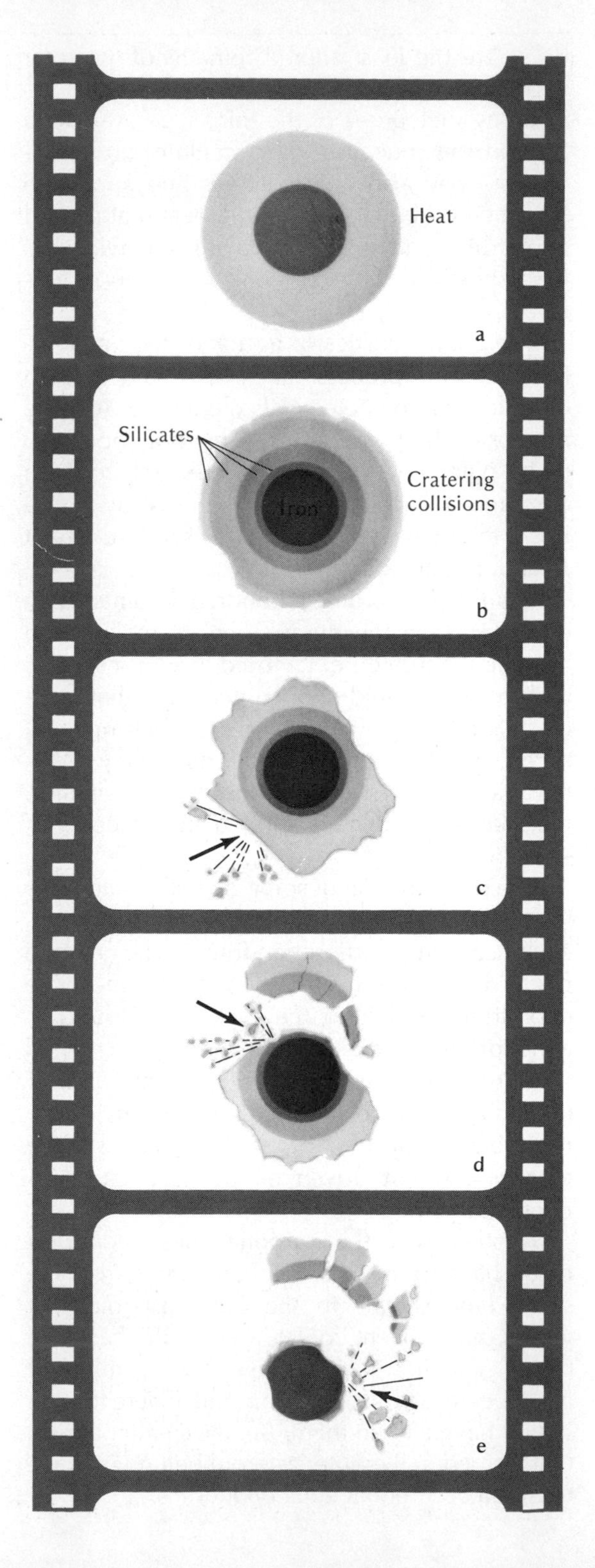


*Figure 13-5. Production of various kinds of meteoroids from asteroids. If it melted, an asteroid (a) could differentiate into an iron core surrounded by layers of different silicate compositions (b). Cratering, collisions, and fragmentation (c, d, e) could break these layers up and produce various kinds of meteorites. (Based on "The Nature of Asteroids" by C. R. Chapman. Copyright © 1975 by Scientific American, Inc. All rights reserved.)*

These theories trace the origin of meteorites to planetesimal-like parent bodies, but they produce a mystery. The small meteoroids in the solar system cannot be fragments of the planetesimals that formed the planets because small meteoroids would be swept up by the planets in only a billion years or less. They could not have survived for 4.6 billion years. Thus the meteorites now in museums all over the world must have broken off planetesimals somewhere in our solar system within the last billion years. Where are these planetesimals?

*Asteroids.* The small bodies that orbit the sun between Mars and Jupiter are apparently the remains of material that was unable to form a planet. Jupiter, just beyond this region, must have scattered the material, throwing some of it into the inner solar system and ejecting some from the system entirely. Jupiter itself may have swept up some of the bodies incorporating them into its mass or adding them to its swarm of satellites.

If the asteroids are the remains of planet-building material, then they are evidently planetesimals. Most of the original planetesimals in the belt were probably removed by Jupiter, but if only a dozen planetesimals remained, they could have produced the asteroid belt. Collisions between planetesimals would have broken many of them into fragments, and more collisions would eventually have produced the ranges of sizes now found in the asteroid belt—that is, from hundreds of kilometers down to bits of dust. Our suspicion that the asteroids are fragments of larger bodies is supported by their irregular shapes.

However, the three largest asteroids—Ceres, Pallas, and Vesta—are approximately spherical. These may be unbroken planetesimals. A body larger than a few hundred kilometers in diameter would have strong enough gravity to hold on to some of the fragments produced during collisions. Thus the largest planetesimals may not have broken up.

If the asteroids are the last remaining planetesimals, then they carry clues of tremendous value. We cannot visit the asteroid belt, but the Mariner and Viking photographs of the moons of Mars may give us a good look at similar bodies (Box 13-2). At present our only samples of asteroids are meteorites, many of which appear to originate in the belt.

The solar nebula must have contained two kinds of planetesimals—rocky ones near the sun, and icy ones farther out. Meteorites and asteroids give us a picture of the rocky bodies, but what of the icy ones? To search out the last of the icy planetesimals, we must study comets.

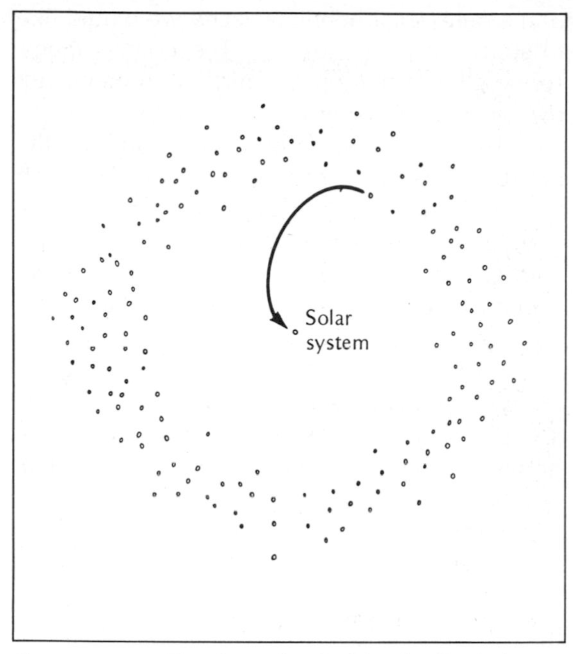


*Figure 13-6. The Oort cloud of icy bodies is believed to extend out to about 50,000 AU from the sun. Collisions among the bodies or the effects of the motions of nearby stars could throw bodies into the inner solar system, where they become comets.*

*Comets.* According to the **Oort cloud theory** (named after the Dutch astronomer Jan Oort), comets are icy bodies a few kilometers in diameter that orbit in a cloud around the sun. This cloud extends out to 50,000 AU, about 20 percent of the way to the nearest star (Figure 13-6). At this distance the ices remain frozen and the comets lack comas or tails. Their orbital velocities are only about 0.13 km/second (about 300 mi/hour), and the slight perturbations caused by the motions of nearby stars could eject a few of these icebergs into long elliptical orbits that carry them into the inner solar system. Thus the Oort cloud theory accounts for the random orientation of cometary orbits—they fall into the solar system from all directions.

The existence of the Oort cloud cannot be confirmed observationally. The icy bodies, mostly less than 10 km in diameter, would be totally undetectable beyond a dozen astronomical units or so. Thus comets can be observed only when they are passing through the inner solar system and the heat of the sun vaporizes some of their ices.

Many astronomers believe that comets originated in the solar nebula among the jovian planets where it was cold enough for ices to solidify. The original planetesimals there must have been rich in ices of water, ammonia, and methane. As the jovian planets grew more massive, they could have swept up many of these planetesimals and ejected others from the solar system, forming the Oort cloud.

Another theory supposes that the comets formed in the distant reaches of the solar nebula before it flattened into a disk. At this stage the nebula was not dense enough to form planetesimals directly, so the theory proposes that regions of the nebula became gravitationally unstable and contracted, flattening into whirling eddies of gas in which icy bodies formed. When the sun cleared the nebula, the icy objects remained as the Oort cloud.

Whether comets formed in the region of the jovian planets or in the distant fringes of the solar system, they represent unaltered samples of the outer solar nebula. Clearly we would like to study their composition. The comets themselves give us that opportunity when they enter the inner solar system.

When a comet falls inward among the planets, it begins a short but showy life. Each time its orbit carries it near the sun, it loses mass as its ices vaporize and the released dust and gas is blown away. The average comet probably cannot survive more than 100 passages through the inner solar system. Halley's comet, for instance, was spectacular in many previous passages, but some believe the 1910 appearance was not as impressive as the earlier passages. If it has begun to run out of ices, its next passage in 1986 may be a disappointment.

Once a comet enters the solar system, it also runs the risk of having its orbit altered by the gravitational influence of the planets (Figure 13-8). If a comet's new orbit keeps it in the inner solar system, the heat rapidly destroys it by vaporizing its ices. A group of comets called "Jupiter's family" is trapped in relatively small orbits that never take them much beyond 5 AU from the sun, and they have lost much of their ice.

When a comet's ices are gone, not much is left but the dirt originally trapped in the ice. Much of the dirt gets scattered along the comet's

---

## BOX 13-2 THE MOONS OF MARS

Deimos and Phobos, the tiny moons of Mars, may be captured asteroids or they may have formed with Mars. In either case, the Mariner and Viking photographs illustrate some of the processes that might affect asteroids as they collide with each other in the asteroid belt.

The photographs reveal a unique set of narrow, parallel grooves on Phobos (Figure 13-7). Averaging 150 m (500 ft) wide and 25 m (80 ft) deep, the grooves run from Stickney, the largest crater, to an oddly featureless region on the opposite side of the satellite. One theory suggests that the grooves are deep fractures produced by the impact that formed the crater. High-resolution photographs show that the grooves are lines of pits, suggesting that the pulverized rock material on the surface has drained into the fractures, or that gas, liberated by the heat of the impact, escaped through the fractures and blew the dusty soil away.

Deimos not only has no grooves; it also looks smoother because of a thicker layer of dust on its surface. This material partially fills craters and covers minor surface irregularities. It seems certain that Deimos experienced collisions in its past, so fractures may be hidden below the debris.

The debris on the surfaces of the moons raises an interesting question. How can the weak gravity of small bodies hold any fragments from meteorite impacts? The escape velocity on Phobos is only about 12 m/second (40 ft/second). An athletic astronaut who could jump 2 m (6 ft) high on earth could jump 2.8 km (1.7 mi) on Phobos. Certainly most of the fragments from an impact should escape, but the slowest particles could fall back in the weak gravity and accumulate on the surface.

Since Deimos is smaller than Phobos, its escape velocity is smaller, so it seems surprising that it has more debris on its surface. This may be related to Phobos's orbit close to Mars. The Martian gravity is almost strong enough to pull loose material off of Phobos's surface, so the moon may be able to retain little of its cratering debris.

The gravitational influence of Phobos on the Viking space craft reveals that the satellite has a low mass. Its average density is only about 2 gm/cm$^3$, which is about the density of carbonaceous chondrites. This plus the dark gray color of the moons leads some astronomers to suggest that they originated in the outer asteroid belt, where it was cool enough for carbonaceous chondrites to form.

---

orbit, forming a swarm of meteoroids (Figure 13-9). If the earth passes through this swarm, the result is a meteor shower, during which an observer might see 60 to 100 meteors per hour instead of the usual 3 to 15.

Since all of the meteoroids belonging to a particular swarm move in nearly identical orbits, they enter earth's atmosphere from the same direction and the meteors they produce appear to radiate from a particular region of the sky. This leads to the custom of naming meteor showers after the constellation from which they appear to radiate. The Perseids, for example, appear to radiate from the constellation Perseus.

The connection between meteor showers and comets is well established. Many meteor showers have been identified with cometary orbits. The Perseid shower, for example, happens every August 12 when the earth crosses the orbit of Comet 1862 III, and the Orionid shower, on October 21, is caused by debris in the orbit of Halley's comet. (See Appendix C.)

If some of these cometary meteors reached the ground, they would bring us samples of icy planetesimals, but, unfortunately, most cometary meteors burn up in the atmosphere. Analysis of the speed and the rate at which they burn suggests they are very low density bits of matter. Collectors carried into the upper atmosphere by rockets and balloons or into space by orbiting

*Figure 13-7. (a) Long parallel grooves about 150 m (500 ft) wide and 25 m (80 ft) deep mark the surface of Phobos. (b) A map of the grooves shows their relation to Stickney, the largest crater. ((a) NASA/JPL; (b) from "Grooves on Phobos" by P. Thomas, J. Vereka, A. Bloom, and T. Duxbury,* Journal of Geophysical Research, *December 19, 1979, Vol. 84, No. B14 Copyright © 1979 by The American Geophysical Union.)*

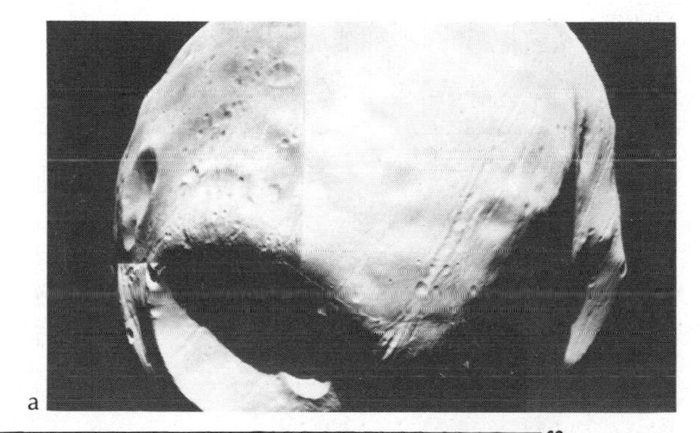

a

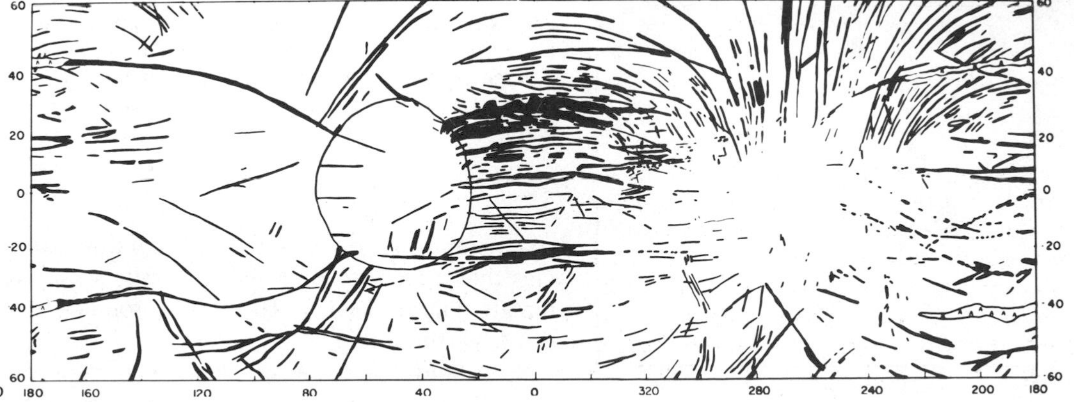


b

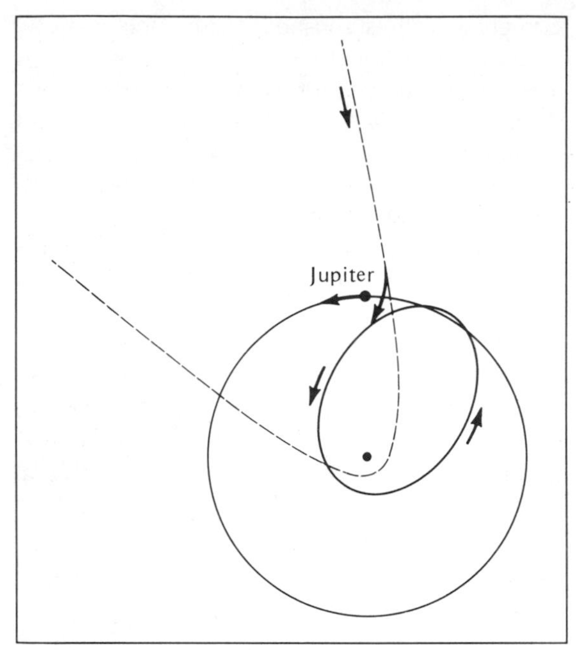


*Figure 13-8.* Massive Jupiter can alter the orbit of a comet (dashed) into a smaller orbit (ellipse) that keeps it in the inner solar system.

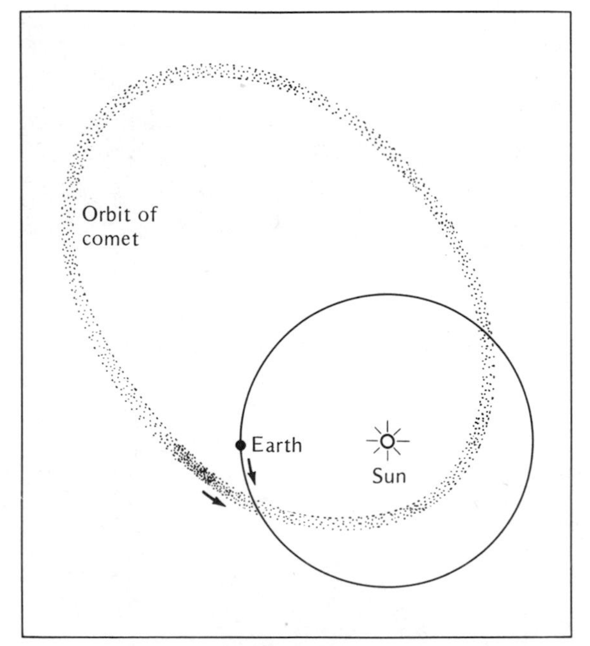


*Figure 13-9.* As a comet's ices evaporate, it releases rocky and metallic bits of material that spread along its orbit. If the earth passes through such material, it experiences a meteor shower.

*Figure 13-10.* Micrometeorites constantly bombard unprotected surfaces in space. Metal plates carried into space aboard rockets, satellites, and spacecraft, and later recovered, are dotted with microscopic craters produced by the impact of micrometeorites no larger than bits of dust. The lines on the metal surface in these drawings are 100 $\mu$ apart. ($1\ \mu = 10^{-6}$m) (Adapted from Dudley Observatory photographs.)

spacecraft have recovered samples of dust particles that appear to be tiny bits of silicates and metals (Figure 13-10). These are probably the dirt from the dirty icebergs.

Our study of comets, asteroids, meteorites, and planetary compositions and motions point to an origin in the same nebula that gave birth to the sun. However, once the nebula cleared the planets stopped growing and began evolving, slowly changing as new processes became important. How these processes molded our beautiful planet and created the cloudy hell of Venus and the deadly dryness of Mars is the subject of the next chapter.

## SUMMARY

The modern theory of the solar system's origin is evolutionary. It proposes that the planets formed from the same cloud of gas that formed the sun.

Planet building probably began as dust grains grew by condensation and accretion into planetesimals ranging from a centimeter to a few kilometers in diameter. These planetesimals settled into a thin plane around the sun and accumulated into larger bodies, the largest of which grew the fastest and eventually became protoplanets. The terrestrial planets must have melted as they formed, differentiating into dense cores and lower-density crusts. The jovian planets probably grew rapidly from icy material and became massive enough to attract and hold vast amounts of nebular gas. Heat of formation raised their temperatures very high when they were young, and Jupiter and Saturn still radiate more heat than they absorb from the sun.

Once the sun became a luminous object, it cleared the nebula, as its light and solar wind pushed material out of the system. The planets helped by absorbing some planetesimals and ejecting others from the system. Once the solar system was clear of debris, planet building ended.

Among the mysteries of the solar system's origin, two stand out. First, the origin of the ring systems around Jupiter, Saturn, and Uranus is not understood. Various processes destroy planetary rings in relatively short intervals, so the rings that now circle at least three of the four jovian planets cannot have lasted since the birth of the planets 4.6 billion years ago. The rings may have been produced later by the capture of debris or the tidal disruption of satellites.

The second mystery is the origin of the earth's moon. Fission, condensation, and capture hypotheses all have flaws. The formation of the moon from material that condensed elsewhere in the solar nebula and was later captured by earth is a compromise theory.

Although the solar nebula must have vanished billions of years ago, it did leave traces. The asteroids, for example, appear to be the remains of material that was never able to produce a planet. Jupiter probably captured or ejected many of the planetesimals in this region. Those that remained collided, broke up, and formed the irregularly shaped asteroids.

The Oort cloud of comets is another fossil of the nebula. These giant icebergs are apparently icy planetesimals. They were either ejected from the outer solar system by the jovian planets, or formed at their present location in the Oort cloud. In either case, they are only observable when they fall into the inner solar system and their ices vaporize in the sun's heat.

The meteoroids are also fossils of the nebula. Most come from the asteroids, and the iron and stony composition suggests that some of the original asteroids differentiated into iron cores and rocky surfaces. The meteors seen in showers are debris scattered along the orbits of comets and consist of small bits of silicate from the dirty icebergs.

## NEW TERMS

planetesimal
condensation
accretion
protoplanet
differentiation
outgassing
Roche limit
radiation pressure
Widmanstätten pattern
chondrite
chondrule
carbonaceous chondrite
achondrite
Oort cloud

## QUESTIONS

1. Describe the processes that formed the planetesimals and protoplanets.
2. Why do some astronomers joke that the moon doesn't exist?
3. What caused the clearing of the solar nebula? Why did that end planet building?
4. Why might we describe meteors, asteroids, and comets as fossils of the solar nebula?
5. Give two possible origins of the objects in the Oort cloud.
6. Why does a comet's tail always point away from the sun?

## PROBLEMS

1. Suppose the earth grew to its present size in a million years through the accretion of planetesimals averaging 100 gm each. On the average, how many planetesimals did earth capture per second?
2. What is the Roche limit for Mars? Do its satellites lie inside this limit? (Hint: See Appendix C.)
3. The velocity of the solar wind is roughly 600 km/second. How long does it take to travel from the sun to the earth?

## RECOMMENDED READING

Cameron, A. G. W. "The Origin and Evolution of the Solar System." *Scientific American* 233 (Sept. 1975), p. 32.

Chapman, C. R. "The Nature of Asteroids." *Scientific American* 232 (Jan. 1975), p. 24.

Grossman, L. "The Most Primitive Objects in the Solar System: Carbonaceous Chondrites." *Scientific American* 232 (Feb. 1975), p. 30.

Hartmann, W. K. *Moons and Planets*. Belmont, Calif.: Wadsworth, 1972.

———. "In the Beginning." *Astronomy* 4 (June 1976), p. 6.

———. "Cratering in the Solar System." *Scientific American* 236 (Jan. 1977), p. 84.

Van de Kamp, P. "Barnard's Star: The Search for Other Solar Systems." *Natural History* 79 (April 1970), p. 38.

Veverka, J. "Phobos and Deimos." *Scientific American* 236 (Feb. 1977), p. 30.

Wetherill, G. W. "Apollo Objects." *Scientific American* 240 (March 1979), p. 54.

Wood, J. A. *The Solar System*. Englewood Cliffs, N. J.: Prentice-Hall, 1979.

*The surface of Mars is marked by streambeds, apparently due to flowing water. Since liquid water cannot now exist on the planet's surface, the streambeds hint that Mars once had a more comfortable climate that might have supported life. (NASA/JPL.)*

# Chapter 14 COMPARATIVE PLANETOLOGY

As soon as a planet forms, its surface begins to change. Only in the last decade have astronomers begun to understand how these changes occur because only recently have space probes returned photographs and measurements from the surfaces of other planets. Until the last few years, comparative planetology, the study of planets by the comparison of different planetary surfaces, was nearly impossible.

Even though we now have detailed information about other planets, we begin our study with the familiar earth. One reason is that the earth is the planet we know best. We can study it carefully and then draw comparisons to other worlds. But another reason is that our home is a planet of extremes. Its interior is molten and generates a magnetic field. Its crust is active, with moving sections that push against each other and trigger earthquakes, volcanoes, and mountain building. Even the earth's atmosphere is extreme. Processes have altered the earth's air from its original composition to a highly unusual, oxygen-rich sea of gas. Once we understand earth's complex properties, the remaining planets in our system should be easier to comprehend.

From our investigation of the earth, we will extract a four-stage history through which all terrestrial planets are presumed to pass. In addition, we will find the same processes at work on other planets. Thus as we explore the solar system, we will not discover entirely new processes, but rather familiar effects working in slightly different ways.

## PLANET EARTH

Like all terrestrial planets, the earth passed through four developmental stages (Figure 14-1). The first, differentiation, occurred during the formation of the planet and allowed the heavy iron and nickel to drain to the center, producing a dense metallic core. The lighter silicates floated to the top and formed a thin, brittle crust.

The second stage, cratering, began as the crust formed. Because the solar nebula was filled with rocky debris, the young earth was heavily battered by meteorites that pulverized the newly formed crust and created a moonlike landscape. The last of the large meteorites blasted out crater basins hundreds of kilometers in diameter. As the solar nebula cleared, the amount of debris decreased, and the level of cratering fell rapidly to its present slow rate.

The third stage, flooding of the basins, began as the decay of radioactive elements heated earth's interior. Lava welled up through fissures in the crust and flooded the deeper basins. Later, as the atmosphere cooled, water condensed and fell as rain, filling the basins and forming the first oceans.

*Figure 14-1. The four stages of planetary development. First: Differentiation into core and crust. Second: Cratering. Third: Flooding of lowlands by lava or water or both. Fourth: Slow surface evolution.*

The fourth stage, slow surface evolution, has continued for the last 3.5 billion years or more. The earth's surface is in constant motion as sections of crust slide over each other, build mountains, and shift continents. In addition, moving water and air erode the surface and wear away geological features.

All terrestrial planets are believed to have passed through these four stages, but some planets, because of their mass or temperature, have emphasized some stages over others. Our goal in this section is to study the earth's interior, crust, and atmosphere in order to establish a base for comparison with other planets. Only by understanding the earth in detail can we understand planets in general.

*The Earth's Interior.* The theory of the origin of planets from the solar nebula predicts that the earth should have melted and differentiated into a dense metallic core with a low-density silicate crust. But did it differentiate? Clearly the earth's surface is made of silicates, but what of the interior?

High temperature and tremendous pressure in the earth's interior make any direct exploration impossible. Even the deepest oil wells extend only a few kilometers down and don't reach through the crust. It is quite impossible to drill far enough to sample the earth's core. Yet earth scientists have studied the interior and found clear proof that the earth did differentiate. This analysis of the earth's interior is possible because earthquakes produce seismic waves that travel through the interior and eventually register on seismographs all over the world.

Such studies show that the interior consists of three parts: a central core, a thick mantle, and a thin crust. Analysis of the kind of seismic waves that can penetrate the core show that it is molten and accounts for about 55 percent of the earth's radius. Theoretical calculations predict that the core is hot (about 4000°K), dense (about 14 gm/cm$^3$), and composed of iron and nickel.

Though the core is hot, it is not entirely molten. Nearer the center the material is under higher pressure, which in turn raises the melting point so high the material cannot melt at the existing temperature. Thus there is an inner core of solid iron and nickel. Estimates suggest the inner core's radius is about 22 percent that of the earth.

The **mantle** is the layer of dense rock and metal oxides that lies between the molten core and the surface. The paths of seismic waves in the mantle show that it is not molten, but it is not precisely solid either. Mantle material behaves as a **plastic,** a material with the properties of a solid but capable of flowing under pressure. The asphalt used in paving roads is a common example of a plastic. It shatters if struck with a sledgehammer, but it bends under the weight of a heavy truck. Just below the earth's crust, where the pressure is less than at greater depths, the mantle is most plastic.

The earth's rocky crust is quite thin, about 70 km (44 mi) under the continents and only 10 km (6 mi) under the oceans. Owing to its lower density (2.5 to 3.5 gm/cm$^3$), this material floats on the denser mantle (3.5 to 5.8 gm/cm$^3$). Unlike the mantle, the crust is brittle and breaks much more easily than the plastic mantle. We will see that this is a characteristic property of many planetary crusts.

Seismic exploration of the earth's interior shows that it did differentiate, but additional phenomena show that the earth has a molten, metallic core. Apparently the earth's magnetic field is a direct result of its rapid rotation and its molten core. The origin of planetary magnetic fields is not yet well understood, but the best current theory is the **dynamo effect.** It supposes that the liquid core is stirred by convection. The rotation of the earth couples this motion into a circulation that generates electric currents throughout the core. Since the highly dense, molten, iron-nickel alloy is a better electrical conductor than copper, the material commonly used for electrical wiring, the currents can flow freely and generate a magnetic field (Figure 14-2).

Though this description of earth's magnetism seems adequate, many mysteries remain. For example, rocks retain traces of the magnetic field in which they solidify, and some contain fields that point backwards. That is, they imply that the earth's magnetic field was reversed at the time they solidified. Careful analysis of such rocks indicates that the earth's field has reversed itself every million years or so, with the north magnetic pole becoming the south magnetic pole and vice versa. These reversals are poorly understood, but may be related to changes in the core convection.

Convection in the earth's core is important because it generates the magnetic field. As we will see in the next section, convection in the mantle constantly remakes the earth's surface.

*Figure 14-2. The dynamo effect couples convection in the liquid core with the earth's rotation and produces electric currents that are believed responsible for the earth's magnetic field.*

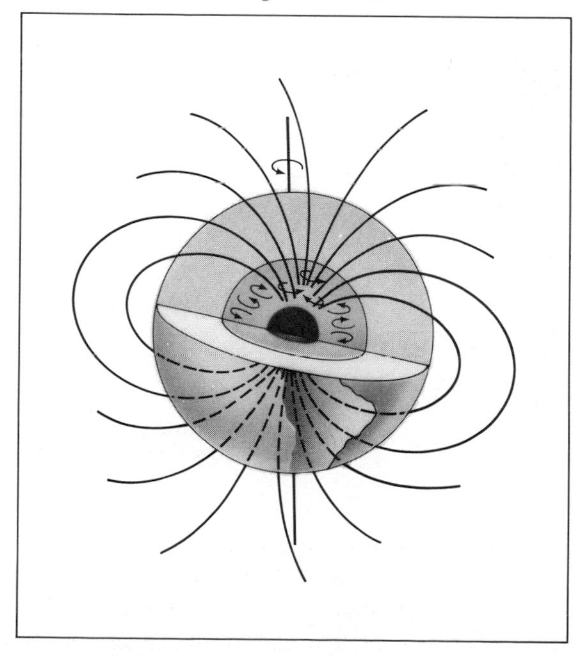

*Plate Tectonics.* The earth's surface is active. It is constantly destroyed and renewed as large sections of crust move about. Geologists refer to this phenomenon as **plate tectonics.** *Plate* refers to the sections of moving crust, and *tectonics* comes from the Greek word for "builder." Plate tectonics is the builder of the earth's surface, because interactions between plates destroy old terrain, push up mountains, and create new crust.

The energy that moves the plates comes from the hot interior. Convection currents of hot mantle material rise from deep layers and spread out under the crust, while other cooler regions sink. This circulation in the mantle apparently drags along large crustal sections at speeds of a few centimeters per year (Figure 14-3).

The key to plate tectonics lies hidden in the ocean floor. The crust there is thin and composed primarily of dense **basalt,** rock characteristic of solidified lava. These seafloors are not at all flat. Running down the center of the ocean beds are undersea mountain ranges called **midocean rises,** and splitting these rises are chasms called **midocean rifts** (Figure 14-3). The rocks of the midocean rises are young, but that only hints at the explanation. The real evidence appeared when scientists measured the residual magnetism of the seafloor and found that some sections retained fields that were the reverse of the earth's present field. Evidently these regions solidified from molten rock during the earth's periodic magnetic reversals. That these regions show alternating bands of magnetism running parallel to the midocean rifts indicates that the earth is creating new crust along the rifts while the seafloors spread outward.

The Atlantic Ocean serves as a good example (Figure 14-4a). Molten material wells up along the midocean rift and solidifies to form young basalt. As the crustal plates spread apart at about 2 to 4 cm/year, the midocean rift opens and new material rises to form more crust. Thus midocean rises are composed of young rock. Parallel magnetic bands form because the earth's field reverses every million years or so, and the solidifying rock records these changes in alternating strips (Figure 14-4b) just as the moving tape in a tape recorder makes a record of the changing magnetic field in the recording head.

If the earth is adding crust in one region, it must be destroying it in another. Crust vanishes in trenches along the coasts of some continents where the spreading seafloor slides downward (Figures 14-3 and 14-7). The collision of moving plates often crumples the crust and pushes up mountains. In addition, the descending seafloor

*Figure 14-3. Currents of moving mantle material rise at midocean and spread the seafloor, moving the continents apart.*

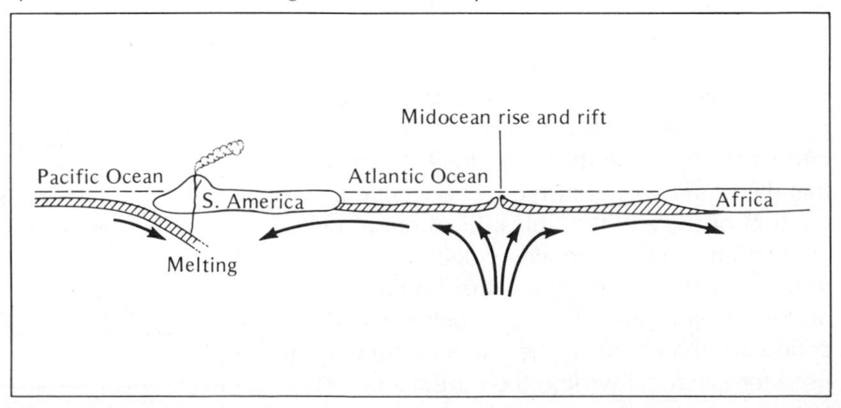

*Figure 14-4. (a) The continents move apart a few centimeters per year as the seafloor spreads and new floor forms along the mid-Atlantic rift (dashed line). Measurements of the residual magnetism of the seafloor near Iceland reveal alternating bands of normal magnetic field (dark) and reversed field (white). (b) The spreading ocean floor records the changing direction of the earth's magnetic field in parallel bands.*

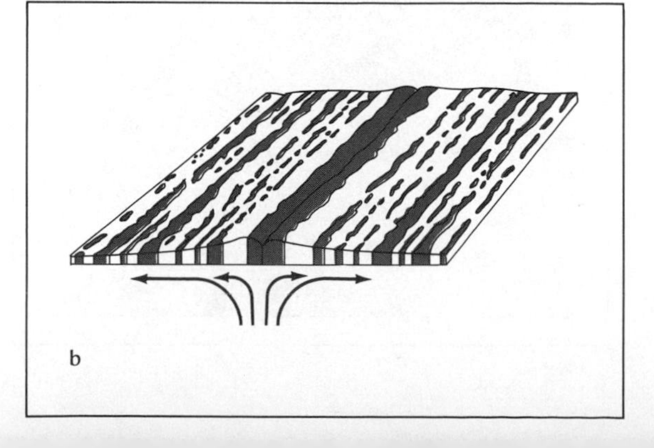

melts along with its accumulated sediment and releases low-density molten rock that rises to the surface and produces volcanism. The Andes Mountains along the west coast of South America and the associated volcanism are due to the descent of the Pacific Ocean floor beneath the continent. The same process produces the volcanoes (including Mt. St. Helens) and earthquakes of Washington State, western Canada, Alaska, and Japan.

The seafloor does not always slip below a continent. The floor of the Atlantic Ocean is locked to North and South America and is pushing the continents westward. Tracing this continental drift backward in time, earth scientists find that North and South America were in contact with Europe and Africa only 200 million years ago. Even a quick glance at a map shows how well these continents fit together (Figure 14-5).

Understanding seafloor spreading and continental drift permits us to construct a history of the earth's surface. The early motions of the earth's crust are unknown, but about 200 million years ago all of the landmasses came together into two large continents. Laurasia lay in the northern hemisphere and consisted of Asia, Europe, and North America. The other land mass was Gondwanaland, composed of South American, Africa, India, Australia, and Antarctica. Rising convection currents in the mantle broke these landmasses into smaller segments that drifted apart, opening the oceans and eventually reaching the present configuration.

*Figure 14-5. About 200 million years ago the continents were joined in the primitive continents Laurasia and Gondwanaland. Since then the motion of crustal plates has moved the continents to their present positions.*

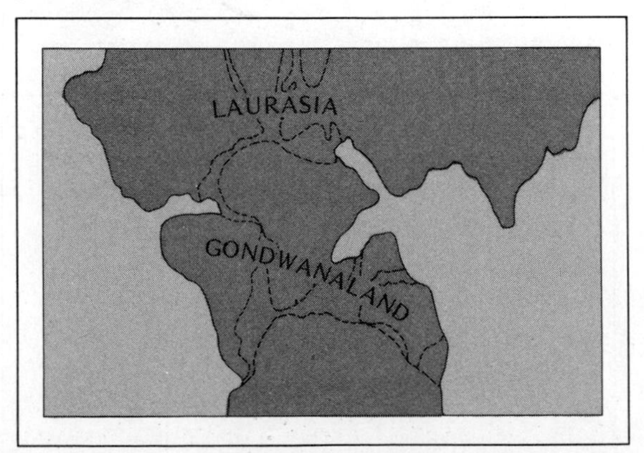


This motion continues today. The plates drift at the urging of the currents in the mantle. They split apart to form new seas, come together to crumple the crust into mountains, and slip against each other to generate earthquakes. The first sign of a continent's splitting is a long, straight, deep depression called a **rift valley.** Africa has recently split from Arabia, opening a rift valley now filled by the Red Sea. Other rift valleys extend southward from the Red Sea across eastern Africa (Figure 14-6). In contrast, the collision of Africa with the east coast of North America roughly 250 million years ago folded the crust into the Appalachian Mountains, just as the collision of India with southern Asia is now building the Himalaya Mountains. In addition, sections of crust in direct contact may slip past each other, as in the case of the Pacific plate carrying part of southern California northward along the San Andreas Fault and generating frequent earthquakes.

Plotting the location of earthquakes and volcanism on a world map reveals the edges of the moving plates (Figure 14-7). The Pacific plate, bounded by a ring of earthquake zones, descends under the Eurasian plate along the North Pacific and Japanese islands. This active zone of circum-Pacific earthquakes and volcanoes is often called the "ring of fire." Earthquakes and volcanoes also occur along the midocean rises where new crust forms and the seafloor spreads.

Because the earth's crust is active, all geological features gradually change. The oldest existing portions of the crust, the Canadian shield and portions of South Africa and Australia, are only about 3.9 billion years old. The constant churning of the earth's surface has wiped away all record of older crust.

*The Atmosphere.* We cannot complete our four-stage history of the earth without mentioning the atmosphere. Not only is it necessary

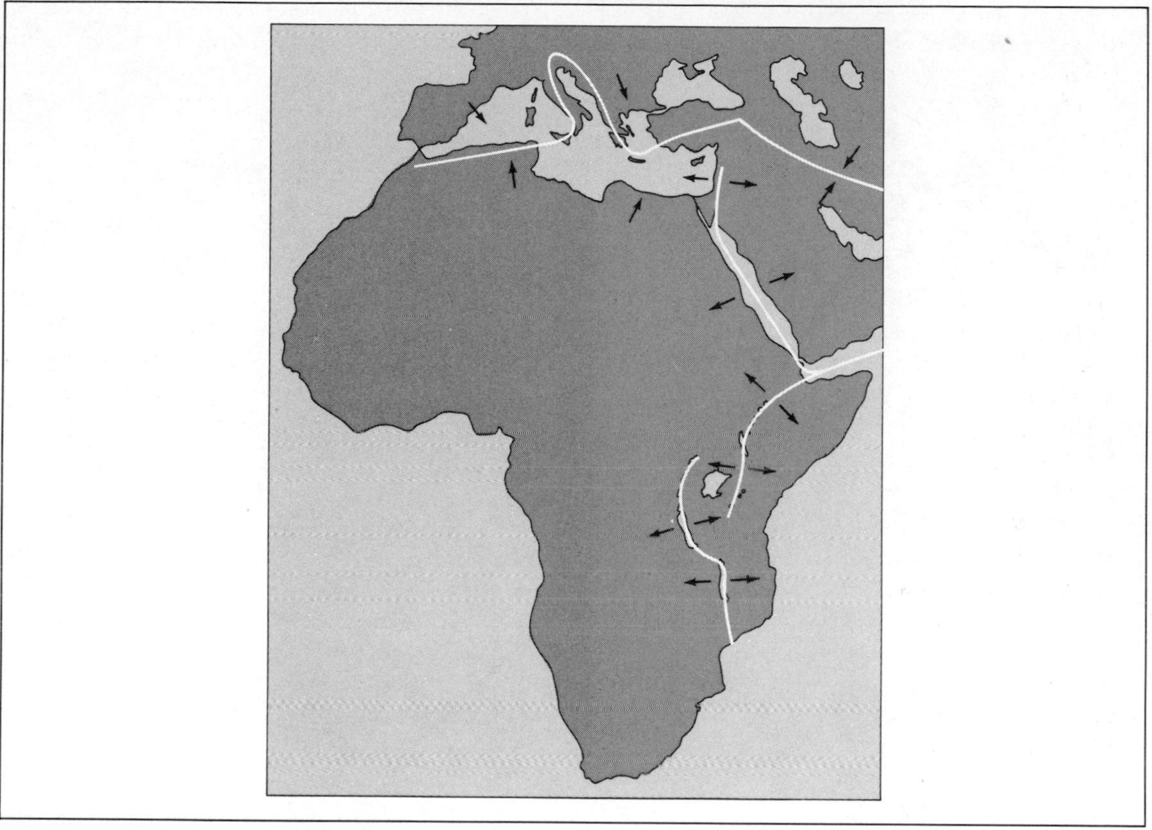

*Figure 14-6. The Red Sea and valleys to the south are rift valleys produced by the separation of crustal plates. The compression of the crust in the area of the Mediterranean accounts for its mountain ranges and volcanism.*

for life, but also it is intimately related to the crust. It affects the surface by eroding geological features through wind and water, and in turn the chemistry of the earth's surface affects the composition of the atmosphere.

Our planet has had three atmospheres—the atmosphere it had when it formed, the atmosphere exhaled from its interior, and its present atmosphere that has developed over the last 3 billion years. The earth's first "air," called the **primeval atmosphere,** consisted of gases from the solar nebula trapped by earth's gravitational field. The composition may have resembled the composition of the solar nebula, being rich in hydrogen and helium and containing traces of methane ($CH_4$), ammonia ($NH_3$), water vapor ($H_2O$), and carbon dioxide ($CO_2$). Any hydrogen and helium, being very light, would have drifted away in the earth's relatively low gravity, leaving an atmosphere containing carbon dioxide, methane, ammonia, and a little water vapor.

As soon as the earth formed, outgassing began to transform the primeval atmosphere. Volcanism is the most impressive example of outgassing, but geysers and the slow seepage of gases through cracks in the crust are also important. Such gases are rich in carbon dioxide, nitrogen, and water vapor, and they also contain methane, ammonia, hydrogen, helium, and other gases. As these gases accumulated and the hydrogen and helium leaked into space, the original primeval atmosphere was replaced, and

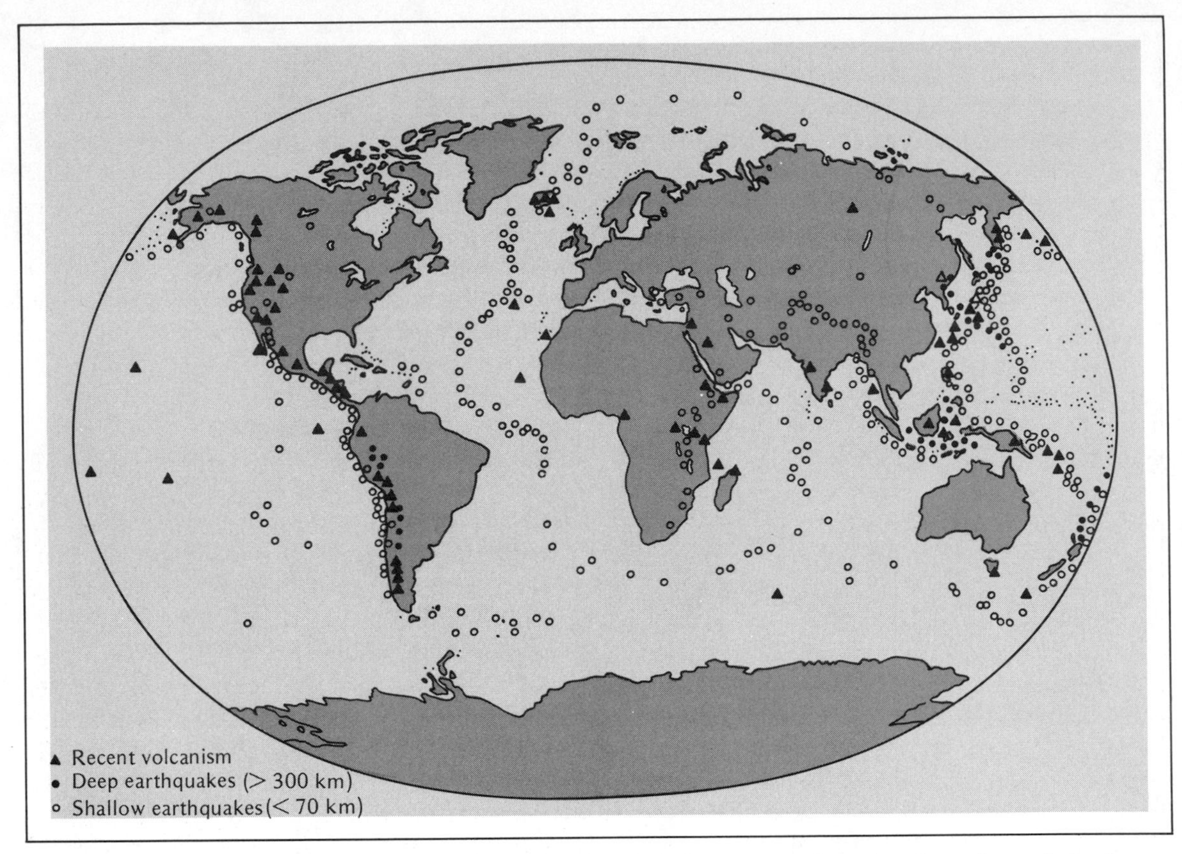


*Figure 14-7. Earthquakes and volcanism outline the crustal plates. Note the volcanism where plates descend below Japan and the west coast of the Americas.*

when the water vapor condensed and formed the oceans, this **secondary atmosphere** was left rich in carbon dioxide, methane, and ammonia.

The first oceans were small, but as more gases escaped and the earth cooled, more and more water rained out of the atmosphere. As the oceans grew larger, carbon dioxide began to dissolve in the water. Carbon dioxide is highly soluble in water—the reason why carbonated beverages are inexpensive—but the primitive oceans could not have absorbed all the atmospheric carbon dioxide had the gas not reacted with dissolved substances such as calcium and magnesium to form silicon dioxide, limestone, and other mineral sediments. Thus the oceans transferred the carbon dioxide from the atmosphere to the seafloor, and left the air rich in methane, ammonia, and water vapor. This removal of carbon dioxide marks the beginning of the transition from the secondary atmosphere to the air we presently breathe.

In the first step of this transition, ultraviolet light eliminated the methane and ammonia. The earth's lower atmosphere is now protected from ultraviolet radiation by a layer of ozone ($O_3$) 15 to 30 km (10 to 20 mi) above the surface. However, the secondary atmosphere did not at first contain free oxygen, so an ozone layer could not form, and the sun's ultraviolet radiation penetrated deep into the atmosphere. The energetic ultraviolet photons broke up water, methane, and ammonia molecules, and the hydrogen escaped to space. The carbon formed carbon dioxide and dissolved in the oceans, leaving nitrogen to accumulate in the atmosphere. By the time the earth was about 2.5 billion years old, the outgassed atmosphere had been cleared of methane and ammonia and was composed mostly of nitrogen.

The origin of the atmospheric oxygen is linked to the origin of life, the subject of Chapter 16, but it is sufficient here to note that life must have originated within a billion years of the earth's formation. This life did not significantly alter the atmosphere, however, until nature invented photosynthesis about 3.3 billion years ago. Photosynthesis absorbs carbon dioxide from the air and utilizes it for plant growth, releasing oxygen back to the atmosphere. Because oxygen is a very reactive element, it combines easily with other compounds, and thus the oxygen abundance grew slowly at first. Apparently the development of large, shallow seas along the continental margins half a billion years ago allowed ocean plants to manufacture oxygen faster than chemical reactions could consume it. Atmospheric oxygen then increased rapidly, and it is still increasing at the rate of about 1 percent every 36 million years.

In this section we have studied the processes that are responsible for the four-step history of the earth. The differentiation of the crust and interior, plate tectonics, and the development of the atmosphere have shaped the surface of our planet and made it a comfortable home for life. But what of other planets? In the next section we will explore the surfaces of other worlds, and then we will study their atmospheres.

## PLANETARY SURFACES

On other planets the balance between processes differs from that on earth. For instance, plate motion on earth moves large sections of crust. Some other planets may have begun plate formation but never reached the stage of plate motion, while others, like the moon, have solid crusts that appear never to have broken into plates.

We have good reason to begin our study of other planets with the moon. The moon is the only astronomical body that humans have visited. It is also the only body beyond the earth from which scientists have samples of known origin that can be analyzed and dated in terrestrial laboratories. In addition, the moon is large enough to pass through the stages of planetary evolution.

*The Moon.* A critical fact about the moon is that it is small, only one-quarter the diameter of the earth. In general, the smaller a body is, the more rapidly it loses its internal heat. The earth's core is large, contains a great deal of heat, and is surrounded by 6378 km (3984 mi) of insulating material. By contrast, the moon's core is small and surrounded by only 1738 km (1080 mi) of insulation. Thus the moon has lost much of its internal heat and is probably now solid rock from its surface well down toward its center. Seismographs left on the moon by the Apollo astronauts have detected moonquakes originating deep within the lunar interior, which suggest it may still have a small, partially molten core (Figure 14-8).

Because the moon's mantle is not plastic, its surface is not broken into plates. In fact, the lunar surface has never been affected by moving plates of crust. There are no lunar mountain ranges marking places where the crust has buckled because of colliding plates. The only mountains are those blasted out of the surface by the impact of large meteorites and small, cold volcanoes dating from the moon's youth.

The moon's small mass means it was unable to retain an atmosphere for any length of time, so erosion has never been important. Wind has not disturbed the dusty surface, and water has never formed ocean beds or river channels. The only erosion comes from heating and cooling of the surface rocks and cratering. The constant bombardment by billions of microscopic meteorites has sandblasted the surface into dust.

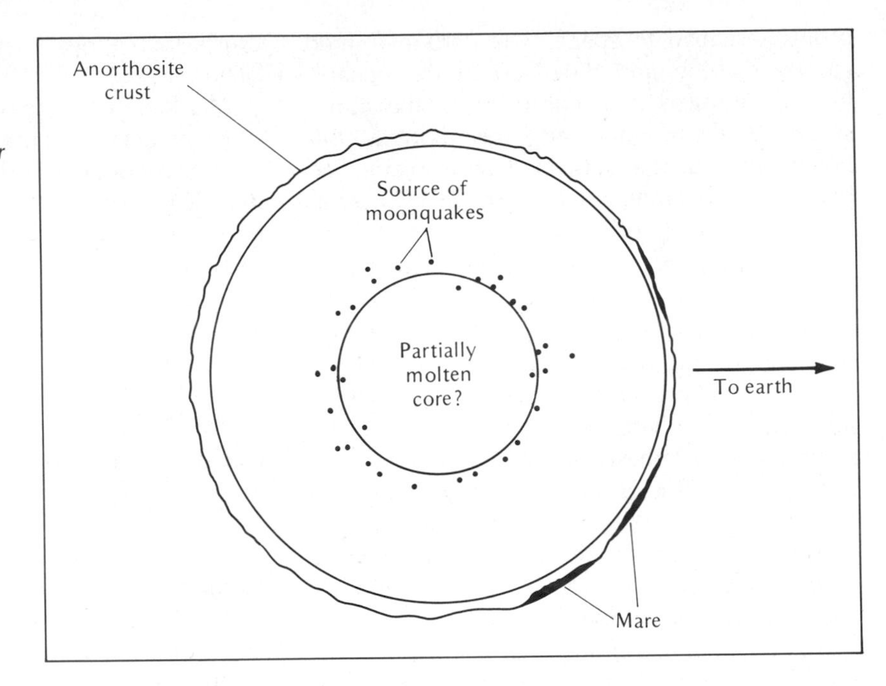


*Figure 14-8. Cross-section of moon shows deep location of moonquakes near what may be a partially molten core. Note that the lunar crust is thinner on the side toward earth.*

Also, the impact of larger meteorites scatters pulverized rock called **ejecta** over large areas of the surface. This is especially obvious where this debris forms **rays,** white streamers radiating from craters (Figure 14-9).

Because the moon's surface has never been modified by moving plates or strongly affected by erosion, it preserves a record of the first three stages of planetary formation unaltered by the passage of 3 billion years. The Apollo explorations recovered samples of basaltic lava, so volcanism and lava flows were once active, but the interior cooled so rapidly the lava flows ended long ago. Since then the only influence on the surface has been cratering.

The lunar terrain divides into two strikingly different regions—the lowlands and the highlands (Figure 14-10, 14-11 and Color Plate 20). The lowlands are smooth, dark plains with generally circular outlines and few craters. These lowlands are called **maria** (singular **mare**), meaning seas, because the first astronomers to examine the moon with telescopes thought they were oceans. In fact, they are low regions filled by successive flows of dark lava. The first Apollo missions landed on these plains and found the surface covered by a thick layer of pulverized lava. The samples returned to earth indicate that the lava solidified about 3.2 to 3.3 billion years ago.

The lunar highlands are the lighter-colored, heavily cratered regions that lie about 3 km (2 mi) higher than the lowlands. Because of their height they were not flooded by the lava that formed the maria, and thus they represent an earlier stage in the moon's history. Because of the interest in the older highlands, the last two Apollo missions risked landing among the jumbled craters. They found that the rock there was not the basalts of the maria but aluminum and calcium silicates called **anorthosite,** a rock type that is lighter in color and lower in density than basalt. As in the case of the maria, the surface in the highlands was covered by a layer of dust and bits of rock debris from the extensive cratering. Nowhere did the explorers find unbroken pieces of the original lunar crust.

The Apollo missions returned 800 pounds of rock samples, of which only 15 percent has been analyzed so far, but that analysis permits

*Figure 14-9. This composite photo of the moon (inverted as it appears in most telescopes) shows the rayed craters Copernicus and Kepler, the partially flooded crater Plato, and the circular outlines of the maria. (Lick Observatory photograph.)*

*Figure 14-10. The lighter-colored lunar highlands have never been flooded by the darker lava flows that formed the maria. Apollo 11 landed in Mare Tranquillitatis. Apollo 17 landed in a small region of highlands near Mare Serenitatis. (Lick Observatory photograph.)*

a

b

*Figure 14-11. (a) Apollo 11 astronaut Edward Aldrin, Jr., stands on the dusty surface of Mare Tranquillitatis in the lunar lowlands, site of the first lunar landing. The flat lava plain is almost featureless and the horizon is straight. (b) In the lunar highlands at Taurus-Littrow the surface is mountainous and irregular. Note the horizon. Apollo 17 scientist-astronaut Harrison Schmitt is shown passing huge boulders. (NASA.)*

us to summarize the four-stage history of the moon. When it formed, it melted and differentiated. A crust of low-density anorthosite solidified between 4.6 and 4.1 billion years ago, as shown by the radioactive ages of the oldest rocks from the highlands.

The second stage, cratering, began as soon as the crust solidified, and the record of the older highlands shows that the cratering was intense during the first billion years. The anorthosite crust was shattered to a depth of about 2 km (1.2 mi) by the constant impact of meteorites. Between 4.1 and 3.9 billion years ago, the rate of cratering declined and the last big impacts formed gigantic crater basins hundreds of kilometers in diameter. The basin that became Mare Imbrium, for instance, was blasted out by the impact of an object about the size of Rhode Island. This Imbrium event occurred about 4.0 billion years ago and hurled ejecta 1400 km (870 mi) away from the site in all directions, blanketing 16 percent of the lunar surface.

The tremendous impacts that formed the lunar basins cracked the anorthosite crust to great depths and led to the third stage. Though the moon cooled rapidly after its formation, some process, perhaps radioactive decay, heated the subsurface material and part of it melted, producing lava that followed the cracks in the crust up into the basins. The basins were flooded by successive lava flows of dark basalts from 3.8 to 3.2 billion years ago (Figure 14-12).

Studies of the lunar crust show that it is thinner on the side toward the earth. This may be due to the earth's tidal effects on the moon. The lava flooded the basins on the earthward side where the crust was thinner, but was unable to flood the lowlands on the far side. Mare Orientale is a gigantic impact basin nearly invisible at the extreme western edge of the moon as seen from earth. It contains only small areas filled with lava (Figure 14-13).

On earth, the deeper basins were flooded first with lava, then with water, but liquid water seems never to have existed on the moon. Thus when the moon cooled further and volcanism retreated into its interior, the airless surface stopped changing. With no water there has been almost no erosion. A few large craters have scarred the smooth maria, and the bombardment of micrometeorites has pulverized the surface, but the moon is geologically dead, frozen between stages three and four.

*Mercury.* Mercury is only 4878 km (3047 mi) in diameter, about 40 percent larger than the

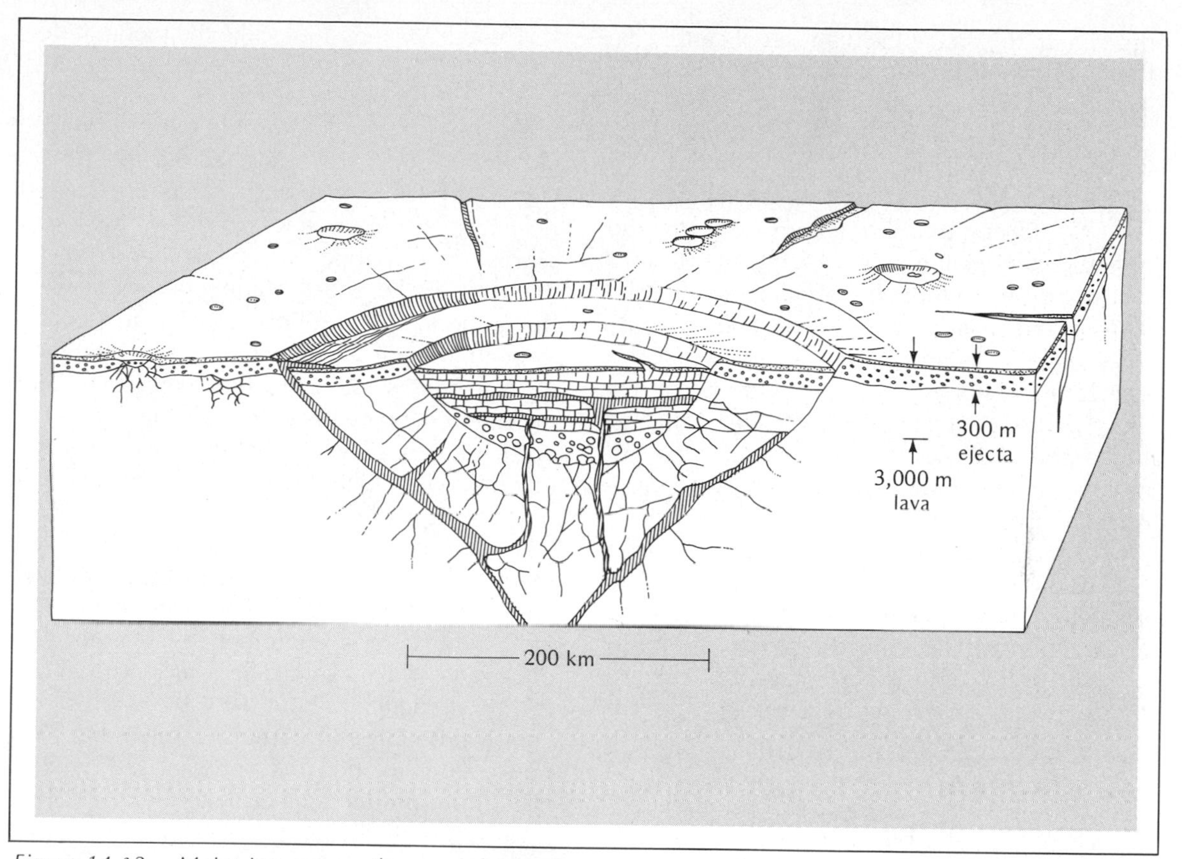


*Figure 14-12.* *Major impacts on the moon broke the crust to great depth and produced large basins. Lava flowed up through the broken crust and flooded the basins to produce maria. (Adapted from William Hartmann, Astronomy: The Cosmic Journey. Belmont, Calif.: Wadsworth Publishing Company, 1978.)*

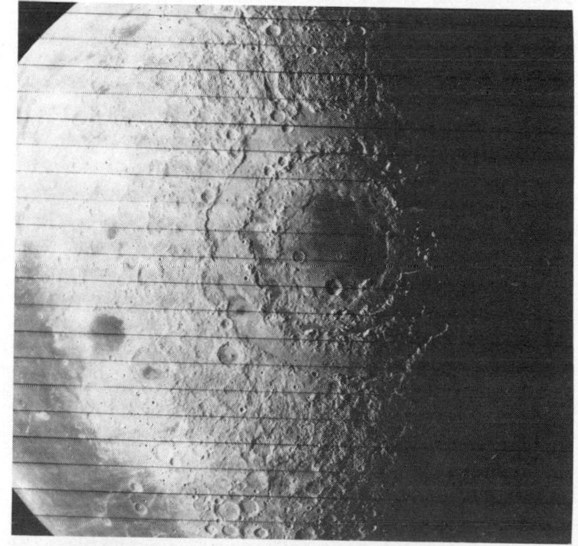

*Figure 14-13.* *Mare Orientale, an impact basin nearly 1000 km (630 mi) in diameter, lies just beyond the western edge of the moon as seen from earth. Note that lava flows have flooded small areas of the basin. (NASA/JPL.)*

moon and 38 percent the diameter of earth. Like the moon, it has been unable to retain an atmosphere and has cooled too fast to have developed plates in motion on a plastic mantle. Thus it too is a dead planet.

Because Mercury's orbit keeps it near the sun, it is difficult to observe from the earth, and little was known about its surface until 1974–75 when Mariner 10 looped through the inner solar system and visited Venus once and Mercury three times. The photos of Mercury revealed a planet whose surface is heavily cratered, much like the moon (see Figure 12-4). Careful analysis of the photos shows that large areas of the surface have been flooded by lava and subsequently cratered.

The largest impact feature on Mercury is the Caloris Basin, a ringed area 1300 km (800 mi) in diameter (Figure 14-14) which looks much like Mare Orientale, the ringed basin on the moon (Figure 14-13). Both features consist of concentric rings of cliffs formed by impact.

*Figure 14-14. Caloris Basin on Mercury lies partially in darkness to the left in this Mariner 10 photo. The basin is a vast ringed impact feature much like Mare Orientale on earth's moon. (NASA/JPL.)*

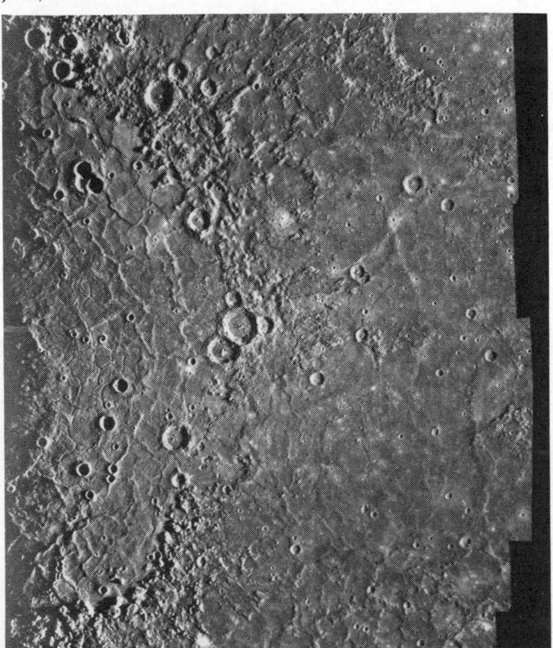

Though Mercury looks like the moon, it does have one characteristic feature the moon lacks. Mariner 10 photos revealed great cliffs called scarps up to 3 km (1.9 mi) high and curving as far as 500 km (310 miles) across the surface (Figure 14-15). Because the cliffs are curved, planetary scientists refer to them as **lobate scarps.** The Mariner 10 encounters did not photograph the entire planet, but everywhere the probe looked it found lobate scarps cutting through the landscape. The scarps even cut through craters, indicating that they formed after the craters. Apparently Mercury's interior cooled and shrank by a few kilometers, forcing the crust to wrinkle into lobate scarps, much like the skin of a drying apple.

Data from Mariner 10 permit us to sketch Mercury's four-stage history. In the first stage, Mercury differentiated. The existence of a nickel-iron core is confirmed by Mariner 10's discovery of a magnetic field around Mercury with a strength of about $10^{-4}$ the earth's. The core is probably still partially molten, but Mercury's slow rotation may prevent the dynamo effect from producing a stronger field.

*Figure 14-15. The lobate scarp (arrows) crosses craters, indicating that Mercury cooled and shrank, wrinkling its crust after many of its craters had formed. (NASA/JPL.)*

The last three stages were similar to those of the moon. Cratering battered the crust and lava flows welled up, filling some of the lowlands just as they did on the moon. Lacking an atmosphere to provide erosion, Mercury has changed very little since the last lava hardened.

*Venus.* Venus is nearly 95 percent the diameter of the earth and has a similar average density. Thus it should have cooled slowly, as the earth did, and may still have an active surface. Unfortunately the surface of Venus is perpetually hidden below thick clouds (Figure 14-16). Our only data on the planet's history come from space probe measurements of its magnetic field, radar maps of its surface, and photographs of small regions of the surface taken by two Russian spacecraft that landed there in 1975.

In 1962, Mariner 2 became the first spacecraft to visit Venus. Since Venus is similar to earth, it should have differentiated and formed a sizable nickel-iron core. Detecting a magnetic field would have confirmed the existence of such a core, but Mariner 2 found no field. Subsequent spacecraft sent to Venus have tried and failed to detect any magnetic field. This cannot be entirely due to the planet's slow rotation. It turns on its axis once in 243 days, but even this slow rotation should produce some magnetic field. The absence of any detectable field is a mystery and illustrates how poorly we understand planetary magnetism.

Almost everything we know about the surface of Venus comes from radar maps. Beginning in 1965, radio astronomers have explored the surface below the clouds by using some of the largest radio telescopes as radar antennas. A normal radio telescope is merely a highly sensitive receiver. A few, however, are equipped to transmit a short burst of radio energy, and then listen for the reflected signal. If the antenna points toward Venus, the signal travels to Venus and back in about five minutes (when the planets are at their closest). By measuring the intensity, timing, and frequencies of the reflected signal, radar astronomers can construct

*Figure 14-16. These ultraviolet photographs of Venus enhance the cloud patterns in the atmosphere. Even at ultraviolet wavelengths the surface is invisible below the clouds. (NASA.)*

maps of the surface of Venus, revealing details as small as 4 km (2.5 mi) in diameter (Figure 14-17). These maps show regions of craters, volcanoes, flooded basins, chains of mountains, and deep valleys.

The craters on Venus appear to be impact features. They range from a few kilometers up to 1000 km (620 mi) in diameter. The craters are shallower than those on the moon, suggesting that erosion is an important process on Venus. The thick atmosphere must protect the surface from smaller meteorites, but large ones smash through the atmosphere with tremendous momentum. Thus we should not be surprised that a few impact craters exist on Venus. What is surprising is that erosion has not erased them as fast as on earth.

The radar maps also show features that appear to be volcanic. A number of volcanoes have been found, some as large as 1000 km across at their base. One feature, called Beta, seems to be a pair of volcanoes, Rhea Mons and Theia Mons, each about 4 km high and 750 km across at the base. These volcanoes appear to be shield volcanoes—wide, low-profile cones produced by highly liquid lava (Figure 14-18). On earth, shield volcanoes develop where molten material breaks through the central portions of crustal plates, as in the case of the Hawaiian Islands. Volcanoes along plate boundaries, such as Mount Fuji in Japan and Mount St. Helens in Washington State, are steeper. Thus the volcanoes on Venus do not necessarily imply plate motion.

Another feature on Venus is Maxwell (Figure 14-17b), a great mountain reaching a height of 11 km (6.9 mi). It tops Mount Everest by 2 km. Some observers believe Maxwell is volcanic. On the west edge of Maxwell lie great ranges of mountains reaching heights from 3 to 6 km (2 to 4 mi). The largest of these ranges is as large as earth's Himalayas. This dramatic region of Venus has been named Terra Ishtar after the Babylonian goddess of love. The landscape might be even more dramatic if the volcanism is still in progress, but radar mapping cannot dis-

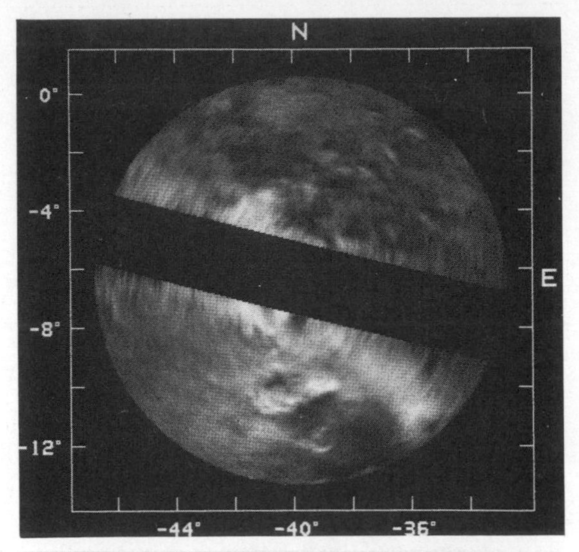


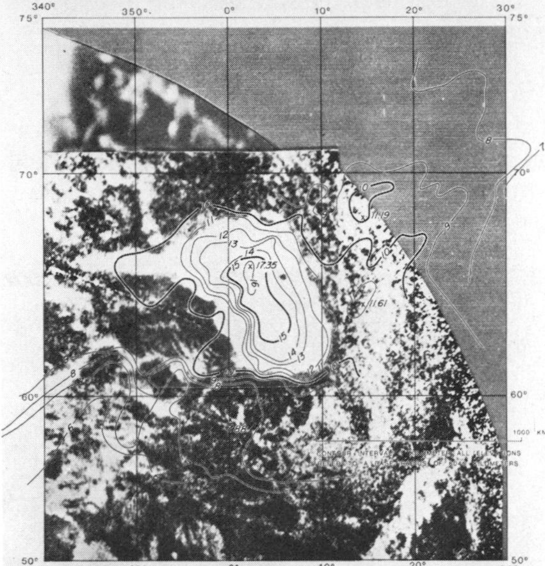


*Figure 14-17. (above) This radar map shows features in a circular region 1500 km (930 mi) in diameter near the equator of Venus. The round feature in the lower half of the image is apparently a caldera (volcanic crater) at the top of a volcanic peak about 500 km (300 mi) in diameter. The dark bar across the map is a region where the radar mapping cannot detect surface details. (below) Radar instruments aboard the Pioneer Venus Orbiter make it possible to add altitude contours to earth-based radar maps. This has revealed that the bright area called Maxwell is a 37,000-foot mountain massif towering 8000 feet higher than Mount Everest. (NASA.)*

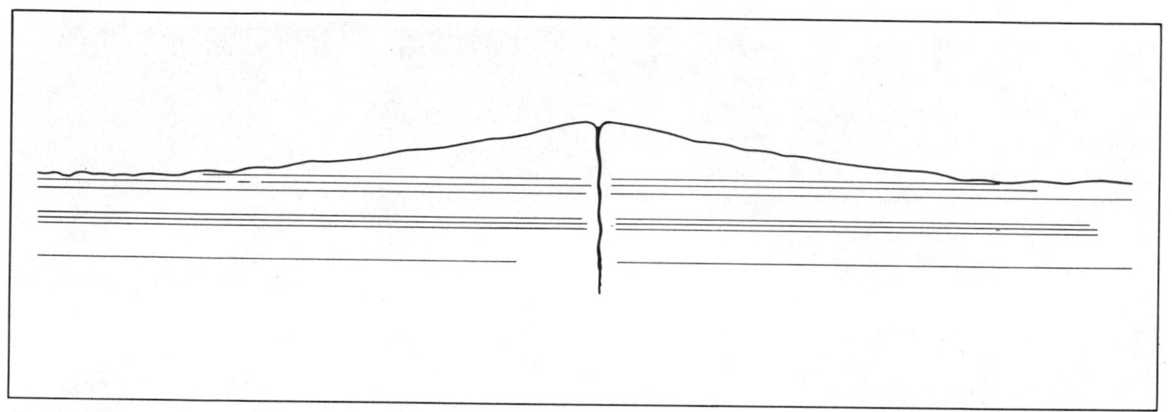

*Figure 14-18. Shield volcanoes may be very large, but, because they form from highly fluid lava, their slopes are not steep.*

tinguish active from inactive volcanism.

About 15 to 20 percent of the planet's surface is covered by relatively flat basins. About 70 percent of the earth is covered by ocean basins produced by seafloor spreading. This difference suggests that whatever the tectonic processes at work on Venus, they are not the same as those on earth. Some scientists have suggested that vertical movement is dominant on Venus, while lateral movement is dominant on earth. Such vertical tectonics would not produce great basins.

The radar studies have found a long deep valley that is more characteristic of plate motion. The valley is about 150 km (90 mi) wide, 2 km (1.2 mi) deep, and about 1400 km (900 mi) long. Such a feature resembles a rift valley, where surface plates move away from each other. Further studies of the surface may eventually show whether this deep feature connects with other valleys to form the boundaries of moving plates.

Only two photographs have ever been taken on the surface of Venus. In October 1975 two Russian spacecraft, Venera 9 and 10, descended through the hot atmosphere and landed on the surface. They found a surface temperature of 745°K (880°F). Venera 9 survived the heat for only 53 minutes and Venera 10 lasted only 65, but during those minutes they took photographs and transmitted them back to earth (Figure 14-19). Venera 9 landed in a region of sharp angular rocks up to 40 cm (16 in) in diameter. The sharp edges show that erosion has not yet rounded the rocks and suggests that the region is geologically young. That is, the rocks were not broken up long ago. A few of the rocks appear to contain holes characteristic of solidified lava. Venera 10 landed 2200 km (1400 mi) away and photographed more rounded rock fragments, suggesting that its region was geologically older. In addition to photographing the surface, the Venera spacecraft also analyzed the surface material for potassium, uranium, and thorium. The measured abundances of these elements show that the surface rocks are probably basalts typical of solidifed lava.

The four-stage history of Venus is difficult to analyze because of its concealing atmosphere. It must have differentiated like the earth and formed a core and crust. The thick atmosphere would protect the surface from small meteorites, but larger objects could penetrate and crater the surface. Volcanism seems to have filled at least some of these basins with lava, and the Venera measurements suggest that lava flows were common. Whether the volcanism is still active and whether the surface ever possessed moving plates will remain mysteries until more detailed radar maps improve our knowledge of the sur-

*Figure 14-19. The surface of Venus, photographed in 1975 by Venera 9 (a) and Venera 10 (b). Portions of the spacecraft are visible at the bottom of each photograph. (Novosti from Sovfoto.)*

face features. Thus the last of the four stages is the most uncertain.

*Mars.* Mercury and the moon are small. Venus and the earth are, for terrestrial planets, large. But Mars occupies an intermediate position. It is twice the diameter of the moon but only 53 percent the earth's diameter. Its small size has allowed it to cool faster than the earth, and much of its atmosphere has leaked away. Its present carbon dioxide atmosphere is only 1 percent as dense as earth's.

In other ways Mars is much like earth. A day on Mars is nearly the same length as an earth day—24 hours and 40 minutes—and its year lasts 1.88 earth years. Also, just as earth's axis is tipped 23½°, Mars's is tipped 24°. As the northern and southern hemispheres turn alternately toward the sun, seasonal changes are visible even through a small telescope. As spring comes to the southern hemisphere, the white polar cap shrinks and the grayish surface markings grow darker, and, according to some observers, greener. At one time these seasonal changes led some to believe that plant life flourished on Mars when spring thawed the polar cap. We will see later that this is not so.

Plants weren't the only thing supposed to flourish on Mars. Because of a discovery made in 1877 by the Italian astronomer Virginio Giovanni Schiaparelli, many people assumed that Mars was inhabited by intelligent life. While he was drawing Mars, Schiaparelli noted straight lines visible in moments when the earth's atmosphere was unusually still. When he reported his discovery he used the Italian word for "channel," the bed of a stream. Newspapers mistranslated it as "canal," an artificially constructed channel, and the canals of Mars were born. Many astronomers of the time could not see any canals, while others drew maps showing hundreds (Figure 14-20). The American astronomer Percival Lowell was especially interested in the canals because of his conviction that they had been built to transport melting

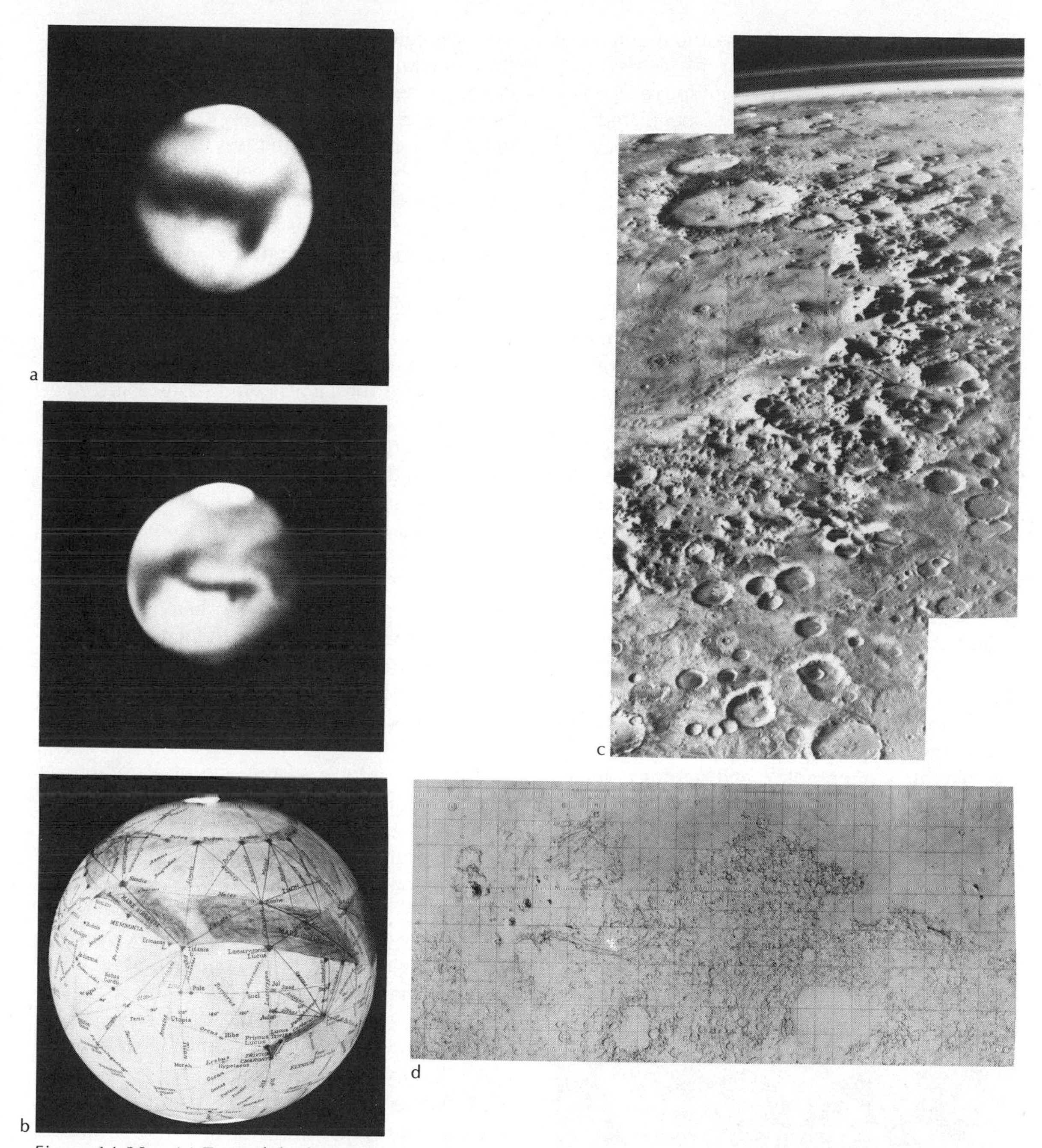


Figure 14-20. (a) Two of the best earth-based photographs of Mars ever taken show little more than a polar cap and general markings. Visual observers mapped hundreds of "canals" (b), but spacecraft revealed Mars is a cratered world (c). Compare the modern map (d) below with the globe (b). ((a) Lick Observatory photograph; (b) Lowell Observatory photograph; (c and d) NASA/JPL.)

water from the polar caps to the arid deserts near the equator.

For nearly 100 years, newspapers, magazines, books, and movies speculated on creatures from Mars. Then in July 1965, Mariner 4 swept past Mars, photographed its surface, and found no canals. The surface was a vast desert, pocked by hundreds of shallow craters, eroded and partially filled with dust (Figure 14-21). Apparently the canals were merely optical illusions produced by the eye-brain process that connects individual dots into complete images.

Not only are the canals nonexistent, but also the seasonal changes on Mars do not appear to be related to plant life. Apparently outcroppings of dark rock are covered by dust in yearly dust storms that sometimes cover the entire planet. As the spring season begins winds gradually strip away the dust and the markings appear to grow darker. The entire planet has a red coloration, probably because of iron oxides (Color Plate 21). The green color of the dark markings is apparently an optical illusion caused by the reddish background.

Though canals and seasonal changes proved to be unrelated to life, later missions to Mars, beginning with Mariner 9 in 1971, revealed that the planet is a complex and geologically interesting place. The photographs showed that the southern hemisphere is old and heavily cratered, giving it a lunar appearance. The northern hemisphere, however, has few craters, and those few are much sharper and less eroded, and thus younger. The region has been smoothed by repeated lava flows that have buried the original surface, and two areas of volcanic cones have been found. In addition, the northern hemisphere is the site of deformed crustal sections in the form of uplifted blocks and collapsed depressions that suggest geological activity.

*Figure 14-21. The Martian surface seen by Viking 1 is a desert of dust and broken rock fragments. The large rock at the left is about 1 by 3 m (3 by 10 ft). (NASA/JPL.)*

Martian volcanoes are of the shield type, showing that the lava flowed easily. Nevertheless, the largest volcano, Olympus Mons, is a vast structure (Figure 14-22). Its base is 600 km (370 mi) in diameter and it towers 25 km (16 mi) above the surface. In contrast, the largest volcano on earth is Mauna Loa in Hawaii. It rises only 10 km (6 mi) above its base on the Pacific ocean floor, and its base is only 225 km (140 mi) in diameter. Mauna Loa is so heavy that it has sunk into the earth's crust, producing an undersea moat around its base, but Olympus Mons, 2.5 times higher, has no moat and is supported entirely by the Martian crust. Evidently the crust of Mars is much thicker than earth's.

Olympus Mons is only one of a number of volcanoes that lie in the Tharsis region near the Martian equator. This region is about 10 km (6 mi) higher than the surrounding surface and appears to have been pushed up by subsurface activity. A similar uplifted volcanic plain, the Elysium region, appears to be older than the Tharsis region. It is more heavily cratered and eroded. The Elysium region was volcanically active about 1.5 billion years ago, but the Tharsis volcanoes may have been active as recently as 0.2 billion years ago.

When the crust of a planet is strained, it may break, producing faults. The Tharsis region is marked by a system of faults, and near it is a great valley, Valles Marineris, named after the Mariner spacecraft that first photographed it (Figure 14-23). The valley is apparently a block of crust that has dropped downward along

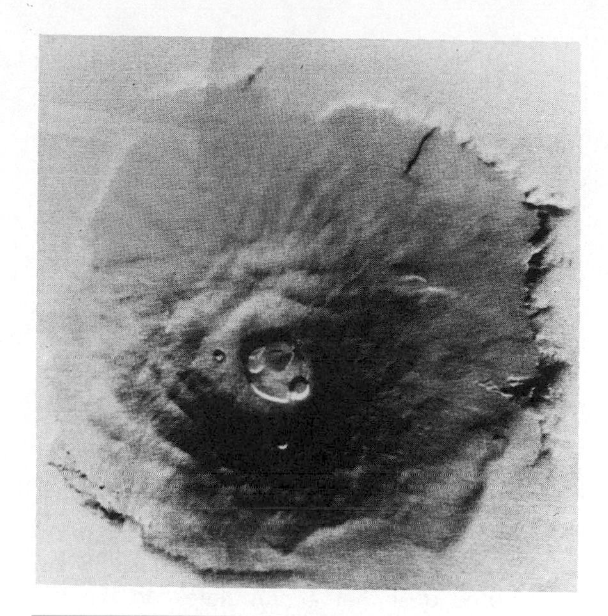

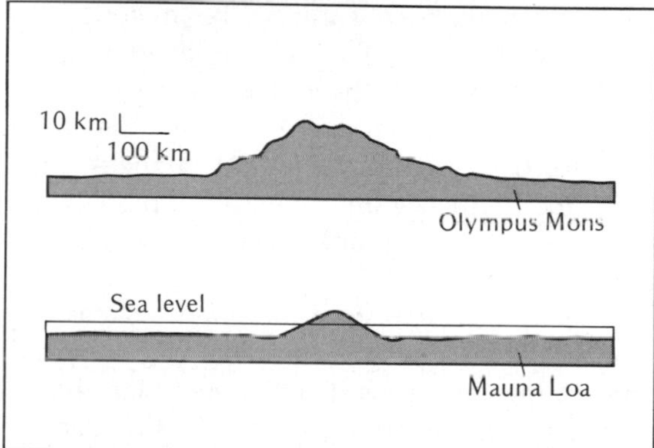


*Figure 14-22.* *Olympus Mons, the tremendous shield volcano on Mars, is much larger than Mauna Loa, largest volcano on earth. Mauna Loa has sunk into the crust, producing a moat around its base. Olympus Mons has no moat, suggesting the Martian crust is stronger than the earth's. (NASA.)*

parallel faults. Erosion and landslides have further modified the structure into a valley nearly 4000 km (2500 mi) long, stretching almost 19 percent of the way around the planet. On earth, it would reach from New York to Los Angeles. At its widest it is 200 km (120 mi) wide, and at its deepest it reaches down 6 km (4 mi), four times the depth of the Grand Canyon.

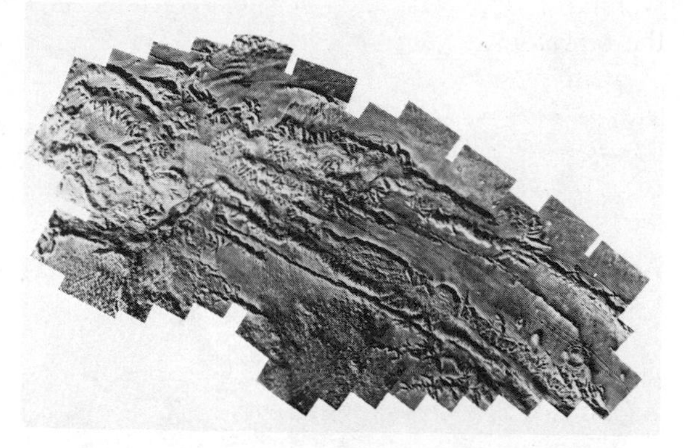

*Figure 14-23.* *A photomosaic of a portion of Valles Marineris, a long, complex valley that stretches 4000 km (2500 mi). The segment shown would reach from one side of Texas to the other. (NASA/JPL.)*

Where Valles Marineris begins, near the Tharsis region, it is marked by numerous faults, suggesting that the valley is related to the uplifted volcanic region. The number of craters in the valley indicates that it is 1 to 2 billion years old, placing its formation sometime before the end of volcanism in the Tharsis region. Some astronomers identify the Valles Marineris as the beginning of a crustal plate boundary. It looks like a rift valley, but it does not connect with other valleys to outline an entire plate. It is as if the crust began to break into plates, but cooled and thickened before the plates could separate entirely and begin to move. Valles Marineris may be the frozen beginnings of this process.

While geologists were fascinated to see signs of an active surface on Mars, they were especially excited to see dry streambeds (Figure 14-24) much like the dry arroyos common in America's Southwest. The streambeds show many of the characteristic forms taken by riverbeds formed on earth—sandbars, tributary streams, braided riverbeds cut by changing currents, and the meandering path characteristic of a river flowing through nearly flat terrain. The air pressure on Mars is now too low to keep water from boiling away, so liquid water could not have cut the streambeds recently. Some aspects

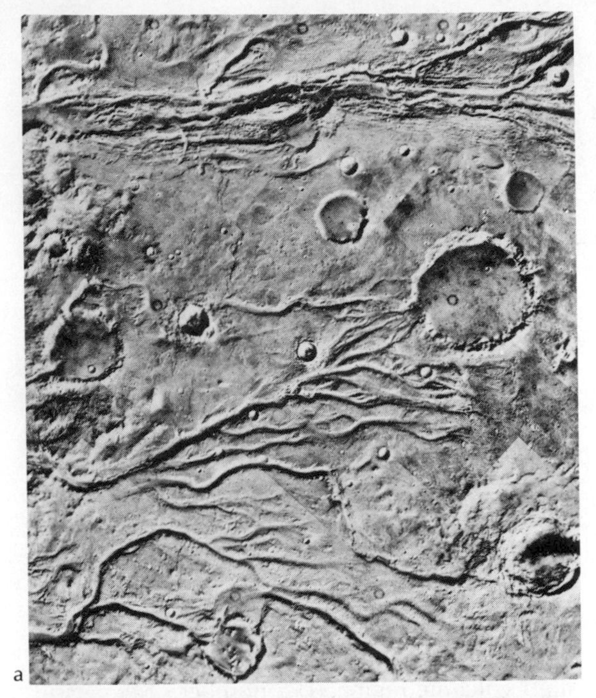

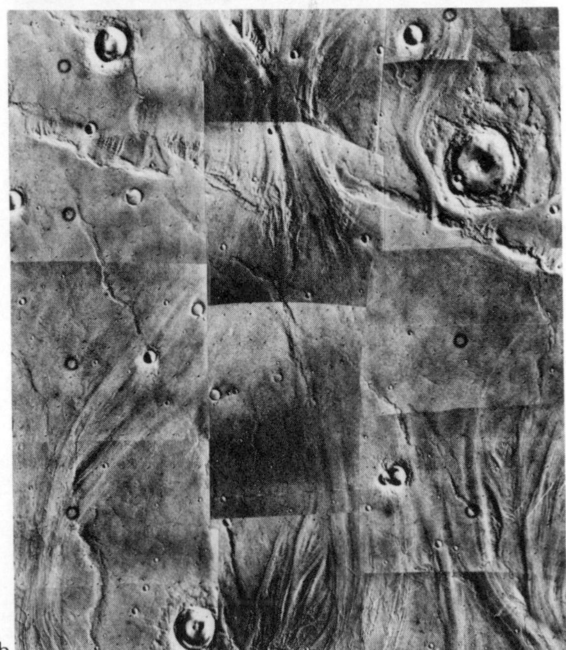

*Figure 14-24. (a) Dry river channels flow through a cratered Martian plain. (b) Signs of erosion due to a moving liquid (probably water) mark this cratered region of Mars. (NASA/JPL.)*

of the stream channels suggest that they are quite old, perhaps as old as 3.5 billion years. Thus Mars appears to have once had a denser atmosphere with liquid water flowing over the surface.

The streambeds suggest that the Martian climate has changed, and careful photographic studies of the polar caps reveal more evidence of climatic variation (Figure 14-25). The polar caps consist of frozen carbon dioxide (dry ice) with frozen water beneath. When spring begins in a hemisphere, the corresponding polar cap begins to shrink as the carbon dioxide turns directly into gas. The water, however, never melts, but stays behind in a smaller permanent ice cap. The region around this permanent cap is marked by layered terrain, evidently deposited by wind-borne dust that accumulates on the ice cap. When the cap shrinks, the material accumulates in layers. Each of the layers is 50 to 100 m (31 to 62 ft) thick. The variation in the layering suggests that some periodic change in climate, perhaps due to orbital changes, may affect the frequency and intensity of planetwide dust storms, and thus alter the rate at which material is deposited.

The four-stage history of Mars is a case of arrested development. The planet began by differentiating into a crust and core. Mars has a magnetic field no stronger than 0.03 percent of earth's, even though it spins rapidly, so it cannot have a molten, conducting core. Since Mars is large enough to remain molten at its center, we must conclude that its core contains not metallic iron, but some less conductive compound.

The crust of Mars is now quite thick, as shown by the mass of Olympus Mons, but it was thinner in the past. Cratering may have broken or at least weakened the crust, triggering lava flows that flooded some basins. Why most of the flows occurred in the northern hemisphere is unknown. Volcanism may have pushed up the Tharsis and Elysium regions and broken the crust to form Valles Marineris, but apparently crustal plates never developed because the planet cooled rapidly and the crust grew thick.

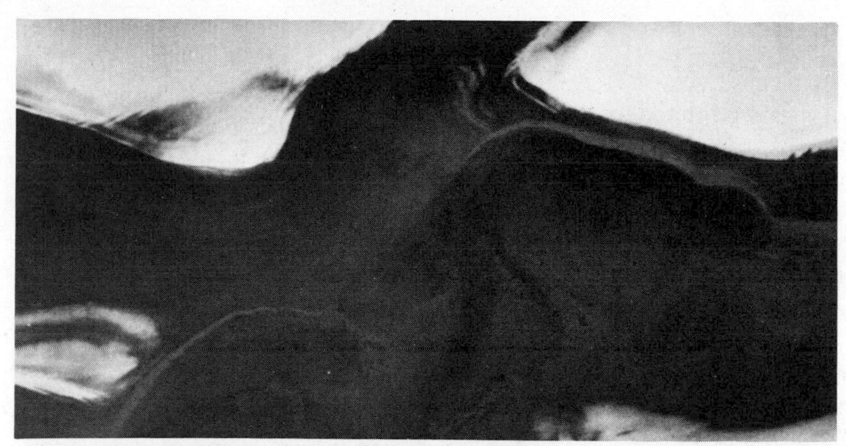

*Figure 14-25. (a) At midsummer the carbon dioxide has evaporated from the north polar cap of Mars, leaving frozen water behind. This Viking photo of the boundary of the polar cap shows layered terrain, apparently due to periodic changes in the Martian climate. (b) Winter on Mars brings a frost of water frozen out of the thin atmosphere. The frost layer may be no more than one-thousandth of an inch thick. (NASA/JPL.)*

The large size of the volcanoes on Mars may be evidence that the plates do not move. On earth, volcanoes like those that formed the Hawaiian Islands occur over rising currents of hot material beneath the crustal plate. Because the plate moves, the hot material breaks through the crust repeatedly and forms a chain of volcanoes instead of a single large feature. The Hawaiian Islands are merely the most recent of a series of volcanic islands called the Hawaiian-Emperor island chain. It stretches nearly 3800 km (2300 mi) across the Pacific Ocean. If this idea is correct, then the lack of plate motion on Mars could have allowed Martian volcanoes to grow to gigantic proportions.

The fourth stage in the history of Mars has been one of slow decline. Volcanic activity probably reached a maximum about 1 billion years ago, and since then the planet has cooled and the crust has thickened. The plates appear never to have pushed against each other and built mountains, and now slow erosion is wearing away the volcanic and impact features.

Our discussion has isolated various geological influences on planetary surfaces, but we have ignored an important factor, planetary atmospheres. An atmosphere that permits water to exist as a liquid can produce rapid erosion. If a planet's atmosphere is thick enough, as is Venus's, it can trap heat and raise the surface temperature. Clearly any discussion of planets would be incomplete without an explanation of the origin and evolution of the gases that surround them.

## PLANETARY ATMOSPHERES

Our study of planetary atmospheres is limited to three terrestrial planets—Venus, Earth, and Mars. The moon and Mercury have no atmosphere, and the jovian planets are so different they deserve separate discussion. We have already discussed the origin and evolution of earth's atmosphere, so only Mars and Venus remain.

The planets illustrate two important principles. Mars has lost much of its atmosphere, demonstrating that a planet's mass and temperature determine how much and what kind of gas it can retain. Venus, being nearly as large as earth, has been able to retain most gases, but it has suffered an atmospheric heat imbalance that has made its surface a sweltering wasteland.

*The Loss of the Martian Atmosphere.* The present atmosphere of Mars is 95 percent carbon dioxide, 2 to 3 percent nitrogen, and about 2 percent argon. Its density is only about 1 percent the earth's. This thin atmosphere is evidently the remains of a thicker blanket that allowed liquid water to carve the streambeds on the Martian surface.

The gases that now cover the surface of Mars were outgassed from its interior. Since Mars formed farther from the sun than the earth did, it may have incorporated more volatiles and then outgassed more. However, most outgassing occurred during the first billion years when the planet's surface was active. Mars has cooled and now releases little gas.

How much atmosphere a planet has depends on how rapidly it releases internal gas and how rapidly it loses gas to space. Since Mars is no longer producing gas rapidly, its losses to space have thinned its atmosphere.

How rapidly a planet loses gas depends on its mass and temperature. The more massive the planet, the higher its escape velocity (Chapter 7) and the more difficult it is for gas atoms to leak into space. Mars has a mass less than 11 percent of earth's and its escape velocity is only 5 km/second, less than half earth's.

A planet's temperature is also important. If the gas is hot, its molecules have a higher average velocity and are more likely to exceed escape velocity. Thus a hot planet is less likely to retain an atmosphere than a cooler planet. However, the velocity of a molecule in a gas also depends on the mass of that molecule. On the average, a low-mass molecule travels faster than a high-mass molecule. Thus a planet loses its lightest gases more easily because they travel the fastest.

Applying these factors to individual planets, we find that all of the terrestrial planets are too small to retain hydrogen and helium. Those molecules are light and leak into space. Though Venus and earth can hold onto their remaining gases, Mercury and the moon are too small and have lost all of their air. Mars is barely able to retain water vapor, methane, and ammonia, but it has a better gravitational grip on carbon dioxide.

Mars should have been able to retain some water and methane but such molecules can be broken up by ultraviolet photons in sunlight. On earth, the ozone layer protects the atmosphere from ultraviolet radiation, but Mars never developed an oxygen-rich atmosphere so it never had an ozone layer. Without that kind of protection, water and methane molecules absorbed ultraviolet photons and broke up. The hydrogen escaped to space, and the oxygen formed oxides in the soil or carbon dioxide in the air. Thus molecules that are too heavy to leak into space can be lost if they dissociate into lighter fragments.

*The Venusian Greenhouse.* Although Venus is the earth's twin in size, the composition, temperature, and density of its atmosphere make it the most inhospitable of planets. About 97 percent of its atmosphere is carbon dioxide, and 2 percent is nitrogen. The remaining 1 percent is water, sulfuric acid ($H_2SO_4$), hydrochloric acid (HCl), and hydrofluoric acid (HF). In fact, the thick clouds that hide the surface are believed to be composed of sulfuric acid droplets and microscopic sulfur crystals. American and Russian spacecraft that have reached the surface report that the temperature is 745°K (880°F) and the atmospheric pressure is 90 times earth's. The air is so dense that if humans could survive its unpleasant composition, intense heat, and high pressure, they could strap wings to their arms and fly.

The present atmosphere of Venus is the result of the planet's proximity to the sun. It is 30 percent closer than the earth, and receives twice as much solar energy. Thus when it formed, it was warmer than the earth, and either this prevented liquid water from condensing and forming oceans, or it evaporated the oceans soon after they began to form. In either case, Venus lacked oceans to absorb carbon dioxide and convert it into mineral deposits. While the earth was removing carbon dioxide from its atmosphere via its oceans, Venus was adding more and more to its atmosphere via its volcanism. When its atmosphere grew rich in carbon dioxide, the fate of Venus was sealed.

A carbon dioxide atmosphere can trap heat by a process called the **greenhouse effect.** When sunlight shines through the glass roof of a greenhouse, it heats the benches and plants inside (Figure 14-26). The warmed interior radiates heat in the form of infrared radiation, but the infrared photons cannot get out through the glass. Heat is trapped within the greenhouse and the temperature climbs until the glass itself grows warm enough to radiate heat away as fast as the sunlight enters.* In the case of a planet, carbon dioxide in the atmosphere admits sunlight and the surface grows warm. However, carbon dioxide is not transparent to infrared radiation, so heat is trapped and the planet's surface temperature rises.

Venus was caught by a runaway greenhouse effect. The rising temperature baked more carbon dioxide out of the surface, and the atmosphere became even less transparent to infrared, which forced the temperature even higher. The surface is now so hot even chlorine and fluorine have baked out of the rock and formed hydrochloric and hydrofluoric acid vapor.

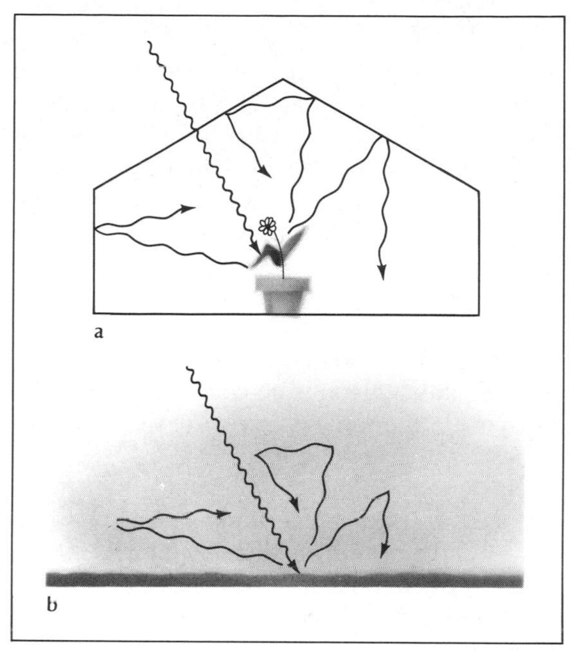


*Figure 14-26. The greenhouse effect. Short-wavelength light can enter a greenhouse (a) and heat its contents, but the longer wavelength infrared cannot get out. The same process heats Venus (b) because its atmosphere of $CO_2$ is not transparent to infrared.*

The earth avoided this runaway greenhouse effect because it was farther from the sun and cooler. Thus it could form and preserve liquid-water oceans to absorb the carbon dioxide, which left a nitrogen atmosphere that was relatively transparent in some parts of the infrared. This points out the risk we run by burning fossil fuels and adding carbon dioxide to the earth's air. Astronomers have measured a gradual increase in carbon dioxide in earth's atmosphere over the last decade, but it does not appear that we are as yet triggering a runaway greenhouse effect.

* A greenhouse also grows warm because the walls prevent the warm air from mixing with the cooler air outside.

## SUMMARY

All terrestrial planets pass through a four-stage history: (1) differentiation; (2) cratering; (3) flooding of the crater basins by lava, water, or both; (4) slow surface evolution. The importance of each of these processes in the evolution of a planet depends on the planet's mass and temperature.

The earth is the largest terrestrial planet in our solar system, and it has passed through all four stages. Studies show that the earth has differentiated into a metallic core and a silicate crust. Currents in the molten portion of the core produce earth's magnetic field.

Because the earth is still partially molten, its surface is active. Plate tectonics refers to the motion of large crustal sections. As new crust appears in the midocean rifts, it pushes the plates apart and destroys old crust where plates slide over each other.

Earth's atmosphere has changed drastically since its formation. The first atmosphere consisted of gas from the nebula that formed the planets. Outgassing produced a secondary atmosphere rich in $CO_2$. As the $CO_2$ dissolved in the oceans and became incorporated in the bottom sediment, surface plant life produced oxygen.

When the moon formed, it differentiated into a dense core and a low-density crust of anorthosite. Cratering broke up this crust, and lava flows of dark basalt filled the lowlands, producing the maria. Cratering has pulverized the surface of the highlands and the lowlands, covering the lunar surface with ejecta.

Mercury, like the moon, has no atmosphere and is covered by craters, but unlike the moon, its surface is marked by lobate scarps. The long curved cliffs suggest that, after formation, Mercury cooled, contracted, and wrinkled its crust.

Little is known about the surface of Venus because it is perpetually hidden by its thick atmosphere. Radar maps show that it is marked by craters, volcanoes, mountains, and flooded basins. One large valley resembles a rift valley, which may be related to surface activity.

Because Mars is small, it has cooled relatively rapidly and is no longer active. Nevertheless, its surface contains lava plains, volcanoes, and a long feature, possibly a rift valley. These suggest that the surface was active in the past.

Venus and Mars illustrate two principles of planetary atmospheres. A thick atmosphere that is opaque to infrared can trap heat via the greenhouse effect and produce a high surface temperature. The earth avoided this by dissolving its $CO_2$ in the oceans, but Venus was too warm for liquid water to form. The second principle is that the density and composition of a planet's atmosphere depend on the mass and temperature of the planet. A warm, low-mass planet has difficulty retaining an atmosphere. Also, since light gases leak away most easily, a planet may be able to retain heavier gases like $CO_2$, while $H_2$ and He leak into space.

## NEW TERMS

mantle
plastic
dynamo effect
plate tectonics
basalt
midocean rise
midocean rift
rift valley
primeval atmosphere
secondary atmosphere
ejecta
rays
maria
anorthosite
lobate scarps
shield volcano
greenhouse effect

## QUESTIONS

1. Summarize the four stages in the development of a terrestrial planet.
2. What evidence do we have that the earth differentiated?
3. How does plate tectonics create and destroy the earth's crust?
4. Describe the evolution of earth's atmosphere.
5. What has produced all of the oxygen in our atmosphere?
6. Summarize the stages in the development of the moon.
7. What kinds of erosion are now active on the moon?
8. What are lobate scarps?
9. How did the earth avoid the greenhouse effect that made Venus so hot?
10. What evidence do we have that Venus and Mars had active crusts?
11. Why doesn't Mars have mountain ranges like earth's?
12. Why is the atmosphere of Venus so rich in carbon dioxide?

## PROBLEMS

1. If the Atlantic seafloor is spreading at 3 cm/year and is now 6400 km (4000 mi) wide, how long ago were the continents in contact?
2. Why do small planets cool faster than large planets?
3. How long would it take for radio signals to travel from earth to Venus and back if Venus were at its farthest point from earth? Why are such observations impractical?
4. Why do low-mass atoms escape from a planet more easily than massive atoms?
5. How does the temperature of a planet affect its ability to retain an atmosphere?

## RECOMMENDED READING


Carr, M. "The Volcanoes of Mars." *Scientific American* 234 (Jan. 1976), p. 32.

Cloud, Preston. *Cosmos, Earth and Man*. New Haven: Yale University Press, 1977.

Hartmann, W. K. "The Early History of Planet Earth." *Astronomy* 6 (Aug. 1978), p. 6. Reprinted in *Astronomy: Selected Readings,* ed. M. A. Seeds. Menlo Park, Calif.: Benjamin/Cummings, 1980.

———. *Moons and Planets*. Belmont, Calif.: Wadsworth, 1972.

———. "The Moon's Early History." *Astronomy* 4 (Sept. 1976), p. 6.

———. "The Significance of the Planet Mercury." *Sky and Telescope* 51 (May 1976), p. 307.

———. "Cratering in the Solar System." *Scientific American* 236 (Jan. 1977), p. 84.

Heirtzler, J. R. and Bryan, W. B. "The Floor of the Mid-Atlantic Rift." *Scientific American* 233 (Aug. 1975), p. 79.

"A History of the Earth's Atmosphere." *Sky and Telescope* 53 (April 1977), p. 266.

Loudon, J. "Pioneer Venus: A First Report." *Sky and Telescope* 57 (Feb. 1979), p. 119.

Oberg, J. E. "Venus." *Astronomy* 4 (Aug. 1976), p. 6.

Rasool, S. I., Hunten, D. M., and Kaula, W. M. "What the Exploration of Mars Tells us about Earth." *Physics Today* 30 (July 1977), p. 23.

Sagan, C. "The Canals of Mars." *Astronomy* 2 (April 1974), p. 4.

*The Solar System*. San Francisco: W. H. Freeman, 1975.

Wilson, J. T., ed. *Continents Adrift*. San Francisco: W. H. Freeman, 1973.

Wood, J. A. "The Moon." *Scientific American* 233 (Sept. 1975), p. 92.

———. *The Solar System*. Englewood Cliffs, N. J.: Prentice-Hall, 1979.

*Jupiter, largest of the jovian planets, dominates the outer solar system. This 200-inch photograph shows the planet's belts of clouds; the Great Red Spot; Ganymede, one of its larger satellites; and the shadow of the satellite on the cloud tops. (Hale Observatories.)*

# Chapter 15 PLANETS OF THE OUTER SOLAR SYSTEM

The sulfuric acid fogs of Venus seem totally alien, but compared to the planets of the outer solar system, Venus is a tropical oasis. Of the five planets beyond the asteroid belt, four have no solid surface and one is so far from the sun its atmosphere lies frozen on its surface.

The four jovian planets—Jupiter, Saturn, Uranus, and Neptune—are strikingly different from the terrestrial planets. Their thick atmospheres of hydrogen, helium, methane, and ammonia are filled with clouds of ammonia and methane crystals. Explorers descending into the atmosphere of a jovian planet would find the atmosphere and its clouds blending gradually with the planet's liquid hydrogen interior. Thus the jovian planets have no solid surface and are not subject to the four-stage development described in the previous chapter. In addition, the thick atmosphere of Jupiter is governed by weather patterns unlike those on earth, and similar processes probably operate on the other jovian planets.

Beyond the jovian planets lies Pluto, a planetary enigma. It is not a jovian planet, but it is so distant we know little more about it. No one is even sure it is a true planet—it may be an escaped moon of Neptune.

## THE JOVIAN PLANETS

Our analysis of the outer planets will concentrate primarily on Jupiter. It is the nearest jovian planet and the most easily observed. Four spacecraft have flown past it and radioed back detailed measurements and photographs. Also, Jupiter is the most massive of the jovian planets, containing 70 percent of all planetary matter in our solar system. This high mass accentuates some processes that are less obvious or nearly absent on the other jovian planets.

*Interiors.* The high mass of the jovian planets compresses their interior to the point that hydrogen becomes a liquid (Figure 15-1). Only the outer 1.3 percent of the planet's radius is gaseous hydrogen, helium, and other compounds (Table 15-1).

Careful measurements of the amount of energy leaving Jupiter reveal that it emits about twice as much energy as it absorbs from the sun. This is evidently heat left over from the formation of the planet. In Chapter 13 we concluded that Jupiter should have grown very hot when it formed, and it might still be as hot as 30,000°K at its center. As this heat leaks away, the planet will contract slightly, producing more heat. In a sense, Jupiter's formation is still in progress—it is still contracting. Similar observations of Saturn show that it too radiates more than it receives.

Theoretical models of Jupiter predict that it contains a core of dense material whose composition resembles earth rocks and metals. Though this is commonly referred to as a rocky core, the high temperature and pressure make

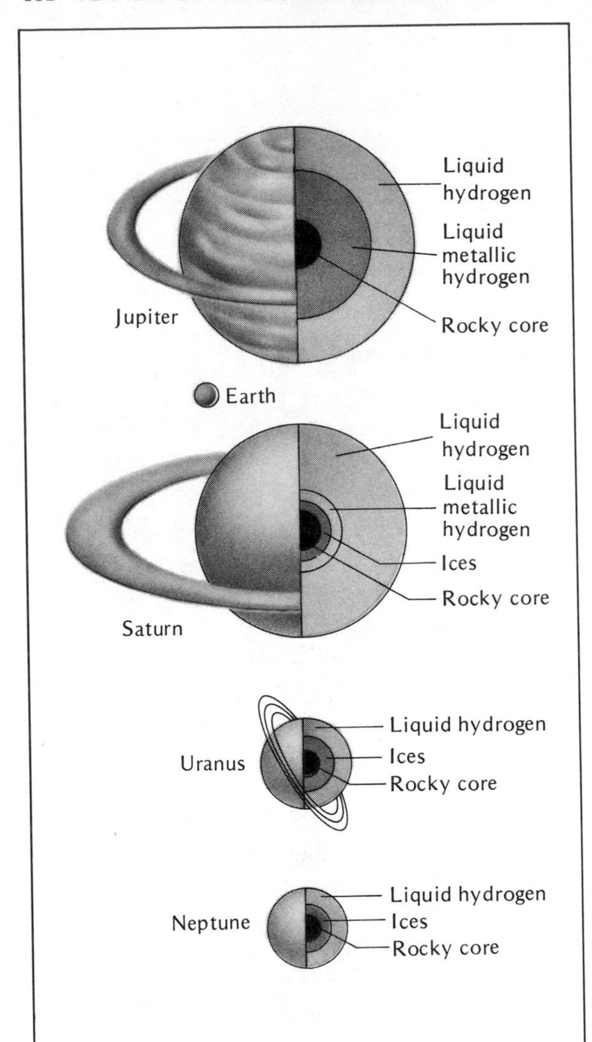


*Figure 15-1. According to recent theoretical models, Jupiter and Saturn contain liquid metallic hydrogen, but Uranus and Neptune do not.*

the material unlike any rock on earth's surface. Jupiter's rocky core occupies less than 1 percent of the planet's total volume and thus has little effect on the planet, even though the core itself is larger than the earth. The other jovian planets probably have rocky cores of similar size or smaller.

Jupiter's internal pressure is as high as 300 million times the pressure of earth's atmosphere. This pressure converts the liquid hydrogen into a different form called liquid metallic hydrogen. It still behaves as a liquid, but it is a very good conductor of electricity. This vast globe of conducting liquid coupled with the rapid rotation of Jupiter generates a powerful magnetic field through the dynamo effect (Chapter 14). The Pioneer and Voyager spacecraft detected a field over 10 times stronger than the earth's extending through a vast region around the planet.

This magnetic field traps charged particles from the solar wind into large doughnut-shaped radiation belts that surround the planet, just as the earth is surrounded by its Van Allen radiation belts (Figure 15-2). Because Jupiter's magnetic field is stronger, its trapped radiation is $10^6$ times more intense than the earth's. The

**Table 15-1 The Composition of Jupiter**

| Element | Abundance | Method of Observation |
|---|---|---|
| Hydrogen | 60% | Infrared—Earth-based observation |
| Helium | 36% | Ultraviolet—Pioneer 10 |
| Neon | 2% | Estimate |
| Water | 0.9% | Infrared—Earth-based observation |
| Ammonia | 0.5% | Infrared—Earth-based observation |
| Argon | 0.3% | Estimate |
| Methane | 0.2% | Infrared—Earth-based observation |

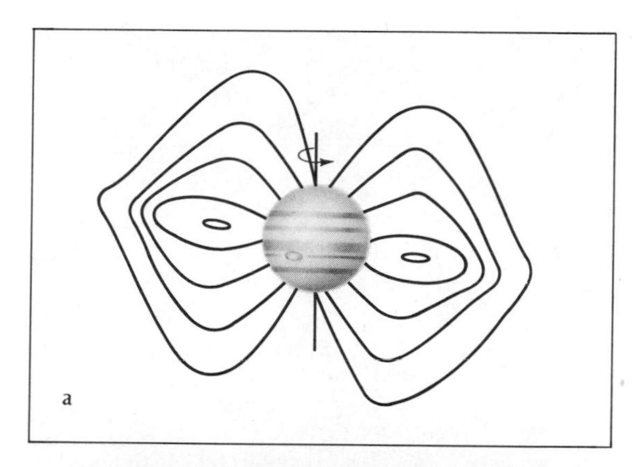


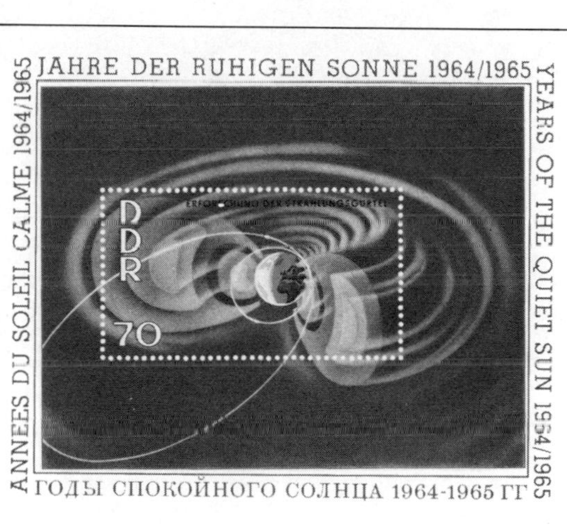


b

*Figure 15-2. (a) Jupiter's intense magnetic field traps particles from the solar wind in radiation belts. (b) A German postage stamp carries a diagram of the earth's Van Allen radiation belts.*

radiation is strong enough to kill an astronaut, and the robot spacecraft that penetrated the belts barely escaped severe damage to their electronic circuits. When Pioneer 11 flew past Saturn in 1979, it found a similar though less intense radiation belt surrounding the planet and rings.

The earth's magnetic field guides charged particles toward the north and south magnetic poles where they strike the upper atmosphere and excite atoms to emit photons. The glowing layers are familiar to residents of high latitudes as **aurora,** and are especially impressive when a storm on the sun ejects bursts of charged particles. When Voyager 1 passed Jupiter, it looked back at Jupiter's night side and saw great sheets of glowing aurora (Figure 15-3). Ultraviolet observations by Pioneer 11 suggest that Saturn too may have aurora near its magnetic poles.

Jupiter's magnetic field is so powerful it emits radio waves. The radiation has a wavelength of about 10 cm (4 in) and originates from electrons trapped in the radiation belts. Radio waves with wavelengths of about 10 m (32 ft) appear to come from interactions between the atmosphere and the particles in the radiation belts. The long-wave radio signals depend in a complex way on the position of Io, the innermost of Jupiter's four large moons. Io's orbit lies inside the inner radiation belt, and it apparently interacts with the field to beam radio waves in specific directions. Thus only when Io is in certain positions in its orbit does one of the radio beams point toward earth.

Saturn is a smaller version of Jupiter. It is less

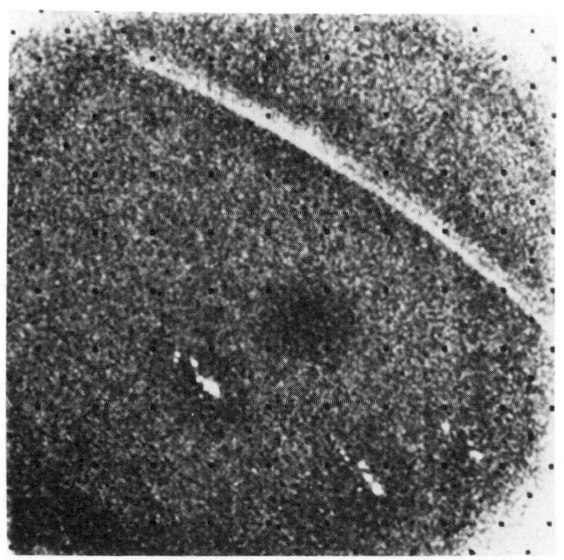

*Figure 15-3. After passing close to Jupiter, the Voyager 1 spacecraft looked back at its dark side and detected aurora along the edge of the disk near the planet's north magnetic pole and lightning bolts among the cloud belts (lower center). (NASA.)*

dense, a fact that leads some to believe it does not have a rocky core, though some calculations suggest a small rocky core coated by a layer of ices. In either case, its smaller mass means it has a lower internal pressure and cannot contain as much liquid metallic hydrogen as can Jupiter. Saturn rotates almost as fast as Jupiter does, but, lacking a large conducting core, the dynamo effect may not produce a strong magnetic field. When Pioneer 11 visited Saturn in 1979, it found a magnetic field only 20 percent as strong as Jupiter's.

Calculations predict that Uranus and Neptune are not massive enough to form a liquid metallic hydrogen, so they may lack strong magnetic fields. Instead they are presumed to be composed of layers of liquid hydrogen, ices, and rocky cores. Because they are smaller than Jupiter and Saturn, they have lost more gases and their rocky cores make them relatively dense for jovian planets.

*Atmospheres.* The outer layers of the jovian planets contain a complex mixture of gases, liquid droplets, and solid crystals. For example, the atmosphere of Jupiter is largely hydrogen and helium in which we see layers of clouds composed of ammonia and ammonia hydrosulfide crystals. Through gaps in these clouds, earth-based telescopes and space probes have seen deeper, warmer layers of water droplet clouds. Below that the atmosphere merges with the liquid hydrogen interior.

The distinctive features of the jovian atmospheres are the **belts** and **zones** that circle the planets parallel to their equators (Figure 12-6 and 15-4). These are most prominent on Jupiter. Even a small telescope can reveal the dark belts and lighter zones. The belts generally appear brown or red, though they may shade into blue-green, and the zones are yellow-white (Color Plate 23). These colors may arise from molecules that are produced by sunlight or lightning interacting with ammonia and other compounds in Jupiter's atmosphere.

The cloud belts of Jupiter are not yet well understood. One popular theory suggests that the zones are high-pressure regions of rising gas, while the belts are low-pressure regions where the gas sinks. On earth, the temperature difference between the equator and poles drives a wavelike wind pattern that organizes such high- and low-pressure regions into cyclonic circulations familiar from weather maps (Figure 15-5). On Jupiter, the equator and poles appear to be about the same temperature, perhaps because of heat rising from the interior. Thus there is no temperature difference to drive the wave circulation, and Jupiter's rapid rotation draws the high- and low-pressure regions into bands that circle the planet. On earth, high- and low-pressure regions are bounded by winds induced by the wave circulation. On Jupiter, the same circulation appears as high-speed winds that blow around the planet at the boundaries of the belts and

*Figure 15-4. Jupiter's swirling cloud belts and zones. The Great Red Spot circulates counterclockwise, as do the smaller spots. (NASA.)*

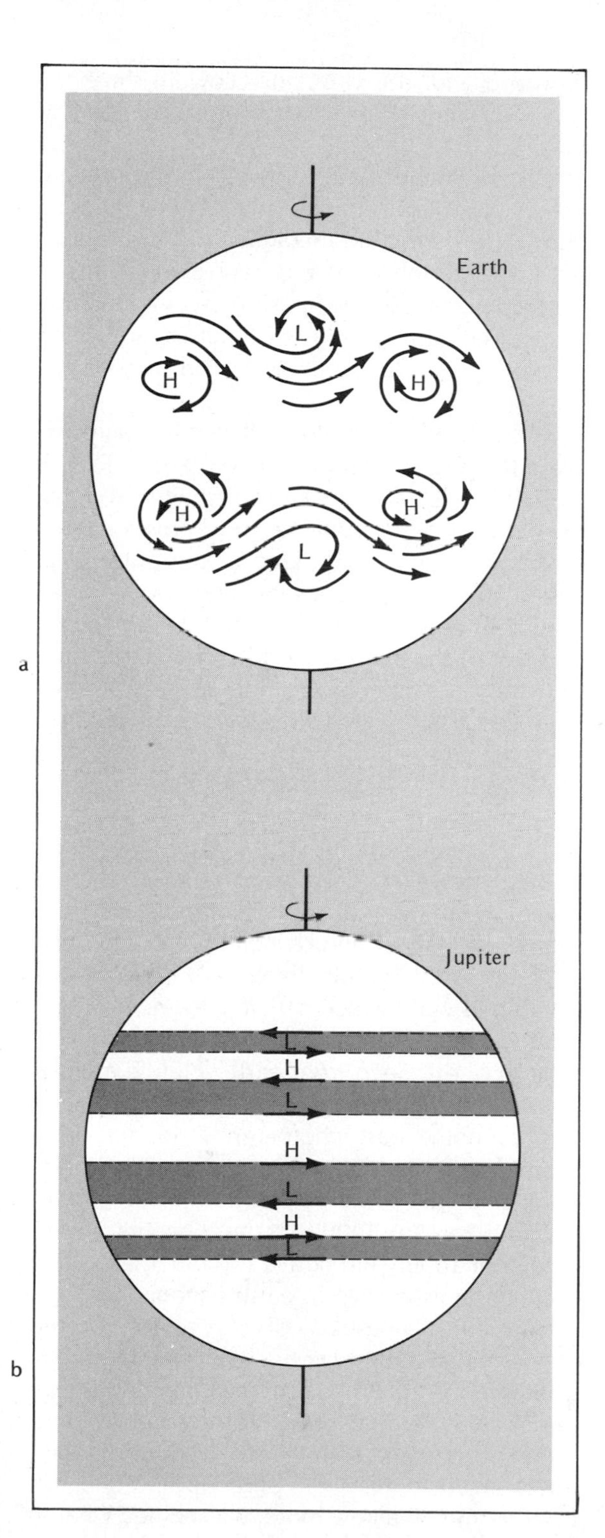


*Figure 15-5. Circulation in earth's atmosphere (a) is dominated by a wave. In Jupiter's atmosphere (b) the circulation is in belts and zones.*

zones. Above about 45° latitude the belt-zone pattern becomes unstable and the circulation decays to turbulent whirlwinds.

Mixed among the belts and zones are light and dark spots, a few larger than the earth. Some appear and disappear in a few days, while others last for years. The largest is the Great Red Spot, a reddish oval that has been in one of the southern zones for at least 300 years (Figure 15-4). Pioneer and Voyager photographs show that the Great Red Spot is a vast cyclonic storm where winds between adjacent belt and zone meet. How such a storm could survive for 300 years, when the smaller spots, apparently also cyclonic disturbances, last only a few years, is unknown.

The belts in Saturn's atmosphere are less distinct than Jupiter's, perhaps because of Saturn's cooler atmosphere (Figure 15-6). Nearly all ammonia in Saturn's atmosphere is frozen into crystals, forming the cloud layers we see. With all the ammonia frozen, the molecules that give Jupiter's belts and zones their colors may not form as easily in Saturn's uppermost cloud layers.

*Figure 15-6. Pioneer 11 photo of Saturn shows faint traces of cloud belts. The satellite Titan is in the background. (NASA/Ames.)*

There is no conclusive proof that Uranus and Neptune have atmospheric belts and zones. They are smaller than Jupiter and Saturn and lie much farther from the sun, making them very difficult to study. The best photographs show little detail (Figure 12-6). The upper layers of their atmospheres are certainly cold enough to freeze all water and ammonia, so the cloud surfaces visible from earth may be methane crystals.

## PLUTO

Pluto is an intriguing anomaly in the solar system. It was discovered by a phenomenal accident, and in some ways it does not seem to belong in the outer solar system. Even its origin is a mystery.

*Discovery.* The story of the discovery of Pluto really begins with the accidental discovery of Uranus on March 13, 1781 by William Herschel. As astronomers studied the orbital motion of Uranus over the following decades, they found small variations they attributed to an as yet undiscovered planet. Through highly complex calculations they predicted the planet's position, and the German astronomer Galle, acting on the predictions, discovered Neptune on September 23, 1846.

Even when corrected for the influence of Neptune, the motion of Uranus contained unexplained variations, and a number of astronomers tried to extend the calculations to search for yet another planet beyond Neptune. One of these, Percival Lowell, predicted the position of the planet and searched for it from 1906 to his death in 1916. The search was resumed in the 1920s and finally in February 1930, after studying photographic plates for nearly a year, Clyde Tombaugh found a faint moving object beyond Neptune. The new planet's discovery was announced on March 13, 1930, which was the 149th anniversary of the discovery of Uranus and the 75th anniversary of Lowell's birth. The planet was named Pluto after the god of the underworld and in a way, after Lowell, since the first two letters in Pluto are the initials of Percival Lowell.

Recent studies have shown that the original observations of Uranus were not accurate enough to permit the prediction of Pluto's position. The discovery of the new planet only 6° from Lowell's predicted position was apparently an accident and only proves that if you search long enough, you are likely to find something.

*The Origin of Pluto.* It is difficult to explain the origin of Pluto for two reasons. First, it is so distant we actually know very little about it. Second, much of what we do know of Pluto makes it seem out of place in the outer solar system. Its orbit is unusual, and it is a small planet among the jovian giants.

Most of the planetary orbits in our solar system are nearly circular, but Pluto's is quite elliptical (see Figure 12-1). Pluto's average distance from the sun is 39.52 AU, but it can come as close as 29.64 AU. In fact, from January 21, 1979 to March 14, 1999, Pluto will be closer to the sun than Neptune. The planets will never collide, however, because Pluto's orbit is inclined 17° to the plane of the solar system. Most of the planetary orbits are inclined by only a few degrees, so the high inclination of Pluto's orbit, combined with its eccentricity, seems peculiar.

Because Pluto is so distant, we know very little about it. Seen from earth, Pluto's angular diameter is slightly larger than 0.1 second of arc, and even the best photographs do not show more than a tiny dot whose diameter is not apparent. Highly sophisticated analysis of Pluto's image in the 5-m (200-in) telescope yields a diameter of about 3600 km (2200 mi), making the planet about the size of earth's moon.

In 1978, James W. Christy of the United States Naval Observatory, while examining a photographic plate, discovered a faint image beside Pluto's. The image turned out to be a moon orbiting the planet once every 6.387 days at an average distance of about 20,000 km (12,000 mi). The new moon was named Charon after the mythological ferryman who transports souls across the river Styx into the underworld.

The discovery of a new object in the solar system is always exciting, but the discovery of Charon was especially important because it made it possible to determine Pluto's mass. The same analysis that reveals the total mass of binary star systems reveals that the total mass of Pluto and its moon is about $7 \times 10^{-9}$ solar masses or about 0.002 earth masses. If astronomers assume that Charon and Pluto have similar surfaces that reflect similar amounts of light, then they can estimate the relative size of the two objects. Thus it seems that Charon is about one-half the diameter of Pluto and about one-tenth its mass.

Combining the mass of Pluto with its observed diameter reveals that its density is quite low, about 0.5 gm/cm$^3$. This suggests that Pluto is composed mostly of frozen volatiles such as water, ammonia, and methane. In fact, spectroscopic observations in 1976 suggested that its surface is covered by frozen methane, and laboratory studies indicate that most gases at Pluto's low surface temperature of 43°K (−230°C) would freeze. Only hydrogen, helium, and neon would remain gaseous, and since hydrogen and helium would leak away from such a small planet and since neon is a relatively rare gas, it seems likely that whatever atmosphere Pluto has is lying frozen on its surface.

Pluto's low density and the frozen methane on its surface are consistent with its formation in the cold outer fringes of the solar nebula. However, its eccentric, highly inclined orbit and its small size have led some astronomers to suggest that Pluto is not really a planet, but an escaped satellite of Neptune. Indeed, studies of the satellites of the jovian planets reveal that some of them are low-density, icy bodies like Pluto. Yet the presence of Charon orbiting Pluto raises an objection to this satellite theory. It does not seem likely that Pluto could have held its moon while it was itself a satellite of Neptune, so we must suppose that it captured its moon after it escaped from Neptune or that Charon is a fragment ripped from Pluto by some cosmic collision that knocked Pluto from its orbit around Neptune. All of these possibilities seem unattractive, so Pluto's origin remains one of the mysteries of the solar system.

Clyde Tombaugh searched for more planets beyond Pluto and found none even though he should have been able to detect any planet as large as Neptune out to a distance of 270 AU. There may be no planets beyond Pluto, but even now spacecraft are revealing new worlds unsuspected within our solar system. The satellites of the planets, little more than dots as seen from earth, are now recognized as complicated worlds themselves (Color Plate 22).

## THE JOVIAN SATELLITES: NEW WORLDS

Since the invention of the telescope in the early 1600s, astronomers have been discovering and studying the moons in the outer solar system, but until the last few years they knew almost nothing about them. Small and distant, they looked like tiny, featureless dots through all but the largest telescopes. Now robot spacecraft are exploring these satellites and showing us that they are as complex as the major planets.

*Triton and Titan.* The largest of the moons of Neptune and Saturn share more than similar names. They are about the same size, could have atmospheres, and lie so far from the sun that we know little about them.

Triton's size is poorly known, but it is larger than Mercury and could be almost as large as Mars. In its orbit around Neptune, it is far enough from the sun to remain cool and hold on to a methane atmosphere. Triton is especially interesting because its orbit is clockwise and unstable. It is spiraling into Neptune, where, in a billion years or so, it will probably be ripped apart by tidal forces and distributed around the planet to form a ring. Unfortunately, no spacecraft has explored as far as Neptune, so little else is known about Triton.

Titan, the largest satellite of Saturn, may be an ocean world. It is 5800 km (3600 mi) in diameter, making it larger than Mercury, and only 15 percent smaller than Mars. Thus it is able to retain an atmosphere containing methane and perhaps hydrogen. Some observations have suggested that it is covered by a layer of reddish clouds. This atmosphere could be as dense as earth's, and infrared observations suggest that a greenhouse effect may warm its surface. However, in the outer reaches of the solar system, far from the sun, *warm* does not mean pleasant. Infrared observations combined with theoretical calculations predict that Titan's surface temperature is about 120°K (−243°F). If this temperature is correct, the clouds may hide a planet-wide ocean of methane and ammonia. Since its average density is quite low, it cannot contain much rocky material, and its oceans may extend nearly to its center.

Unlike Titan and Triton, the four largest satellites of Jupiter are not large enough or cold enough to retain much atmosphere. Nevertheless, they are some of the most interesting bodies in the solar system. The outer two are cold, dead, icy worlds. The inner two are denser and carry signs of activity. In the case of the innermost, that activity is spectacular.

*Callisto.* The outermost of Jupiter's large satellites, Callisto, is superficially similar to earth's moon. It is 44 percent larger, heavily cratered, and, like all of Jupiter's large satellites, tidally locked to its planet. Unlike our moon, it is not solid rock.

The average density of Callisto is only 1.79 gm/cm$^3$, which means it cannot contain much rock. Its interior may be liquid water or a slush of ices and water. The dark crust appears to be a layer of dirty ice, perhaps only a few hundred kilometers thick. The whitish material surrounding many of its craters is apparently cleaner ice ejected by impacts (Figure 15-7 and Color Plate 26). Photographs taken by the Voyager spacecraft show no high relief—that is, no mountains or other high features. This is understandable since an icy crust is physically weak and a

*Figure 15-7. Callisto's dark surface is probably dirty ice broken by heavy cratering. The ringed plane is the largest impact feature known in the solar system. (NASA.)*

mountain of ice would gradually slump lower and lower. Even large crater rims could not last long in such a surface.

Callisto is the site of the largest known impact feature in the solar system. No crater marks the impact, owing to the plastic nature of the surface, but concentric rings reach 1500 km (930 mi) across the surface in an unmistakable bull's-eye similar to those around Mare Orientale on the moon and Caloris Basin on Mercury. Two similar though smaller ringed plains mark other regions of Callisto.

Callisto is now a dead world. The number of craters on its surface suggests that it formed a frozen crust over 4 billion years ago, and that most of the craters date from that time. Since then only occasional cratering has altered its ancient terrain.

*Ganymede.* The largest of Jupiter's satellites, Ganymede, is slightly closer to Jupiter. It is larger than Mercury and has had a complex history. Its low density of 1.9 gm/cm$^3$ and the in-

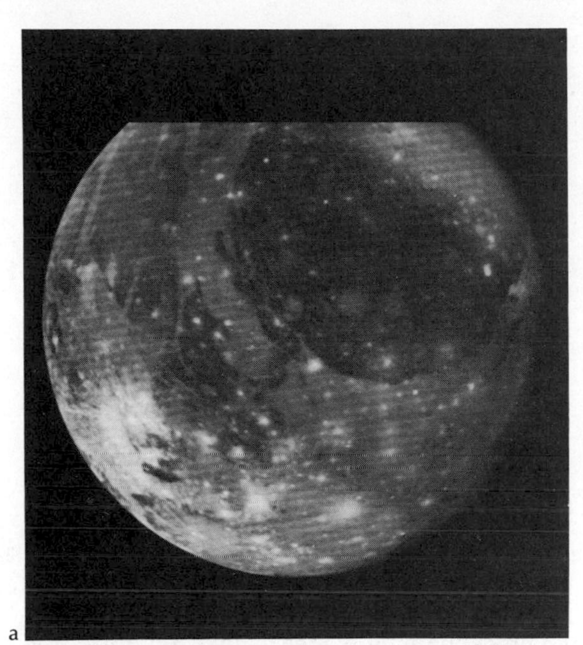

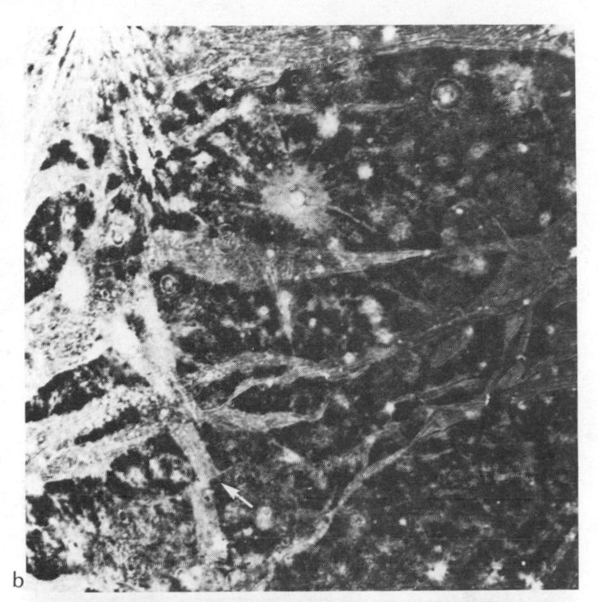

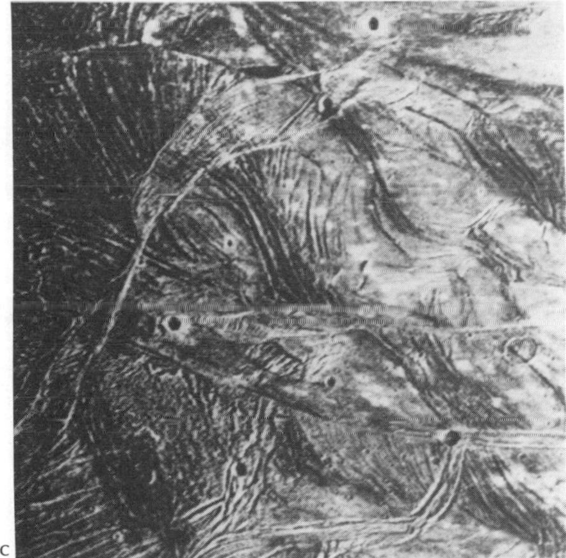

*Figure 15-8. (a) Ganymede's surface is divided into dark cratered terrain and lighter grooved regions. (b) Note the fault cutting across the grooves (arrow). (c) The grooves may be ancient breaks in the icy crust. (NASA.)*

frared detection of water suggest that it is, like Callisto, an ice-liquid-rock mixture. As in the case of Callisto, the Voyager spacecraft photographed no high relief features, indicating that its surface is icy and cannot support the weight of high features like mountains.

Though Ganymede resembles Callisto, it is clearly unique. Its surface is divided into cratered regions and grooved areas (Figure 15-8). These grooves are 5 to 15 km (3 to 9 mi) wide and a few hundred meters deep. Groups of a dozen parallel grooves reach up to 1000 km (600 miles) across the surface. These may be breaks in the frozen surface caused by the motion of crustal plates. A surface fault in one photograph shows where crustal motion has shifted a grooved section about 50 km (30 mi) to one side (Figure 15-8b).

Studies of the number of craters in different regions suggest that Ganymede retained a liquid interior longer than Callisto, perhaps because it was closer to Jupiter and was heated by tidal squeezing. The crust was free to move, and as it broke and refroze, the grooved surface features developed. The number of craters in the grooved areas suggests that the grooves formed over 2 billion years ago. Future exploration of Ganymede may tell us more about the geology of ice planets.

*Europa.* The density of Europa, the next satellite inward, is 3.03 gm/cm$^3$, which means that it contains more rock than Ganymede and Callisto. Earth-based infrared observations have detected water ice, and the Voyager spacecraft found no surface features higher than about 50 m (160 ft), making Europa the smoothest object in the solar system (Figure 15-9a). This leads some to suspect that Europa has an icy crust only about 100 km (60 mi) thick covering an interior of liquid and rock.

Voyager photographs show Europa's surface crossed by lines 10 to 50 km (6 to 30 mi) wide and 1000 km (600 mi) long (Figure 15-9b and Color Plate 27). Some wind more than halfway around the planet. These lines are not rays of ejecta from craters. In fact, Europa is nearly clear of large craters—only three have been found. Instead the lines suggest fractures in the crust, perhaps due to internal convection or tidal forces causing tectonic motion in the surface and opening breaks in a planetwide ice sheet. The nearly complete lack of craters suggests that they have been erased by an active surface, that the crust is much too weak to retain any trace of an impact, or that the craters have been covered over by ice.

a

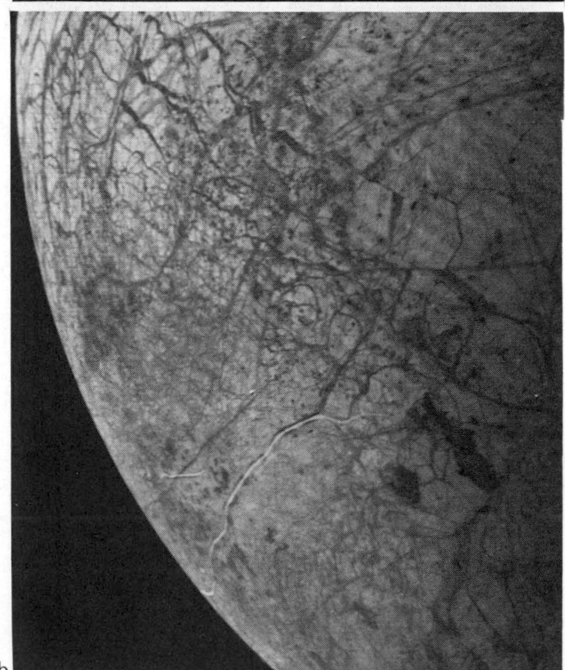

b

*Figure 15-9. (a) The satellite Europa. (b) The satellite's surface is marked by long, narrow lines that may be the scars of past fractures in a planetwide ice sheet. Few craters are evident, suggesting an active surface. (NASA.)*

*Io.* The innermost of the four large moons of Jupiter is also the most peculiar. Earth-based observations have revealed that Io is surrounded by a cloud of sodium and sulfur, and that it occasionally grows brighter in the infrared. However, few astronomers had any idea how complex the little planet really is until the Voyager spacecraft flew past it in 1979.

Voyager photographs showed the surface mottled with red, orange, brown, and black like a badly made pizza (Color Plate 24). Careful analysis of the photographs revealed volcanoes spewing out gas and solid material hundreds of kilometers above the surface (Figure 15-10 and Color Plate 25). The escape velocity from Io's surface is only 2.5 km/second, so some of this material probably escapes and forms the sodium-sulfur cloud. Much of the solid ejecta falls back, blanketing the surface with about 1 mm of material per year. Thus no craters are visible because they have been buried.

The volcanism on Io differs from earthly volcanism in some mysterious ways. The power to eject ash from earth's volcanoes comes primarily from water vapor in the lava flashing into steam, but observations of Io reveal no water. Clearly some other gas is involved. Also, three

a

b

c

*Figure 15-10. (a) Io seen against the cloud belts of Jupiter. (b) Io's surface is free of craters due to the presence of active volcanoes that spew ash in complex patterns and bury new features. (c) Seen from Io's night side, this volcano blasts volcanic material over 100 km above the surface. (NASA.)*

regions of Io's surface have temperatures 150°K hotter than the rest of the surface. These were first identified as lava pools, but they are too cool for that. Their temperatures are only about 60°F, much too cool for normal lava, and even too cool for molten sulfur, a common product of volcanism. Exactly what produces the volcanoes, the ejecta, the hot spots, and the multicolored lava flows is unknown, though astronomers suspect that sulfur and sulfur dioxide are important factors.

Io is much too small to have remained hot since it formed, or to have grown hot from internal radioactivity. Its orbit lies inside a doughnut-shaped belt of radioactivity around Jupiter, so it may gain some heat from that source, but the best theory attributes the heat to changing tides. Europa and Ganymede force Io's orbit into an ellipse. As its distance from Jupiter varies, the size of its tidal bulge changes and its surface rises and falls by several hundred feet. The friction is sufficient to melt its interior. If that is true, Io may have a molten silicate interior topped by a crust of silicates, frozen sulfur, and liquid sulfur dioxide.

The satellites of our solar system only whet our appetite to explore more planets. Have we now seen examples of all types of planets and planetary surfaces, or do other solar systems contain planets where surface features are unlike anything in our system? In fact, the new features seen on Jupiter's satellites are not really new. The earth has craters, moving sheets of ice, and volcanoes. What is surprising about the satellites in our solar system is the unfamiliar aspect these familiar processes wear. Perhaps the geology of other planets orbiting other stars is similar to ours after all.

## SUMMARY

The jovian planets are so massive they have been able to retain much of their initial hydrogen and helium. This excess of light elements gives the jovian planets densities much lower than those of the terrestrial planets. In their interiors, the tremendous pressure forces the hydrogen into a liquid state, and in Jupiter and Saturn the pressure is so high the hydrogen becomes a liquid metal that is a good conductor of electricity. Electric currents within this material produce magnetic fields that trap particles from the solar wind to form radiation belts around the planets. Both Jupiter and Saturn are known to have radiation belts similar to the earth's Van Allan belts, and Jupiter's belts are very intense.

The atmosphere of Jupiter is marked by high- and low-pressure regions in the form of belts and zones that encircle the planet. Chemical reactions in the clouds are believed to produce compounds that color the belts. Saturn's belts and zones are not as well defined as those on Jupiter, and those of Uranus and Neptune are barely visible.

The Great Red Spot on Jupiter is believed to be a cyclonic disturbance much like an earthly hurricane. It has lasted for at least 300 years. The Voyager spacecraft revealed that Jupiter's atmosphere contains a number of similar spots that are smaller and shorter-lived.

Pluto is very distant and little is known about it. It is about the size of earth's moon and has a satellite of its own called Charon. This satellite makes it possible to determine Pluto's mass, which, combined with its size, yields a density of about 0.5 gm/cm$^3$. This low density and the spectroscopically observed presence of frozen methane on its surface are consistent with its formation in the cold outer fringes of the solar system.

Pluto's orbit is peculiar for being eccentric and inclined 17° to the plane of the solar system. In fact, Pluto will be closer to the sun than Neptune from January 21, 1979 to March 14, 1999, though the planets will never collide because Pluto's orbit is so highly inclined. This peculiar orbit and Pluto's small size have led some astronomers to suggest that Pluto is an escaped satellite of Neptune, though the moon orbiting Pluto makes this explanation less likely.

Exploration of the outer solar system is revealing that the satellites of the planets are complex worlds themselves. A few astronomers have speculated that Titan, largest moon of Saturn, could be an ocean planet. The outer two of Jupiter's four largest satellites are ice planets heavily battered by meteorite impacts. The inner two have very few impact features suggesting that their surfaces are active and that impact craters do not survive for long periods. In fact, the innermost of the four moons, Io, has a highly active crust on which volcanoes blast ash high above the surface.

## NEW TERMS

aurora
belts
zones

## QUESTIONS

1. Describe the interior of Jupiter. How does it compare with the interior of the other jovian planets?
2. How does the dynamo effect account for the magnetic fields of Jupiter and Saturn?
3. Why does the dynamo theory predict that Uranus and Neptune should have weaker magnetic fields than Jupiter and Saturn?
4. How do Jupiter's belts and zones resemble earthly weather patterns?
5. Why is it difficult to explain Pluto's origin?
6. What evidence do we have that the surfaces of Jupiter's satellites have been geologically active?
7. Why do Callisto, Ganymede, and Europa lack high mountains?

## PROBLEMS

1. How does the central temperature of Jupiter compare with the earth's central temperature?
2. How long would Io's volcanoes take to bury a new crater under a 1-km thick layer of volcanic debris?
3. Draw and label a diagram to show Pluto, Triton, Titan, the four largest moons of Jupiter, and earth's moon to scale. (Hint: See Appendix C.)

## RECOMMENDED READING

Beatty, J. K. "The Far Out Worlds of Voyager 1." *Sky and Telescope* 57 (May/June 1979), pp. 423 and 516.

———. "Voyager's Encore Performance." *Sky and Telescope* 58 (Sept. 1979), p. 206.

Hartmann, W. K. *Moons and Planets.* Belmont, Calif.: Wadsworth, 1972.

Ingersoll, A. "The Meteorology of Jupiter." *Scientific American* 234 (March 1976), p. 46.

Soderblom, L. A. "The Galilean Moons of Jupiter." *Scientific American* 242 (Jan. 1980), p. 88.

*The Solar System.* San Francisco: W. H. Freeman, 1975.

Wood, J. A. *The Solar System.* Englewood Cliffs, N. J.: Prentice-Hall, 1979.

See also *Science* 1 June 1975, 23 February 1979, 6 July 1979, 23 November 1979, 25 January 1980.

*At least once in the history of the universe a planet's surface has kindled the fire of life. Extensions of that planet are now exploring the universe looking for other planets where the spark exists. (NASA.)*

# Chapter 16 LIFE ON OTHER WORLDS

Are there intelligent beings living on other planets? That is the last and perhaps the most challenging question in our study of astronomy. We will try to answer it in three steps, each dealing with a different aspect of life.

First, we must decide what we mean by life. A living thing is not so much a physical object as a process. We are not simply the matter that forms our bodies, but rather a tremendously complex system that has the ability to duplicate and protect itself. Thus life is based on information that contains the directions for the processes of duplication and preservation.

Our second step is to study the origin of life. Direct investigation is limited to the earth, but if we can understand how life began here, then we may better estimate the chances that it occurred elsewhere. We will find that life on earth probably began with simple chemical reactions that happened naturally. If these gave rise to life on earth, then similar reactions may have provided the spark on other worlds.

Our third step is to study evolution, the process by which life improves its ability to survive. The survival of stable species has transformed the simple organisms that began in the earth's oceans into a wide variety of creatures with special adaptations. The rose's thorn, the rabbit's long ears, and the human's intelligence are protective adaptations. Evolution is so natural a process that it surely works on any planet where life begins, and if we assume that intelligence is a valuable trait, then intelligent beings may eventually emerge.

If life is common in the universe, where might we look for it, and how might we communicate with other intelligent beings? Certainly the prospects of finding life, intelligent or otherwise, on any of the other planets in our solar system are bleak. If we are to find extraterrestrial life, we must go beyond our solar system and search among any planets that may orbit other stars.

Communication with intelligent races on other worlds may be possible, but we cannot expect to travel between solar systems. Interstellar distances are so great that only in science fiction do spaceships flit from star to star. However, it may be possible to communicate via radio. If civilizations can survive for long periods of time at a technological level at which they can build large radio telescopes, then we may be able to send and receive messages. Such messages would mark a turning point in the history of humanity. If life is common in the universe, such signals may arrive during our lifetime.

## THE NATURE OF LIFE

What is life? Philosophers have struggled with that question for thousands of years, so it is unlikely that we will answer it here. But we must agree on a working model of life before we can speculate on its occurrence on other worlds. To that end we will identify in living things three properties—a process, a physical base, and a unit of controlling information.

The life process is aimed at survival. Living things extract energy from their environment and use that energy to modify their surroundings to make their own preservation more likely. For example, human beings obtain energy by eating and breathing and they use that energy to build houses, cities, and stable societies to protect themselves. The same can be said of a bacterium absorbing food and reinforcing its cell structure.

This apparently selfish protection of the individual is aimed at the preservation of the race through safe reproduction. The ability to reproduce is one of the distinguishing characteristics of living organisms, and any organism that does not, in some way, ensure safe reproduction will not survive many generations. The entire life process is aimed at safe reproduction because any other target is self-destructive.

The physical basis of life on earth is the chemistry of the carbon atom (Figure 16-1). Because of the way this atom bonds to other atoms, it can form long, complex, stable chains that are capable of extracting, storing, and utilizing energy. Other chemical bases may exist. Science fiction stories and movies abound with silicon creatures, living things whose body chemistry is based on silicon rather than carbon. However, silicon forms weaker bonds than carbon does, and it cannot form double bonds as easily. Consequently it cannot form the long, complex, stable chains that carbon can. Silicon is 135 times more common on earth than carbon, yet there are no silicon people among us. All earth life is carbon based. Thus the likelihood that distant planets are inhabited by silicon people seems small, but we cannot rule out life based on noncarbon chemistry.

In fact, nonchemical life might be possible. All that nature requires is some physical base capable of supporting the extraction and utilization of energy that we have identified as life. One could at least imagine life based on electromagnetic fields and ionized gas. No one has ever met such a creature, but science fiction writers conjure up all sorts.

Clearly we could range far in space and time theorizing about alien life, but to make progress

a

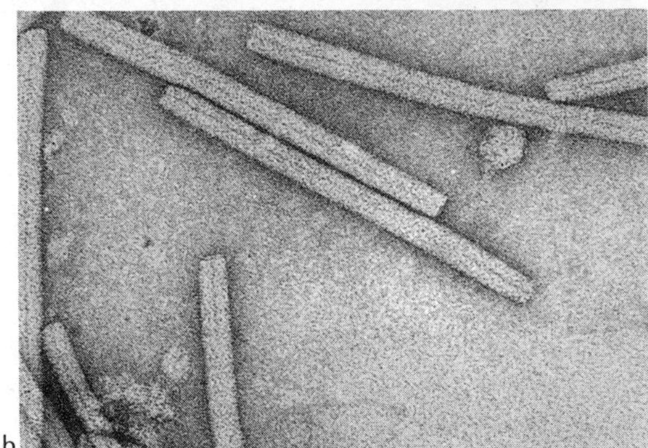

b

*Figure 16-1. All living things on earth are based on carbon chemistry. (a) Elsa, a complex mammal containing about 30 AU of DNA. (b) Tobacco mosaic virus. Each rod is a single spiral strand of RNA about 0.01 mm long surrounded by a protein coat. ((a) Janet Michael Seeds; (b) L. D. Simon)*

we must discuss what we know best, carbon-based life on earth. We must try to understand how it works and how it came to exist. Only then can we consider life on other worlds.

*The DNA Code.* The key to understanding life is information—the information the organism uses to control its utilization of energy. We must discover how life stores and uses that information and how the information changes and thus preserves the species.

The unit of life on earth is the cell (Figure 16-2), the self-contained factory capable of absorbing nourishment from its surroundings, maintaining its own existence, and performing its task within the larger organism. The foundation of the cell's activity is a set of patterns that describe how it is to function. This information must be stored in the cell in some safe location, yet it must be passed on easily to new cells and be used readily to guide the cell's activity. To understand how matter can be alive, we must understand how the cell stores, reproduces, and uses this information.

The information is stored in long carbon-chain molecules called **DNA** (**deoxyribonucleic acid**), most of which reside in the cell nucleus. The structure of DNA resembles a long twisted ladder. The rails of the ladder are made of alternating phosphates and sugars, while the rungs are made of pairs of molecules called bases (Figure 16-3). There are only four kinds of bases in DNA, and the order in which they appear on the DNA ladder represents the information the cell needs to function. One human cell stores about 1.5 m of DNA, containing about 4.5 billion pairs of bases. Thus 4.5 billion pieces of information are available to run a human cell. That is enough to record all of the works of Shakespeare over 200 times. Since the human

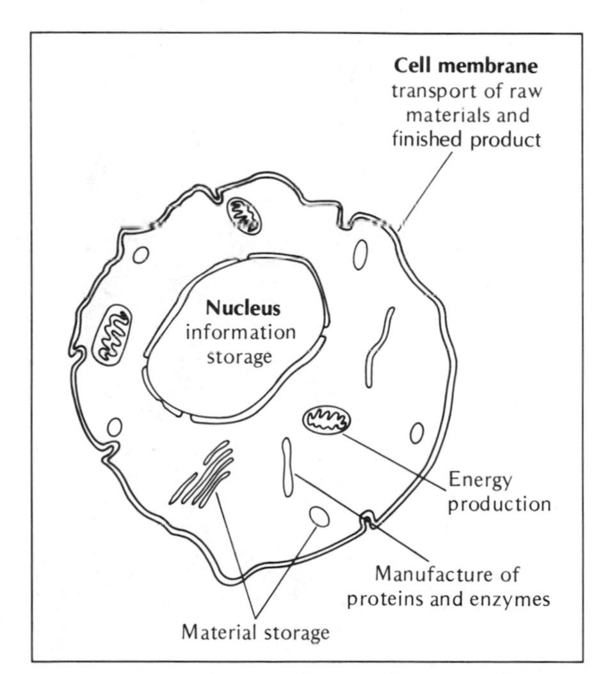


*Figure 16-2. A living cell is a self-contained factory that absorbs raw materials from its surroundings and uses them to maintain itself and manufacture finished products for the use of the organism as a whole.*

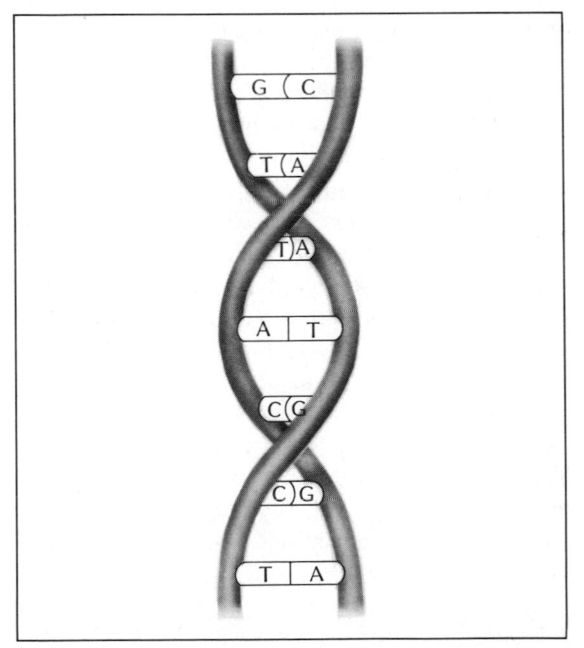


*Figure 16-3. The DNA molecule consists of two rails of sugars and phosphates (dark) and rungs made of bases [adenine (A), cytosine (C), guanine (G), thymine (T)].*

body contains about $60 \times 10^{12}$ cells, the total DNA in a single human would stretch $9 \times 10^{13}$ m, about 600 AU.

Storing all this data in each cell does the organism no good unless the data can be reproduced and passed on to new cells. The DNA molecule is especially adapted for duplicating itself by splitting its ladder down the center of the rungs, producing two rails with protruding bases (Figure 16-4). These quickly bond with the proper bases, phosphates, and sugars to reconstruct the missing part of the molecule, and, presto, the cell has two complete copies of the critical information. One set goes to each of the newly forming cells. Thus the DNA is the genetic information passed from parent to offspring.

Segments of the DNA molecules are patterns for the production of **proteins**. Many proteins are structural molecules—the cell might make protein to repair its cell wall, for example. **Enzymes** are special proteins that control other processes, growth for example. Thus the DNA molecule contains the recipes to make all of the different molecules required in an organism.

Actually, the cell does not risk its precious DNA patterns by involving them directly in the manufacture of protein. The DNA stays safely in the cell nucleus where it produces a copy of the patterns by assembling a long carbon-chain molecule called **RNA (ribonucleic acid)**. This RNA carries the information out of the nucleus and then assembles the proteins from simple molecules called **amino acids,** the basic building blocks of protein. Thus the RNA acts as a messenger, carrying copies of the necessary plans from the central office to the construction site.

Though the information coded on the DNA

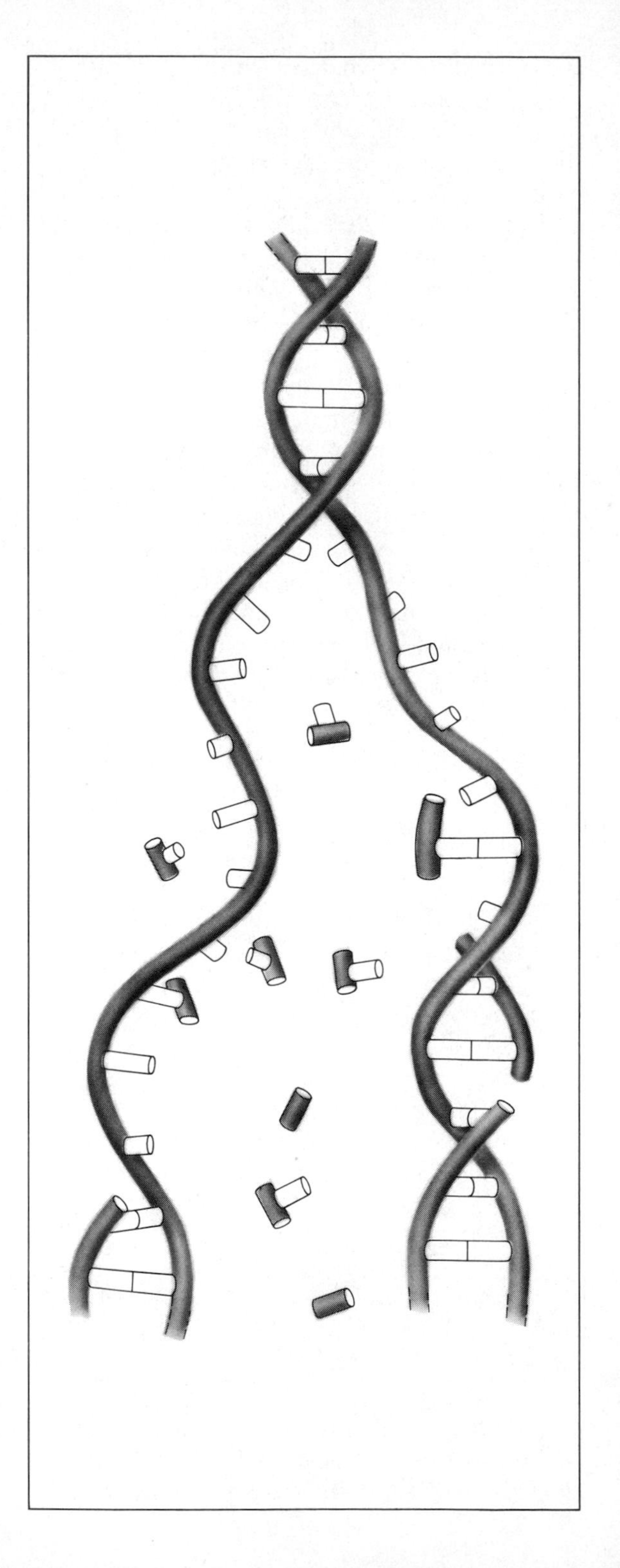

*Figure 16-4. The DNA molecule can duplicate itself by splitting in half (top), assembling matching bases, sugars, and phosphates (center), and thus producing two DNA molecules (bottom). The actual duplication process is significantly more complex than in this schematic diagram.*

must be preserved for the survival of the organism, it must be changeable or the species will become extinct. To see why, we must study evolution, the process that rewrites the data in the DNA.

*The Evolution of Life Forms.* Every living thing on earth is part of a web of interdependence. Not only now, but ever since life began, life forms have depended on each other for food and shelter. This exposes life to a serious danger in that gradual changes in climate may destroy one life form and endanger hundreds of others. A slight warming of the climate, for example, might kill a species of plant, starving the rabbits, deer, and other herbivores, and leaving the hawks, wolves, and mountain lions with no prey. If a species is to survive in such a world, it must be able to adapt to changing conditions, and that means the data coded in the DNA must change. The species must evolve.

Species evolve because of a process called **natural selection**. Each time an organism reproduces, its offspring receive the data stored in the DNA, but some variation is possible. For example, most of the rabbits in a litter may be normal, but it is possible for one to get a DNA recipe that gives it stronger teeth. If it has stronger teeth, it may be able to eat something other than the plant the others depend on, and if that plant is becoming scarce, the rabbit with stronger teeth has a survival advantage. It can eat other plants and will be healthier than its litter mates and have more offspring. Some of these offspring may also have stronger teeth as the altered DNA data are handed down to the new generation. Thus nature selects and preserves those attributes that contribute to the survival of the species. Those that are unfit die. Natural selection is merciless to the individual, but it gives the species the best possible chance to survive.

The only way nature can obtain new DNA patterns from which to select the best is to alter actual DNA molecules. This can happen through chance mismatching of base pairs—errors—in the reproduction of the DNA molecule. Another way this could occur is through damage to reproductive cells from exposure to radioactivity. Cosmic rays or natural radioactivity in the soil might play this role. In any case, an offspring born with altered DNA is called a **mutant.** Most mutations are fatal, and the individual dies long before it can have offspring of its own. But rarely a mutation may give a species a new survival advantage. Then natural selection makes it likely that the new DNA message will survive and be handed down, making the species more capable of surviving.

## THE ORIGIN OF LIFE

Clearly the carbon chemistry of life on earth is extremely complex. How could it have ever gotten started? Obviously, 4.5 billion chemical bases didn't just happen to drift together to form the DNA formula for a human. The key is evolution. Once a life form begins to reproduce itself, natural selection preserves the most advantageous traits. Over long periods of time spanning thousands, perhaps millions, of generations, the life form becomes more fit to survive. This nearly always means the life form becomes more complex. Thus life could have begun as a very simple process that gradually became more sophisticated as it was modified by evolution.

We begin our search on earth, where fossils and an intimate familiarity with carbon-based life give us a glimpse of the first living matter. Once we discover how earthly life could have begun, we can look for signs that life began on other planets in our solar system. Finally, we can speculate on the chances that other planets, orbiting other stars, have conditions that give rise to life.

*The Origin of Life on Earth.* The oldest fossils hint that life began in the oceans. The oldest easily identified fossils appear in sedimentary rocks that formed between 0.6 and 0.5 billion years ago—the **Cambrian period.** Such Cambrian fossils were simple ocean crea-

*Figure 16-5. Trilobites made their first appearance in the Cambrian oceans about 600 million years ago. These fossils, about 400 million years old, came from Ontario. (Smithsonian Institution Photo No. 76-17821.)*

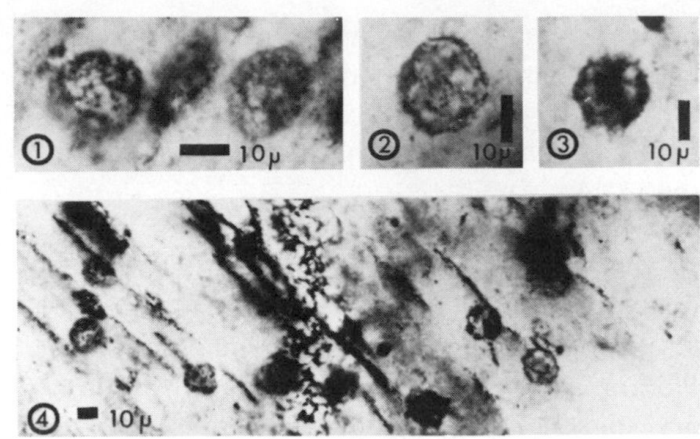


*Figure 16-6. Among the oldest fossils known, these microscopic spheres lie in the Precambrian fig-tree chert near Barberton, South Africa. They are at least 3.1 billion years old. The black bars are 10 μ long (1 μ = $10^{-6}$m). (E. S. Barghoorn.)*

tures, the most complex of which were trilobites (Figure 16-5), but there are no Cambrian fossils of land planets or animals. Evidently land surfaces were totally devoid of life until only 400 million years ago.

Precambrian deposits contain no obvious fossils, but microscopes reveal microfossils that were the ancestors of the Cambrian creatures. Fig-tree chert in South Africa is 3.0 to 3.3 billion years old, and the Onverwacht shale, also found in South Africa, may be as old as 3.6 billion years. Both contain structures that appear to be microfossils of bacteria or simple algae such as those that live in water (Figure 16-6). Apparently life was already active in the earth's oceans a billion years after the earth formed.

The key to the origin of this life may lie in an experiment performed by Stanley Miller and Harold Urey in 1952. This **Miller experiment** sought to reproduce the conditions on earth under which life began (Figure 16-7). In a closed glass container, the experimenters placed water (to represent the oceans), the gases hydrogen, ammonia, and methane (to represent the primitive atmosphere), and an electric arc (to represent lightning bolts). The apparatus was sterilized, sealed, and set in operation.

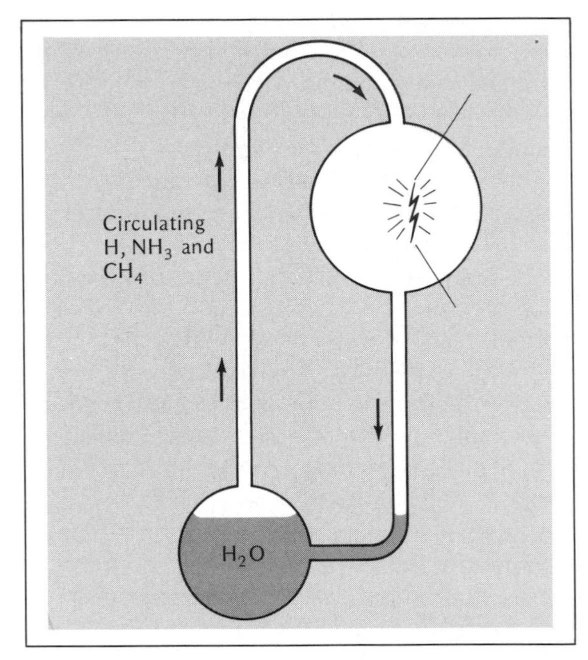


*Figure 16-7. The Miller experiment circulated gases through water in the presence of an electric arc. This simulation of primitive conditions on earth produced amino acids, the building blocks of protein.*

After a week, Miller and Urey stopped the experiment and analyzed the material in the flask. Among the many compounds the experiment produced, they found four amino acids that are common building blocks in protein, various fatty acids, and urea, a molecule common to many life processes. Evidently the energy from the electric arc had molded the atmospheric gases into some of the basic components of living matter. Other energy sources such as hot silica (to simulate hot lava spilling into the sea) and ultraviolet radiation (to simulate sunlight) give similar results. The Miller experiment did not create life, but it did show that energy released in the earth's primitive atmosphere could have created the chemical compounds found in living matter. Today such compounds occur naturally only in organic material, so the fact that they could have formed in the earth's earliest atmosphere suggests that life must have begun through such simple reactions.

The next step on the journey toward life is for the compounds dissolved in the oceans to link up and form larger molecules. Amino acids, for example, can link together to form proteins. This linkage occurs when amino acids join together end to end and release a water molecule (Figure 16-8). Among the ways this could have happened, retreating ocean tides might have left behind pools of water to evaporate in the sun. As the water evaporated, it would leave a concentrated broth of amino acids, facilitating linkups between molecules. Sunlight would drive off water molecules from between the amino acids, and long protein chains would grow until the returning tide swept them all back into the sea.

Although these proteins might have contained hundreds of amino acids, they were not alive. Not yet. Such molecules did not reproduce, but merely linked together and broke apart at random. However, because some molecules are stronger than others and because some molecules bond together more readily than others, this blind **chemical evolution** led to the concentration of the varied smaller molecules into the most stable larger forms. Eventually, somewhere in the oceans, a molecule took shape that could reproduce a copy of itself. At that point the chemical evolution of molecules became the biological evolution of living things.

Which came first, reproducing molecules or the cell? Because we think of the cell as the basic unit of life, this question seems to make no sense, but in fact the cell may have originated during chemical evolution. If a dry mixture of amino acids is heated, the acids form long, proteinlike molecules that, when poured into water, collect to form microscopic spheres that function in ways similar to cells (Figure 16-9).

*Figure 16-8. Amino acids can link together through the evaporation of a water molecule to form long carbon-chain molecules. The amino acid in this hypothetical example is alanine, one of the simplest.*

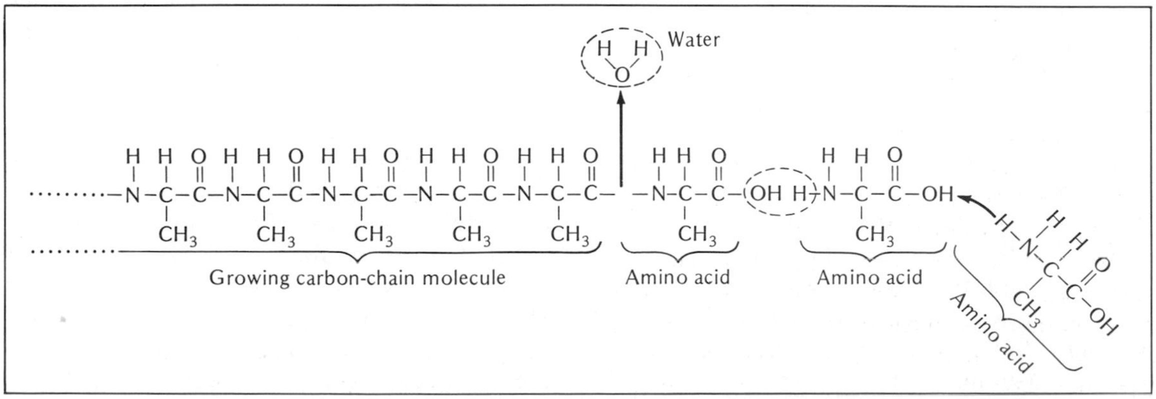

Water

Growing carbon-chain molecule

Amino acid

Amino acid

Amino acid

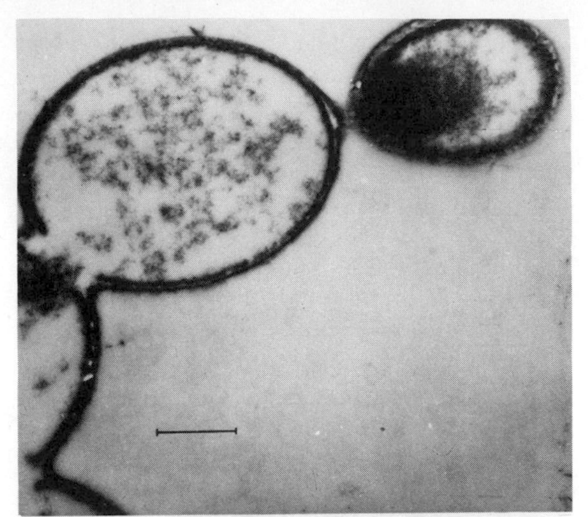

*Figure 16-9. Proteinlike material added to water forms microspheres. Although these bodies do not contain DNA or related genetic information, they have many of the properties of cells such as a double-layered boundary similar to a cell membrane. Thus the first cell structures may have originated through the self-ordering properties of the first complex molecules. The horizontal bar is 1 $\mu$ long. (Courtesy of S. W. Fox from S. W. Fox and K. Dose,* Molecular Evolution and the Origin of Life, *rev. ed. New York: Marcel Dekker, 1977.)*

They appear to have a thin membrane surface, they can absorb material from their surroundings, they grow in size, and they can divide and bud just as cells do. However, they contain no large molecule that copies itself. Thus the structure of the cell may have originated first and the reproducing molecules later.

An alternative theory supposes that the replicating molecule developed first. Such a molecule would be exposed to damage if it were bare, so the first to manufacture or attract a protective coating of protein would have a significant survival advantage. If this is the case, then the protective cell membrane is a later development of biological evolution.

The first living things must have been single-celled organisms much like modern bacteria and simple algae. How evolution shaped these creatures to live in the ancient oceans, molded them into multicellular organisms, and developed sexual reproduction, photosynthesis, and respiration is a fascinating story (see Box 16-1), but we cannot explore it in detail here. We can see that life could begin through simple chemical reactions building complex molecules, and that once some DNA-like molecule formed, it protected its own survival with selfish determination. Over billions of years the genetic information stored in living things kept those qualities that favored survival and discarded the rest. As Samuel Butler said, "The chicken is the egg's way of making another egg." In that sense, all living matter on earth is merely the physical expression of DNA's mindless determination to continue its existence.

Perhaps this seems harsh. Human experience goes far beyond mere reproduction. *Homo sapiens* has art, poetry, music, philosophy, religion, science—all of the great, sensitive accomplishments of our intelligence. Perhaps that is more than mere reproduction of DNA, but intelligence, the ability to analyze complex situations and respond with appropriate action, must have begun as a survival mechanism. For example, a fixed escape strategy stored in the DNA is a disadvantage for a creature that frequently moves from one environment to another. A rodent that always escapes from predators by automatically climbing the nearest tree would be in serious jeopardy if it met a hungry fox in a treeless clearing. Even a faint glimmer of intelligence might allow the rodent to analyze the situation and, finding no trees, to choose running over climbing. Thus intelligence, of which *Homo sapiens* is so proud, may have developed in ancient creatures as a way of making them more versatile.

If life could originate on earth and develop into intelligent creatures, perhaps the same thing could have happened on other planets. This raises three questions. First, could life originate if conditions were suitable? Second, if life begins on a planet, will it evolve toward intelligence? The answer to both questions seems to be yes. The direction of chemical and biological evolution is directed toward survival, which should lead to versatility and intelligence. But what of the third question: Are suitable conditions so rare that life almost never gets started? The only way to answer that is to search for life on other planets. We begin, in the next section, with the other planets in our solar system.

*Life in Our Solar System.* Though life based on other than carbon chemistry may be possible, we must limit our discussion to life as we know it and the conditions it requires. The most important condition is the presence of liquid water, not only as part of chemical reactions, but also as a medium to transport nutrients and wastes within the organism. This means the temperature must be moderate. Thus our search for life in the solar system must look for a planet where liquid water could exist.

The water requirement automatically eliminates a number of worlds. The moon and Mercury are airless, and thus liquid water cannot exist on their surfaces. Venus has some water vapor but it is much too hot for liquid water, and in the outer regions of the solar system the temperature is much too low.

Mars gives us good reason to hope for life. It is a temperate planet with summer temperatures at high noon that are not unlike a pleasant autumn day (17°C = 62°F). But the thin atmosphere provides little insulatation so that temperatures at night drop to about −73°C (−100°F). This is unpleasantly cold, but life forms on earth have evolved to survive under harsh circumstances, so life forms on Mars might have evolved ways of coping with the midnight cold. Even if there is no life there now, Mars may once have had a thicker atmosphere, liquid water, and a more moderate temperature range. Life might have developed under such circumstances, but withered as the atmosphere leaked away. Such life may have left seeds or spores that would germinate with the return of acceptable conditions.

Searching for life on Mars is difficult as long as humans cannot visit the planet. The solution is robots that land on the surface and perform automatic experiments under remote guidance from earth. The Viking 1 and Viking 2 spacecraft (Figure 16-10) landed safely on the Martian surface in the summer of 1976, each carrying three experiments designed to search for life in the Martian soil by exposing soil samples to water, light, and nutrients.

The labeled-release experiment (Figure 16-11) placed a thimbleful of soil in a container and dampened it with a rich broth of nutrients that any earth organism would have loved. Some of the carbon atoms in this broth were radioactive carbon-14, and the assumption was that any living things in the soil would absorb the nutrients and release carbon dioxide containing radioactive carbon. Counters measuring the amount of radioactivity in the gases escaping from the container could thus measure the level of biological activity. As a control, the experiment could be repeated with a soil sample first heated to 160°C (320°F) to kill any life. The radioactivity in the control phase would show the scientists what to expect if no life were present.

Both of the Viking 1 and Viking 2 landers functioned properly, and their remote-controlled arms scooped up soil and deposited

*Figure 16-10. Viking 1 and Viking 2 landers reached the Martian surface in 1976. Among other experiments, they searched for signs of life in soil samples collected by a remote-controlled arm. (NASA.)*

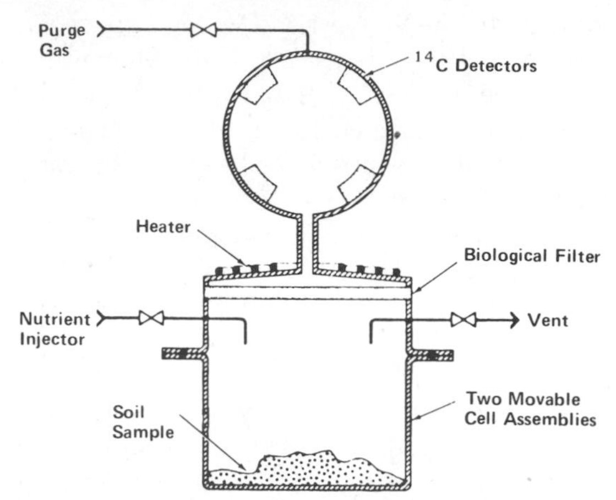


*Figure 16-11. The labeled-release experiment aboard the Viking landers exposed soil samples to a nutrient containing radioactive carbon-14. Radioactivity in the gas given off by the sample indicates the level of biological activity. (Martin Marietta Aerospace.)*

it in the experimental chambers. When the nutrients were added, the radioactivity went up rapidly, and when the sample was heated, the radioactivity declined. This sounds encouraging but the radioactivity release came much too soon. Living things should have taken some time to absorb and process the molecules in the nutrient solution. The experiment was repeated a number of times with new samples, but the results were always inconsistent with life. Evidently the released gases came from simple chemical reactions in the soil.

Two other automatic experiments on the landers searched for signs of photosynthesis and respiration (the absorption of carbon dioxide gas and the release of oxygen). All three experiments were repeated a number of times but in every case the results were negative or borderline. There was no indication of anything more

## BOX 16-1 GEOLOGIC TIME

Humanity is a very new experiment on planet earth. All of the evolution that leads from the primitive life forms in the oceans of the Cambrian period 0.6 billion years ago, to fishes, amphibians, reptiles, and mammals can fit on a single chart like Figure 16-12. But the 3-million-year history of humanity is so short that it appears as a single, thin line at the top of the diagram. If we tried to represent the entire geologic history of the earth in such a chart (Figure 16-13), the portion describing the rise of life on the land would be an unreadably small segment.

One way to represent the evolution of life is to compress the 4.6-billion-year history of earth into a one-year-long film. In such a film, the earth forms as the film begins on January 1, and through all of January and February it cools and is cratered and the first oceans form. But those oceans remain lifeless until sometime in March or early April, when the first living things develop. The 4-billion-year history of precambrian evolution lasts until the film reaches mid-November, when primitive ocean life begins to evolve into complex organisms such as trilobites.

If we examine the land instead of the oceans, we find a lifeless waste. But once our film shows plant and animal life on the land, about November 28, evolution proceeds rapidly. Dinosaurs, for example, appear about December 12 and vanish by Christmas evening, to be followed by the mammals and birds.

Throughout the one-year-run of our film there are no humans, and even during the last days of the year as the mammals rise and dominate the landscape, there are no people. In the early evening of December 31, vaguely human forms move through the grasslands, and by late evening they begin making stone tools. The Stone Age lasts till about 11:45 PM, and the first signs of civilization, towns and cities, do not appear until 11:54 PM. The Christian era begins only 14 seconds before the New Year, and the Declaration of Independence is signed with but one second to spare.

*Figure 16-12. When the ages of life on earth are plotted on a single chart, the age of humanity becomes no more than a thick line at the top. (Compare with Figure 16-13.)*

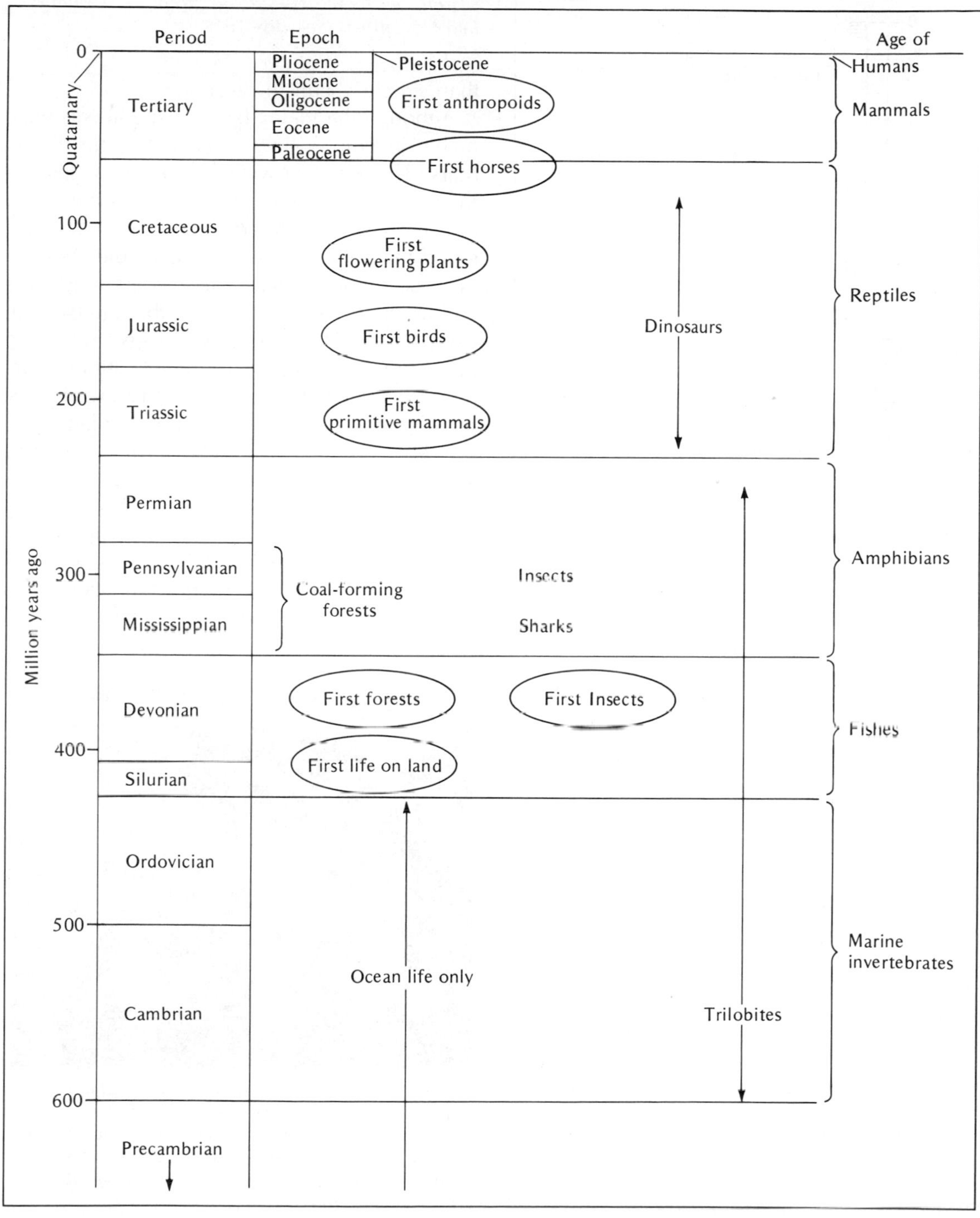

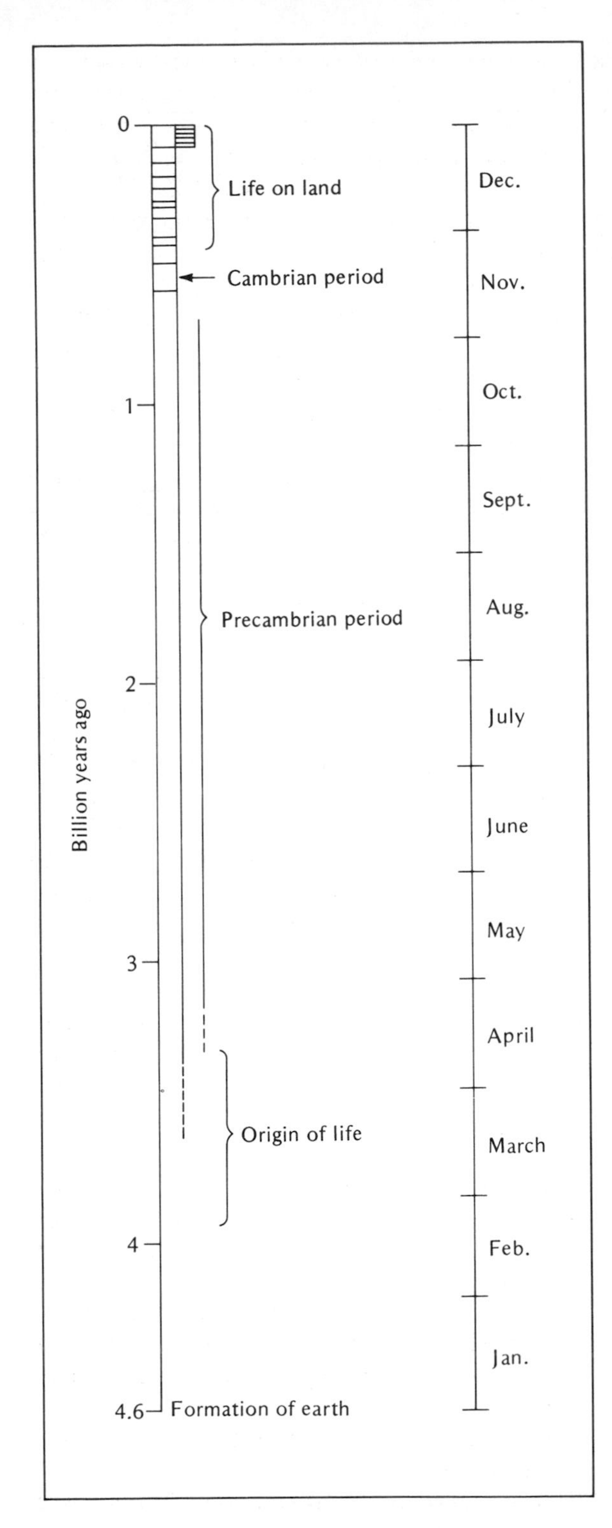


*Figure 16-13. If the entire history of the earth were compressed into one year, life would originate in March, but would not emerge from the sea until late November. (See Box 16-1.)*

than simple chemical reactions.

Although the Viking lander experiments were highly ingenious, they could detect no life on Mars. There are, however, two hopes. Life on Mars may be so unlike us that it did not respond to the experimental conditions. It may have found the nutrient broth unacceptable, perhaps even poisonous. Alternatively, Mars may have had life in the distant past, but that life became extinct when the atmosphere leaked away and the water vanished. An astronaut visiting Mars may someday find fossils in the dry streambeds, proving that life can begin on other planets.

The disappointing news from Mars leaves us with one remaining candidate in our solar system, Jupiter. We might hope for life on Jupiter since its present atmosphere is composed of gases quite similar to the primeval atmosphere of earth (Figure 16-14). Energy sources such as solar ultraviolet radiation and tremendous lightning bolts could make complicated molecules

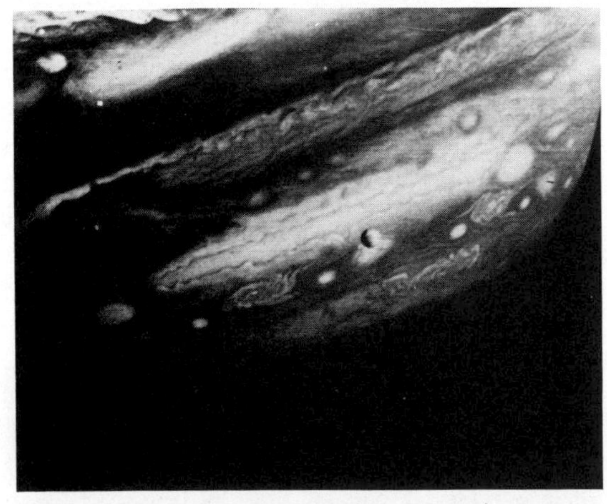

*Figure 16-14. Because Jupiter's atmospheric composition is much like earth's primeval atmosphere, some astronomers have speculated that sunlight or lightning may have generated life among the jovian clouds. The satellites in the foreground are Io (left) and Europa (right). (NASA.)*

in the atmosphere. In some of the deeper layers, water exists as vapor and as droplets, and the temperature 50 km (30 mi) below the cloud tops is a pleasant 27°C (80°F). Under these circumstances, life may have begun and evolved into forms like the plankton that lives in earth's oceans. Some have even speculated about jovian equivalents to fish and sharks—floating creatures that feed on the plankton and on each other. This is obviously speculation, and we will have no evidence until space probes descend into the jovian clouds. Such a mission is planned for the mid-1980s.

A few of the satellites of the jovian planets might have conditions that could support life. Saturn's moon Titan has an atmosphere and may have a moderate temperature. The same may be true of Neptune's moon Triton. The only way to search for life on these satellites is to visit them or send automatic probes to obtain samples. Because of the great distance involved, it is unlikely that these moons can be explored in such detail in the near future.

Slightly nearer to earth, Jupiter's moon Europa has been mentioned as a possible abode of life. Its icy crust may conceal a liquid water mantle, and if that water has never been frozen, living things may have developed beneath the ice. Again, the only way to be sure is to drill through the ice to reach any water that may lie there. The chance of life on Europa is probably slim, but it will be many years before that chance can be tested conclusively.

So far as we know now, the solar system is bare of life except for our planet. Consequently, our search for life in the universe takes us to other planetary systems.

*Life in Other Planetary Systems.* Might life exist in other solar systems? To consider this question let us try to decide how common planets are and what conditions a planet must fulfill for life to originate and evolve to intelligence. The first question is astronomical, while the second is biological. Our ability to discuss the problem of life outside our solar system is severely limited by our lack of experience.

In Chapter 12 we concluded that planets form as a natural result of star formation. In addition, the process that gives rise to planets is probably related to the process that forms binary star systems. Had Jupiter been 50 times more massive, it would have been a star instead of a planet. Since about half of all stars are members of binary or multiple star systems, it seems that this process is very common, implying that planetary systems are also common.

If a planet is to become a suitable home for life, it must have a stable orbit around its sun. This is simple in a solar system like our own, but in a binary system most planetary orbits are unstable. Most planets in such systems would not last long before they were swallowed up by one of the stars or ejected from the system.

Thus it seems that single stars are the most likely to have planets suitable for life. Since our galaxy contains about $10^{11}$ stars, half of which are single, there should be roughly $5 \times 10^{10}$ planetary systems in which we might look for life.

A few million years of suitable conditions does not seem to be enough time to originate life. Our planet required at least 0.5 to 1 billion years to create the first cells and 4.6 billion to create intelligence. Clearly conditions on a planet must remain acceptable over a long time. This eliminates giant stars that change their luminosity rapidly as they evolve. It also eliminates massive stars that remain stable on the main sequence for only a few million years. If life requires a few billion years to originate and evolve to intelligence then no star hotter than about F5 will do. This is not really a serious restriction, since upper main-sequence stars are rare anyway.

In previous sections we decided that life, as we know it at least, requires liquid water. That requirement defines a **life zone** (or ecosphere) around each star, a region within which a planet has temperatures that permit the existence of liquid water.

The size of the life zone depends on the temperature of the star (Figure 16-15). Hot stars have larger life zones because the planets must be more distant to remain cool. But the short main-sequence lives of these stars make them unacceptable. M stars have small life zones because they are extremely cool—only planets very near the star receive sufficient warmth. However, planets that are close to a star would probably become tidally coupled, keeping the same side toward the star. This might allow the water and atmosphere to freeze in the perpetual darkness of the planet's night side and end all chance of life. Thus the life zone restricts our search for life to main-sequence G and K stars. Some of the cooler F stars and warmer M stars might also be good candidates.

Even a star on the main sequence is not perfectly stable. Main-sequence stars gradually grow more luminous as they convert their hydrogen to helium, and thus the life zone around a star gradually moves outward. A planet might form in the life zone and life might begin and evolve for billions of years only to be destroyed as the slowly increasing luminosity of its star moved the life zone outward and evaporated the planet's oceans and drove off its atmosphere. If a planet is to remain in the life zone for 4 to 5 billion years, it must form on the outer edge of the zone. This may be the most serious restriction we have yet discussed.

If all of these requirements are met, will life begin? Early in this chapter we decided that life could begin through simple chemical reactions, so perhaps we should change our question and ask, What could prevent life from beginning? Given what we know about life, it should arise wherever conditions permit, and our galaxy

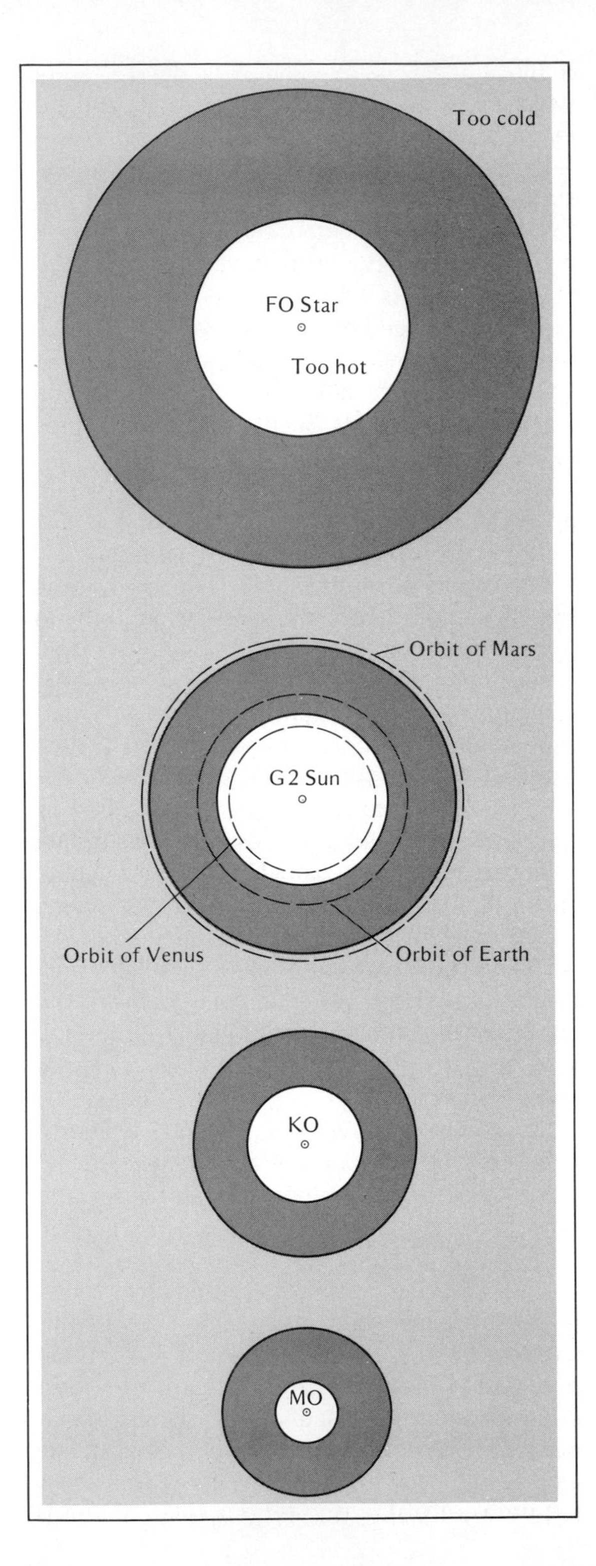


*Figure 16-15. The size of the life zone around a star depends on the temperature of the star. Stars hotter than about F5 do not remain stable long enough for life to develop. Stars cooler than about M0 may not support life because their life zone is too near the star and the planet's rotation becomes locked to the star.*

should be filled with planets that are inhabited with living creatures. Then why haven't we heard from them?

## COMMUNICATION WITH DISTANT CIVILIZATIONS

If other civilizations exist, perhaps we can communicate with them in some way. Sadly, travel between the stars appears more difficult in real life than in science fiction. It may in fact be almost impossible. If we can't physically visit, perhaps we can communicate by radio. Again, nature places restrictions on such conversations, but the restrictions are not too severe. The real problem lies with the nature of civilizations.

*Travel Between the Stars.* Roaming among the stars is in practice tremendously difficult because of three limitations: distance, speed, and fuel. The distances between stars are almost beyond comprehension. It does little good to explain that if we represent the sun by a ping-pong ball in New York City, the nearest star would be another ping-pong ball in Chicago. It is only slightly better to note that the fastest commercial jet would take about 4 million years to reach the nearest star.

The second limitation is a speed limit—we may not travel faster than the speed of light. Though science fiction writers invent hyperspace drives so their heros can zip from star to star, the speed of light is a natural and unavoidable limit that we cannot exceed. This, combined with the large distances between stars, makes interstellar travel very time consuming.

The third limitation says that we can't even approach the speed of light without using a fantastic amount of fuel. Even if we ignore the problem of escaping from the earth's gravity, we must still use energy stored in fuel to accelerate to high speed and to decelerate to a stop when we reach our destination. To return to earth, assuming we wish to, we have to repeat the process. These changes in velocity require a tremendous amount of fuel. If we flew a spaceship as big as a large yacht to a star 5 ly (1.5 pc) away and wanted to get there in only 10 years, we would use 40,000 times as much energy as the United States consumes in a year.

Travel for a few individuals might be possible if we accept very long travel times. That would require some form of suspended animation (currently unknown) or colony ships that carry a complete, though small, society in which people are born, live, and die generation after generation. Whether the occupants of such a ship would retain the social characteristics of humans over a long voyage is questionable.

These three limitations not only make it difficult for us to leave our solar system, but they would also make it difficult for aliens to visit earth. Reputable scientists have studied UFOs and related phenomena and have never found any evidence that the earth is being visited or has ever been visited by aliens from other worlds. Thus it seems unlikely that humans will ever meet an alien face to face. The only way we can communicate with other civilizations is via radio.

*Radio Communication.* Nature places two restrictions on our ability to communicate with distant societies by radio. One has to do with simple physics, is well understood, and merely makes the communication difficult. The second has to do with the fate of technological civilizations, is still unresolved, and may severely limit the number of societies we can detect by radio.

Radio signals are electromagnetic waves that travel at the speed of light. Since even the nearest civilizations must be a few light-years away, this limits our ability to carry on a conversation with distant beings. If we ask a question of a creature 4 ly away, we will have to wait 8 years for a reply. Clearly the give and take of normal conversation will be impossible.

Instead we could simply broadcast a radio beacon of friendship to announce our presence. Such a beacon would have to consist of a pattern of pulses obviously designed by intelligent beings to distinguish it from natural radio signals emitted by nebulae, pulsars, and so on. For example, pulses counting off the first dozen prime numbers would do. In fact, we are already broadcasting a recognizable beacon. Short-wavelength radio signals, such as TV and FM, have been leaking into space for the last 20 years or so. Any civilization within 20 ly might already have detected us.

If we intentionally broadcast such a signal, we could give listening aliens a good idea of what humanity is like by including coded data in the signal. For example, in 1974 at the rededication of the 1000-ft radio telescope at Arecibo, Puerto Rico, radio astronomers transmitted a series of pulses toward the globular cluster M 13 in Hercules (Figures 16-16). The number of data points in the message was 1679, a number selected because it can only be factored into 23 and 73. When the signal arrives at the globular cluster 26,000 years from now, any aliens who detect it will be able to arrange the data in only two ways—23 rows of 73 data points each, or 73 rows of 23 points each. The first way yields nonsense, but the second produces a picture that describes our solar system, the chemical basis of our life form, the general shape and size of the human body, and the number of humans on earth. Whether there will still be humans on earth in 52,000 years when any reply to our message returns remains to be discovered.

It took 30 minutes to transmit the Arecibo message. If more time were taken, a more detailed picture could be sent, and if we were sure our radio telescope were pointed at a listening civilization, we could send a long series of pictures. With pictures we could teach them our language and tell them all about our life, our difficulties, and accomplishments.

If we can think of sending such signals, then aliens can think of it too. If we point our radio telescopes in the right direction and listen at the right wavelength, we might hear other intelligent races calling out to each other. This raises two questions. Which stars are the best candidates, and what wavelengths are most likely? We have already answered the first question. Main-sequence G and K stars have the most favorable characteristics. But the second question is more complex.

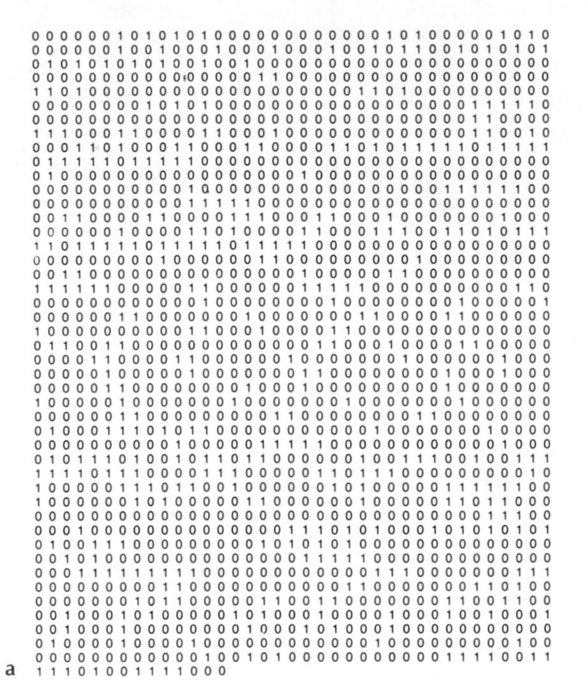


*Figure 16-16. (a) The Arecibo message of pulses transmitted toward the globular cluster M 13 is shown as a series of 0's and 1's. (b) Arranged in 73 rows of 23 pulses each and represented as light and dark squares, the message would tell aliens about human life. (From "The Search for Extraterrestrial Intelligence" by Carl Sagan and Frank Drake. Copyright © 1975 by Scientific American, Inc. All rights reserved.)*

Only certain wavelengths would be useful for communication. We cannot use wavelengths longer than about 30 cm because the signal would be lost in the background radio noise from our galaxy. Nor can we go to wavelengths much shorter than 1 cm because of absorption within our atmosphere. Thus only a certain range of wavelengths, a radio window, is open for communication (Figure 16-17).

This communications window is very wide,

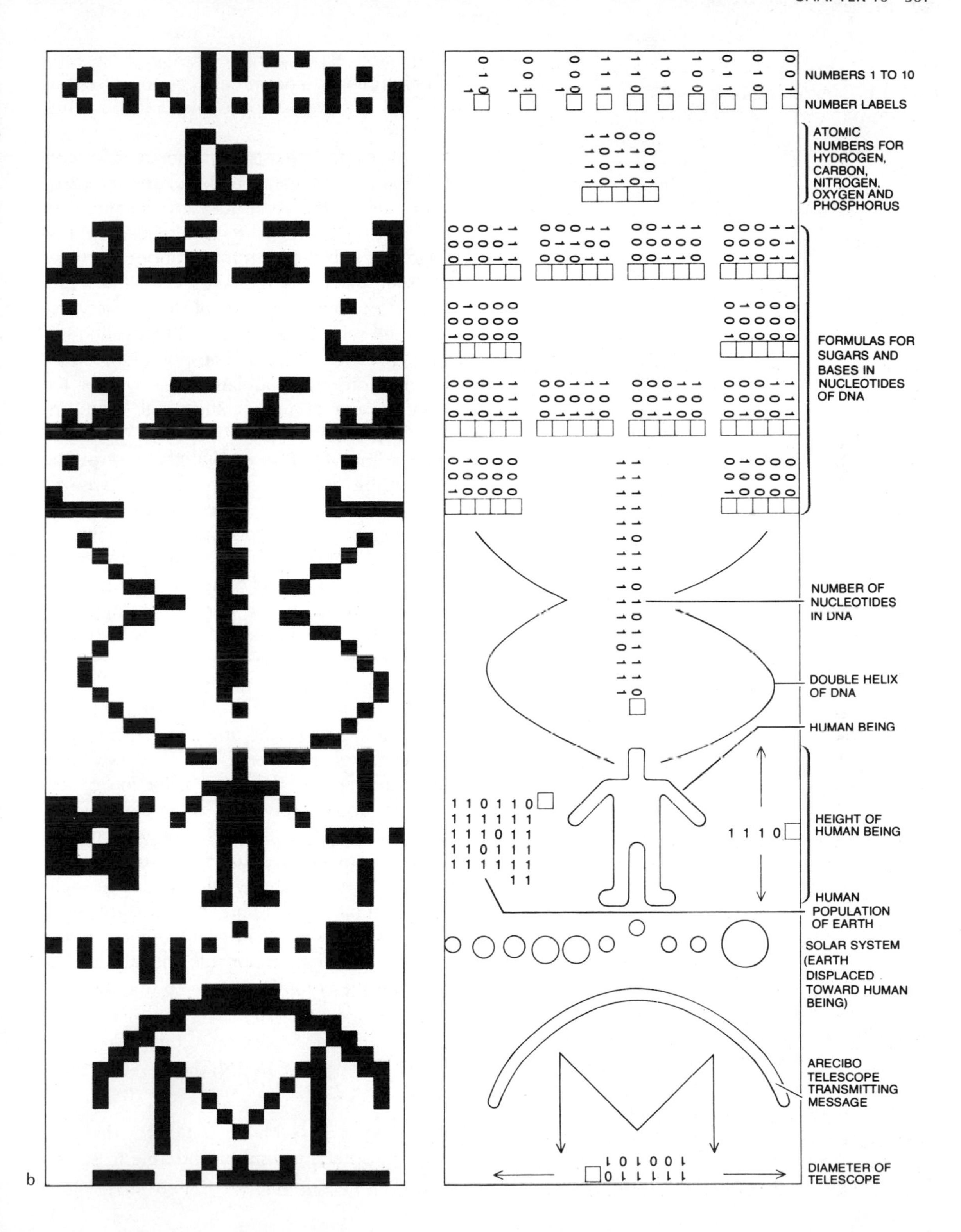
NUMBERS 1 TO 10
NUMBER LABELS
ATOMIC NUMBERS FOR HYDROGEN, CARBON, NITROGEN, OXYGEN AND PHOSPHORUS
FORMULAS FOR SUGARS AND BASES IN NUCLEOTIDES OF DNA
NUMBER OF NUCLEOTIDES IN DNA
DOUBLE HELIX OF DNA
HUMAN BEING
HEIGHT OF HUMAN BEING
HUMAN POPULATION OF EARTH
SOLAR SYSTEM (EARTH DISPLACED TOWARD HUMAN BEING)
ARECIBO TELESCOPE TRANSMITTING MESSAGE
DIAMETER OF TELESCOPE
b

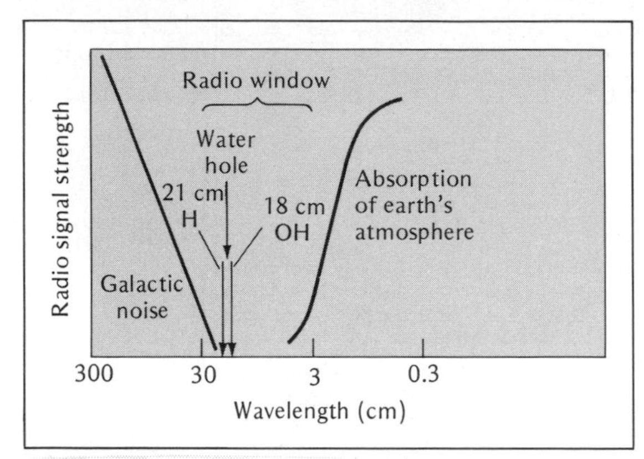


*Figure 16-17. The wavelength range between 30 cm and 3 cm is a window for possible communication between civilizations. The "water hole" between the radio emissions of H and OH is an especially likely wavelength range.*

so a radio relescope would take a long time to tune over all the wavelengths searching for intelligent signals. However, nature may have given us a way to narrow the search. Within the communications window lie the 21-cm line of neutral hydrogen and the 18-cm line of OH. The interval between these two lines has been dubbed the **water hole** because the combination of H and OH yields water ($H_2O$). Water is the fundamental solvent in our life form, so it might seem natural for similar water creatures to call out to each other at wavelengths in the water hole. But even a silicon creature would be familiar with the 21-cm line of hydrogen. Thus they too might select wavelengths near the water hole.

This discussion is not just speculation. In the summer of 1971 NASA and the American Society for Engineering Education assembled two dozen scientists and engineers to study the feasibility of detecting distant civilizations. The study, called Project Cyclops, concluded that the technology already existed to search for radio signals from extraterrestrial beings. The search would take a few billion dollars and might expect success in a few decades. Although the Cyclops search has not been funded, smaller, less expensive searches have already been completed and others are in progress. None has detected alien communication thus far.

The technology exists, but the most serious restriction on the search is the unanswered social question. How long does a civilization remain at a high enough technological level to engage in galactic communication? If other planets are like earth, life takes 4.6 billion years to reach a technological level. If a society destroys itself within 100 years of the invention of radio (by nuclear war, nuclear pollution, chemical pollution, overpopulation, etc.), then the chances of our communicating with them are very small. Most life forms in the galaxy would be on the long road up to civilization or on the path leading downward from the collapse of their technology. But if a technological society can solve its internal problems and remain stable for a million years, at least some would be at that stage at the proper time to communicate with us. Estimates of the number of communicative civilizations in our galaxy range from 10 million to 1 (see Box 16-2).

Are we the only thinking race? If we are, then we bear the sole responsibility to understand and admire the universe. Then we are the sole representatives of that state of matter called intelligence. The mere detection of signals from another civilization would demonstrate that we share the universe with others. Though we might never leave our solar system, such communication would end the self-centered isolation of humanity and stimulate a reevaluation of the meaning of our existence. We may never realize our full potential as humans until we communicate with nonhumans.

## PERSPECTIVE: COMING OF AGE IN THE GALAXY

If our galaxy is rich with life, if it is crowded with civilizations, then we should be able to search out the nearest and begin a dialogue. Since we are just entering the communicative stage, we are beginners, and most of these civilizations would be far more advanced than we and should be able to help us in various ways. However, before we announce our presence, we should recall that there are many cities on earth where we would not venture out alone after dark, and there may be similar rough neighborhoods in our galaxy.

Communication is fraught with both external hazards, such as invasion, and internal risks. Communication with an alien civilization may induce a difficult and irrevocable transformation for humanity, tearing us from our isolation and

BOX 16-2 THE NUMBER OF COMMUNICATIVE CIVILIZATIONS IN OUR GALAXY

Simple arithmetic can give us an estimate of the number of technological civilizations with which we might communicate. The formula for the number of communicative civilizations in a galaxy, $N_c$, is

$$N_c = N^* \cdot f_P \cdot n_{LZ} \cdot f_L \cdot f_I \cdot F_S$$

$\mathbf{N^*}$ is the number of stars in a galaxy, and $\mathbf{f_P}$ represents the probability that a star has planets. If all single stars have planets, then $\mathbf{f_P}$ is about 0.5. The factor $\mathbf{n_{LZ}}$ is the average number of planets in a solar system suitably placed in the life zone, $\mathbf{f_L}$ is the probability that life will originate if conditions are suitable, and $\mathbf{f_I}$ is the probability that the life form will evolve to intelligence. These factors can be roughly estimated, but the remaining factor is much more uncertain.

$\mathbf{F_S}$ is the fraction of a star's life during which the life form is communicative. Here we assume a star lives about 10 billion years. As explained in the text, if a society survives at a technological level for only 100 years, our chances of communicating with it are small. But a society that stabilizes and remains technological for a long time is much more likely to be in the communicative phase at the proper time to signal to us. If we assume technological societies destroy themselves in about 100 years, then $\mathbf{F_S}$ is 100 divided by 10 billion, or $10^{-8}$. But if societies can remain technological for a million years, then $\mathbf{F_S}$ is $10^{-4}$. The influence of this factor is shown in Table 16-1.

If the optimistic estimates are true, there may be a communicative civilization within a few dozen light-years of us and we could locate it by searching through only a few thousand stars. On the other hand, if the pessimistic view is correct, we may be the only planet in our galaxy capable of communication. We may never know until we understand how technological societies function.

**Table 16-1 The Number of Technological Civilizations per Galaxy**

| | | Estimates | |
|---|---|---|---|
| | | Pessimistic | Optimistic |
| $N^*$ | Number of stars per galaxy | $2 \times 10^{11}$ | $2 \times 10^{11}$ |
| $f_P$ | Fraction of stars with planets | 0.01 | 0.5 |
| $n_{LZ}$ | Number of planets per star that lie in life zone for longer than 4 billion years | 0.01 | 1 |
| $f_L$ | Fraction of suitable planets on which life begins | 0.01 | 1 |
| $f_I$ | Fraction of life forms that evolve to intelligence | 0.01 | 1 |
| $F_S$ | Fraction of star's life during which a technological society survives | $10^{-8}$ | $10^{-4}$ |
| $N_c$ | Number of communicative civilizations per galaxy | $2 \times 10^{-5}$ | $10 \times 10^{6}$ |

thrusting us into a larger society of civilized races. Like all comings of age it could be both painful and beneficial.

*Hazards.* The most obvious, though least likely, hazard of announcing our presence is invasion, that calamity so dear to science fiction fans. Transmission of signals marks the earth as habitable, identifies us as technological, and pinpoints our location. A fleet of armed interstellar starships would find earth easy pickings.

Nature, however, seems to have given us natural safety factors that protect us from such invasion. Interstellar travel is at least difficult and tremendously expensive, and it may be impossible. In addition, the earth might not be attractive to aliens based on other body chemistries and far advanced in technology.

Some have suggested that we ensure our safety by not broadcasting signals, but merely listening. If we eavesdrop on the rest of the galaxy, we could test their good intentions. But we already leak TV and FM radio signals into space, and it is unlikely that we can stop the leak in the near future. We have no choice but to trust to the isolation of planetary systems.

The worst hazard we might face would be **cultural shock**, the bewildering impact of one society on another. The history of earth is filled with examples of smaller, weaker societies that have crumbled when forced into contact with larger, more powerful cultures. Our only defense is the insulation of vast distance that makes conversation impossible and permits us time for thoughtful consideration of each new piece of information.

One aspect of cultural shock would be the impact of the discovery of nonhumans on our religious beliefs. Many of earth's religions teach that humanity is chosen, especially blessed, even made in the image of God. TV pictures of a race of intelligent squids with glowing antennae would challenge many of our most sincerely held beliefs.

Another aspect of cultural shock is the dependence we might develop if we began to receive all the answers to our problems via radio. George Wald, Nobel-prize-winning biologist has said, "I can conceive of no nightmare as terrifying as establishing such communication with a so-called superior (or if you wish, advanced) technology in outer space." His point, that the human enterprise of trying to understand nature might wither, is a serious consideration. However, we should remember that interstellar distances are vast and conversations are impossible. The information we might receive would be minimal compared to the questions it would generate, questions we could not put to the aliens because we could not wait for the long delayed answers. Communication with superior cultures might act as a stimulus to our human quest for understanding rather than as a suppressant. We should also note that refraining from communication for this reason is in effect preferring ignorance to knowledge. We would abandon our human enterprise in trying to preserve it.

The last and most dangerous hazard we might face is internal. Our planet is divided into nations that barely coexist. If one nation detected signals from a superior culture and began to learn advanced technology, it might enjoy a tremendous military advantage over its fellows. If it kept its discovery secret, no other nation could discover it without precise knowledge of the proper star and radio frequency. Unfortunately, there is no defense against ourselves except international cooperation to assure free access to such signals by all nations.

These hazards may seem so severe that we should close our ears and refuse to end our isolation on earth. But the most chilling aspect of the venture is inevitability. If the galaxy is populated by many civilizations communicating with each other, it is only a matter of time until we discover them, on purpose or by accident, and then all these theoretical hazards become real.

*Benefits.* If the galaxy is richly populated with technological civilizations, then we will discover them eventually and we might as well look at the bright side. We could benefit tremendously. Such communication would open a galaxy of new ideas and might give us new insight into what it means to be human.

Probably the most extensive benefit would be the end of the isolation of earth. The process that Copernicus began when he said the earth was not at the center of the universe would be brought to completion by the realization that humanity is not the only life form. If we can survive the cultural shock, the discovery of other intelligent beings should be a growing experience for all humans.

One obvious benefit of interstellar communication would be survival information. If some societies destroy themselves, we are unlikely to discover them. The civilizations we might find would be the survivors, the ones that solved the critical problems and reached stability. We can hope their signals would contain survival information. They might warn us from certain technological, economic, political, or social systems that lead inevitably to collapse. Our civilization faces so many critical situations it is hard to guess which might be fatal. Just knowing that some societies have solved their problems would be encouraging.

With the critical survival questions answered, a distant society might broadcast examples of their aesthetic expression. We might receive examples of painting and sculpture, music, literature, philosophy, and more. Are the laws of artistic composition the same for aliens? Is the theme of *Hamlet* universal?

Of course, the most obvious and perhaps most useful benefits of such communication would be technological. Such messages might contain directions for new methods of making steel, plastics, electronics circuits, and more. After all, a distant civilization might be alien, but it would work with the same chemical elements and the same laws of physics, so its technological processes should be applicable to earthly problems. Medical technology might be another matter, however. Even if a distant race was carbon based, they would probably have significantly different body chemistry and their medical technology would not be immediately applicable to human ills.

Science, however, might benefit even more than technology. Communication with an advanced race might give us a much more sophisticated understanding of nature than we now have. Even if they were no more advanced than we, communication would allow us to compare our world with theirs. Thus we would no longer be confined to the study of a single planet and its life. We could compare the earth's geology, for example, with that of other planets in minute detail. We could create whole new areas of study such as comparative evolution, comparative ecology, comparative psychology, and others.

If we assume that communication between civilizations is common, then our galaxy may be filled with a network of signals carrying beneficial information. Some were transmitted long ago and some recently. Some may be rebroadcasts received and retransmitted over and over as civilized races preserve our galactic heritage. To join that network would be a coming of age for humanity, an assumption of our heritage as living, intelligent beings.

*Where Are They?* If the galaxy is filled with such a hum of communication, why haven't we heard it? To date, radio astronomers have tested about 1000 stars for radio signals and found none. But even if technological civilizations are common, we might have to test 10,000 to 100,000 stars before we found a civilization transmitting a signal. The search has just begun.

Nevertheless, some astronomers feel that life on other worlds is very unlikely. The process that transforms chemical evolution into organic evolution is very complex, and it may require such special circumstances that biological evolution almost never gets started. If that is the case, we may be the only life in our galaxy.

Even if life begins easily, evolution may not lead to intelligence except in very special circumstances. In that case, there may be many planets where life exists, but we may be the only life that has the intelligence to grasp the mystery of existence. Or life may inevitably progress to intelligence, but intelligence may be a self-destroying trait. Intelligent species may destroy themselves by altering their planet so drastically they cannot survive. This is the most frightening prospect of all. If it is correct, we are not only alone, but we also face an unavoidable catastrophe in the near future.

Before we abandon all hope, we should consider the comment of astronomer Martin Rees: "Absence of evidence is not evidence of absence." We have hardly begun to look for fellow beings, and until we search many stars and many different radio wavelengths, we can hardly conclude that we are alone.

## SUMMARY

To discuss life on other worlds, we must first understand something about life in general, life on earth, and the origin of life. In general we can identify three properties in living things—a process, a physical basis, and a controlling unit of information. The process must extract energy from the surroundings, maintain the organism, and modify the surroundings to promote the organism's survival. The physical basis is the arrangement of matter and energy that implements the life process. On earth all life is based on carbon chemistry. The controlling information is the data necessary to maintain the organism's function. Data for earth life is stored in long carbon-chain molecules called DNA.

The DNA molecule stores information in the form of chemical bases linked together like the rungs of a ladder. When these patterns are copied by RNA molecules, they can direct the manufacture of proteins and enzymes. Thus the DNA information is the chemical formulae the cell needs to function. When a cell divides, the DNA molecule splits lengthwise and duplicates itself so that each of the new cells has a copy of the information. Errors in the duplication or damage to the DNA molecule can produce mutants, organisms that contain new DNA information and have new properties. Natural selection determines which of these new organisms are most suited to survive, and the species evolves to fit its environment.

The Miller experiment duplicated conditions in the earth's primitive environment and suggests that energy sources such as lightning could have formed amino acids and other complex molecules. Chemical evolution would have connected these together in larger and more complex, but not yet living, molecules. When a molecule acquired the ability to produce copies of itself, natural selection perfected the organism through biological evolution. Though this may have happened in the first billion years, life did not become diverse and complex until the Cambrian period about 0.6 billion years ago. Life emerged from the oceans about 0.4 billion years ago, and humanity developed only a few million years ago.

It seems unlikely that there is life on other planets in our solar system. Most of the planets are too hot or too cold. Mars may have had life long ago if its atmosphere was thicker and liquid water existed on its surface, but the Viking landers performed three kinds of experiments to look for life and found none. We can imagine how lightning bolts in Jupiter's atmosphere might have spawned life, but there are no data available on complex molecules in the jovian atmosphere.

To find life we must look beyond our solar system. Since we suspect that planets form from the left over debris of star formation, we suspect that most stars have planets. The rise of intelligence may take billions of years, however, so short-lived massive stars and binary stars with unstable planetary orbits must be discarded. The best candidates are G and K main-sequence stars.

The distances between stars are too large to

permit travel, but communication by radio could be possible. A certain wavelength range called a radio window is suitable, and a small range between the radio signals of H and OH, the so-called water hole, is especially likely.

Humanity now has the technology to search for other intelligent life in the universe. Though there are some hazards such as cultural shock, such communication would probably be beneficial. In any case, if life is common in the universe, and if technological societies do not destroy themselves too quickly, the discovery of radio signals from extraterrestrial life is inevitable.

## NEW TERMS

DNA (deoxyribonucleic acid)
protein
enzyme
RNA (ribonucleic acid)
amino acid
natural selection
mutant
Cambrian period
Miller experiment
chemical evolution
life zone
water hole
cultural shock

## QUESTIONS

1. What three properties do we associate with living things?
2. Why must the genetic information coded in a species' DNA be changeable?
3. Why do we believe life on earth began in the sea?
4. What is the difference between chemical evolution and biological evolution?
5. In what way is intelligence a survival mechanism?
6. Discuss the prospects of finding life on other planets in our solar system.
7. What role did the control play in the Viking labeled-release experiment?
8. Why are upper main-sequence stars unlikely sites for intelligent civilizations?
9. Describe the content of the Arecibo message.
10. What risks might we face if we detected distant civilizations?
11. What limitations make interstellar travel difficult?
12. Make as strong an argument as you can that we are alone in our galaxy.

## PROBLEMS

1. A single human cell encloses about 1.5 m of DNA containing 4.5 billion base pairs. What is the spacing between these base pairs in Angstroms? That is, how far apart are the rungs of the DNA ladder?
2. If we represent the history of the earth by a line 1 m long, how long a segment would represent the 400 million years since life moved onto the land? How long a segment would represent the 3-million-year history of humanity?
3. If a star must remain on the main sequence for at least 5 billion years for life to evolve to intelligence, how massive could a star be and still harbor intelligent life on one of its planets? (Hint: See Box 6-2.)
4. If there are about $1.4 \times 10^{-4}$ stars like the sun per cubic light-year, how many lie within 100 ly of earth (Hint: The volume of a sphere is $^4/_3\pi r^3$.)
5. Calculate the number of communicative civilizations per galaxy from your own estimates of the factors in Box 16-2.

## RECOMMENDED READING

Bracewell, R. N. *The Galactic Club: Intelligent Life in Outer Space.* San Francisco: W. H. Freeman, 1975.

Cloud, Preston. *Cosmos, Earth and Man.* New Haven: Yale University Press, 1978.

Dawkins, R. *The Selfish Gene.* New York: Oxford University Press, 1976.

Goldsmith, D., and Owen, T. *The Search of Life in the Universe.* Menlo Park, Calif.: Benjamin/Cummings, 1980.

Horowitz, N. "The Search for Life on Mars." *Scientific American* 237(Oct. 1977), p. 52.

Kurten, B. "Continental Drift and Evolution." *Scientific American* 220(March 1969) p. 54.

Lehninger, A. L. *Biochemistry.* 2nd ed. Chapter 37. New York: Worth, 1975.

Parker, B. "Are We the Only Intelligent Life in Our Galaxy?" *Astronomy* 7(Jan. 1979), p. 6. Reprinted in *Astronomy: Selected Readings.* ed. M. A. Seeds. Menlo Park, Calif.: Benjamin/Cummings, 1980.

Ponnamperuma, C. *The Origins of Life.* New York: Dutton, 1972.

*Project Cyclops: A Design Study of a System for Detecting Extraterrestrial Intelligent Life.* NASA CR 114445. Available from Dr. John Billingham, NASA/Ames Research Center, Code LT, Moffett Field, CA 94035.

Ridpath, I. *Messages From the Stars.* New York: Harper & Row, 1978.

Sagan, C. *The Cosmic Connection.* New York: Dell, 1973.

———. *The Dragons of Eden: Speculations on the Evolution of Human Intelligence.* New York: Random House, 1977.

————, and Page, T., eds. *UFO's: A Scientific Debate.* New York: W. W. Norton, 1972.

Shklovskii, I. S., and Sagan, S. *Intelligent Life in the Universe.* San Francisco: Holden-Day, 1966.

Stebbins, G. L. *Processes of Organic Evolution.* 2nd ed. Englewood Cliffs, N.J.: Prentice-Hall, 1971.

Stern, D. K. "First Contact with Non-human Cultures." *Mercury* 4(Sept./Oct. 1975), p. 14.

*Planet Earth seen from the Apollo 17 spacecraft. The Mediterranean lies at the extreme top of the photograph, with the Arabian Peninsula at the top center. The outline of Africa extends south toward Antarctica at the bottom. (NASA.)*

# Chapter 17 AFTERWORD

Our journey is over, but before we part company, there is one last thing to discuss—the place of humanity in the universe. Astronomy gives us some comprehension of the workings of stars, galaxies, and planets, but its greatest value lies in what it teaches us about ourselves. Now that we have surveyed astronomical knowledge, we can better understand our own position in nature.

To some, the word *nature* conjures up visions of furry rabbits hopping about in a forest glade dotted with pastel wildflowers. To others, nature is the blue-green ocean depths filled with creatures swirling in a mad struggle for survival. Still others think of nature as windswept mountain tops of gray stone and glittering ice. As diverse as these images are, they are all earth bound. Having studied astronomy, we can view nature as a beautiful mechanism composed of matter and energy interacting according to simple rules to form galaxies, stars, planets, mountain tops, ocean depths, and forest glades.

Perhaps the most important astronomical lesson is that we are a small but important part of the universe. Most of the universe is lifeless. The vast reaches between the galaxies appear to be empty of all but the thinnest gas, and the stars, which contain most of the mass, are much too hot to preserve the chemical bonds that seem necessary for life to survive and develop. Only on the surfaces of a few planets, where temperatures are moderate, could atoms link together to form living matter.

If life is special, then intelligence is precious. The universe must contain many planets devoid of life, planets where the wind has blown unfelt for billions of years. There may also exist planets where life has developed but has not become complex, planets on which the wind stirs wide plains of grass and rustles dark forests. On some planets, insects, fish, birds, and animals watch the passing days unaware of their own existence. It is intelligence, human or alien, that gives meaning to the landscape.

As the primary intelligent species on this planet, we are the custodians of a priceless gift—a planet filled with living things. This is especially true if life is rare in the universe. In fact, if the earth is the only inhabited planet, then our responsibility is overwhelming. In any case, we are the only creatures who can take action to preserve the existence of life on earth, and, ironically, it is our own actions that are the most serious hazards.

The future of humanity is not secure. We are trapped on a tiny planet with limited resources and a population growing faster than our ability to produce food. In our efforts to survive, we have already driven some creatures to extinction and now threaten others. If our civilization collapses because of starvation, or if our race destroys itself somehow, the only bright spot is that the rest of the creatures on earth will be better off for our absence.

But even if we control our population and conserve and recycle our resources, life on earth is doomed. In five billion years, the sun will leave the main sequence and swell into a red giant, incinerating the earth. However, earth will be lifeless long before that. Within the next few billion years the growing luminosity of the sun will first alter the earth's climate, and then boil its atmosphere and oceans. Our earth is, like everything else in the universe, only temporary.

To survive, humanity must leave the earth and search for other planets. Colonizing the moon and other planets of our solar system will not save us, since they will face the same fate as the earth when the sun dies. But travel to other stars is tremendously difficult, and may be impossible with the limited resources we have in our small solar system. We and all of the living things that depend on us for survival may be trapped.

This is a depressing prospect, but a few factors are comforting. First, everything in the universe is temporary. Stars die, galaxies die, perhaps the entire universe will fall back in a "big crunch" and die. That our distant future is limited only assures us that we are a part of a much larger whole. Second, we have a few billion years to prepare and a billion years is a very long time. Only a few million years ago our ancestors were learning to walk erect and communicate with each other. A billion years ago our ancestors were microscopic organisms living in the primeval oceans. To suppose that a billion years hence we humans will still be human, or that we will still be the dominant species on earth, or that we will still be the sole intelligence on earth is the ultimate conceit.

Our responsibility is not to save our race for all eternity, but to behave as dependable custodians of our planet, preserving it, admiring it, and trying to understand it. That will call for drastic changes in our behavior toward other living things and a revolution in our attitude toward our planet's resources. Whether we can change our ways is debatable—humanity is far from perfect in its understanding, abilities, or intentions. However, we must not imagine that we and our civilization are less than precious. We have the gift of intelligence, and that is the finest thing this planet has ever produced.

# Appendix A ASTRONOMICAL COORDINATES AND TIME

The rotation of the earth on its axis is of great importance in astronomy. It defines a system of latitude and longitude used to locate places on earth, and that system, projected into the sky, provides an astronomical coordinate system. In addition, the rotation of the earth is the basis of our timekeeping.

## ASTRONOMICAL COORDINATES

The earth is girded by a mesh of imaginary lines that define the latitude and longitude of every spot on the earth's surface. This system, projected on the sky, is the basis for the astronomical coordinate system.

An observer's latitude is measured north or south from the earth's equator. People living on the earth's equator, in Quito, Ecuador for instance, have latitude 0°, while an encampment of explorers at the earth's north pole are at latitude 90°. The people of New York City live at latitude 40°45′ and those in Los Angeles at 34°05′. Because latitude is measured north from the equator, people living south of the equator have negative latitudes. Thus we can draw lines around the earth parallel to the equator and refer to them as circles of latitude (Figure A-1a).

Longitude is measured east or west of Greenwich, England. Thus we can say that New York is 74° west of Greenwich. A north-south line through New York City extending from the north pole to the south pole is a line of longitude, and anyone on that line has the same longitude as New York City. Consequently we can cover the earth with north-south lines of longitude (Figure A-1b).

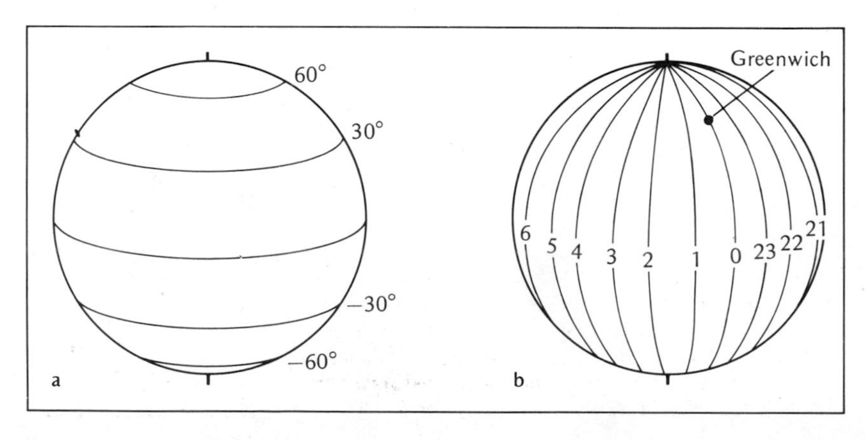


*Figure A-1. (a) Lines on the earth parallel to earth's equator are lines of constant latitude. (b) North-south lines extending from the earth's north pole to its south pole are lines of constant longitude. Greenwich, England, is the reference point from which longitude is measured.*

It is common to express a longitude as a time instead of an angle. This arises because the earth rotates through 360° in 24 hours. Thus 15° is equivalent to 1 hour. Instead of saying New York City is 74° west of Greenwich, we could say it is $4^h56^m$ west.

A similar system of lines on the sky defines the celestial coordinate system but instead of latitude we refer to a star's **declination** and instead of its longitude we refer to its **right ascension**. These are often abbreviated Dec. and R. A. In some books and tables, you may find declination abbreviated $\delta$ and right ascension abbreviated $\alpha$.

Declination is measured north or south from the celestial equator. A star on the celestial equator has declination 0°, and a hypothetical star at the north celestial pole has declination 90°. Polaris, the North Star, is not precisely at the pole, so its declination, 89°11′, is slightly less than 90°. Stars located south of the celestial equator have negative declinations, such as Rigel at −8°13′. Thus lines drawn around the celestial sphere parallel to the celestial equator are lines of constant declination (Figure A-2a).

Before we can measure right ascension, we must have some reference mark on the sky from which to measure. Longitude is measured east or west from Greenwich, but there are no cities on the sky, and the stars are all moving slowly as they orbit the center of our galaxy. Faced with this dilemma, astronomers choose as their reference point the vernal equinox, the point on the celestial equator where the sun crosses into the northern sky in the spring. As we will see later, even the vernal equinox is moving slowly.

Unlike longitude, right ascension is always measured eastward from the reference mark and

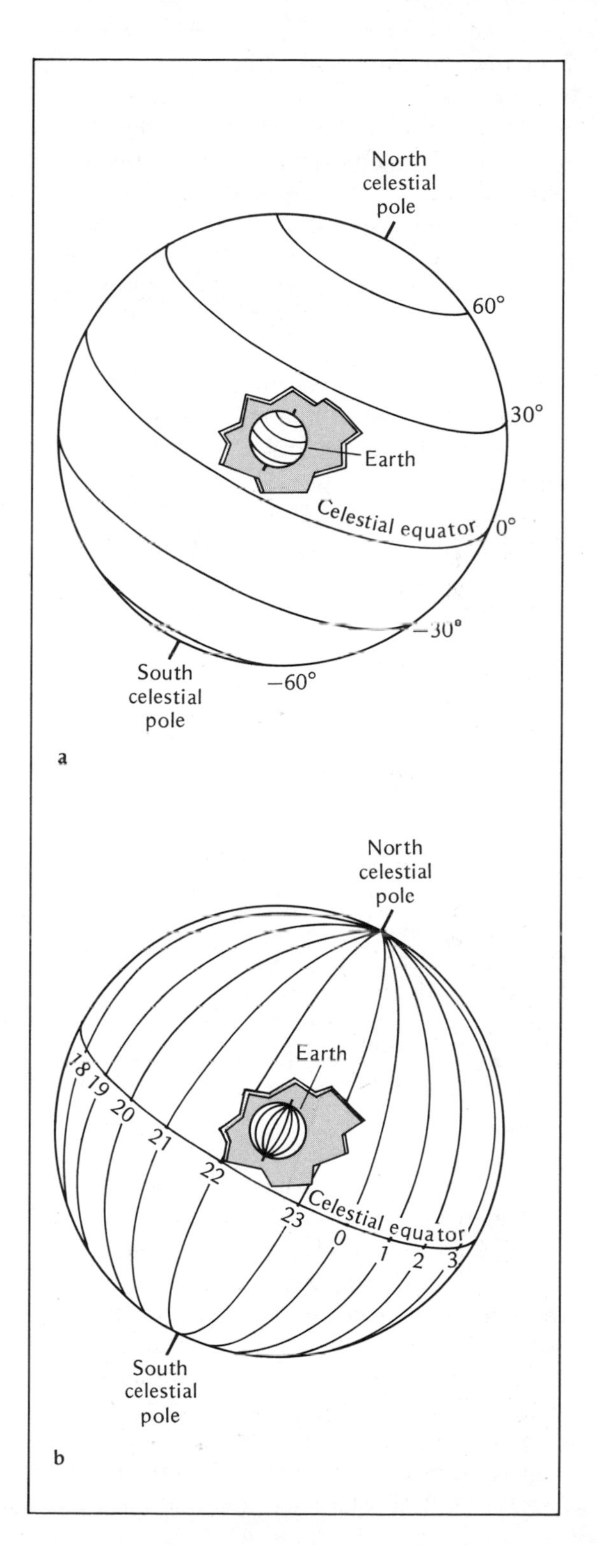


*Figure A-2. The celestial coordinate system is the projection on the sky of latitude and longitude. (a) On the celestial sphere, lines of constant declination are parallel to the celestial equator. (b) Lines of constant right ascension run north-south from celestial pole to celestial pole.*

is nearly always expressed in units of time. We can say the star Sirius has a right ascension of $6^h44^m$, meaning it lies 6 hours and 44 minutes east of the vernal equinox. Thus we divide the celestial equator into 24 equal divisions called hours of right ascension, and we draw 24, equally spaced north-south lines from the north celestial pole to the south celestial pole to serve as lines of constant right ascension (Figure A-2b).

Using this mesh of declination and right ascension lines, we can define the declination and right ascension of any object. For instance, the nearest star to the sun, Alpha Centauri, is located at R. A. $14^h38^m$, Dec. $-60°46'$. We could also give the right ascension and declination of the sun, moon, and planets, but since these objects move relatively rapidly, their coordinates change from day to day. Thus we would have to list them in an almanac giving the location of the object on specific dates.

This coordinate system is based on the rotation of the earth, but the earth's axis of rotation is not fixed in space. It precesses in a conical motion, taking 26,000 years for one circuit. Because this motion changes the orientation of the earth's axis, it also changes the direction toward the celestial poles and the celestial equator. Thus the mesh of imaginary lines that define the celestial coordinate system slips slowly across the sky. As a result, the vernal equinox, the intersection of the celestial equator and the ecliptic, moves slowly westward along the celestial equator.

This is a slow process, but it significantly affects the coordinates of a celestial body. For example, in 1970 the star Alpheratz in the constellation Andromeda (Figure 1-2) had the coordinates R. A. $0^h6^m48^s$, Dec. $+28°55'$, but by 1980 they had changed to R. A. $0^h7^m18^s$, Dec. $+28°58'$. Consequently, precise astronomical tables list the epoch of the coordinates—the date for which the coordinates are given. If the coordinates are used for some other date, they will be in error by some small amount.

## ASTRONOMICAL TIME SYSTEMS

Though an object's coordinates specify the direction toward that object, the rotation of the earth makes the entire sky seem to rotate around the earth. Consequently we cannot use an object's coordinates to point a telescope at it until we know which direction the earth is facing in its daily rotation.

Since the sky is divided into 24 equal intervals of right ascension, it is simple to set up a 24-hour-clock to read zero hours when the vernal equinox is on our local celestial meridian—the north-south line that goes through the observer's zenith. As time goes by, the rotation of the earth will cause the sky to move westward and the right ascension of objects on the local celestial meridian will gradually increase. If the clock is properly adjusted, it will keep pace with this change, and we can tell by glancing at its dial the right ascension of objects currently on the local celestial meridian. This kind of clock would keep **sidereal time**—time based on the motion of the stars across the sky.

However, we would not want to strap a sidereal clock to our wrist to govern our daily activities. To see why, imagine that we set our clocks at noon on the day of the vernal equinox. Both the sun and the vernal equinox would be on the local celestial meridian, so we would set a normal clock to read 12:00 noon and we would set our sidereal clock to read 0 hours. The next day when our sidereal clock read 0 hours again, the vernal equinox would be back on the local celestial meridian. However, during the interval of one day, the earth would have moved along its orbit and consequently the sun would have moved eastward about 1° along the ecliptic. It would not be at the vernal equinox, but about 1° east. Since the earth requires four minutes to rotate 1°, our normal clock would read 11:56 AM (Figure A-3). Thus a sidereal clock that keeps track of the stars must run four minutes a day faster than a normal clock that keeps track of the sun. If we tried to run our lives

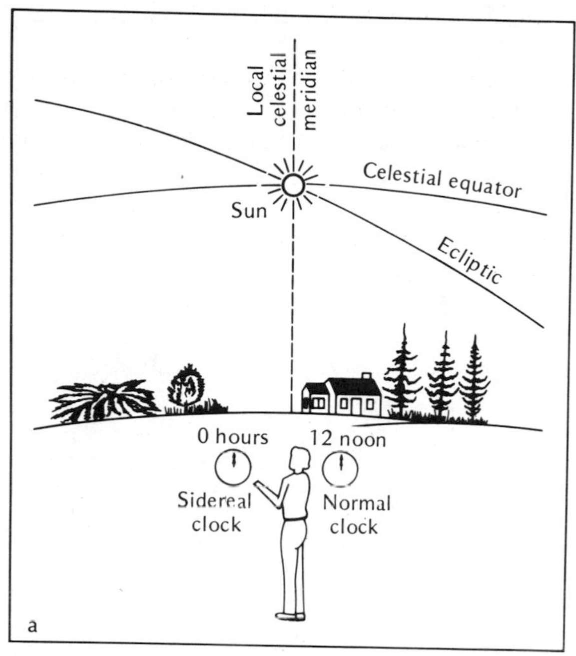


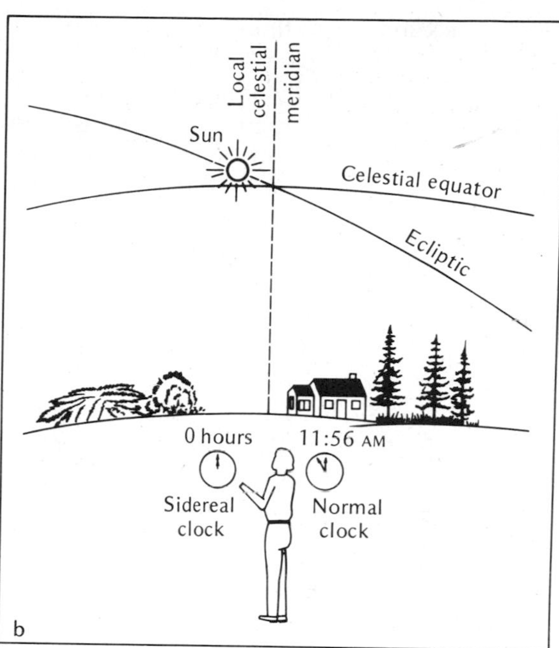


Figure A-3. *(a) About noon on a day in late March when the vernal equinox and the sun are on the local celestial meridian, we set a normal clock to 12:00 noon and a sidereal clock to 0:00 hours. (b) One day later the earth has rotated once on its axis and the vernal equinox is back on the local celestial meridian and our sidereal clock reads 0:00 hours. However, the sun has moved eastward along the ecliptic, and our normal clock reads 11:56* AM.

according to a sidereal clock, it would gain four minutes a day and in six months we would be eating supper at sunrise and breakfast at sunset.

## ZONE TIME

It would, in fact, be impossible to regulate a wristwatch to keep precise time according to the sun. Because the earth's orbit is slightly elliptical, it moves more rapidly when it is closer to the sun and more slowly when it is farther away. This makes the sun appear to move more rapidly along the ecliptic in January than in July. This and other factors make the sun's motion around the sky slightly nonuniform. A wristwatch, running at a steady rate, could not keep track of the exact position of the sun.

This is one reason a sundial does not usually agree with our wristwatch. The sundial reads the true apparent solar time. Our wristwatch gives the time assuming the sun moves at a constant rate around the sky. The two can differ by as much as 15 minutes.

Another reason sundials don't agree with our wristwatches is related to time zones. If one person lived 50 miles west of another person, their local celestial meridians would differ. The sun would first cross the local celestial meridian of the eastern observer, and his sundial would read noon. However, the western observer would see the sun slightly east of his meridian, and his sundial would say it was a few minutes before noon. Thus every person with a different longitude has a different time called local time.

This made no difference in the distant past, but with the advent of rapid communication and transportation it became very inconvenient if the clocks in one town were a few minutes different from the clocks in the next town west. To solve this problem, time zones were established, and every clock in a given time zone was set to the same time. Philadelphia and Pittsburgh have local times that differ by about 20 minutes, but they both lie in the Eastern Time Zone so the clocks in both cities read the same time. The standard time adopted in each time zone is the time kept by clocks on its central meridian (Figure A-4). Thus when it is 12 noon in New York, it is 11:00 AM in Kansas, 10:00 AM in Colorado, and 9:00 AM in California.

When we compare our wristwatches with a sundial, we compare zone time to local time. The sundial is slow by four minutes for every degree of longitude it lies west of the central meridian of the time zone. This combined with the uneven motion of the earth along its orbit can add up to an appreciable difference between zone time and a sundial.

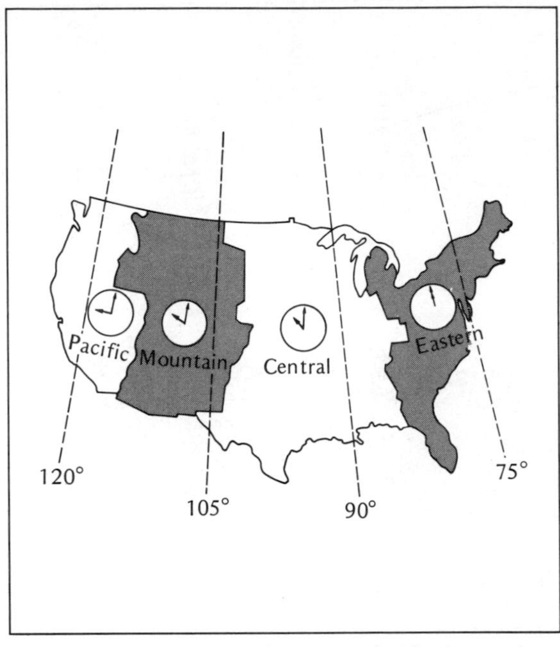


*Figure A-4.* *For convenience we divide the earth into time zones and all clocks within a zone keep the same time as a clock on the zone's central meridian. Adjacent zones differ by one hour.*

## NEW TERMS

declination
right ascension
sidereal time

# Appendix B NUMBERS AND UNITS

## POWERS OF 10 NOTATION

It is often convenient to write very large numbers with powers of 10. For example, the nearest star is about 43,000,000,000,000 km from the sun. It is easier to write this number as $4.3 \times 10^{13}$ km.

Very small numbers can similarly be written with powers of 10. For example, the wavelength of visible light is about 0.0000005 m. In powers of 10 this becomes $5 \times 10^{-7}$ m.

The powers of ten used in this notation are as follows.

$$
\begin{aligned}
&\vdots \\
10^{5} &= 100{,}000 \\
10^{4} &= 10{,}000 \\
10^{3} &= 1{,}000 \\
10^{2} &= 100 \\
10^{1} &= 10 \\
10^{0} &= 1 \\
10^{-1} &= 0.1 \\
10^{-2} &= 0.01 \\
10^{-3} &= 0.001 \\
10^{-4} &= 0.0001 \\
&\vdots
\end{aligned}
$$

The exponent tells us how to move the decimal point. If the exponent is positive, we move the decimal point to the right. If the exponent is negative, we move the decimal point to the left. Thus $2 \times 10^{3}$ equals 2000.0 and $2 \times 10^{-3}$ equals 0.002.

## METRIC UNITS

| Length | | | English units |
|---|---|---|---|
| 1 meter (m) | = 100 cm | = 1000 mm | = 39.36 inches |
| | | | = 1.0933 yards |
| 1 kilometer (km) | = 1000 m | | = 0.6214 miles |
| 1 centimeter (cm) | = 10 mm | = 1/100 m | = 0.39 inches |
| 1 millimeter (mm) | = 1/1000 m | | = 0.039 inches |
| 1 micron ($\mu$) | = $10^{-6}$ meters | | = 0.000039 inches |
| Mass | | | |
| 1 gram (gm) | = 1/1000 kg | | = 1/453.9 pounds |
| 1 kilogram (kg) | = 1000 g | | = 2.205 pounds |

## UNITS OF MEASUREMENT

| | | | |
|---|---|---|---|
| 1 Ångstrom (Å) | = $10^{-8}$ cm | = $10^{-10}$ m | |
| 1 astronomical unit (AU) | = $1.495979 \times 10^{13}$ cm | = $92.95582 \times 10^{6}$ miles | |
| 1 light-year (ly) | = $6.3240 \times 10^{4}$AU | = $9.46053 \times 10^{17}$ cm | = $5.9 \times 10^{12}$ miles |
| 1 parsec (pc) | = 206265 AU | = $3.085678 \times 10^{18}$ cm | = 3.261633 ly |
| 1 kiloparsec (kpc) | = 1000 pc | | |
| 1 megaparsec (Mpc) | = 1,000,000 pc | | |

## CONSTANTS

| | | |
|---|---|---|
| astronomical unit (AU) | = | $1.495979 \times 10^{13}$ cm |
| parsec (pc) | = | 206265 AU |
| | = | $3.085678 \times 10^{18}$ cm |
| | = | 3.261633 ly |
| light-year (ly) | = | $9.46053 \times 10^{17}$ cm |
| velocity of light (c) | = | $2.997925 \times 10^{10}$ cm/second |
| gravitational constant (G) | = | $6.67 \times 10^{-8}$ dyne cm$^2$/gm$^2$ |
| mass of earth | = | $5.976 \times 10^{27}$ gm |
| equatorial radius of earth | = | 6378.164 km |
| mass of sun ($M_\odot$) | = | $1.989 \times 10^{33}$ gm |
| radius of sun ($R_\odot$) | = | $6.9599 \times 10^{10}$ cm |
| solar luminosity ($L_\odot$) | = | $3.826 \times 10^{33}$ erg/second |
| mass of H atom | = | $1.67352 \times 10^{-24}$ gm |

# TEMPERATURE SCALES

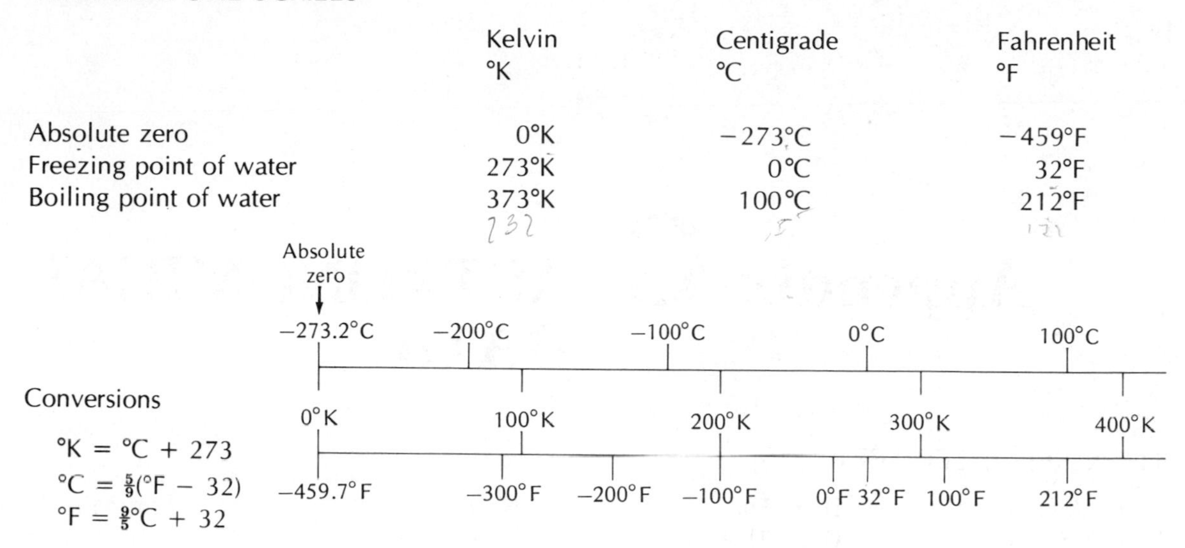

| | Kelvin °K | Centigrade °C | Fahrenheit °F |
|---|---|---|---|
| Absolute zero | 0°K | −273°C | −459°F |
| Freezing point of water | 273°K | 0°C | 32°F |
| Boiling point of water | 373°K | 100°C | 212°F |



Conversions

$°K = °C + 273$

$°C = \frac{5}{9}(°F - 32)$

$°F = \frac{9}{5}°C + 32$

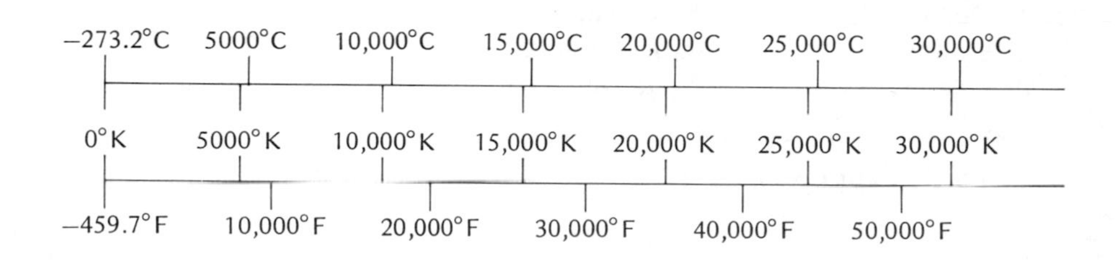

# Appendix C ASTRONOMICAL DATA

## THE GREEK ALPHABET

| | | | |
|---|---|---|---|
| A, α Alpha | H, η Eta | N, ν Nu | T, τ Tau |
| B, β Beta | Θ, θ Theta | Ξ, ξ Xi | Υ, υ Upsilon |
| Γ, γ Gamma | I, ι Iota | O, ο Omicron | Φ, ϕ Phi |
| Δ, δ Delta | K, κ Kappa | Π, π Pi | X, χ Chi |
| E, ϵ Epsilon | Λ, λ Lambda | P, ρ Rho | Ψ, ψ Psi |
| Z, ζ Zeta | M, μ Mu | Σ, σ Sigma | Ω, ω Omega |

## THE CONSTELLATIONS

| | | Approximate Position | |
|---|---|---|---|
| | | **R. A. (h)** | **Dec. (°)** |
| Andromeda (And) | The Princess | 1 | +40 |
| Antlia (Ant) | The Air Pump | 10 | −35 |
| Apus (Aps) | The Bird of Paradise | 16 | −75 |
| Aquarius (Aqr) | The Water Bearer | 23 | −15 |
| Aquila (Aql) | The Eagle | 20 | +5 |
| Ara (Ara) | The Altar | 17 | −55 |
| Aries (Ari) | The Ram | 3 | +20 |
| Auriga (Aur) | The Charioteer | 6 | +40 |
| Boötes (Boo) | The Bear Driver | 15 | +30 |
| Caelum (Cae) | The Sculptor's Chisel | 5 | −40 |
| Camelopardus (Cam) | The Giraffe | 6 | +70 |
| Cancer (Cnc) | The Crab | 9 | +20 |
| Canes Venatici (CVn) | The Hunting Dogs | 13 | +40 |
| Canis Major (CMa) | The Greater Dog | 7 | −20 |
| Canis Minor (CMi) | The Smaller Dog | 8 | +5 |

| | | Approximate Position | |
|---|---|---|---|
| | | R. A. (h) | Dec. (°) |
| Capricornus (Cap) | The Sea Goat | 21 | −20 |
| Carina (Car) | The Keel | 9 | −60 |
| Cassiopeia (Cas) | The Seated Queen | 1 | +60 |
| Centaurus (Cen) | The Centaur | 13 | −50 |
| Cepheus (Cep) | The King | 22 | +70 |
| Cetus (Cet) | The Whale | 2 | −10 |
| Chamaeleon (Cha) | The Chameleon | 11 | −80 |
| Circinus (Cir) | The Compasses | 15 | −60 |
| Columba (Col) | The Dove | 6 | −35 |
| Coma Berenices (Com) | Berenice's Hair | 13 | +20 |
| Corona Australis (CrA) | The Southern Crown | 19 | −40 |
| Corona Borealis (CrB) | The Northern Crown | 16 | +30 |
| Corvus (Crv) | The Crow | 12 | −20 |
| Crater (Crt) | The Cup | 11 | −15 |
| Crux (Cru) | The Southern Cross | 12 | −60 |
| Cygnus (Cyg) | The Swan | 21 | +40 |
| Delphinus (Del) | The Dolphin | 21 | +10 |
| Dorado (Dor) | The Swordfish | 5 | −65 |
| Draco (Dra) | The Dragon | 17 | +65 |
| Equuleus (Equ) | The Foal | 21 | +10 |
| Eridanus (Eri) | The River | 3 | −20 |
| Fornax (For) | The Laboratory Furnace | 3 | −30 |
| Gemini (Gem) | The Twins | 7 | +20 |
| Grus (Gru) | The Crane | 22 | −45 |
| Hercules (Her) | Hercules | 17 | +30 |
| Horologium (Hor) | The Clock | 3 | −60 |
| Hydra (Hya) | The Water Serpent | 10 | −20 |
| Hydrus (Hyi) | The Water Snake | 2 | −75 |
| Indus (Ind) | The American Indian | 21 | −55 |
| Lacerta (Lac) | The Lizard | 22 | +45 |
| Leo (Leo) | The Lion | 11 | +15 |
| Leo Minor (LMi) | The Lion Cub | 10 | +35 |
| Lepus (Lep) | The Hare | 6 | −20 |
| Libra (Lib) | The Scales | 15 | −15 |
| Lupus (Lup) | The Wolf | 15 | −45 |
| Lynx (Lyn) | The Lynx | 8 | +45 |
| Lyra (Lyr) | The Lyre | 19 | +40 |
| Mensa (Men) | The Table Mountain | 5 | −80 |
| Microscopium (Mic) | The Microscope | 21 | −35 |
| Monoceros (Mon) | The Unicorn | 7 | −5 |

| | | Approximate Position | |
|---|---|---|---|
| | | R. A. (h) | Dec. (°) |
| Musca (Mus) | The Fly | 12 | −70 |
| Norma (Nor) | The Carpenter's Square | 16 | −50 |
| Octans (Oct) | The Octant | 22 | −85 |
| Ophiuchus (Oph) | The Serpent Holder | 17 | 0 |
| Orion (Ori) | The Great Hunter | 5 | 0 |
| Pavo (Pav) | The Peacock | 20 | −65 |
| Pegasus (Peg) | The Winged Horse | 22 | +20 |
| Perseus (Per) | The Hero | 3 | +45 |
| Phoenix (Phe) | The Phoenix | 1 | −50 |
| Pictor (Pic) | The Painter's Easel | 6 | −55 |
| Pisces (Psc) | The Fishes | 1 | +15 |
| Piscis Austrinus (PsA) | The Southern Fish | 22 | −30 |
| Puppis (Pup) | The Stern | 8 | −40 |
| Pyxis (Pyx) | The Compass Box | 9 | −30 |
| Reticulum (Ret) | The Net | 4 | −60 |
| Sagitta (Sge) | The Arrow | 20 | +10 |
| Sagittarius (Sgr) | The Archer | 19 | −25 |
| Scorpius (Sco) | The Scorpion | 17 | −40 |
| Sculptor (Scl) | The Sculptor's Workshop | 0 | −30 |
| Scutum (Sct) | The Shield | 19 | −10 |
| Serpens (Ser) | The Serpent | 17 | 0 |
| Sextans (Sex) | The Sextant | 10 | 0 |
| Taurus (Tau) | The Bull | 4 | +15 |
| Telescopium (Tel) | The Telescope | 19 | −50 |
| Triangulum (Tri) | The Triangle | 2 | +30 |
| Triangulum Australe (TrA) | The Southern Triangle | 16 | −65 |
| Tucana (Tuc) | The Toucan | 0 | −65 |
| Ursa Major (UMa) | The Greater Bear | 11 | +50 |
| Ursa Minor (UMi) | The Smaller Bear | 15 | +70 |
| Vela (Vel) | The Sail | 9 | −50 |
| Virgo (Vir) | The Maiden | 13 | 0 |
| Volans (Vol) | The Flying Fish | 8 | −70 |
| Vulpecula (Vul) | The Fox | 20 | +25 |

# THE NEAREST STARS

| Name | Absolute Magnitude $M_v$ | Distance (ly) | Spectral Type | Apparent Visual Magnitude $m_v$ |
|---|---|---|---|---|
| Sun | 4.83 | | G2 | −26.8 |
| α Cen A | 4.38 | 4.3 | G2 | 0.1 |
| B | 5.76 | 4.3 | K5 | 1.5 |
| Barnard's Star | 13.21 | 5.9 | M5 | 9.5 |
| Wolf 359 | 16.80 | 7.6 | M6 | 13.5 |
| Lalande 21185 | 10.42 | 8.1 | M2 | 7.5 |
| Sirius A | 1.41 | 8.6 | A1 | −1.5 |
| B | 11.54 | 8.6 | white dwarf | 7.2 |
| Luyten 726-8A | 15.27 | 8.9 | M5 | 12.5 |
| B (UV Cet) | 15.8 | 8.9 | M6 | 13.0 |
| Ross 154 | 13.3 | 9.4 | M5 | 10.6 |
| Ross 248 | 14.8 | 10.3 | M6 | 12.2 |
| ϵ Eri | 6.13 | 10.7 | K2 | 3.7 |
| Luyten 789-6 | 14.6 | 10.8 | M7 | 12.2 |
| Ross 128 | 13.5 | 10.8 | M5 | 11.1 |
| 61 CYG A | 7.58 | 11.2 | K5 | 5.2 |
| B | 8.39 | 11.2 | K7 | 6.0 |
| ϵ Ind | 7.0 | 11.2 | K5 | 4.7 |
| Procyon A | 2.64 | 11.4 | F5 | 0.3 |
| B | 13.1 | 11.4 | white dwarf | 10.8 |
| Σ 2398 A | 11.15 | 11.5 | M4 | 8.9 |
| B | 11.94 | 11.5 | M5 | 9.7 |
| Groombridge 34 A | 10.32 | 11.6 | M1 | 8.1 |
| B | 13.29 | 11.6 | M6 | 11.0 |
| Lacaille 9352 | 9.59 | 11.7 | M2 | 7.4 |
| τ Ceti | 5.72 | 11.9 | G8 | 3.5 |
| BD +5° 1668 | 11.98 | 12.2 | M5 | 9.8 |
| L 725-32 | 15.27 | 12.4 | M5 | 11.5 |
| Lacaille 8760 | 8.75 | 12.5 | M0 | 6.7 |
| Kapteyn's Star | 10.85 | 12.7 | M0 | 8.8 |
| Kruger 60 A | 11.87 | 12.8 | M3 | 9.7 |
| B | 13.3 | 12.8 | M4 | 11.2 |

## THE BRIGHTEST STARS

| Star | Name | Apparent Visual Magnitude ($m_v$) | Spectral Type | Absolute Visual Magnitude ($M_v$) | Distance (ly) |
|---|---|---|---|---|---|
| α CMa A | Sirius | −1.47 | A1 | 1.4 | 8.7 |
| α Car | Canopus | −0.72 | F0 | −3.1 | 98 |
| α Cen | Rigil Kentaurus | −0.01 | G2 | 4.4 | 4.3 |
| α Boo | Arcturus | −0.06 | K2 | −0.3 | 36 |
| α Lyr | Vega | 0.04 | A0 | 0.5 | 26.5 |
| α Aur | Capella | 0.05 | G8 | −0.6 | 45 |
| β Ori A | Rigel | 0.14 | B8 | −7.1 | 900 |
| α CMi A | Procyon | 0.37 | F5 | 2.7 | 11.3 |
| α Ori | Betelgeuse | 0.41 | M2 | −5.6 | 520 |
| α Eri | Achernar | 0.51 | B3 | −2.3 | 118 |
| β Cen AB | Hadar | 0.63 | B1 | −5.2 | 490 |
| α Aql | Altair | 0.77 | A7 | 2.2 | 16.5 |
| α Tau A | Aldebaran | 0.86 | K5 | −0.7 | 68 |
| α Cru | Acrux | 0.90 | B2 | −3.5 | 260 |
| α Vir | Spica | 0.91 | B1 | −3.3 | 220 |
| α SCO A | Antares | 0.92 | M1 | −5.1 | 520 |
| α PsA | Fomalhaut | 1.15 | A3 | 2.0 | 22.6 |
| β Gem | Pollux | 1.16 | K0 | 1.0 | 35 |
| α Cyg | Deneb | 1.26 | A2 | −7.1 | 1600 |
| β Cru | Beta Crucis | 1.28 | B0.5 | −4.6 | 490 |

## PROPERTIES OF MAIN-SEQUENCE STARS

| Spectral Type | Absolute Visual Magnitude ($M_v$) | Luminosity* | Temperature (°K) | λ max (Å) | Mass* | Radius* | Average Density ($gm/cm^3$) |
|---|---|---|---|---|---|---|---|
| O5 | −5.8 | 501,000 | 40,000 | 724 | 40 | 17.8 | 0.01 |
| B0 | −4.1 | 20,000 | 28,000 | 1000 | 18 | 7.4 | 0.1 |
| B5 | −1.1 | 790 | 15,000 | 1900 | 6.4 | 3.8 | 0.2 |
| A0 | +0.7 | 79 | 9900 | 2900 | 3.2 | 2.5 | 0.3 |
| A5 | +2.0 | 20 | 8500 | 3400 | 2.1 | 1.7 | 0.6 |
| F0 | +2.6 | 6.3 | 7400 | 3900 | 1.7 | 1.4 | 1.0 |
| F5 | +3.4 | 2.5 | 6600 | 4400 | 1.3 | 1.2 | 1.1 |
| G0 | +4.4 | 1.3 | 6000 | 4800 | 1.1 | 1.0 | 1.4 |
| G5 | +5.1 | 0.8 | 5500 | 5200 | 0.9 | 0.9 | 1.6 |
| K0 | +5.9 | 0.4 | 4900 | 5900 | 0.8 | 0.8 | 1.8 |

* Luminosity, mass, and radius are given in terms of the sun's luminosity, mass, and radius.

| Spectral Type | Absolute Visual Magnitude ($M_v$) | Luminosity* | Temperature (°K) | λ max (Å) | Mass* | Radius* | Average Density (gm/cm³) |
|---|---|---|---|---|---|---|---|
| K5 | +7.3 | 0.2 | 4100 | 7000 | 0.7 | 0.7 | 2.4 |
| M0 | +9.0 | 0.1 | 3500 | 8300 | 0.5 | 0.6 | 2.5 |
| M5 | +11.8 | 0.01 | 2800 | 10,000 | 0.2 | 0.3 | 10.0 |
| M8 | +16 | 0.001 | 2400 | 12,000 | 0.1 | 0.1 | 63 |

## PROPERTIES OF THE PLANETS AND SATELLITES

### Planets: Physical Properties

| Planet | Equatorial (km) | Radius (Earth=1) | Mass (Earth=1) | Average Density (gm/cm³) | Surface Gravity (Earth=1) | Escape Velocity (km/sec) | Period of Revolution | Inclination of Equator to Orbit |
|---|---|---|---|---|---|---|---|---|
| Mercury | 2425 | 0.38 | 0.0554 | 5.44 | 0.378 | 4.2 | $59^{d}$ | <28° |
| Venus | 6070 | 0.95 | 0.815 | 5.24 | 0.894 | 10.3 | $244.3^{d}$ | 177° |
| Earth | 6378 | 1.00 | 1.00 | 5.497 | 1.00 | 11.2 | $23^{h}56^{m}04.1^{s}$ | 23°27′ |
| Mars | 3395 | 0.53 | 0.1075 | 3.9 | 0.379 | 5.0 | $24^{h}37^{m}22.6^{s}$ | 23°59′ |
| Jupiter | 71,300 | 11.18 | 317.83 | 1.34 | 2.54 | 61 | $9^{h}50^{m}30^{s}$ | 3°5′ |
| Saturn | 60,100 | 9.42 | 95.147 | 0.7 | 1.07 | 37 | $10^{h}14^{m}$ | 26°44′ |
| Uranus | 26,230 | 4.11 | 14.54 | 1.58 | 0.919 | 22 | $15^{h}$? | 97°55′ |
| Neptune | 25,100 | 3.93 | 17.23 | 2.3 | 1.19 | 25 | $22^{h}$? | 28°48′ |
| Pluto | 1500 | 0.24 | 0.002? | 0.5? | 0.05? | 1.2? | $6^{d}9^{h}17^{m}$ | 54°? |

### Planets: Orbital Properties

| Planet | Semimajor Axis (a) (AU) | (10⁶ km) | Orbital Period (P) (yr) | (Days) | Average Orbital Velocity (km/sec) | Orbital Eccentricity | Inclination to Ecliptic |
|---|---|---|---|---|---|---|---|
| Mercury | 0.3871 | 57.9 | 0.24084 | 87.96 | 47.89 | 0.2056 | 7°0′26″ |
| Venus | 0.7233 | 108.2 | 0.61515 | 224.68 | 35.03 | 0.0068 | 3°23′40″ |
| Earth | 1 | 149.6 | 1 | 365.26 | 29.79 | 0.0167 | 0°0′14″ |
| Mars | 1.5237 | 227.9 | 1.8808 | 686.95 | 24.13 | 0.0934 | 1°51′09″ |
| Jupiter | 5.2028 | 778.3 | 11.867 | 4334.3 | 13.06 | 0.0484 | 1°18′29″ |
| Saturn | 9.5388 | 1427.0 | 29.461 | 10,760 | 9.64 | 0.0560 | 2°29′17″ |
| Uranus | 19.1914 | 2871.0 | 84.014 | 30,685 | 6.81 | 0.0461 | 0°46′23″ |
| Neptune | 30.0611 | 4497.1 | 164.793 | 60,189 | 5.43 | 0.0100 | 1°46′27″ |
| Pluto | 39.44 | 5900 | 247.7 | 90,465 | 4.74 | 0.2484 | 17°9′3″ |

## Satellites of the Solar System

| Planet | | Satellite | Radius (km) | Distance from Planet ($10^3$ km) | Orbital Period (days) | Orbital Eccentricity | Orbital Inclination |
|---|---|---|---|---|---|---|---|
| Earth | | Moon | 1738 | 384 | 27.322 | 0.055 | 23° |
| Mars | | Phobos | 14 | 9 | 0.3189 | 0.018 | 1°.0 |
| | | Deimos | 8 | 23 | 1.262 | 0.002 | 1°.3 |
| Jupiter | XV | 1979J1* | 20 | 57 | .30 | — | 0° |
| | V | Amalthea | 80 | 181 | .4982 | 0.003 | 0°.4 |
| | I | Io | 1810 | 422 | 1.769 | 0.000 | 0° |
| | II | Europa | 1480 | 671 | 3.551 | 0.000 | 0° |
| | III | Ganymede | 2600 | 1070 | 7.155 | 0.001 | 0° |
| | IV | Callisto | 2360 | 1883 | 16.69 | 0.01 | 0° |
| | XIII | Leda | 8 | 11,110 | 239 | 0.147 | 26°.7 |
| | VI | Himalia | 50 | 11,476 | 250.6 | 0.158 | 27°.6 |
| | VII | Elara | 12 | 11,737 | 259.7 | 0.207 | 24°.8 |
| | X | Lysithea | 8 | 11,700 | 263.6 | 0.130 | 29°.0 |
| | XII | Ananke | 8 | 21,200 | 631.1 | 0.169 | 147° |
| | XI | Carme | 9 | 22,600 | 692.5 | 0.207 | 164° |
| | VIII | Pasiphae | 10 | 23,500 | 738.9 | 0.378 | 145° |
| | IX | Sinope | 9 | 23,600 | 758 | 0.275 | 153° |
| Saturn | | 1979 S1† | 100 | 152 | .700 | — | 0° |
| | | Janus | 150 | 159 | .749 | 0.0 | 0°.0 |
| | | Mimas | 270 | 186 | .942 | 0.020 | 1°.5 |
| | | Enceladus | 300 | 238 | 1.370 | 0.004 | 0°.0 |
| | | Tethys | 500 | 295 | 1.888 | 0.000 | 1°.1 |
| | | Dione | 480 | 377 | 2.737 | 0.002 | 0°.0 |
| | | Rhea | 650 | 527 | 4.518 | 0.001 | 0°.4 |
| | | Titan | 2440 | 1222 | 15.95 | 0.029 | 0°.3 |
| | | Hyperion | 220 | 1483 | 21.28 | 0.104 | 0°.4 |
| | | Iapetus | 550 | 3560 | 79.33 | 0.028 | 14°.7 |
| | | Phoebe | 120 | 12,950 | 550.5 | 0.163 | 150° |
| Uranus | | Miranda | 120 | 130 | 1.414 | 0.017 | 0° |
| | | Ariel | 350 | 192 | 2.520 | 0.003 | 0° |
| | | Umbriel | 250 | 267 | 4.144 | 0.004 | 0° |
| | | Titania | 500 | 438 | 8.706 | 0.002 | 0° |
| | | Oberon | 450 | 586 | 13.46 | 0.001 | 0° |
| Neptune | | Triton | 1900 | 355 | 5.877 | 0.000 | 160° |
| | | Nereid | 120 | 5562 | 359.4 | 0.76 | 27°.4 |
| Pluto | | Charon | 700? | 20,000 | 6.3861 | 0? | 54° |

* Discovered by Voyager 2 spacecraft.
† Discovered by Pioneer 10 spacecraft.

## METEOR SHOWERS

| Shower | Dates | Hourly Rate | Radiant R. A. | Radiant Dec. | Associated Comet |
|---|---|---|---|---|---|
| Quadrantids | Jan. 2–4 | 30 | $15^h24^m$ | 50° | |
| Lyrids | April 20–22 | 8 | $18^h\ 4^m$ | 33° | 1861 I |
| $\eta$ Aquarids | May 2–7 | 10 | $22^h24^m$ | 0° | Halley? |
| $\delta$ Aquarids | July 26–31 | 15 | $22^h36^m$ | −10° | |
| Perseids | Aug. 10–14 | 40 | $3^h\ 4^m$ | 58° | 1982 III |
| Orionids | Oct. 18–23 | 15 | $6^h20^m$ | 15° | Halley? |
| Taurids | Nov. 1–7 | 8 | $3^h40^m$ | 17° | Encke |
| Leonids | Nov. 14–19 | 6 | $10^h12^m$ | 22° | 1866 I Temp |
| Geminids | Dec. 10–13 | 50 | $7^h28^m$ | 32° | |

## THE MESSIER OBJECTS

| M | NGC* | Right Ascension (1950) (h) | (m) | Declination (1950) (°) | (′) | Apparent Visual Magnitude | Description |
|---|---|---|---|---|---|---|---|
| 1 | 1952 | 5 | 31.5 | +21 | 59 | 8.4 | Crab Nebula in Taurus; remains of supernova |
| 2 | 7089 | 21 | 30.9 | −1 | 02 | 6.4 | Globular cluster in Aquarius |
| 3 | 5272 | 13 | 39.8 | +28 | 38 | 6.3 | Globular cluster in Canes Venatici |
| 4 | 6121 | 16 | 20.6 | −26 | 24 | 6.5 | Globular cluster in Scorpius |
| 5 | 5904 | 15 | 16.0 | +2 | 16 | 6.1 | Globular cluster in Serpens |
| 6 | 6405 | 17 | 36.8 | −32 | 10 | 5.3 | Open cluster in Scorpius |
| 7 | 6475 | 17 | 50.7 | −34 | 48 | 4.1 | Open cluster in Scorpius |
| 8 | 6523 | 18 | 00.1 | −24 | 23 | 6.0 | Lagoon Nebula in Sagittarius |
| 9 | 6333 | 17 | 16.3 | −18 | 28 | 7.3 | Globular cluster in Ophiuchus |
| 10 | 6254 | 16 | 54.5 | −4 | 02 | 6.7 | Globular cluster in Ophiuchus |
| 11 | 6705 | 18 | 48.4 | −6 | 20 | 6.3 | Open cluster in Scutum |
| 12 | 6218 | 16 | 44.7 | −1 | 52 | 6.6 | Globular cluster in Ophiuchus |
| 13 | 6205 | 16 | 39.9 | +36 | 33 | 5.9 | Globular cluster in Hercules |
| 14 | 6402 | 17 | 35.0 | −3 | 13 | 7.7 | Globular cluster in Ophiuchus |
| 15 | 7078 | 21 | 27.5 | +11 | 57 | 6.4 | Globular cluster in Pegasus |
| 16 | 6611 | 18 | 16.1 | −13 | 48 | 6.4 | Open cluster with nebulosity in Serpens |
| 17 | 6618 | 18 | 17.9 | −16 | 12 | 7.0 | Swan or Omega Nebula in Sagittarius |
| 18 | 6613 | 18 | 17.0 | −17 | 09 | 7.5 | Open cluster in Sagittarius |
| 19 | 6273 | 16 | 59.5 | −26 | 11 | 6.6 | Globular cluster in Ophiuchus |
| 20 | 6514 | 17 | 59.4 | −23 | 02 | 9.0 | Trifid Nebula in Sagittarius |

* New General Catalog number.

| M | NGC | Right Ascension (1950) | | Declination (1950) | | Apparent Visual Magnitude | Description |
|---|---|---|---|---|---|---|---|
| | | (h) | (m) | (°) | (′) | | |
| 21 | 6531 | 18 | 01.6 | −22 | 30 | 6.5 | Open cluster in Sagittarius |
| 22 | 6656 | 18 | 33.4 | −23 | 57 | 5.6 | Globular cluster in Sagittarius |
| 23 | 6494 | 17 | 54.0 | −19 | 00 | 6.9 | Open cluster in Sagittarius |
| 24 | 6603 | 18 | 15.5 | −18 | 27 | 11.4 | Open cluster in Sagittarius |
| 25 | IC4725* | 18 | 28.7 | −19 | 17 | 6.5 | Open cluster in Sagittarius |
| 26 | 6694 | 18 | 42.5 | −9 | 27 | 9.3 | Open cluster in Scutum |
| 27 | 6853 | 19 | 57.5 | +22 | 35 | 7.6 | Dumb-bell Planetary Nebula in Vulpecula |
| 28 | 6626 | 18 | 21.4 | −24 | 53 | 7.6 | Globular cluster in Sagittarius |
| 29 | 6913 | 20 | 22.2 | +38 | 21 | 7.1 | Open cluster in Cygnus |
| 30 | 7099 | 21 | 37.5 | −23 | 24 | 8.4 | Globular cluster in Capricornus |
| 31 | 224 | 0 | 40.0 | +41 | 00 | 4.8 | Andromeda galaxy |
| 32 | 221 | 0 | 40.0 | +40 | 36 | 8.7 | Elliptical galaxy; companion to M31 |
| 33 | 598 | 1 | 31.0 | +30 | 24 | 6.7 | Spiral galaxy in Triangulum |
| 34 | 1039 | 2 | 38.8 | +42 | 35 | 5.5 | Open cluster in Perseus |
| 35 | 2168 | 6 | 05.7 | +24 | 21 | 5.3 | Open cluster in Gemini |
| 36 | 1960 | 5 | 33.0 | +34 | 04 | 6.3 | Open cluster in Auriga |
| 37 | 2099 | 5 | 49.1 | +32 | 33 | 6.2 | Open cluster in Auriga |
| 38 | 1912 | 5 | 25.3 | +35 | 47 | 7.4 | Open cluster in Auriga |
| 39 | 7092 | 21 | 30.4 | +48 | 13 | 5.2 | Open cluster in Cygnus |
| 40 | — | 12 | 20.0 | +58 | 20 | — | Close double star in Ursa Major |
| 41 | 2287 | 6 | 44.9 | −20 | 41 | 4.6 | Loose open cluster in Canis-Major |
| 42 | 1976 | 5 | 32.9 | −5 | 25 | 4.0 | Orion Nebula |
| 43 | 1982 | 5 | 33.1 | −5 | 19 | 9.0 | Northeast portion of Orion Nebula |
| 44 | 2632 | 8 | 37.0 | +20 | 10 | 3.7 | Praesepe; open cluster in Cancer |
| 45 | — | 3 | 44.5 | +23 | 57 | 1.6 | The Pleiades; open cluster in Taurus |
| 46 | 2437 | 7 | 39.5 | −14 | 42 | 6.0 | Open cluster in Puppis |
| 47 | 2422 | 7 | 34.3 | −14 | 22 | 5.2 | Loose group of stars in Puppis |
| 48 | 2458 | 8 | 11.0 | −5 | 38 | 5.5 | Open cluster in Hydra |
| 49 | 4472 | 12 | 27.3 | +8 | 16 | 8.5 | Elliptical galaxy in Virgo |
| 50 | 2323 | 7 | 00.6 | −8 | 16 | 6.3 | Loose open cluster in Monoceros |
| 51 | 5194 | 13 | 27.8 | +47 | 27 | 8.4 | Whirlpool spiral galaxy in Canes Venatici |
| 52 | 7654 | 23 | 22.0 | +61 | 20 | 7.3 | Loose open cluster in Cassiopeia |
| 53 | 5024 | 13 | 10.5 | +18 | 26 | 7.8 | Globular cluster in Coma Berenices |
| 54 | 6715 | 18 | 51.9 | −30 | 32 | 7.3 | Globular cluster in Sagittarius |
| 55 | 6809 | 19 | 36.8 | −31 | 03 | 7.6 | Globular cluster in Sagittarius |
| 56 | 6779 | 19 | 14.6 | +30 | 05 | 8.2 | Globular cluster in Lyra |
| 57 | 6720 | 18 | 51.7 | +32 | 58 | 9.0 | Ring Nebula; planetary nebula in Lyra |
| 58 | 4579 | 12 | 35.2 | +12 | 05 | 8.2 | Barred spiral galaxy in Virgo |
| 59 | 4621 | 12 | 39.5 | +11 | 56 | 9.3 | Elliptical spiral galaxy in Virgo |

* Index Catalogue (IC) number.

| M | NGC | Right Ascension (1950) | | Decli-nation (1950) | | Apparent Visual Magnitude | Description |
|---|---|---|---|---|---|---|---|
| | | (h) | (m) | (°) | (′) | | |
| 60 | 4649 | 12 | 41.1 | +11 | 50 | 9.0 | Elliptical galaxy in Virgo |
| 61 | 4303 | 12 | 19.3 | +4 | 45 | 9.6 | Spiral galaxy in Virgo |
| 62 | 6266 | 16 | 58.0 | −30 | 02 | 6.6 | Globular cluster in Ophiuchus |
| 63 | 5055 | 13 | 13.5 | +42 | 17 | 10.1 | Spiral galaxy in Canes Venatici |
| 64 | 4826 | 12 | 54.2 | +21 | 57 | 6.6 | Spiral galaxy in Coma Berenices |
| 65 | 3623 | 11 | 16.3 | +13 | 22 | 9.4 | Spiral galaxy in Leo |
| 66 | 3627 | 11 | 17.6 | +13 | 16 | 9.0 | Spiral galaxy in Leo; companion to M65 |
| 67 | 2682 | 8 | 48.4 | +12 | 00 | 6.1 | Open cluster in Cancer |
| 68 | 4590 | 12 | 36.8 | −26 | 29 | 8.2 | Globular cluster in Hydra |
| 69 | 6637 | 18 | 28.1 | −32 | 24 | 8.9 | Globular cluster in Sagittarius |
| 70 | 6681 | 18 | 40.0 | −32 | 20 | 9.6 | Globular cluster in Sagittarius |
| 71 | 6838 | 19 | 51.5 | +18 | 39 | 9.0 | Globular cluster in Sagitta |
| 72 | 6981 | 20 | 50.7 | −12 | 45 | 9.8 | Globular cluster in Aquarius |
| 73 | 6994 | 20 | 56.2 | −12 | 50 | 9.0 | Open cluster in Aquarius |
| 74 | 628 | 1 | 34.0 | +15 | 32 | 10.2 | Spiral galaxy in Pisces |
| 75 | 6864 | 20 | 03.1 | −22 | 04 | 8.0 | Globular cluster in Sagittarius |
| 76 | 650 | 1 | 38.8 | +51 | 19 | 11.4 | Planetary nebula in Perseus |
| 77 | 1068 | 2 | 40.1 | −0 | 12 | 8.9 | Spiral galaxy in Cetus |
| 78 | 2068 | 5 | 44.2 | +0 | 02 | 8.3 | Small reflection nebula in Orion |
| 79 | 1904 | 5 | 22.1 | −24 | 34 | 7.5 | Globular cluster in Lepus |
| 80 | 6093 | 16 | 14.0 | −22 | 52 | 7.5 | Globular cluster in Scorpius |
| 81 | 3031 | 9 | 51.7 | +69 | 18 | 7.9 | Spiral galaxy in Ursa Major |
| 82 | 3034 | 9 | 51.9 | +69 | 56 | 8.4 | Irregular galaxy in Ursa Major |
| 83 | 5236 | 13 | 34.2 | −29 | 37 | 10.1 | Spiral galaxy in Hydra |
| 84 | 4374 | 12 | 22.6 | +13 | 10 | 9.4 | S0 type galaxy in Virgo |
| 85 | 4382 | 12 | 22.8 | +18 | 28 | 9.3 | S0 type galaxy in Coma Berenices |
| 86 | 4406 | 12 | 23.6 | +13 | 13 | 9.2 | Elliptical galaxy in Virgo |
| 87 | 4486 | 12 | 28.2 | +12 | 40 | 8.7 | Elliptical galaxy in Virgo |
| 88 | 4501 | 12 | 29.4 | +14 | 42 | 10.2 | Spiral galaxy in Coma Berenices |
| 89 | 4552 | 12 | 33.1 | +12 | 50 | 9.5 | Elliptical galaxy in Virgo |
| 90 | 4569 | 12 | 34.3 | +13 | 26 | 9.6 | Spiral galaxy in Virgo |
| 91* | 4571(?) | — | — | — | — | — | |
| 92 | 6341 | 17 | 15.6 | +43 | 12 | 6.4 | Globular cluster in Hercules |
| 93 | 2447 | 7 | 42.4 | −23 | 45 | 6.0 | Open cluster in Puppis |
| 94 | 4736 | 12 | 48.6 | +41 | 24 | 8.3 | Spiral galaxy in Canes Venatici |
| 95 | 3351 | 10 | 41.3 | +11 | 58 | 9.8 | Barred spiral galaxy in Leo |
| 96 | 3368 | 10 | 44.1 | +12 | 05 | 9.3 | Spiral galaxy in Leo |
| 97 | 3587 | 11 | 12.0 | +55 | 17 | 12.0 | Owl Nebula; planetary nebula in Ursa Major |
| 98 | 4192 | 12 | 11.2 | +15 | 11 | 10.2 | Spiral galaxy in Coma Berenices |

* Items of doubtful identification.

| M | NGC | Right Ascension (1950) | | Declination (1950) | | Apparent Visual Magnitude | Description |
|---|---|---|---|---|---|---|---|
| | | (h) | (m) | (°) | (′) | | |
| 99 | 4254 | 12 | 16.3 | +14 | 42 | 9.9 | Spiral galaxy in Coma Berenices |
| 100 | 4321 | 12 | 20.4 | +16 | 06 | 10.6 | Spiral galaxy in Coma Berenices |
| 101 | 5457 | 14 | 01.4 | +54 | 36 | 9.6 | Spiral galaxy in Ursa Major |
| 102* | 5866(?) | — | — | — | — | — | |
| 103 | 581 | 1 | 29.9 | +60 | 26 | 7.4 | Open cluster in Cassiopeia |
| 104 | 4594 | 12 | 37.4 | −11 | 21 | 8.3 | Spiral galaxy in Virgo |
| 105 | 3379 | 10 | 45.2 | +13 | 01 | 9.7 | Elliptical galaxy in Leo |
| 106 | 4258 | 12 | 16.5 | +47 | 35 | 8.4 | Spiral galaxy in Canes Venatici |
| 107 | 6171 | 16 | 29.7 | −12 | 57 | 9.2 | Globular cluster in Ophiuchus |

* Items of doubtful identification.

# GLOSSARY

**absolute visual magnitude** ($M_V$) Intrinsic brightness of a star. The apparent visual magnitude the star would have if it were 10 pc away.

**absorption line** A dark line in a spectrum. Produced by the absence of photons absorbed by atoms or molecules.

**absorption spectrum (dark line spectrum)** A spectrum that contains absorption lines.

**accretion** The sticking together of solid particles to produce a larger particle.

**accretion disk** The whirling disk of gas that forms around a compact object such as a white dwarf, neutron star, or black hole as matter is drawn in.

**achondrites** Stony meteorites containing no chondrules or volatiles.

**achromatic lens** A telescope lens composed of two lenses ground from different kinds of glass and designed to bring two selected colors to the same focus and correct for chromatic aberration.

**active-core radio galaxy** A galaxy that emits radio energy from its core. Believed to result from an eruption in the core.

**amino acids** Carbon-chain molecules that are the building blocks of protein.

**Angstrom (Å)** A unit of distance. 1 Å = $10^{-10}$ m. Commonly used to measure the wavelength of light.

**annular eclipse** A solar eclipse in which the solar photosphere appears around the edge of the moon in a bright ring or annulus. The corona, chromosphere, and prominences cannot be seen.

**anorthosite** Rock of aluminum and calcium silicates found in the lunar highlands.

**apparent visual magnitude** ($m_v$) The brightness of a star as seen by human eyes on earth.

**associations** Groups of widely scattered stars (10 to 1000) moving together through space. Not gravitationally bound into clusters.

**asteroids** Small rocky worlds most of which lie between Mars and Jupiter in the asteroid belt.

**astronomical unit (AU)** Average distance from the earth to the sun; $1.5 \times 10^8$ km, or $93 \times 10^6$ mi.

**atmospheric windows** Wavelength regions in which our atmosphere is transparent—at visual wavelengths, infrared and at radio wavelengths.

**aurora** The glowing light display that results when a planet's magnetic field guides charged particles toward the north and south magnetic poles, where they strike the upper atmosphere and excite atoms to emit photons.

**autumnal equinox** The point on the celestial sphere where the sun crosses the celestial equator going southward. Also, the time when the sun reaches this point and autumn begins in the northern hemisphere—about September 22.

**Balmer series** Spectral lines in the visible and near ultraviolet spectrum of hydrogen produced by transitions whose lowest orbit is the second.

**Barnard's star** A star located near the sun in space. Consequently it has a high proper motion. Some believe it has planets in orbit around it.

**barred spiral galaxy** A spiral galaxy with an elongated nucleus resembling a bar from which the arms originate.

**basalt** Dark, igneous rock characteristic of solidified lava.

**belts** Dark bands of clouds that circle Jupiter parallel to its equator. Generally red, brown, or blue-green. Believed to be regions of descending gas.

**big bang theory** The theory that the universe began with a violent explosion from which the expanding universe of galaxies eventually formed.

**binary stars** Pairs of stars that orbit around their common center of mass.

**binding energy** The energy needed to pull an electron away from its atom.

**black body radiation** Radiation emitted by a hypothetical perfect radiator. The spectrum is continuous, and the wavelength of maximum emission depends only on the body's temperature.

**black dwarf** The end state of a white dwarf that has cooled to low temperature.

**black hole** A mass that has collapsed to such a small volume that its gravity prevents the escape of all radiation. Also, the volume of space from which radiation may not escape.

**Bok globules** Small, dark clouds only about 1 ly in diameter that contain 10 to 1000 $M_{\odot}$ of gas and dust. Believed related to star formation.

**bright line spectrum** See **emission spectrum.**

**Cambrian period** A geological period 0.6 to 0.5 billion years ago during which life on earth became diverse and complex. Cambrian rocks contain the oldest easily identifiable fossils.

**carbonaceous chondrites** Stony meteorites that contain both chondrules and volatiles. They may be the least altered remains of the solar nebula still present in the solar system.

**carbon detonation** The explosive ignition of carbon burning in some giant stars. A possible cause of some supernova explosions.

**carbon-nitrogen-oxygen (CNO) cycle** A series of nuclear reactions that use carbon as a catalyst to combine four hydrogen atoms to make one helium atom plus energy. Effective in stars more massive than the sun.

**Cassegrain telescope** A reflecting telescope in which the secondary mirror reflects light back down the tube through a hole in the center of the objective mirror.

**celestial equator** The imaginary line around the sky directly above the earth's equator.

**celestial sphere** An imaginary sphere of very large radius surrounding the earth and to which the planets, stars, sun, and moon seem to be attached.

**Cepheid variable stars** Variable stars with a period of 1 to 60 days. Period of variation related to luminosity.

**Chandrasekhar limit** The maximum mass of a white dwarf, about 1.4 $M_{\odot}$. A white dwarf of greater mass cannot support itself and will collapse.

**chemical evolution** The chemical process that led to the growth of complex molecules on the primitive earth. This did not involve the reproduction of molecules.

**chondrite** A stony meteorite that contains chondrules.

**chondrules** Round glassy bodies in some stony meteorites. Believed to have solidified very quickly from molten drops of silicate material.

**chromatic aberration** A distortion found in refracting telescopes because lenses focus different colors at slightly different distances. Images are consequently surrounded by color fringes.

**chromosphere** Bright gases just above the photosphere of the sun. Responsible for the emission lines in the flash spectrum.

**closed universe** A model universe in which the average density is great enough to stop the expansion and make the universe contract.

**cluster method** The method of determining the masses of galaxies based on the motions of galaxies in a cluster.

**cocoon** The cloud of gas and dust around a contracting protostar that conceals it at visible wavelengths.

**coma** The glowing head of a comet.

**comparison spectrum** A spectrum of known spectral lines used to identify unknown wavelengths in an object's spectrum.

**condensation** The growth of a particle by addition of material from surrounding gas, atom by atom.

**condensation sequence** The sequence in which different materials condense from the solar nebula as we move outward from the sun.

**constellation** One of the stellar patterns identified by name, usually of mythological gods, people, animals, or objects. Also, the region of the sky containing that star pattern.

**continuity of energy law** One of the basic laws of stellar structure. The amount of energy flowing out of the top of a shell must equal the amount coming in at the bottom plus whatever energy is generated within the shell.

**continuity of mass law** One of the basic laws of stellar structure. The total mass of the star must equal the sum of the masses of the shells, and the mass must be distributed smoothly through the star.

**continuous spectrum** A spectrum in which there are no absorption or emission lines.

**Copernican principle** The belief that the earth is not in a special place in the universe.

**corona** The faint outer atmosphere of the sun. Composed of low-density, very hot, ionized gas.

**cosmic rays** Atomic nuclei that enter earth's atmosphere at nearly the speed of light. Some originate in solar flares and some may come from supernova explosions, but their true nature is not well understood.

**Cosmological principle** The assumption that any observer in any galaxy sees the same general features of the universe.

**cosmological test** A measurement or observation whose result can help us choose between different cosmological theories.

**cosmology** The study of the nature, origin, and evolution of the universe.

**Coudé focus** The focal arrangement of a reflecting telescope in which mirrors direct the light to a fixed focus beyond the bounds of the telescope's movement, typically in a separate room. Usually used for spectroscopy.

**Coulomb barrier** The electrostatic force of repulsion between bodies of like charge. Commonly applied to atomic nuclei.

**cultural shock** The bewildering impact of an advanced society upon a less sophisticated society.

**dark line spectrum** See **absorption spectrum.**

**declination** A coordinate used on the celestial sphere just as latitude is used on earth. An object's declination is measured from the celestial equator—positive to the north, and negative to the south.

**deferent** In the Ptolemaic theory, the large circle around the earth along which the center of the epicycle moved.

**degenerate matter** Extremely high density matter in which pressure no longer depends on temperature due to quantum mechanical effects.

**density wave theory** Theory proposed to account for spiral arms as compressions of the interstellar medium in the disk of the galaxy.

**differential rotation** The rotation of a body in

which different parts of the body have different periods of rotation. This is true of the sun, the jovian planets, and the disk of the galaxy.

**differentiation** The separation of planetary material according to density.

**diffraction fringe** Blurred fringe surrounding any image caused by the wave properties of light. Because of this, no image detail can be seen smaller than the fringe.

**disk component** All material confined to the plane of the galaxy.

**distance indicators** Objects whose luminosities or diameters are known. Used to find the distance to a star cluster or galaxy.

**distance modulus** The difference between the apparent and absolute magnitude of a star. A measure of how far away the star is.

**DNA (deoxyribonucleic acid)** The long carbon-chain molecule that records information to govern the biological activity of the organism. DNA carries the genetic data passed to offspring.

**Doppler effect** The change in the wavelength of radiation due to relative radial motion of source and observer.

**double galaxy method** A method of finding the masses of galaxies from orbiting pairs of galaxies.

**double-lobed radio galaxy** A galaxy that emits radio energy from two regions (lobes) located on opposite sides of the galaxy.

**dwarf nova** A star that undergoes novalike explosions every few days or weeks. Believed associated with mass transfer onto a white dwarf in a binary system.

**dynamo effect** The theory that the earth's magnetic field is generated in the conducting material of its molten core.

**dyne** A unit of force. One dyne is the force needed to accelerate a mass of one gram one centimeter per second in one second.

**eclipsing binary** A binary star system in which the stars eclipse each other.

**ecliptic** The apparent path of the sun around the sky.

**ejecta** Pulverized rock scattered by meteorite impacts on a planetary surface.

**electromagnetic radiation** Changing electric and magnetic fields that travel through space and transfer energy from one place to another. For example, light, radio waves, etc.

**electrons** Low-mass atomic particles carrying negative charges.

**elliptical galaxy** A galaxy that is round or elliptical in outline, contains little gas and dust, no disk or spiral arms, and few hot, bright stars.

**emission line** A bright line in a spectrum caused by the emission of photons from atoms.

**emission nebula** A cloud of glowing gas excited by ultraviolet radiation from hot stars.

**emission spectrum (bright line spectrum)** A spectrum containing emission lines.

**energy transport** Energy must flow from hot regions to cooler regions by conduction, convection, or radiation.

**enzymes** Special proteins that control processes in an organism.

**epicycle** The small circle followed by a planet in the Ptolemaic theory. The center of the epicycle follows a larger circle (the deferent) around the earth.

**equant** The point off center in the deferent from which the center of the epicycle appears to move uniformly.

**erg** A unit of energy equivalent to a force of 1 dyne acting over a distance of 1 cm. $10^7$ ergs expended in 1 second equals 1 watt of power.

**escape velocity** The initial velocity an object needs to escape from the surface of a celestial body.

**evening star** Any planet visible in the sky just after sunset.

**event horizon** The boundary of the region of a black hole from which no radiation may escape. No event that occurs within the event horizon is visible to a distant observer.

**excited atom** An atom in which an electron has moved from a lower to a higher orbit.

**eyepiece** A short-focal-length lens used to enlarge the image in a telescope. The lens nearest the eye.

**flash spectrum** The emission spectrum of the chromosphere that is visible for the few seconds during a total solar eclipse when the moon has covered the photosphere but has not yet covered the chromosphere.

**focal length** The focal length of a lens is the distance from the lens to the point where it focuses parallel rays of light.

**Galilean satellites** The four largest satellites of Jupiter, named after their discoverer, Galileo.

**geocentric universe** A model universe with the earth at the center, such as the Ptolemaic universe.

**giant stars** Large, cool, highly luminous stars in the upper right of the H-R diagram. Typically 10 to 100 times the diameter of the sun.

**glacial period** An interval when ice sheets cover large areas of the land.

**glitch** A sudden change in the period of a pulsar.

**globular cluster** A star cluster containing 50,000 to 1 million stars in a sphere about 75 ly in diameter. Generally old, metal poor, and found in the spherical component of the galaxy.

**grating** A piece of material in which numerous microscopic parallel lines are scribed. Light encountering a grating is dispersed to form a spectrum.

**gravitational red shift** The lengthening of the wavelength of a photon due to its escape from a gravitational field.

**greenhouse effect** The process by which a carbon dioxide atmosphere traps heat and raises the temperature of a planetary surface.

**ground state** The lowest permitted electron orbit in an atom.

**halo** The spherical region of a spiral galaxy containing a thin scattering of stars, star clusters, and small amounts of gas.

**heliocentric universe** A model of the universe with the sun at the center, such as the Copernican universe.

**helium flash** The explosive ignition of helium burning that takes place in some giant stars.

**Herbig-Haro objects** Small nebulae that vary irregularly in brightness. Believed associated with star formation.

**Hertzsprung-Russell diagram** A plot of the intrinsic brightness versus the surface temperature of stars. It separates the effects of temperature and surface area on stellar luminosity. Commonly absolute magnitude versus spectral type, but also luminosity versus surface temperature or color.

**homogeneity** The assumption that, on the large scale, matter is uniformly spread through the universe.

**H-R diagram** See **Hertzsprung-Russell diagram.**

**H II region** A region of ionized hydrogen around a hot star.

**Hubble constant (H)** A measure of the rate of expansion of the universe. The average value of velocity of recession divided by distance. Presently believed to be about 50 km/second/Mpc.

**hydrostatic equilibrium** The balance between the weight of the material pressing downward on a layer in a star, and the pressure in that layer.

**instability strip** The region of the H-R diagram in which stars are unstable to pulsation. A star passing through this strip becomes a variable star.

**interglacial period** A period when ice sheets melt back and the climate is warmer.

**interstellar medium** The gas and dust distributed between the stars.

**interstellar reddening** The process in which dust scatters blue light out of starlight and makes the stars look redder.

**ion** An atom that has lost or gained one or more electrons.

**ionization** The process in which atoms lose or gain electrons.

**irregular galaxy** A galaxy with a chaotic appearance, large clouds of gas and dust, both population I and population II stars, but without spiral arms.

**isotopes** Atoms that have the same number of protons but a different number of neutrons.

**isotropy** The assumption that in its general properties the universe looks the same in every direction.

**jovian planets** Jupiter-like planets with large diameters and low densities.

**kiloparsec (kpc)** A unit of distance equal to 1000 pc or 3260 ly.

**life zone** A region around a star within which a planet can have temperatures that permit the existence of liquid water.

**light curve** A graph of brightness versus time commonly used in analyzing variable stars and eclipsing binaries.

**light-gathering power** The ability of a telescope to collect light. Proportional to the area of the telescope objective lens or mirror.

**lighthouse theory** The theory that a neutron star produces pulses of radiation by sweeping radio beams around the sky as it rotates.

**light-year** A unit of distance. The distance light travels in one year.

**lobate scarp** A curved cliff such as those found on Mercury.

**look-back time** The amount by which we look into the past when we look at a distant galaxy. A time equal to the distance to the galaxy in light-years.

**luminosity** The total amount of energy a star radiates in one second.

**luminosity class** A category of stars of similar luminosity. Determined by the widths of lines in their spectra.

**lunar eclipse** The darkening of the moon when it moves through the earth's shadow.

**Lyman series** Spectral lines in the ultraviolet spectrum of hydrogen produced by transitions whose lowest orbit is the ground state.

**magnifying power** The ability of a telescope to make an image larger.

**Magellanic clouds** Small, irregular galaxies that are companions to the Milky Way. Visible in the southern sky.

**magnitude scale** The astronomical brightness scale. The larger the number, the fainter the star.

**main sequence** The region of the H-R diagram running from upper left to lower right, which includes roughly 90 percent of all stars.

**mantle** The layer of dense rock and metal oxides that lies between the molten core and the surface of the earth. Also, similar layers in other planets.

**Mare (sea)** One of the lunar lowlands filled by successive flows of dark lava.

**mass-luminosity relation** The more massive a star is, the more luminous it is.

**megaparsec (Mpc)** A unit of distance equal to 1,000,000 pc.

**metals** In astronomical usage, all atoms heavier than helium.

**meteor** A small bit of matter heated by friction to incandescent vapor as it falls into earth's atmosphere.

**meteorite** A meteor that has survived its passage through the atmosphere and strikes the ground.

**meteoroid** A meteor in space before it enters the earth's atmosphere.

**midocean rift** Chasms that split the midocean rises where crustal plates move apart.

**midocean rise** One of the undersea mountain ranges that push up from the seafloor in the center of the oceans.

**Miller experiment** An experiment that reproduced the conditions under which life began on earth and manufactured amino acids and other organic compounds.

**minute of arc** An angular measure. Each degree is divided into 60 minutes of arc.

**missing mass** Unobserved mass in clusters of galaxies believed to provide sufficient gravity to bind the cluster together.

**molecule** Two or more atoms bonded together.

**morning star** Any planet visible in the sky just before sunrise.

**mutant** Offspring born with altered DNA.

**natural selection** The process by which the

best traits are passed on, allowing the most able to survive.

**neap tides** Ocean tides of low amplitude occurring at first- and third-quarter moon.

**Newtonian focus** The focal arrangement of a reflecting telescope in which a diagonal mirror reflects light out the side of the telescope tube for easier access.

**neutrino** A neutral, massless atomic particle that travels at the speed of light.

**neutron** An atomic particle with no charge and about the same mass as a proton.

**neutron star** A small, highly dense star composed almost entirely of tightly packed neutrons. Radius about 10 km.

**node** The points where an object's orbit passes through the plane of the earth's orbit.

**non-Doppler red shift** Proposed cause of the red shift in the spectra of QSOs. Not dependent on the Doppler effect.

**north celestial pole** The point on the celestial sphere directly above the earth's north pole.

**nova** From the Latin "new," a sudden brightening of a star making it appear as a "new" star in the sky. Believed associated with eruptions on white dwarfs in binary systems.

**nucleus (of an atom)** The central core of an atom containing protons and neutrons. Carries a net positive charge.

**objective lens** In a refracting telescope, the long-focal-length lens that forms an image of the object viewed. The lens closest to the object.

**objective mirror** In a reflecting telescope, the principal mirror (reflecting surface) that forms an image of the object viewed.

**Olbers' paradox** The conflict between observation and theory as to why the night sky should or should not be dark.

**135 km/second arm** A receding cloud of neutral hydrogen lying on the far side of the galactic center.

**Oort cloud** The hypothetical source of comets. A swarm of icy bodies believed to lie in a spherical shell 50,000 AU from the sun.

**opacity** The resistance of a gas to the passage of radiation.

**open cluster** A cluster of 10 to 10,000 stars with an open, transparent appearance. The stars are not tightly grouped. Usually relatively young and located in the disk of the galaxy.

**open universe** A model universe in which the average density is less than the critical density needed to halt the expansion.

**oscillating universe** The theory that the universe begins with a big bang, expands, is slowed by its own gravity, and then falls back to create another big bang.

**outgassing** The release of gases from a planet's interior.

**parallax (p)** The apparent change in the position of an object due to a change in the location of the observer. Astronomical parallax is measured in seconds of arc.

**parsec (pc)** The distance to a hypothetical star whose parallax is one second of arc. 1 pc = 206,265 AU = 3.26 ly.

**Paschen series** Spectral lines in the infrared spectrum of hydrogen produced by transitions whose lowest orbit is the third.

**penumbra** The portion of a shadow that is only partially shaded.

**perfect cosmological principle** The belief that, in general properties, the universe looks the same from every location in space at any time.

**perihelion** The orbital point of closest approach to the sun.

**period-luminosity relation** The relation between period of pulsation and intrinsic brightness among Cepheid variable stars.

**permitted orbit** One of the energy levels in an atom that an electron may occupy.

**photometer** An instrument used to measure the intensity and color of starlight.

**photon** A quantum of electromagnetic energy. Carries an amount of energy that depends inversely on its wavelength.

**photosphere** The bright visible surface of the sun.

**planetary nebula** An expanding shell of gas ejected from a star during the latter stages of its evolution.

**planetesimal** One of the small bodies that formed from the solar nebula and eventually grew into protoplanets.

**plastic** A material with the properties of a solid but capable of flowing under pressure.

**plate tectonics** The constant destruction and renewal of earth's surface by the motion of sections of crust.

**poor galaxy cluster** An irregularly shaped cluster that contains fewer than 1000 galaxies, many spiral, and no giant ellipticals.

**population I** Stars rich in atoms heavier than helium. Nearly always relatively young stars found in the disk of the galaxy.

**population II** Stars poor in atoms heavier than helium. Nearly always relatively old stars found in the halo, globular clusters, or the nuclear bulge.

**precession** The slow change in the direction of the earth's axis of rotation. One cycle takes nearly 26,000 years.

**pressure broadening** The blurring of spectral lines due to the gas pressure in a star's atmosphere.

**prime focus** The point at which the objective mirror forms an image in a reflecting telescope.

**primeval atmosphere** Earth's first air, composed of gases from the solar nebula.

**primordial background radiation** Radiation from the hot clouds of the big bang explosion. Because of its large red shift it appears to come from a body whose temperature is only 2.7°K.

**primordial black holes** Low-mass black holes that may have formed during the big bang explosion.

**prominences** Eruptions on the solar surface. Visible during total solar eclipses.

**proper motion** The rate at which a star moves across the sky. Measured in seconds of arc per year.

**proteins** Complex molecules composed of amino acid units.

**proton** A positively charged atomic particle contained in the nucleus of atoms. The nucleus of a hydrogen atom.

**proton-proton chain** A series of three nuclear reactions that builds a helium atom by adding together protons. The main energy source in the sun.

**protoplanet** Massive object resulting from the coalescence of planetesimals in the solar nebula and destined to become a planet.

**protostar** A collapsing cloud of gas and dust destined to become a star.

**pulsar** A source of short, precisely timed radio bursts. Believed to be spinning neutron stars.

**quasi-stellar objects (QSOs)** Small, powerful sources of radio signals. They seem to lie at great distances and therefore must be 10 to 1000 times as energetic as a normal galaxy.

**radial velocity ($V_r$)** That component of an object's velocity directed away from or toward the earth.

**radiation pressure** The force exerted on the surface of a body by its absorption of light. Small particles floating in the solar system can be blown outward by the pressure of the sunlight.

**radio galaxy** A galaxy that is a strong source of radio signals.

**radio interferometer** Two or more radio telescopes that combine their signals to achieve the resolving power of a larger telescope.

**rays** Ejecta from meteorite impacts forming white streamers radiating from some lunar craters.

**recurrent novae** Stars that erupt as novae every few dozen years.

**reflecting telescope** A telescope that uses a concave mirror to focus light into an image.

**refracting telescope** A telescope that forms images by bending (refracting) light with a lens.

**relativistic red shift** The red shift due to the Doppler effect for objects traveling at speeds near the speed of light.

**resolving power** The ability of a telescope to reveal fine detail. Depends on the diameter of the telescope objective.

**retrograde motion** The apparent backward (westward) motion of planets as seen against the background of stars.

**rich galaxy cluster** A cluster containing over 1000 galaxies, mostly elliptical, scattered over a volume about 3 Mpc in diameter.

**rift valley** A long, straight, deep valley produced by the separation of crustal plates.

**right ascension (R. A.)** A coordinate used on the celestial sphere just as longitude is used on earth. An object's right ascension is measured eastward from the vernal equinox.

**ring galaxy** A galaxy that resembles a ring around a bright nucleus. Believed to be the result of a head-on collision of two galaxies.

**RNA (ribonucleic acid)** Long carbon-chain molecules that use the information stored in DNA to manufacture complex molecules necessary to the organism.

**Roche limit** The minimum distance between a planet and a satellite that holds itself together by its own gravity. If a satellite's orbit brings it within its planet's Roche limit, tidal forces will pull the satellite apart.

**Roche lobe** The volume of space a star controls gravitationally within a binary system.

**rotation curve** A graph of orbital velocity versus radius in the disk of a galaxy.

**rotation curve method** A method of determining a galaxy's mass by observing the orbital velocity and orbital radius of stars in the galaxy.

**RR Lyrae variable stars** Variable stars with periods of from 12 to 24 hours. Common in some globular clusters.

**Sagittarius A** The powerful radio source located at the core of the Milky Way galaxy.

**Schmidt camera** A photographic telescope that takes wide-angle photographs.

**Schwarzschild radius** The radius of the event horizon around a black hole.

**second of arc** An angular measure. Each minute of arc is divided into 60 seconds of arc.

**secondary atmosphere** The gases outgassed from a planet's interior. Rich in carbon dioxide.

**secondary mirror** In a reflecting telescope, the mirror that reflects the light to a point of easy observation.

**seeing** Atmospheric conditions on a given night. When the atmosphere is unsteady, producing blurred images, the seeing is said to be poor.

**self-sustained star formation** The process by which the birth of stars compresses the surrounding gas clouds and triggers the formation of more stars. Proposed to explain spiral arms.

**Seyfert galaxy** An otherwise normal spiral galaxy with an unusually bright, small core that fluctuates in brightness. Believed to indicate the core is erupting.

**shield volcanoes** Wide, low-profile volcanic cones produced by highly liquid lava.

**shock wave** A sudden change in pressure that travels as an intense sound wave.

**sidereal time** Time based on the rotation of the earth with respect to the stars. The sidereal time at any moment equals the right ascension of objects on the upper half of the local celestial meridian.

**singularity** The object of zero radius into which the matter in a black hole is believed to fall.

**solar eclipse** The event that occurs when the moon passes directly between the earth and sun, blocking our view of the sun.

**solar granulation** The patchwork pattern of bright areas with dark borders observed on the sun. The tops of rising currents of hot gas in the convective zone.

**solar wind** Rapidly moving atoms and ions that escape from the solar corona and blow outward through the solar system.

**south celestial pole** The point on the celestial sphere directly above the earth's south pole.

**spectral class or type** A star's position in the temperature classification system O B A F G K M. Based on the appearance of the star's spectrum.

**spectral sequence** The arrangement of spectral classes (O, B, A, F, G, K, M) ranging from hot to cool.

**spectrograph** A device that separates light by wavelength to produce a spectrum.

**spectroscopic binary** A star system in which the stars are too close together to be visible separately. We see a single point of light and only by taking a spectrum can we determine that there are two stars.

**spectroscopic parallax** The method of determining a star's distance by comparing its apparent magnitude with its absolute magnitude as estimated from its spectrum.

**spherical component** The part of the galaxy including all matter in a spherical distribution around the center (the halo and nuclear bulge).

**spiral arms** Long spiral patterns of bright stars, star clusters, gas, and dust, that extend from the center to the edge of the disk of spiral galaxies.

**spiral galaxy** A galaxy with an obvious disk component containing gas, dust, hot, bright stars, and spiral arms.

**spiral tracers** Objects used to map the spiral arms (e.g., O and B associations, open clusters, clouds of ionized hydrogen, and some types of variable stars).

**spring tides** Ocean tides of high amplitude that occur at full and new moon.

**stellar density function** A description of the abundance of stars of different types in space.

**stellar model** A table of numbers representing the conditions in various layers within a star.

**summer solstice** The point on the celestial sphere where the sun is at its most northerly point. Also, the time when the sun passes this point about June 22. Summer begins in the northern hemisphere.

**sunspots** Relatively dark spots on the sun that contain intense magnetic fields.

**supergiant stars** Exceptionally luminous stars 10 to 1000 times the sun's diameter.

**supernova remnant** The expanding shell of gas marking the site of a supernova explosion.

**synchrotron radiation** Radiation emitted when high-speed electrons move through a magnetic field.

**T Tauri stars** Young stars surrounded by gas and dust. Believed to be contracting toward the main sequence.

**terrestrial planets** Earthlike planets—small, dense, rocky.

**3-kpc arm** A cloud of neutral hydrogen moving outward from the nucleus of our galaxy at about 53 km/second. It lies 3 kpc from the center of the galaxy.

**Titius-Bode Rule** a simple series of steps that produces numbers approximately matching the sizes of the planetary orbits.

**transition** The movement of an electron from one atomic orbit to another.

**tuning fork diagram** A system of classification for elliptical, spiral, and irregular galaxies.

**turn-off point** The point in an H-R diagram where a cluster's stars turn off of the main sequence and move toward the red giant region, revealing the approximate age of the cluster.

**umbra** The region of a shadow that it totally shaded.

**uncompressed density** The density a planet would have if its gravity did not compress it.

**uniform circular motion** The classical belief that the perfect heavens could only move by the combination of uniform motion along circular orbits.

**universality** The assumption that the physical laws observed on earth apply everywhere in the universe.

**velocity dispersion method** A method of finding a galaxy's mass by observing the range of velocities within the galaxy.

**vernal equinox** The place on the celestial sphere where the sun crosses the celestial equator moving northward. Also, the time of year when the sun crosses this point, about March 21, and spring begins in the northern hemisphere.

**visual binary** A binary star system in which the two stars are separately visible in the telescope.

**water hole** The interval of the radio spectrum between the 21 cm hydrogen radiation and the 18 cm OH radiation. Likely wavelengths to use in the search for extraterrestrial life.

**wavelength** The distance between successive peaks or troughs of a wave. Usually represented by $\lambda$.

**wavelength of maximum ($\lambda_{max}$)** The wavelength at which a perfect radiator emits the maximum amount of energy. Depends only on the object's temperature.

**white dwarf stars** Dying stars that have collapsed to the size of the earth and are slowly cooling off. At the lower left of the H-R diagram.

**Widmanstätten patterns** Bands in iron meteorites due to large crystals of nickel-iron alloys.

**winter solstice** The point on the celestial sphere where the sun is farthest south. Also the time of year when the sun passes this point, about December 22. Winter begins in northern hemisphere.

**zero-age main sequence** The locus in the H-R diagram where stars first reach stability as hydrogen burning stars.

**zones** Yellow-white regions that circle Jupiter parallel to its equator. Believed to be areas of rising gas.

# ANSWERS TO EVEN NUMBERED PROBLEMS

**Chapter 1**

2. 2755
4. 400,000
6. 6790 km (smaller)

**Chapter 2**

2. 192,000
4. (a) Full
   (b) First Quarter
   (c) Waxing Gibbous
   (d) Waxing Crescent

**Chapter 3**

2. 4600 Å, $4.6 \times 10^{-7}$ m
4. 97,000 Å, infrared
6. The 200-inch gathers 100 times more.
8. 0.5 inches
10. The smaller radio telescope gathers 100 times less.

**Chapter 4**

2. 2500 Å
4. (a) 20,000°K
   (b) 7,500°K
   (c) 3,000°K
   (d) 4,500°K
6. 91 km/sec, receding

**Chapter 5**

2. 62.5 pc, 2
4. B
6. 160 pc
8. a, c, c, c, d

**Chapter 6**

4. $9 \times 10^{20}$ ergs
8. 9.8 million years

## Chapter 7

4. about 920 years ago i.e. about 1060 AD
6. about 178 million years
8. about 30 Å, x-ray

## Chapter 8

2. about 11%
4. about 6300 pc
6. 21 kpc

## Chapter 10

2. 0.024 pc or 28 lightdays
4. 0.16
6. 95,000 km/sec

## Chapter 11

2. $17.2 \times 10^9$ years

## Chapter 12

2. The moon. It has the smallest mass.

## Chapter 13

2. Assuming the planet and satellites have the same density the Roche limit is 8284 km. Phobos lies just outside this limit and Deimos is nearly three times farther away.

## Chapter 15

2. about 100,000 years

## Chapter 16

2. 87 mm, 0.6 mm
4. about 600

# INDEX

Page numbers in **boldface** refer to the place in the text where the term is defined.
Page numbers in *italic* refer to figures.

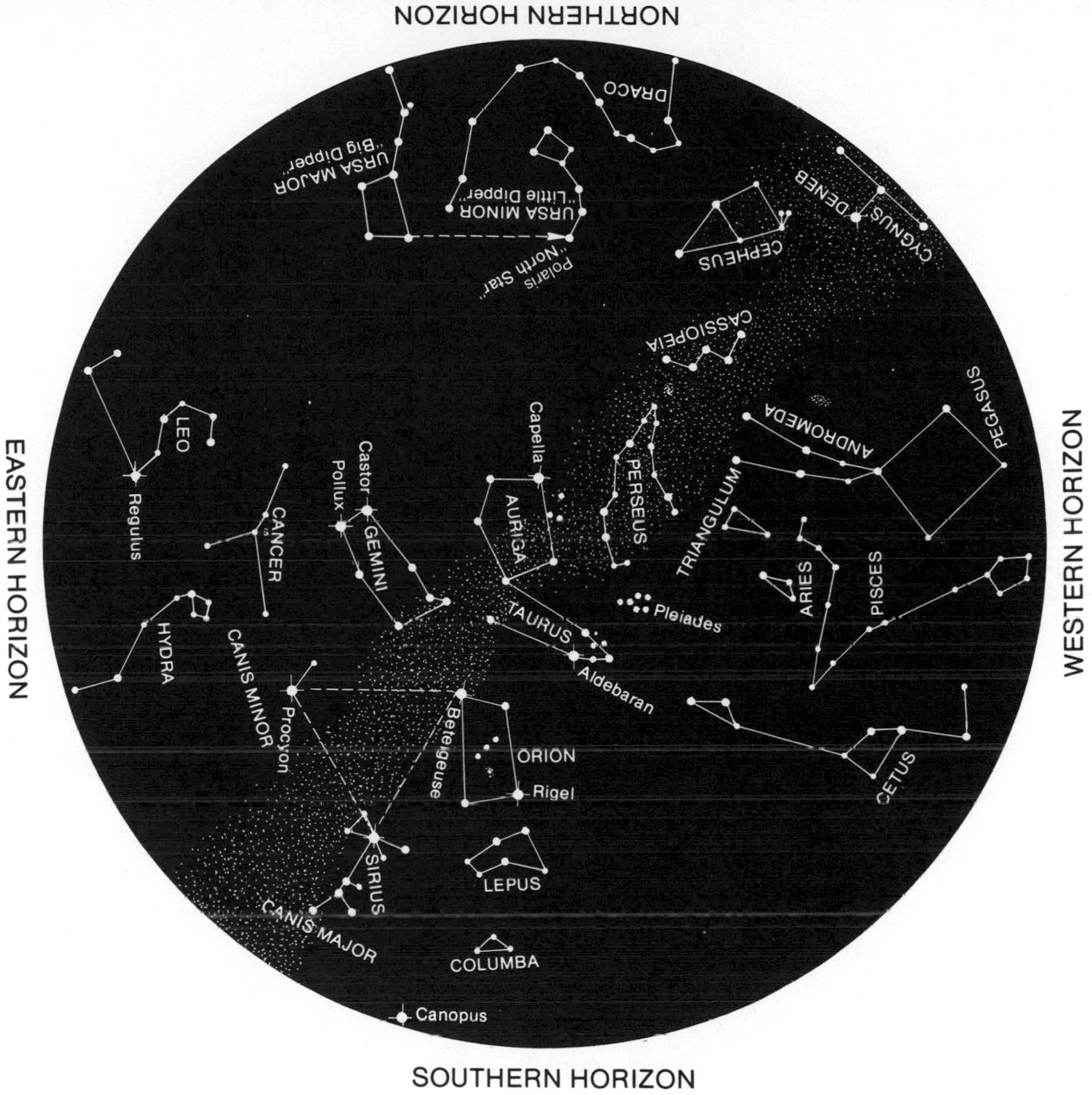

NORTHERN HORIZON

EASTERN HORIZON

WESTERN HORIZON

SOUTHERN HORIZON

## THE NIGHT SKY IN JANUARY

To use: Hold chart vertically and turn it so the direction you are facing shows at the bottom.

Chart Time (local Standard):

| | |
|---|---|
| 10:00 p.m. | First of month |
| 9:00 p.m. | Middle of month |
| 8:00 p.m. | Last of month |

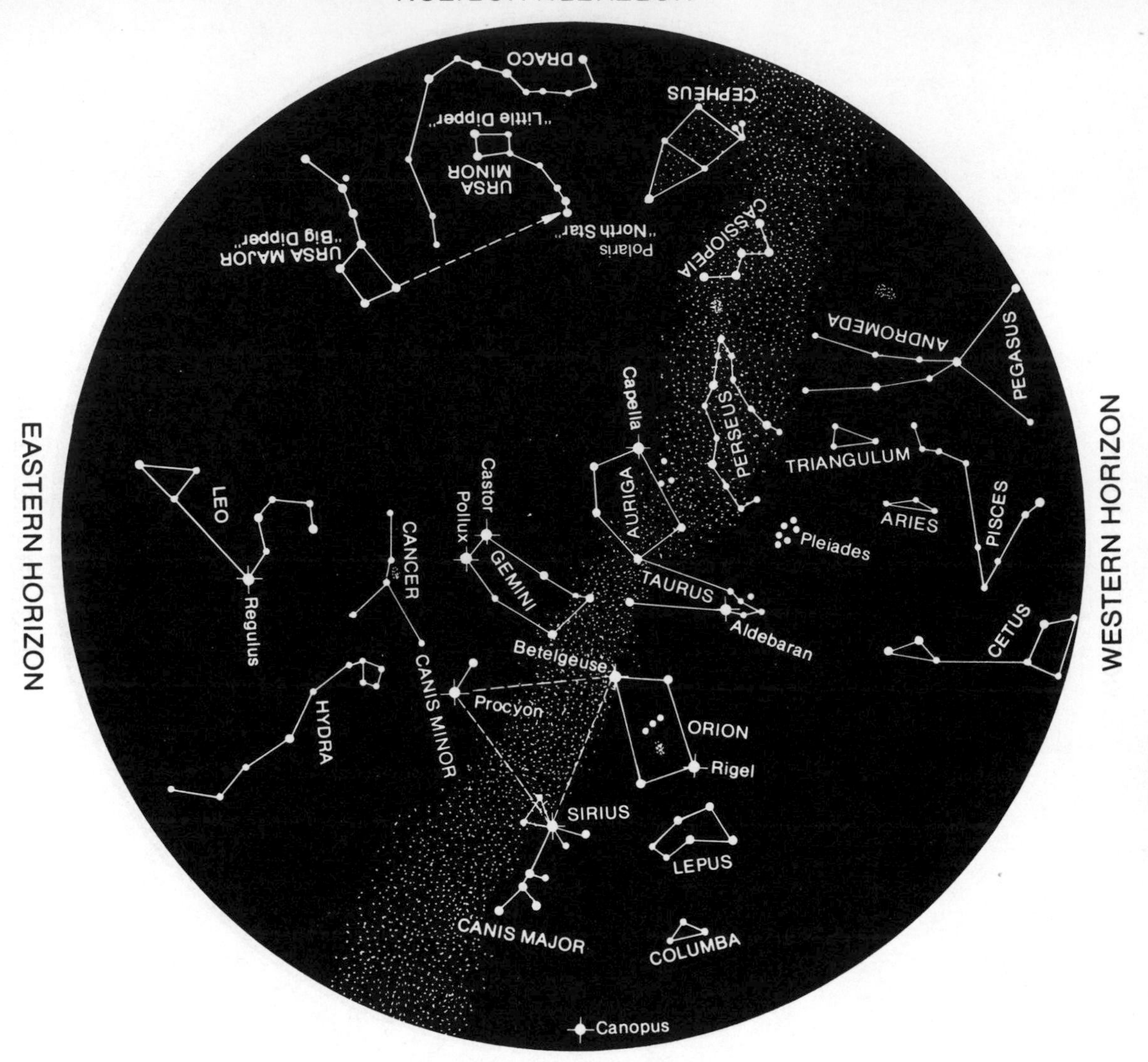


THE NIGHT SKY IN FEBRUARY

To use: Hold chart vertically and turn it so the direction you are facing shows at the bottom.

Chart Time (local Standard):

| | |
|---|---|
| 10:00 p.m. | First of month |
| 9:00 p.m. | Middle of month |
| 8:00 p.m. | Last of month |

## THE NIGHT SKY IN MARCH

To use: Hold chart vertically and turn it so the direction you are facing shows at the bottom.

Chart Time (local Standard):

| | |
|---|---|
| 10:00 p.m. | First of month |
| 9:00 p.m. | Middle of month |
| 8:00 p.m. | Last of month |

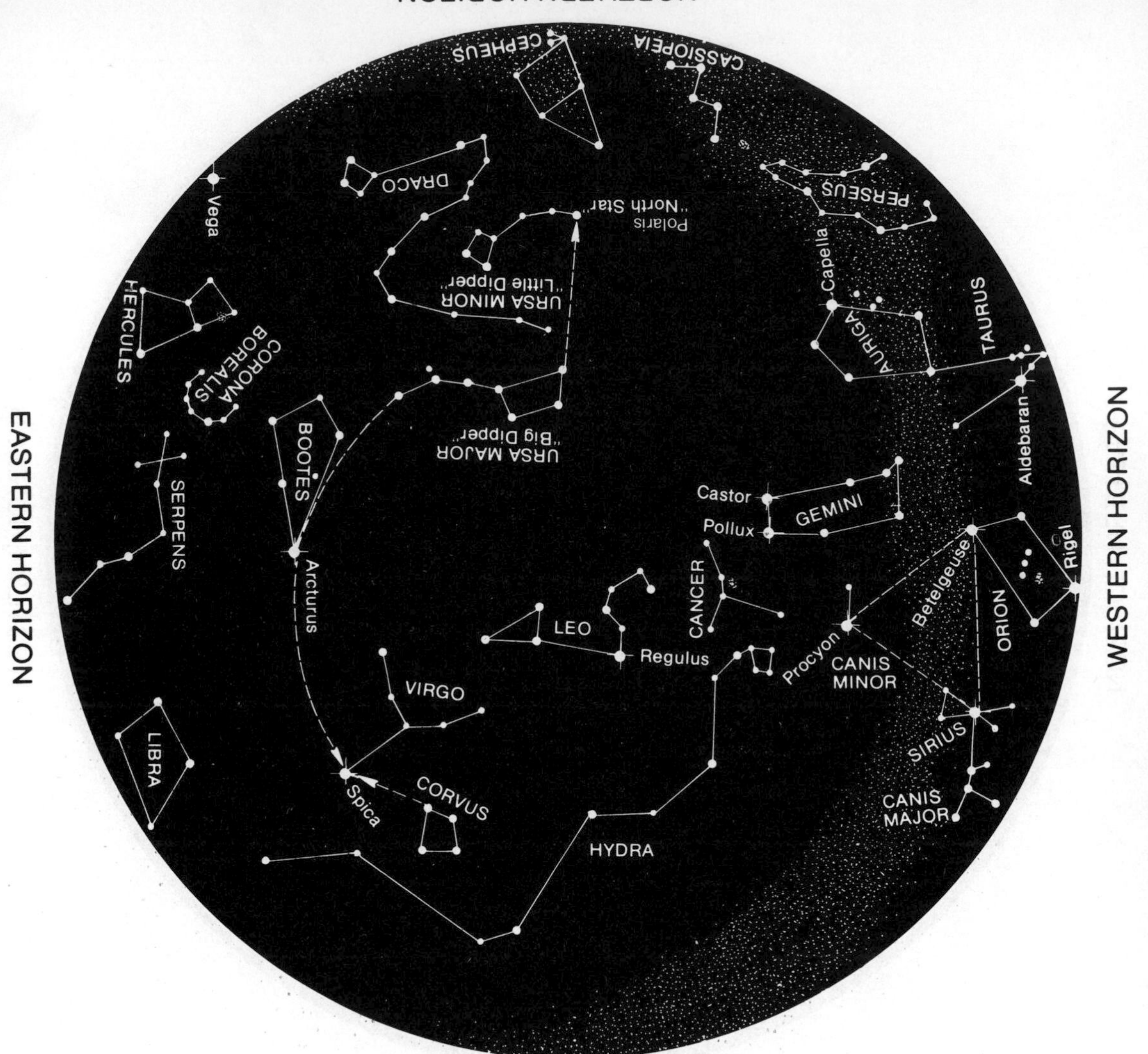


## THE NIGHT SKY IN APRIL

To use: Hold chart vertically and turn it so the direction you are facing shows at the bottom.

Chart Time (local Standard):

| | |
|---|---|
| 10:00 p.m. | First of month |
| 9:00 p.m. | Middle of month |
| 8:00 p.m. | Last of month |

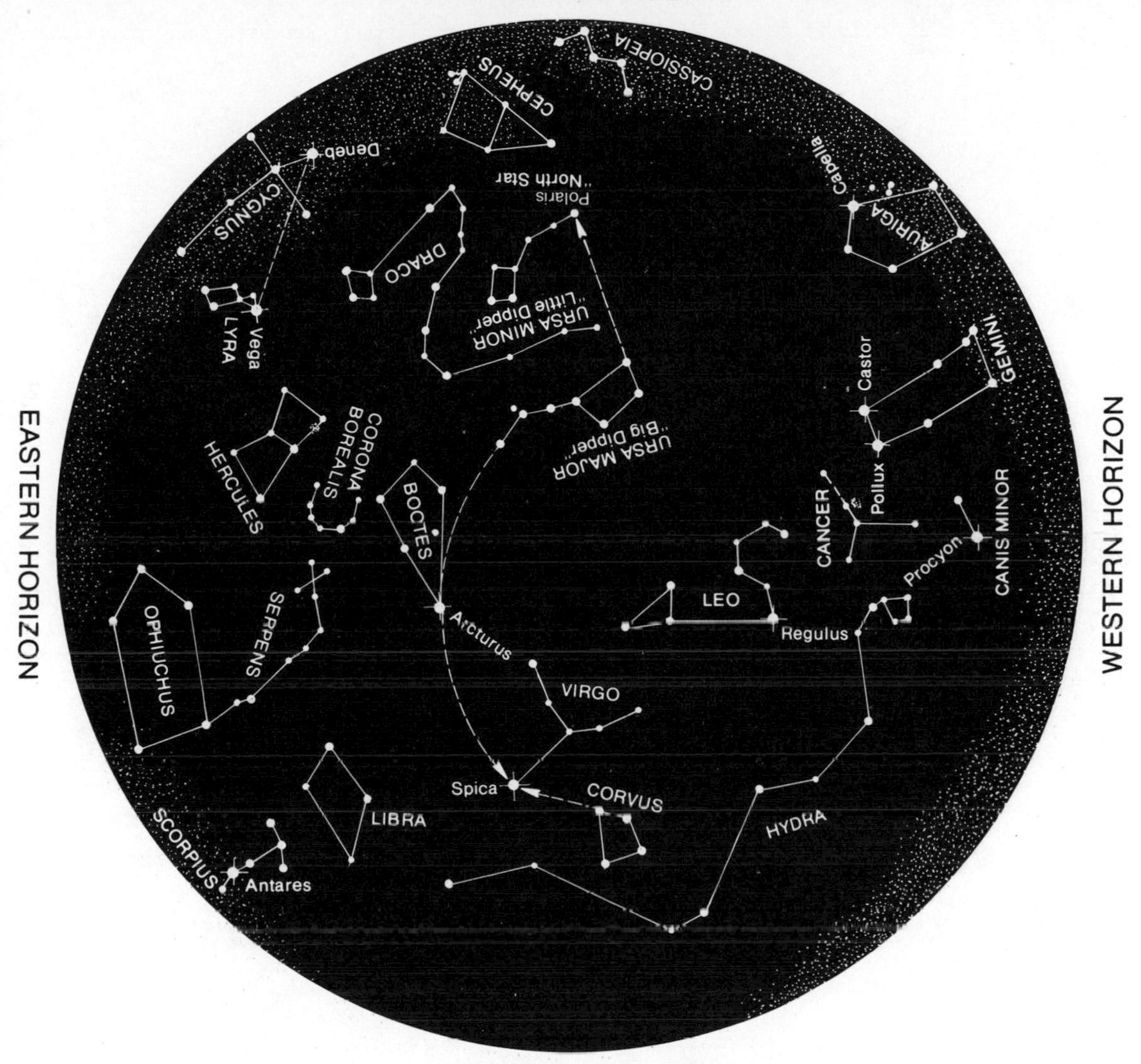


SOUTHERN HORIZON

THE NIGHT SKY IN MAY

To use: Hold chart vertically and turn it so the direction you are facing shows at the bottom.

Chart Time (local Standard):

| | |
|---|---|
| 10:00 p.m. | First of month |
| 9:00 p.m. | Middle of month |
| 8:00 p.m. | Last of month |

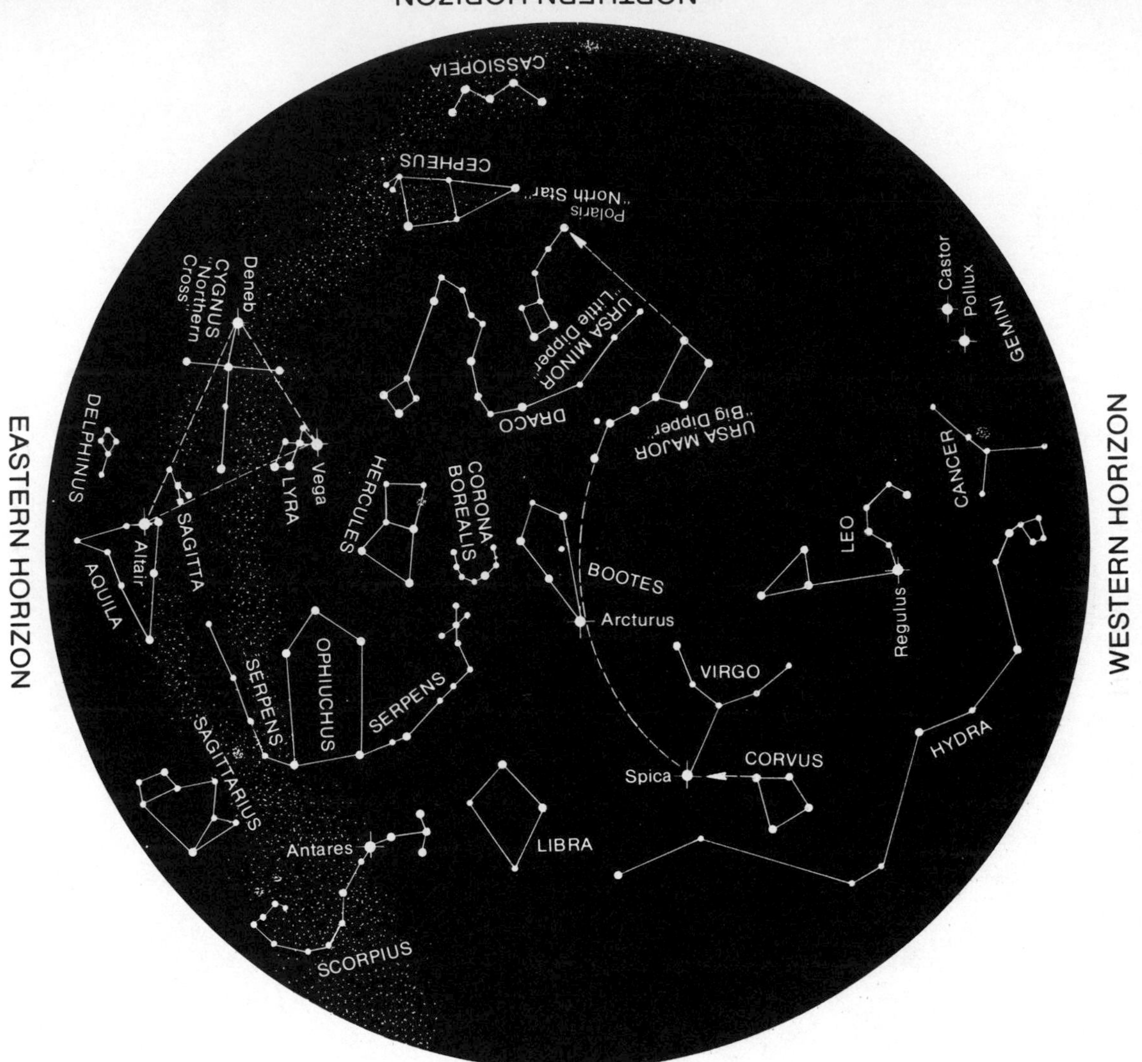


# THE NIGHT SKY IN JUNE

To use: Hold chart vertically and turn it so the direction you are facing shows at the bottom.

Chart Time (local Standard):

| | |
|---|---|
| 10:00 p.m. | First of month |
| 9:00 p.m. | Middle of month |
| 8:00 p.m. | Last of month |

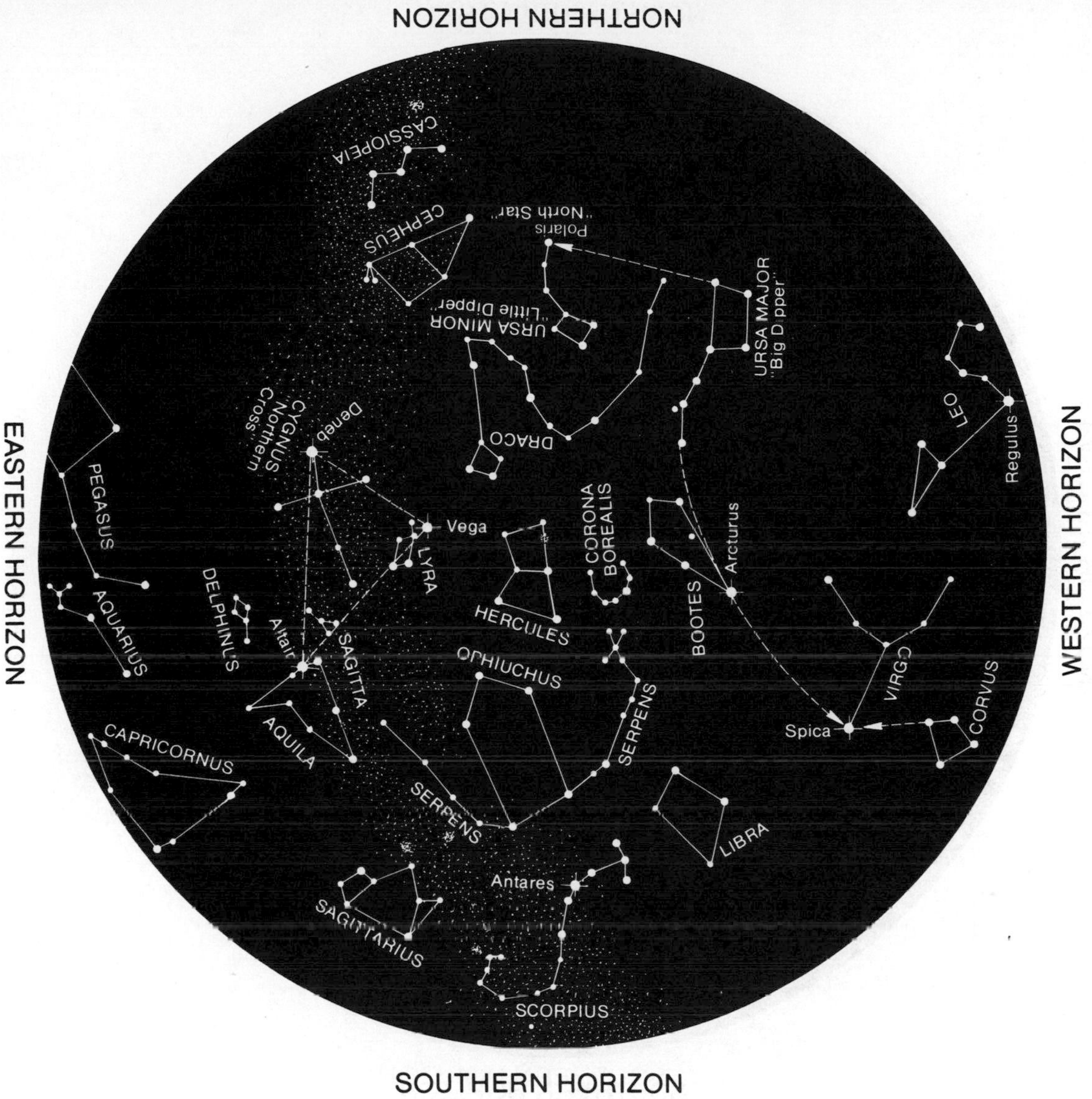

## THE NIGHT SKY IN JULY

To use: Hold chart vertically and turn it so the direction you are facing shows at the bottom.

Chart Time (local Standard):

| | |
|---|---|
| 10:00 p.m. | First of month |
| 9:00 p.m. | Middle of month |
| 8:00 p.m. | Last of month |

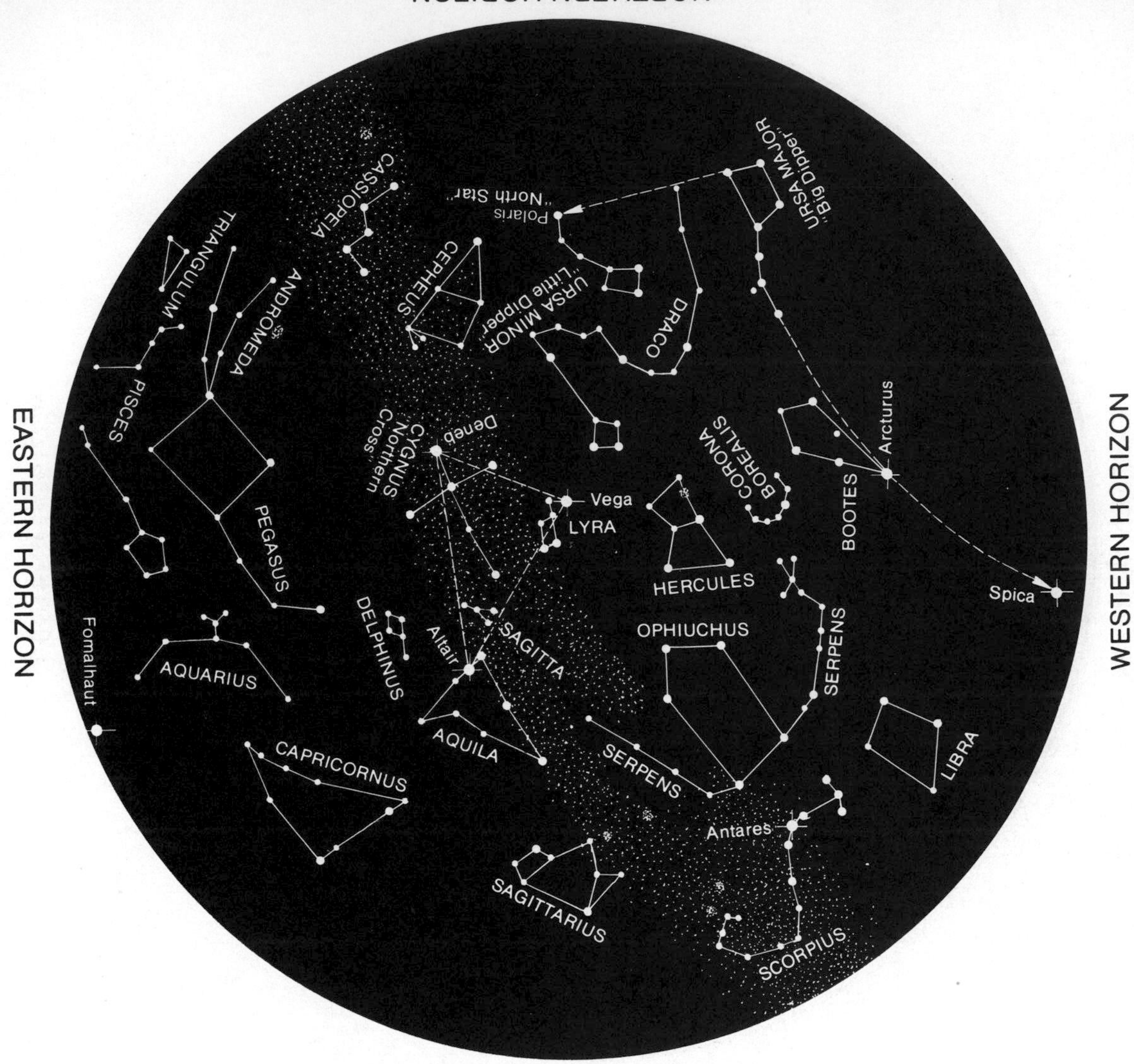


SOUTHERN HORIZON

## THE NIGHT SKY IN AUGUST

To use: Hold chart vertically and turn it so the direction you are facing shows at the bottom.

Chart Time (local Standard):

| | |
|---|---|
| 10:00 p.m. | First of month |
| 9:00 p.m. | Middle of month |
| 8:00 p.m. | Last of month |

NORTHERN HORIZON

EASTERN HORIZON

WESTERN HORIZON

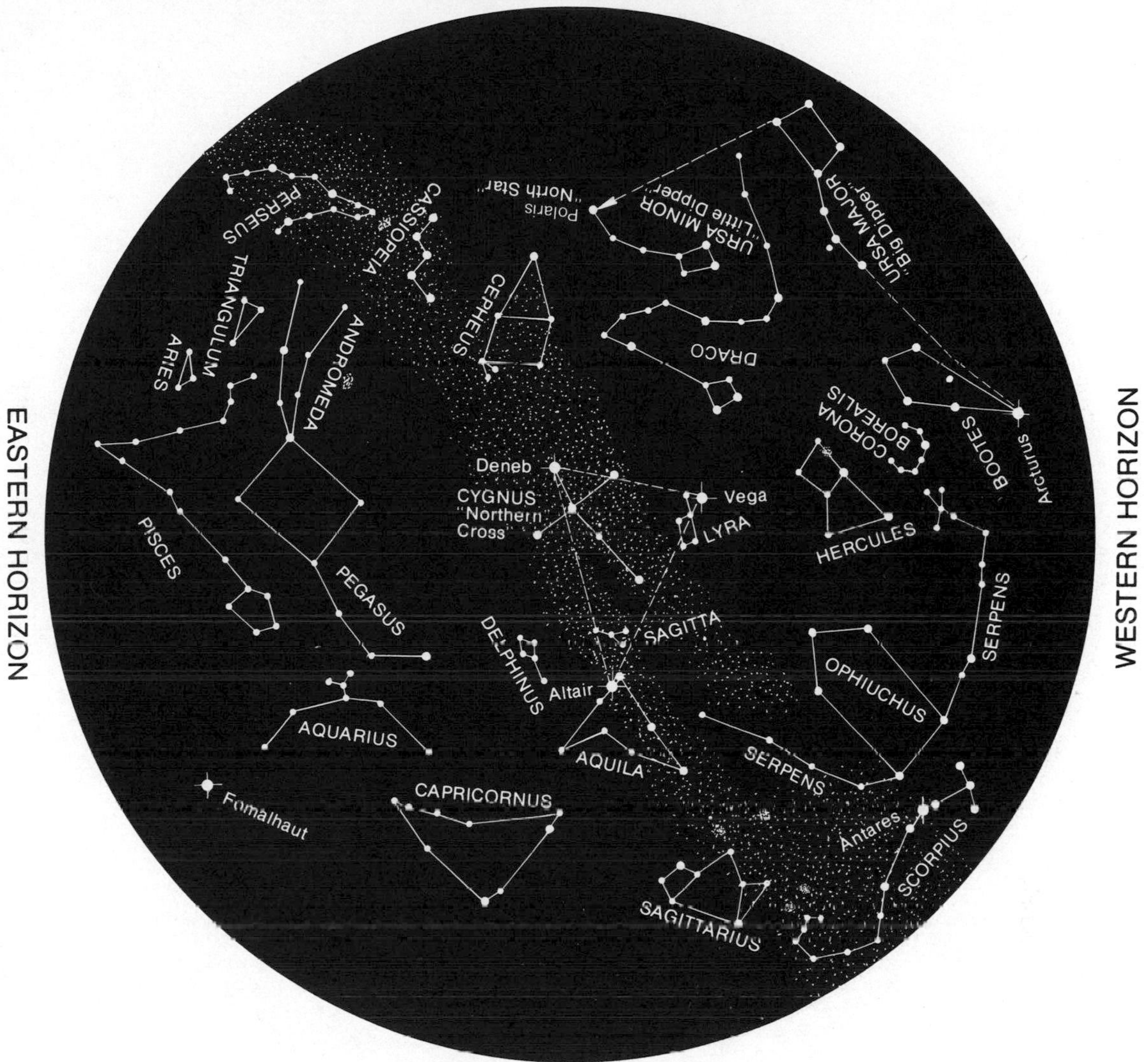


SOUTHERN HORIZON

## THE NIGHT SKY IN SEPTEMBER

To use: Hold chart vertically and turn it so the direction you are facing shows at the bottom.

Chart Time (local Standard):

| | |
|---|---|
| 10:00 p.m. | First of month |
| 9:00 p.m. | Middle of month |
| 8:00 p.m. | Last of month |

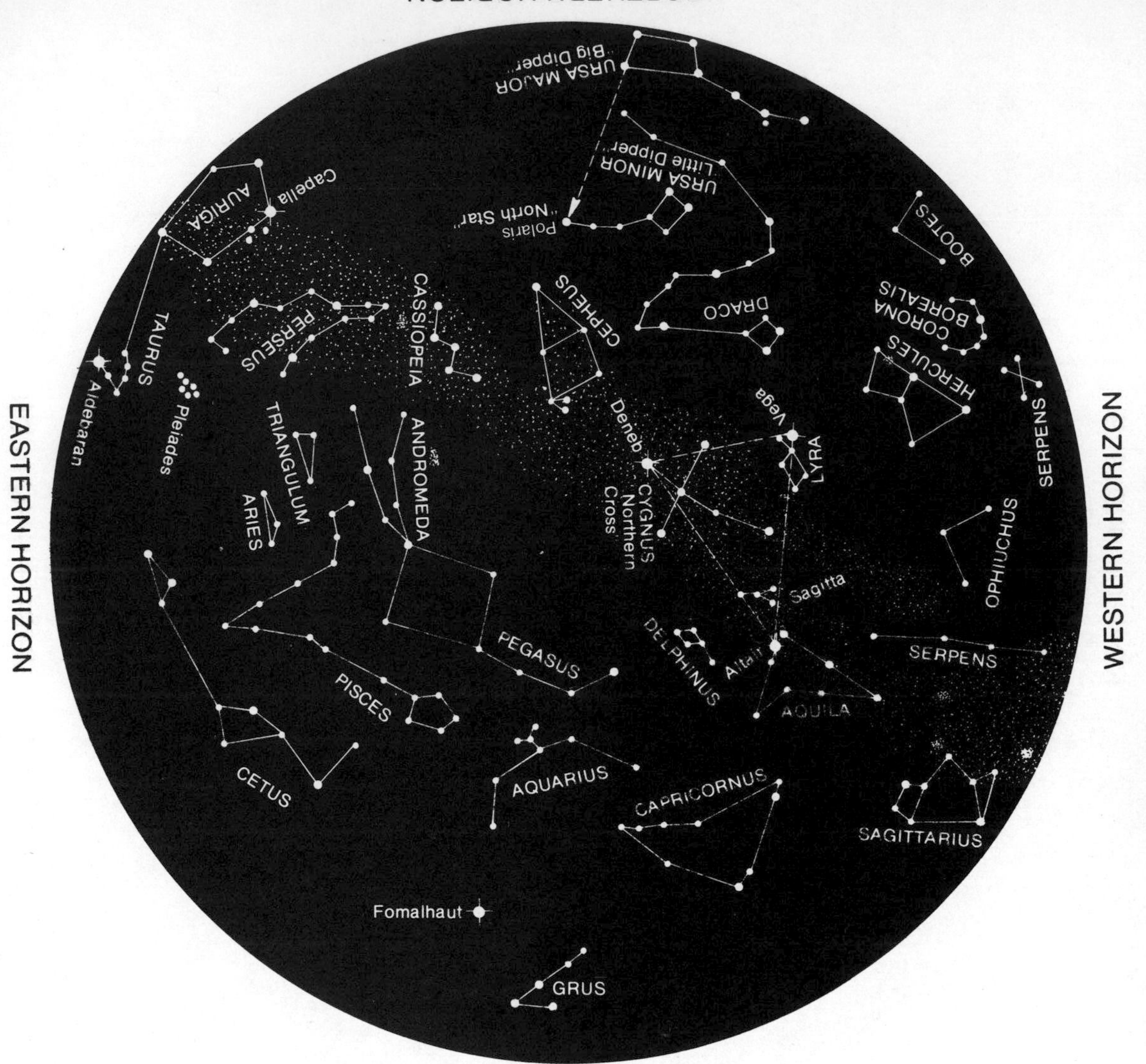


SOUTHERN HORIZON

## THE NIGHT SKY IN OCTOBER

To use: Hold chart vertically and turn it so the direction you are facing shows at the bottom.

Chart Time (local Standard):

| | |
|---|---|
| 10:00 p.m. | First of month |
| 9:00 p.m. | Middle of month |
| 8:00 p.m. | Last of month |

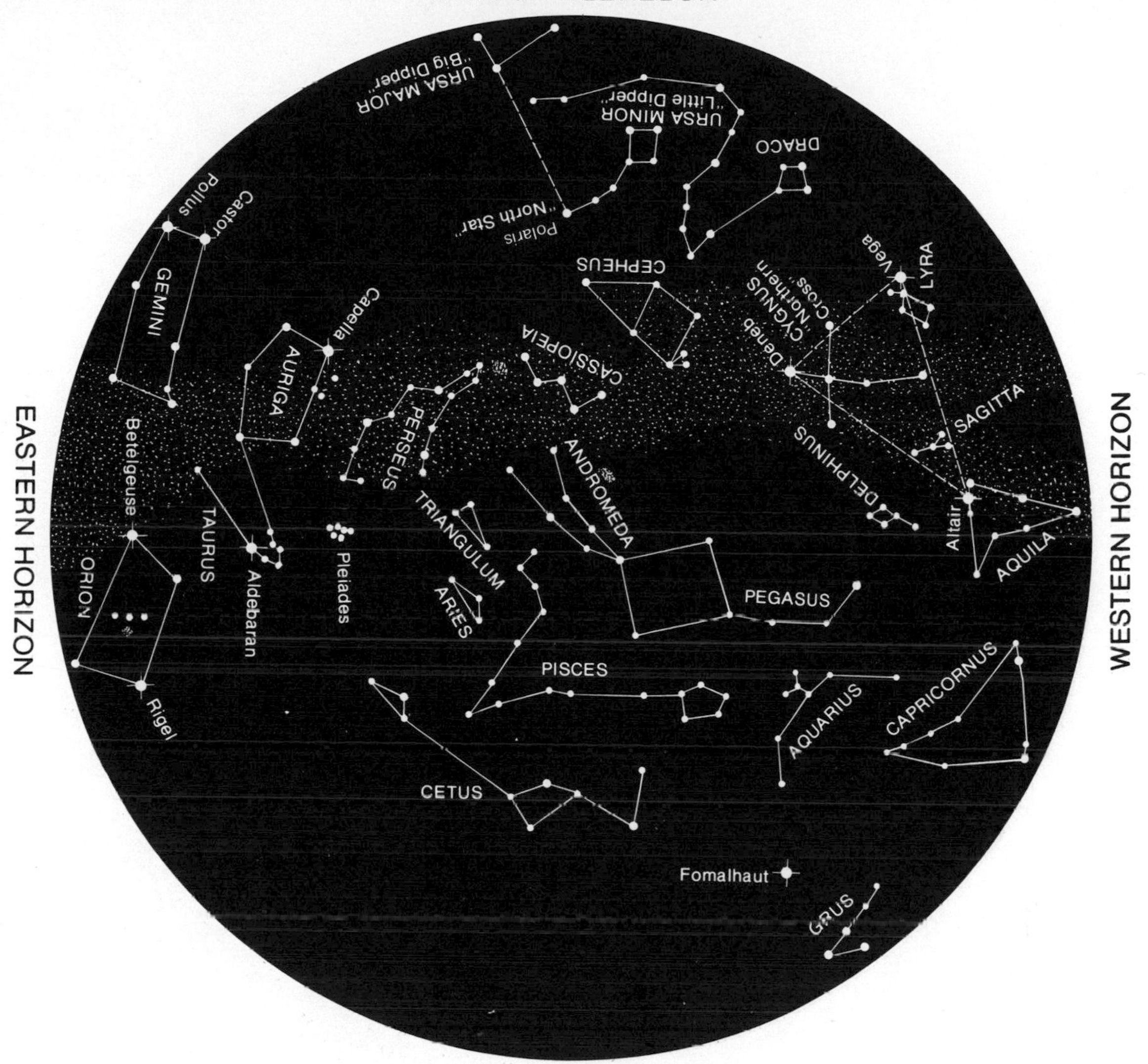


SOUTHERN HORIZON

## THE NIGHT SKY IN NOVEMBER

To use: Hold chart vertically and turn it so the direction you are facing shows at the bottom.

Chart Time (local Standard):

| | |
|---|---|
| 10:00 p.m. | First of month |
| 9:00 p.m. | Middle of month |
| 8:00 p.m. | Last of month |

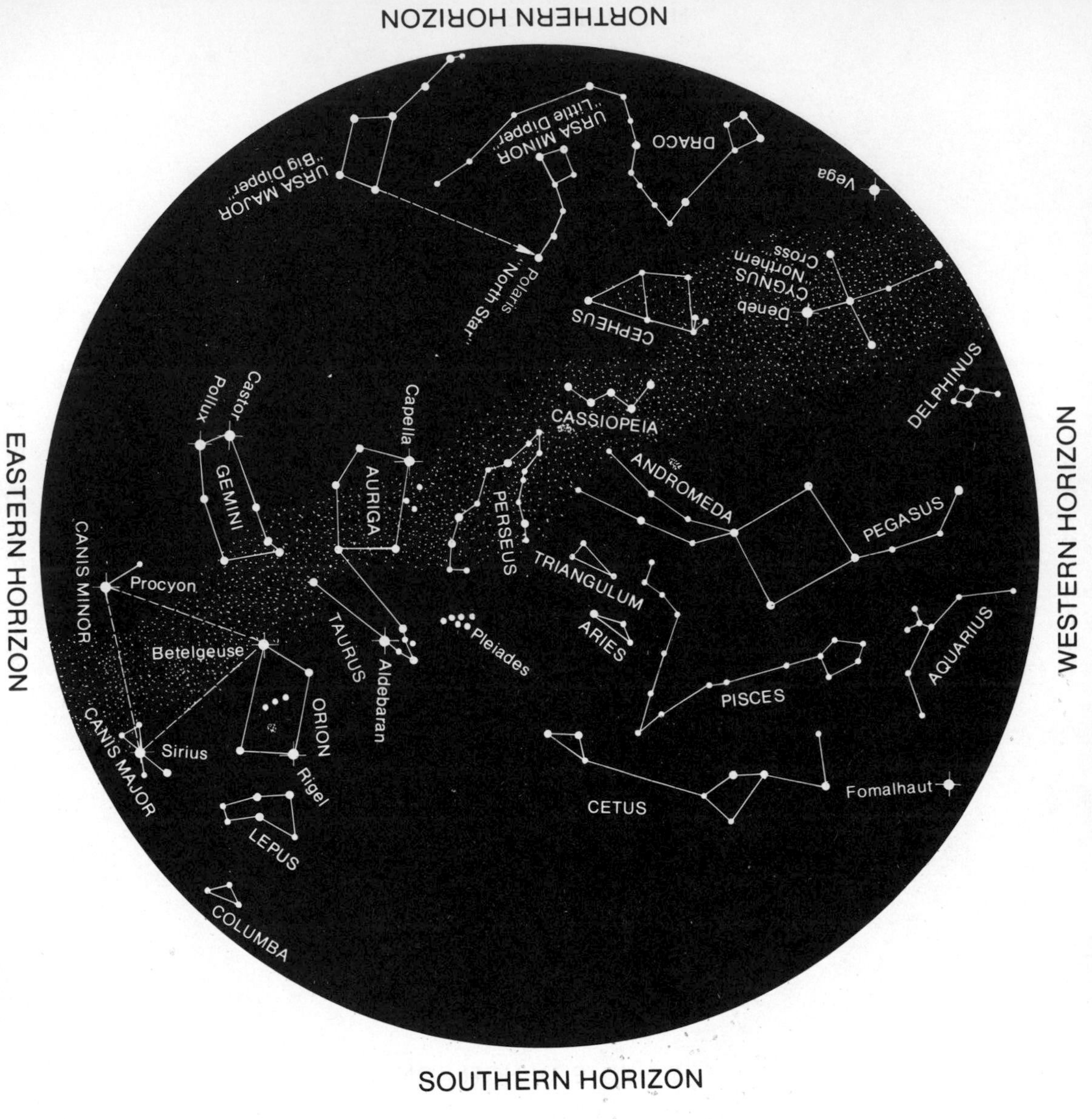

## THE NIGHT SKY IN DECEMBER

To use: Hold chart vertically and turn it so the direction you are facing shows at the bottom.

Chart Time (local Standard):

| | |
|---|---|
| 10:00 p.m. | First of month |
| 9:00 p.m. | Middle of month |
| 8:00 p.m. | Last of month |

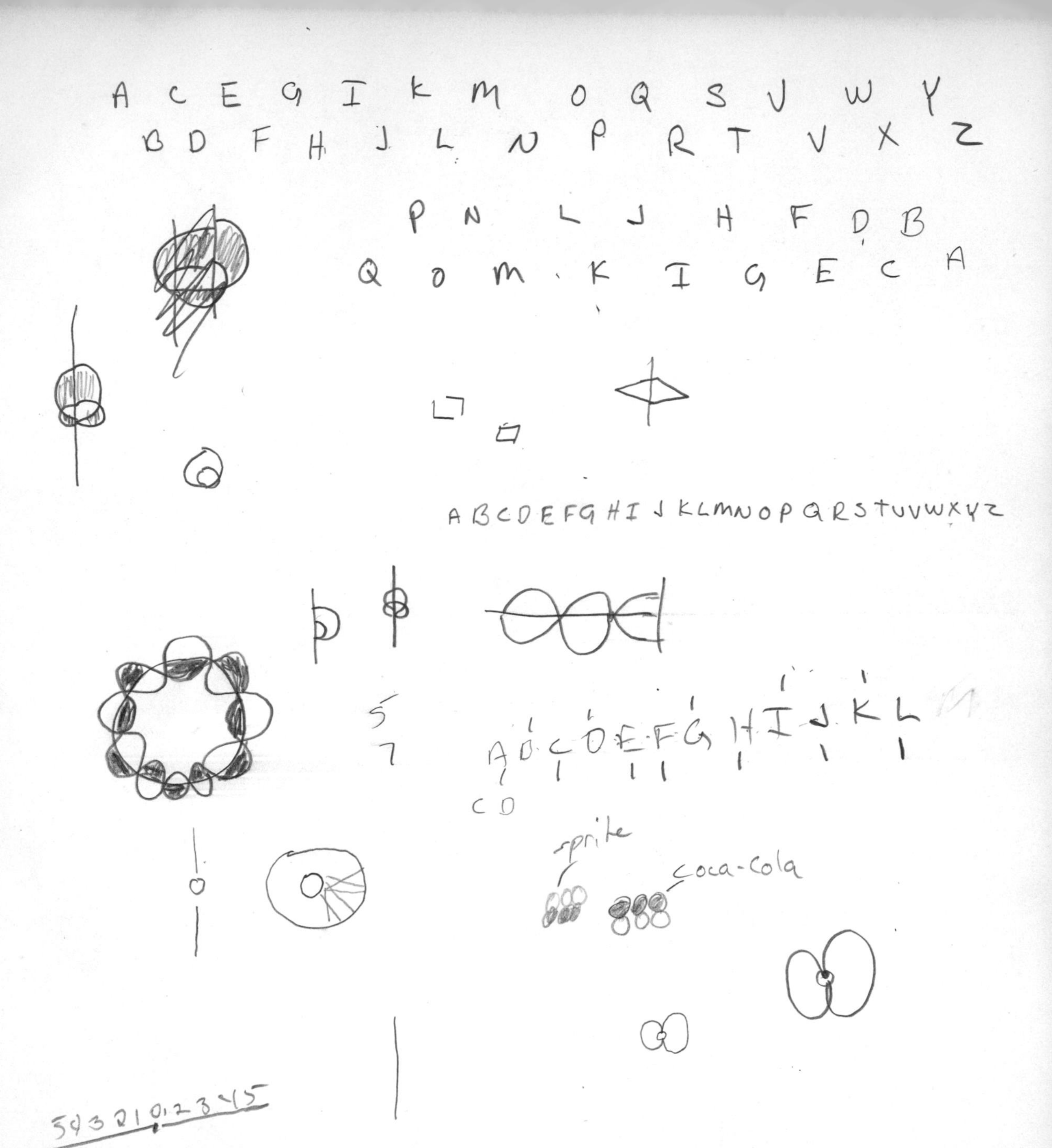

A C E G I K M O Q S U W Y
B D F H J L N P R T V X Z
P N L J H F D B
Q O M K I G E C A
A B C D E F G H I J K L M N O P Q R S T U V W X Y Z
5
7
A B C D E F G H I J K L
C D
sprite
coca-cola
5 4 3 2 1 0 1 2 3 4 5

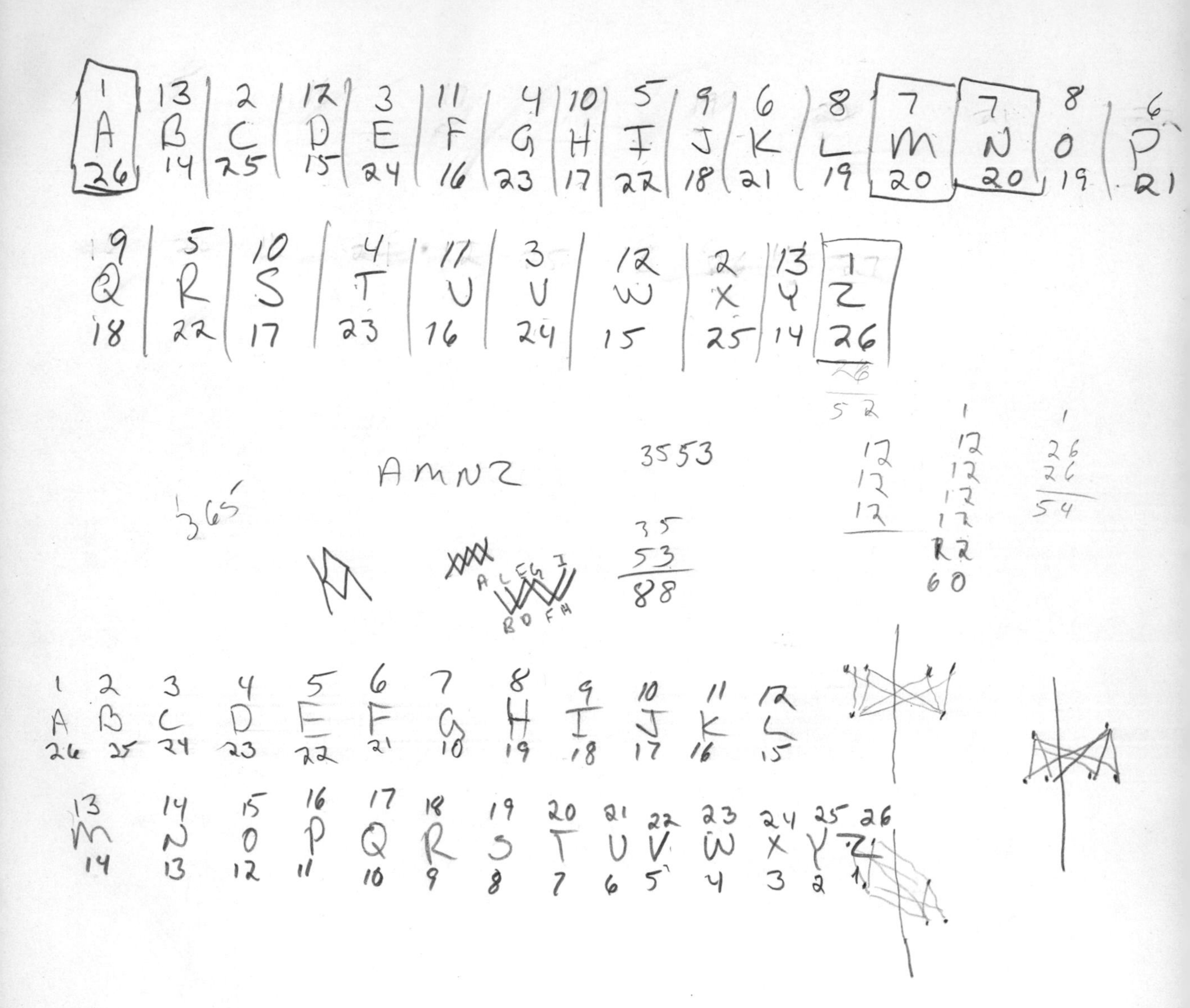

1 13 2 12 3 11 4 10 5 9 6 8 7 7 8 6
A B C D E F G H I J K L M N O P
26 14 25 15 24 16 23 17 22 18 21 19 20 20 19 21
9 5 10 4 11 3 12 2 13 1
Q R S T U V W X Y Z
18 22 17 23 16 24 15 25 14 26
AMNZ
3553
365
35
53
88
1 2 3 4 5 6 7 8 9 10 11 12
A B C D E F G H I J K L
26 25 24 23 22 21 18 19 18 17 16 15
13 14 15 16 17 18 19 20 21 22 23 24 25 26
M N O P Q R S T U V W X Y Z
14 13 12 11 10 9 8 7 6 5 4 3 2 1

# Keep Track of Physics and Astronomy All Year Long!

The editors of Wadsworth Publishing Company hope you will enjoy Seeds's HORIZONS: EXPLORING THE UNIVERSE. In announcing this exciting book, we have produced a handsome physics and astronomy calendar, beautifully illustrated with full-color photos and extensively annotated with important physical and astronomical dates and events occurring during the year.

We are pleased to make available to you a complimentary copy of this useful, attractive calendar. To receive your free calendar, simply fill out and remove the coupon on the bottom, and send to Wadsworth Publishing Company, Attn: Order Department, Box BK-CAL, Ten Davis Drive, Belmont, California 94002.

NOTE: Offer begins with 1982 calendar. Requests accepted for this calendar after May 1, 1981. In subsequent years, requests received before May 1st will be sent the calendar for the current year. Requests received after May 1st will be sent the calendar for the coming year.

YOU MUST USE THE *ORIGINAL* COUPON APPEARING HERE TO RECEIVE YOUR COMPLIMENTARY CALENDAR.
CALENDARS ARE PUBLISHED IN OCTOBER.

---

To: Wadsworth Publishing Company
Attn: Order Dept.
Box BK-CAL
Ten Davis Dr.
Belmont, CA 94002

☐ Please send me my complimentary copy of the PHYSICS AND ASTRONOMY Calendar.

Send to (PRINT CLEARLY, AS THIS IS YOUR MAILING LABEL)

NAME

STREET ADDRESS

CITY

STATE ZIP CODE

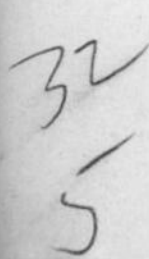